Graduate Texts in Mathematics 111

Graduate Texts in Mathematics

1 TAKEUTI/ZARING. Introduction to Axiomatic Set Theory. 2nd ed.
2 OXTOBY. Measure and Category. 2nd ed.
3 SCHAEFER. Topological Vector Spaces. 2nd ed.
4 HILTON/STAMMBACH. A Course in Homological Algebra. 2nd ed.
5 MAC LANE. Categories for the Working Mathematician. 2nd ed.
6 HUGHES/PIPER. Projective Planes.
7 J.-P. SERRE. A Course in Arithmetic.
8 TAKEUTI/ZARING. Axiomatic Set Theory.
9 HUMPHREYS. Introduction to Lie Algebras and Representation Theory.
10 COHEN. A Course in Simple Homotopy Theory.
11 CONWAY. Functions of One Complex Variable I. 2nd ed.
12 BEALS. Advanced Mathematical Analysis.
13 ANDERSON/FULLER. Rings and Categories of Modules. 2nd ed.
14 GOLUBITSKY/GUILLEMIN. Stable Mappings and Their Singularities.
15 BERBERIAN. Lectures in Functional Analysis and Operator Theory.
16 WINTER. The Structure of Fields.
17 ROSENBLATT. Random Processes. 2nd ed.
18 HALMOS. Measure Theory.
19 HALMOS. A Hilbert Space Problem Book. 2nd ed.
20 HUSEMOLLER. Fibre Bundles. 3rd ed.
21 HUMPHREYS. Linear Algebraic Groups.
22 BARNES/MACK. An Algebraic Introduction to Mathematical Logic.
23 GREUB. Linear Algebra. 4th ed.
24 HOLMES. Geometric Functional Analysis and Its Applications.
25 HEWITT/STROMBERG. Real and Abstract Analysis.
26 MANES. Algebraic Theories.
27 KELLEY. General Topology.
28 ZARISKI/SAMUEL. Commutative Algebra. Vol.I.
29 ZARISKI/SAMUEL. Commutative Algebra. Vol.II.
30 JACOBSON. Lectures in Abstract Algebra I. Basic Concepts.
31 JACOBSON. Lectures in Abstract Algebra II. Linear Algebra.
32 JACOBSON. Lectures in Abstract Algebra III. Theory of Fields and Galois Theory.
33 HIRSCH. Differential Topology.
34 SPITZER. Principles of Random Walk. 2nd ed.
35 ALEXANDER/WERMER. Several Complex Variables and Banach Algebras. 3rd ed.
36 KELLEY/NAMIOKA et al. Linear Topological Spaces.
37 MONK. Mathematical Logic.
38 GRAUERT/FRITZSCHE. Several Complex Variables.
39 ARVESON. An Invitation to C^*-Algebras.
40 KEMENY/SNELL/KNAPP. Denumerable Markov Chains. 2nd ed.
41 APOSTOL. Modular Functions and Dirichlet Series in Number Theory. 2nd ed.
42 J.-P. SERRE. Linear Representations of Finite Groups.
43 GILLMAN/JERISON. Rings of Continuous Functions.
44 KENDIG. Elementary Algebraic Geometry.
45 LOÈVE. Probability Theory I. 4th ed.
46 LOÈVE. Probability Theory II. 4th ed.
47 MOISE. Geometric Topology in Dimensions 2 and 3.
48 SACHS/WU. General Relativity for Mathematicians.
49 GRUENBERG/WEIR. Linear Geometry. 2nd ed.
50 EDWARDS. Fermat's Last Theorem.
51 KLINGENBERG. A Course in Differential Geometry.
52 HARTSHORNE. Algebraic Geometry.
53 MANIN. A Course in Mathematical Logic.
54 GRAVER/WATKINS. Combinatorics with Emphasis on the Theory of Graphs.
55 BROWN/PEARCY. Introduction to Operator Theory I: Elements of Functional Analysis.
56 MASSEY. Algebraic Topology: An Introduction.
57 CROWELL/FOX. Introduction to Knot Theory.
58 KOBLITZ. p-adic Numbers, p-adic Analysis, and Zeta-Functions. 2nd ed.
59 LANG. Cyclotomic Fields.
60 ARNOLD. Mathematical Methods in Classical Mechanics. 2nd ed.
61 WHITEHEAD. Elements of Homotopy Theory.
62 KARGAPOLOV/MERLZJAKOV. Fundamentals of the Theory of Groups.
63 BOLLOBAS. Graph Theory.

(continued after index)

Dale Husemöller

Elliptic Curves

Second Edition

With Appendices by Otto Forster, Ruth Lawrence, and Stefan Theisen

With 42 Illustrations

Dale Husemöller
Max-Planck-Institut für Mathematik
Vivatsgasse 7
D-53111 Bonn
Germany
dale@mpim-bonn.mpg.de

Mathematics Subject Classification (2000): 14-01, 14H52

Library of Congress Cataloging-in-Publication Data
Husemöller, Dale.
Elliptic curves.— 2nd ed. / Dale Husemöller ; with appendices by Stefan Theisen, Otto Forster, and Ruth Lawrence.
p. cm. — (Graduate texts in mathematics; 111)
Includes bibliographical references and index.
ISBN 0-387-95490-2 (alk. paper)
1. Curves, Elliptic. 2. Curves, Algebraic. 3. Group schemes (Mathematics) I. Title. II. Series.
QA567 .H897 2002
516.3′52—dc21 2002067016

ISBN 0-387-95490-2 Printed on acid-free paper.

Printed in the United States of America. (TXB/EB)

9 8 7 6 5 4 3 2

springeronline.com

To
Robert
and the memory of
Roger,
with whom I first learned
the meaning of collaboration

Preface to the Second Edition

The second edition builds on the first in several ways. There are three new chapters which survey recent directions and extensions of the theory, and there are two new appendices. Then there are numerous additions to the original text. For example, a very elementary addition is another parametrization which the author learned from Don Zagier $y^2 = x^3 - 3\alpha x + 2\beta$ of the basic cubic equation. This parametrization is useful for a detailed description of elliptic curves over the real numbers.

The three new chapters are Chapters 18, 19, and 20. Chapter 18, on Fermat's Last Theorem, is designed to point out which material in the earlier chapters is relevant as background for reading Wiles' paper on the subject together with further developments by Taylor and Diamond. The statement which we call the modular curve conjecture has a long history associated with Shimura, Taniyama, and Weil over the last fifty years. Its relation to Fermat, starting with the clever observation of Frey ending in the complete proof by Ribet with many contributions of Serre, was already mentioned in the first edition. The proof for a broad class of curves by Wiles was sufficient to establish Fermat's last theorem. Chapter 18 is an introduction to the papers on the modular curve conjecture and some indication of the proof.

Chapter 19 is an introduction to K3 surfaces and the higher dimensional Calabi–Yau manifolds. One of the motivations for producing the second edition was the utility of the first edition for people considering examples of fibrings of three dimensional Calabi–Yau varieties. Abelian varieties form one class of generalizations of elliptic curves to higher dimensions, and K3 surfaces and general Calabi–Yau manifolds constitute a second class.

Chapter 20 is an extension of earlier material on families of elliptic curves where the family itself is considered as a higher dimensional variety fibered by elliptic curves. The first two cases are one dimensional parameter spaces where the family is two dimensional, hence a surface two dimensional surface parameter spaces where the family is three dimensional. There is the question of, given a surface or a three dimensional variety, does it admit a fibration by elliptic curves with a finite number of exceptional singular fibres. This question can be taken as the point of departure for the Enriques classification of surfaces.

There are three new appendices, one by Stefan Theisen on the role of Calabi–Yau manifolds in string theory and one by Otto Forster on the use of elliptic curves in computing theory and coding theory. In the third appendix we discuss the role of elliptic curves in homotopy theory. In these three introductions the reader can get a clue to the far-reaching implications of the theory of elliptic curves in mathematical sciences.

During the final production of this edition, the ICM 2002 manuscript of Mike Hopkins became available. This report outlines the role of elliptic curves in homotopy theory. Elliptic curves appear in the form of the Weierstasse equation and its related changes of variable. The equations and the changes of variable are coded in an algebraic structure called a Hopf algebroid, and this Hopf algebroid is related to a cohomology theory called topological modular forms. Hopkins and his coworkers have used this theory in several directions, one being the explanation of elements in stable homotopy up to degree 60. In the third appendix we explain how what we described in Chapter 3 leads to the Weierstrass Hopf algebroid making a link with Hopkins' paper.

Max-Planck-Institut für Mathematik — Dale Husemöller
Bonn, Germany

Preface to the First Edition

The book divides naturally into several parts according to the level of the material, the background required of the reader, and the style of presentation with respect to details of proofs. For example, the first part, to Chapter 6, is undergraduate in level, the second part requires a background in Galois theory and the third some complex analysis, while the last parts, from Chapter 12 on, are mostly at graduate level. A general outline of much of the material can be found in Tate's colloquium lectures reproduced as an article in *Inventiones* [1974].

The first part grew out of Tate's 1961 Haverford Philips Lectures as an attempt to write something for publication closely related to the original Tate notes which were more or less taken from the tape recording of the lectures themselves. This includes parts of the Introduction and the first six chapters. The aim of this part is to prove, by elementary methods, the Mordell theorem on the finite generation of the rational points on elliptic curves defined over the rational numbers.

In 1970 Tate returned to Haverford to give again, in revised form, the original lectures of 1961 and to extend the material so that it would be suitable for publication. This led to a broader plan for the book.

The second part, consisting of Chapters 7 and 8, recasts the arguments used in the proof of the Mordell theorem into the context of Galois cohomology and descent theory. The background material in Galois theory that is required is surveyed at the beginnng of Chapter 7 for the convenience of the reader.

The third part, consisting of Chapters 9, 10, and 11, is on analytic theory. A background in complex analysis is assumed and in Chapter 10 elementary results on p-adic fields, some of which were introduced in Chapter 5, are used in our discussion of Tate's theory of p-adic theta functions. This section is based on Tate's 1972 Haverford Philips Lectures.

Max-Planck-Institut für Mathematik
Bonn, Germany

Dale Husemöller

Acknowledgments to the Second Edition

Stefan Theisen, during a period of his work on Calabi–Yau manifolds in conjunction with string theory, brought up many questions in the summer of 1998 which lead to a renewed interest in the subject of elliptic curves on my part.

Otto Forster gave a course in Munich during 2000–2001 on or related to elliptic curves. We had discussions on the subject leading to improvements in the second edition, and at the same time he introduced me to the role of elliptic curves in cryptography.

A reader provided by the publisher made systematic and very useful remarks on everything including mathematical content, exposition, and English throughout the manuscript.

Richard Taylor read a first version of Chapter 18, and his comments were of great use. F. Oort and Don Zagier offered many useful suggestions for improvement of parts of the first edition. In particular the theory of elliptic curves over the real numbers was explained to me by Don.

With the third appendix T. Bauer, M. Joachim, and S. Schwede offered many useful suggestions.

During this period of work on the second edition, I was a research professor from Haverford College, a visitor at the Max Planck Institute for Mathematics in Bonn, a member of the Graduate College and mathematics department in Munich, and a member of the Graduate College in Münster. All of these connections played a significant role in bringing this project to a conclusion.

Max-Planck-Institut für Mathematik
Bonn, Germany

Dale Husemöller

Acknowledgments to the First Edition

Being an amateur in the field of elliptic curves, I would have never completed a project like this without the professional and moral support of a great number of persons and institutions over the long period during which this book was being written.

John Tate's treatment of an advanced subject, the arithmetic of elliptic curves, in an undergraduate context has been an inspiration for me during the last 25 years while at Haverford. The general outline of the project, together with many of the details of the exposition, owe so much to Tate's generous help.

The E.N.S. course by J.-P. Serre of four lectures in June 1970 together with two Haverford lectures on elliptic curves were very important in the early development of the manuscript. I wish to thank him also for many stimulating discussions. Elliptic curves were in the air during the summer seasons at the I.H.E.S. around the early 1970s. I wish to thank P. Deligne, N. Katz, S. Lichtenbaum, and B. Mazur for many helpful conversations during that period. It was the Haverford College Faculty Research Fund that supported many times my stays at the I.H.E.S.

During the year 1974–5, the summer of 1976, the year 1981–2, and the spring of 1986, I was a guest of the Bonn Mathematics Department SFB and later the Max Planck Institute. I wish to thank Professor F. Hirzebruch for making possible time to work in a stimulating atmosphere and for his encouragement in this work. An early version of the first half of the book was the result of a Bonn lecture series on Elliptische Kurven. During these periods, I profited frequently from discussions with G. Harder and A. Ogg.

Conversations with B. Gross were especially important for realizing the final form of the manuscript during the early 1980s. I am very thankful for his encouragement and help. In the spring of 1983 some of the early chapters of the book were used by K. Rubin in the Princeton Junior Seminar, and I thank him for several useful suggestions. During the same time, J. Coates invited me to an Oberwolfach conference on elliptic curves where the final form of the manuscript evolved.

During the final stages of the manuscript, both R. Greenberg and R. Rosen read through the later chapters, and I am grateful for their comments. I would like to thank P. Landweber for a very careful reading of the manuscript and many useful comments.

Ruth Lawrence read the early chapters along with working the exercises. Her contribution was very great with her appendix on the exercises and suggested improvements in the text. I wish to thank her for this very special addition to the book.

Free time from teaching at Haverford College during the year 1985–1986 was made possible by a grant from the Vaughn Foundation. I wish to express my gratitude to Mr. James Vaughn for this support, for this project as well as others, during this difficult last period of the preparation of the manuscript.

Max-Planck-Institut für Mathematik
Bonn, Germany

Dale Husemöller

Contents

Introduction to Rational Points on Plane Curves

This introduction is designed to bring up some of the main issues of the book in an informal way so that the reader with only a minimal background in mathematics can get an idea of the character and direction of the subject.

An elliptic curve, viewed as a plane curve, is given by a nonsingular cubic equation. We wish to point out what is special about the class of elliptic curves among all plane curves from the point of view of arithmetic. In the process the geometry of the curve also enters the picture.

For the first considerations our plane curves are defined by a polynomial equation in two variables $f(x, y) = 0$ with rational coefficients. The main invariant of this f is its degree, a natural number. In terms of plane analytic geometry there is a curve C_f which is the locus of this equation in the x, y-plane, that is, C_f is defined as the set of $(x, y) \in \mathbb{R}^2$ satisfying $f(x, y) = 0$. To emphasize that the locus consists of points with real coordinates (so is in $\mathbb{R}^2$), we denote this real locus by $C_f(\mathbb{R})$ and consider $C_f(\mathbb{R}) \subset \mathbb{R}^2$.

Since some curves C_f, like for example $f(x, y) = x^2 + y^2 + 1$, have an empty real locus $C_f(\mathbb{R})$, it is always useful to work also with the complex locus $C_f(\mathbb{C})$ contained in $\mathbb{C}^2$ even though it cannot be completely pictured geometrically. For geometric considerations involving the curve, the complex locus $C_f(\mathbb{C})$ plays the central role.

For arithmetic the locus of special interest is the set $C_f(\mathbb{Q})$ of rational points $(x, y) \in \mathbb{Q}^2$ satisfying $f(x, y) = 0$, that is, points whose coordinates are rational numbers. The fundamental problem of this book is the description of this set $C_f(\mathbb{Q})$. An elementary formulation of this problem is the question whether or not $C_f(\mathbb{Q})$ is finite or even empty.

This problem is attacked by a combination of geometric and arithmetic arguments using the inclusions $C_f(\mathbb{Q}) \subset C_f(\mathbb{R}) \subset C_f(\mathbb{C})$. A locus $C_f(\mathbb{Q})$ can be compared with another locus $C_g(\mathbb{Q})$, which is better understood, as we illustrate for lines where $\deg(f) = 1$ and conics where $\deg(f) = 2$. In the case of cubic curves we introduce an internal operation.

In terms of the real locus, curves of degree 1, degree 2, and degree 3 can be pictured respectively as follows.

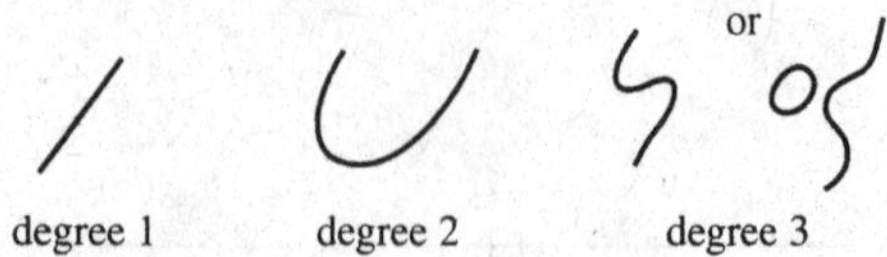

§1. Rational Lines in the Projective Plane

Plane curves C_f can be defined for any nonconstant complex polynomial with complex coefficients $f(x, y) \in \mathbb{C}[x, y]$ by the equation $f(x, y) = 0$. For a nonzero constant k the equations $f(x, y) = 0$ and $kf(x, y) = 0$ have the same solutions and define the same plane curve $C_f = C_{kf}$. When f has complex coefficients, there is only a complex locus defined. If f has real coefficients or if f differs from a real polynomial by a nonzero constant, then there is also a real locus with $C_f(\mathbb{R}) \subset C_f(\mathbb{C})$. Such curves are called real curves.

(1.1) Definition. A rational plane curve or a curve defined over $\mathbb{Q}$ is one of the form C_f where $f(x, y)$ is a polyomial with rational coefficients.

This is an arithmetic definition of rational curve, and it should not be confused with the geometric definition of rational curve or variety. We will not use the geometric concept.

In the case of a rational plane curve C_f we have rational, real, and complex points $C_f(\mathbb{Q}) \subset C_f(\mathbb{R}) \subset C_f(\mathbb{C})$ or loci.

A polynomial of degree 1 has the form $f(x, y) = a + bx + cy$. We assume the coefficients are rational numbers and begin by describing the rational line $C_f(\mathbb{Q})$. For c nonzero we can set up a bijective correspondence between rational points on the line C_f and on the x-axis using intersections with vertical lines.

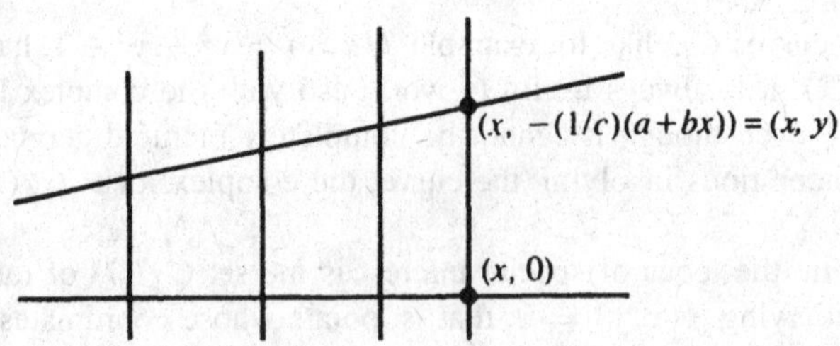

The rational point $(x, 0)$ on the x-axis corresponds to the rational point

$$(x, -(1/c)(a + bx))$$

on C_f. When b is nonzero, the points on the rational line $C_f(\mathbb{Q})$ can be put in bijective correspondence with the rational points on the y-axis using intersections with horizontal lines. Observe that the vertical or horizontal lines relating rational points are themselves rational lines.

Instead of using parallel lines to relate points on two lines $L = C_f$ and $L' = C_{f'}$, we can use a point $P_0 = (x_0, y_0)$ not on either L or L' and relate points using the family of all lines through P_0. The pair P on L and P' on L' correspond when P, P', and P_0 are all on a line.

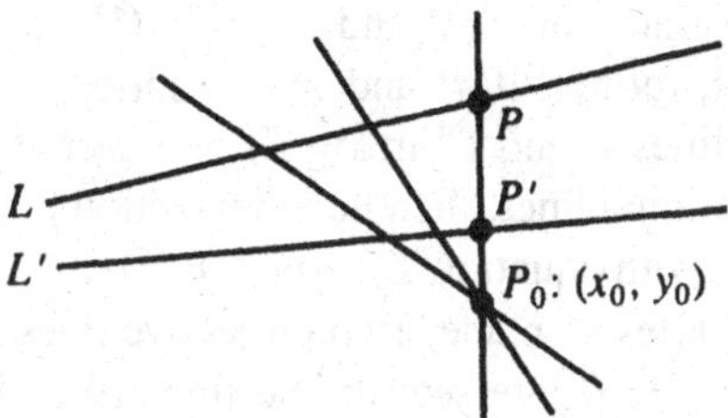

If L and L' are rational lines, and if P_0 is a rational point, then for two corresponding points P on L and P' on L' the point P is rational if and only if P' is rational, and this defines a bijection between $C_f(\mathbb{Q})$ and $C_{f'}(\mathbb{Q})$.

Observe that there are special cases of lines through P_0, i.e., those parallel to L or L', which as matters stand do not give a corresponding pair of points between L and L'. This is related to the fact that the two types of correspondence with parallel lines and with lines through a point are really the same when viewed in terms of the projective plane, for parallel lines intersect at a point on the "line at infinity." As we see in the next paragraphs, the projective plane is the ordinary Cartesian or affine plane together with an additional line called the line at infinity.

(1.2) Definition. The projective plane $\mathbb{P}_2$ is the set of all triples $w : x : y$, where w, x, and y are not all zero and the points $w : x : y$ and $w' : x' : y'$ are considered equal provided there is a nonzero constant k with

$$w' = kw, \quad x' = kx, \quad y' = ky.$$

As with the affine plane and plane curves we have three basic cases

$$\mathbb{P}_2(\mathbb{Q}) \subset \mathbb{P}_2(\mathbb{R}) \subset \mathbb{P}_2(\mathbb{C})$$

consisting of triples proportional to $w : x : y$, where $w, x, y \in \mathbb{Q}$ for $\mathbb{P}_2(\mathbb{Q})$, where $w, x, y \in \mathbb{R}$ for $\mathbb{P}_2(\mathbb{R})$, and where $w, x, y \in \mathbb{C}$ for $\mathbb{P}_2(\mathbb{C})$.

Note $w : x : y \in \mathbb{P}_2(\mathbb{C})$ is also in $\mathbb{P}_2(\mathbb{Q})$ if and only if $w, x, y \in \mathbb{C}$ can be rescaled to be elements of $\mathbb{Q}$.

(1.3) Remarks. A line C_f in $\mathbb{P}_2$ is the locus of all $w : x : y$ satisfying the equation $F(w, x, y) = aw + bx + cy = 0$. The line at infinity L_∞ is given by the equation $w = 0$. A point in $\mathbb{P}_2 - L_\infty$ has the form $1 : x : y$ after multiplying with the factor w^{-1}. The point $1 : x : y$ in the projective plane corresponds to (x, y) in the usual Cartesian plane. For a line L given by $aw + bx + cy = 0$ and L' given by $a'w + b'x + c'y = 0$ we have $L = L'$ if and only if $a : b : c = a' : b' : c'$ in the projective plane. In particular the points $a : b : c$ in the projective plane can be used to parametrize the lines in the projective plane.

From the theory of elimination of variables in beginning algebra we have the following geometric assertions of projective geometry whose verification is left to the reader.

(1.4) Assertion. Two distinct points P and P' in $\mathbb{P}_2(\mathbb{C})$ lie on a unique line L in the projective plane, and, further, if P and P' are rational points, then the line L is rational. Two distinct lines L and L' in $\mathbb{P}_2(\mathbb{C})$ intersect at a unique point P, and further, if L and L' are rational lines, then the intersection point P is rational.

The projective line L with equation $L : aw + bx + cy = 0$ determines the line $a + bx + cy = 0$ in the Cartesian plane. Two projective lines $L : aw + bx + cy = 0$ and $L' : a'w + b'x + c'y = 0$ intersect on the line at infinity $w = 0$ if and only if $b : c = b' : c'$, that is, the pairs (b, c) and (b', c') are proportional. Hence the corresponding lines in the x, y-plane given by

$$a + bx + cy = 0 \quad \text{and} \quad a' + b'x + c'y = 0$$

have the same slope or are parallel exactly when the projective lines intersect at infinity. Now the reader is invited to reconsider the correspondence between rational points on two rational lines L and L' which arises by intersecting L and L' with all rational lines through a fixed point P_0 not on either L or L'.

To define more general plane curves in projective space, we use nonzero homogeneous polynomials $F(w, x, y) \in \mathbb{C}[w, x, y]$. Then we have the relation

$$F(qw, qx, qy) = q^d F(w, x, y),$$

where $q \in \mathbb{C}$ and d is the degree of the homogenous polynomial $F(w, x, y)$. The locus C_F is the set of all $w : x : y$ in the projective plane such that $F(w, x, y) = 0$. The homogeneity of $F(w, x, y)$ is needed for $F(w, x, y) = 0$ to be independent of the scale for $w : x : y \in \mathbb{P}_2$. Again the complex points of C_F are denoted by $C_F(\mathbb{C}) \subset \mathbb{P}_2(\mathbb{C})$, and, moreover, $C_F(\mathbb{C}) = C_{F'}(\mathbb{C})$ if and only if $F(w, x, y)$ and $F'(w, x, y)$ are proportional with a nonzero complex number. This assertion is not completely evident and is taken up again in Chapter 2.

(1.5) Definition. A rational (resp. real) plane curve in $\mathbb{P}_2$ is one of the form C_F where $F(w, x, y)$ has rational (resp. real) coefficients.

As in the x, y-plane for a rational plane curve C_F, we have rational, real, and complex points $C_F(\mathbb{Q}) \subset C_F(\mathbb{R}) \subset C_F(\mathbb{C})$.

(1.6) Remark. The above definition of a rational plane curve is an arithmetic notion, and it means the curve can be defined over $\mathbb{Q}$. There is a geometric concept of rational curve (genus = 0) which should not be confused with (1.5).

§2. Rational Points on Conics

Now we study rational points on rational plane curves of degree 2 which in x, y-coordinates are given by

$$0 = f(x, y) = a + bx + cy + dx^2 + exy + fy^2$$

and in homogeneous form for the projective plane are given by

$$0 = F(w, x, y) = aw^2 + bwx + cwy + dx^2 + exy + fy^2.$$

Observe that the two polynomials are related by $f(x, y) = F(1, x, y)$ and $F(w, x, y) = w^2 f(x/w, y/w)$. More generally, if $f(x, y)$ has degree d, then $F(w, x, y) = w^d f(x/w, y/w)$ is the corresponding homogenous polynomial, and the curve C_f in x, y-space is the curve C_F minus the points on the line at infinity. We will frequently pass between the projective and affine descriptions of conics and plane curves.

Returning to the conic defined by a polynomial f of degree 2, we begin by excluding the case where f factors as a product of two linear polynomials, i.e., C_f is the union of two lines or a single double line. These are exactly the singular conics, and we return later to the general concept of singularity on a curve. One example of such a conic is $xy = 0$, the locus for the x and y axis.

(2.1) Remark. Let $C = C_f$ be a nonsingular rational conic. There are two questions related to the determination of the rational points on C:

(1) Is there a rational point P_0 on C at all? If not, then $C_f(\mathbb{Q})$ is the empty set!
(2) Given a rational point P_0 on C, determine all other rational points P on C in terms of P_0.

The second problem has a particularly simple elegant solution in terms of the ideas introduced in the previous section. To carry out this solution, we need the following intersection result.

(2.2) Remark. If one of the two intersection points of a rational conic with a rational line is a rational point, then the other intersection point is rational.

To see this, we use the equation $aw + bx + cy = 0$ of the line to eliminate one variable in the second-order equation $F(w, x, y) = 0$ of the conic. For intersections off the line at infinity, given by $w = 0$, one is left with a quadratic equation in the x coordinate or in the y coordinate of the intersection points. The equation of the line comes in again here to recover the other coordinate. Thus the intersection points will be rational if and only if the roots of the quadratic equation are rational. In general they are conjugate quadratic irrationalities for rational lines and conics, and an intersection point is rational if and only if its x coordinate is a rational number. Thus (2.2) reduces to the algebraic statement: if a quadratic polynomial with rational coefficients has one rational root, then the other root is rational.

Let C be a rational conic with a rational point O on it. Choose a rational line L not containing O, and project the conic C onto the line L from this point O.

For every point Q on the line L by joining it to O one gets a point P on the conic C, and in the other direction, a line meets the conic C in two points, so to every point P on the conic C there corresponds a point Q on the line L. This sets up a correspondence between points on the conic and points on the line L. Since O is assumed to be rational, we see from (2.2) that the point P is rational if and only if the point Q is rational.

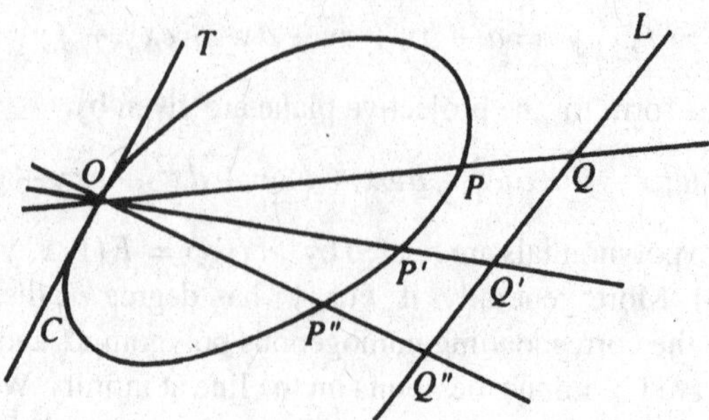

(2.3) Assertion. Assume that L intersects the tangent line T to C at O at a point R. Then the rational points on the conic C different from the rational point O are in one-to-one correspondence with the rational points on the line L different from R. We complete the correspondence between $C(\mathbb{Q})$ and $L(\mathbb{Q})$ by letting O on C correspond to R on L.

Now we return to the first question of whether there is a rational point at all on a rational conic. For example, clearly the circles $x^2 + y^2 = 1$ and $x^2 + y^2 = 2$ have rational points on them. On the other hand, $x^2 + y^2 = 3$ has no rational point; that is, it is impossible for the sum of the squares of two rational numbers to equal three.

To see that there are no rational points on $x^2 + y^2 = 3$, we can introduce homogeneous coordinates $w : x : y$ and clear denominators of the rational numbers x and y to look for integers satisfying $x^2 + y^2 = 3w^2$, where x, y, and w have no common factor. In this case 3 does not divide either x or y. For if $3|x$, then $3|y^2$, and hence $3|y$. From this it would follow that 9 divides $x^2+y^2 = 3w^2$. This would mean that $3|w^2$ and thus $3|w$ which contradicts the fact that x, y, and w have no common factor. This means that $x, y \equiv \pm 1 \pmod 3$. This implies that $x^2 + y^2 \equiv 1 + 1 = 2 \pmod 3$, so that the sum $x^2 + y^2$ cannot be divisible by 3. We conclude that $x^2 + y^2 = 3w^2$ has no solutions. Hence there are no two rational numbers whose squares add to 3.

The argument given for $x^2 + y^2 = 3$ gives an indication of the general method which can also be applied directly to show that there are no rational points on the circle $x^2 + y^2 = n$ for any n of the form $n = 4k + 3$. The reader is invited to carry out the argument.

More generally there is a test by which, in a finite numbers of steps, one can determine whether or not a given rational conic has a rational point. It consists in seeing whether a certain congruence can be satisfied, and the theorem goes back to Legendre.

(2.4) Legendre's Theorem. *For a conic $ax^2 + by^2 = w^2$ there exists a certain number m such that $ax^2 + by^2 = w^2$ has an integral solution if and only if the congruence*

$$ax^2 + by^2 \equiv w^2 \pmod m$$

has a solution in the integers modulo m.

There is a more elegant and general way of stating the theorem which is due to Hasse in its final form and uses p-adic numbers.

(2.5) Hasse–Minkowski Theorem. *A homogeneous quadratic equation in several variables is solvable by rational numbers, not all zero, if and only if it is solvable in the p-adic numbers for each prime p including the infinite prime. The p-adic numbers at the infinite prime are the real numbers.*

From this result the theorem of Legendre about the congruence follows in a very elementary way. The p-adic theorem is the better statement, and for the interested reader a proof can be found in Chapter 4 of J.-P. Serre, *Course in Arithmetic*, or in Appendix 3 of Milnor and Husemöller, *Symmetric Bilinear Forms* (both from Springer-Verlag).

§3. Pythagoras, Diophantus, and Fermat

The simple conic with equation $x^2 + y^2 = 1$ or $x^2 + y^2 = w^2$ has a long history stretching back to Pythagoras in the sixth century B.C. It started with the relation between the lengths of the three sides of a right triangle

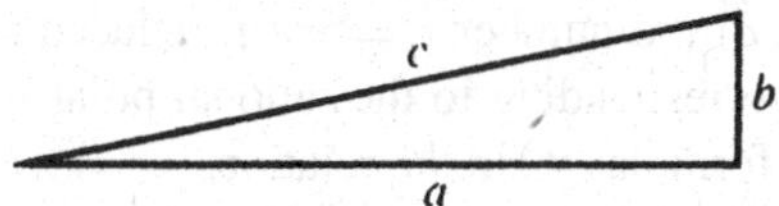

The relation $c^2 = a^2 + b^2$ is attributed to Pythagoras, but it seems to have been known in Babylon at the time of Hammurabi and to the Egyptians, besides to the members of Pythagoras' school in Cortona in southern Italy.

Triples of whole numbers (a, b, c) satisfying $c^2 = a^2+b^2$ are called Pythagorean triples. Some of the first examples known from the time of Pythagoras were $(3,4,5)$, and $(5, 12, 13)$, and $(9, 40, 41)$. Of course, if (a, b, c) is a Pythagorean triple, then so is (ka, kb, kc) for any whole number k. Thus it suffices to determine primitive Pythagorean triples where the greatest common divisor of a, b, and c is 1. The above examples are primitive. The determination of all primitive Pythagorean triples goes back to Diophantus of Alexandria, about 250 A.D.

(3.1) Theorem. *Let m and n be two relatively prime natural numbers such that $n-m$ is positive and odd. Then $(n^2-m^2, 2mn, n^2+m^2)$ is a primitive Pythagorean triple, and each primitive Pythagorean triple arises in this way for some m, n.*

This theorem follows from the considerations of the previous section where a conic was projected onto a line in (2.3). Consider the conic $x^2 + y^2 = 1$. Project from the point $(-1, 0)$ the points on this circle onto the y-axis. The line L_t through

$(-1,0)$ and $(0,t)$ on the y-axis has equation $y = t(x+1)$. If the line L_t intersects the circle $x^2 + y^2 = 1$ at the points $(-1,0)$ and (x,y), then we have

$$x = \frac{1-t^2}{1+t^2} \quad \text{and} \quad y = \frac{2t}{1+t^2}.$$

Observe that t is rational if and only if (x,y) is a rational point on the circle. The value infinity corresponds to the base of the projection $(-1,0)$.

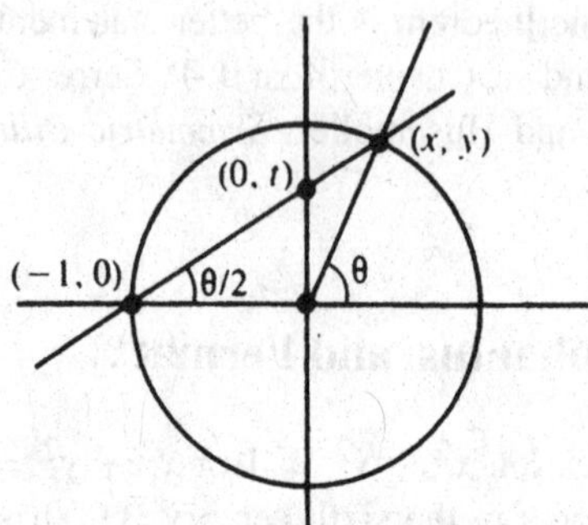

In order to prove the theorem of Diophantus, we consider for any primitive Pythagorean triple (a,b,c) the number $t = m/n$, reduced to lowest terms, giving the point on the y-axis corresponding to the rational point $(a/c, b/c)$ on the circle $x^2 + y^2 = 1$. The above formulas yield the relations

$$a = n^2 - m^2, \quad b = 2mn, \quad c = n^2 + m^2.$$

The first assertion of the theorem follows from the computation

$$\left(n^2 - m^2\right)^2 + (2mn)^2 = \left(n^2 + m^2\right)^2.$$

The above projection of the circle on the y-axis is also related to the following trigonometric identities, left to the reader as an exercise,

$$\tan\left(\frac{\theta}{2}\right) = \frac{\sin\theta}{1+\cos\theta}, \quad \cos\theta = \frac{1-\tan^2\left(\frac{\theta}{2}\right)}{1+\tan^2\left(\frac{\theta}{2}\right)}, \quad \sin\theta = \frac{2\tan\left(\frac{\theta}{2}\right)}{1+\tan^2\left(\frac{\theta}{2}\right)}.$$

If $\int R(\sin\theta, \cos\theta, \tan\theta, \cot\theta, \sec\theta, \csc\theta)\,d\theta$ is an integral whose integrand is a rational function R of the six trigonometric functions, then it transforms into an integral of the form $\int S(t)\,dt$, where $S(t)$ is a rational function of t under the substitution $t = \tan(\theta/2)$. These classical substitutions of calculus come from the previous correspondence between points on the y-axis and on the unit circle $x^2 + y^2 = 1$.

There is a natural generalization of the unit circle.

(3.2) Definition. The Fermat curve F_n of order n is given by the equation in affine x, y-coordinates

$$x^n + y^n = 1,$$

or in projective coordinates by $w^n = x^n + y^n$.

While F_2 has infinitely many rational points on it as given above, Fermat, in 1621, conjectured that the only rational points on F_n for $3 \le n$ were the obvious ones. This is called Fermat's last theorem.

(3.3) Fermat's Last Theorem. *For $3 \le n$, the only rational points on F_n lie on the x-axis and y-axis.*

Fermat stated the theorem in the following form:

Cubum autem in duos cubos, aut quadrato-quadratum in duos quadrato-quadratos, et generaliter nullam in infinitum ultra quadratum potestatem in duas ejusdem nominis fas est dividere; cujus rei demonstrationem mirabilem sane detexi. Hanc marginis exiguitas non caperet.

It is the last comment that has puzzled people for a long time. Proofs were given for special values of n by many mathematicians: For $n = 4$ by Fermat using (3.1), for $n = 3$ by Euler in 1770, for $n = 5$ by Legendre in 1825, and for $n = 7$ by G. Lamé in 1839. The conjecture of Fermat, that is, Fermat's last theorem, had been checked for all n up to a very large six-digit number, and Kummer proved it for all n a regular prime. Only in 1983 as a solution to the more general Mordell conjecture was given by Gerd Faltings, did we know that $F_n(\mathbb{Q})$ has at most finitely many points. We will return to the Mordell conjecture in §6. Finally in 1995 through the effort of A. Wiles and others can we say Fermat's Last Theorem is established, see Chapter 18.

Again we return to a problem related to the unit circle. Recently J. Tunnell has considered the problem of the existence of Pythagorean triples (a, b, c) of positive rational numbers where the area $A = (1/2)ab$ of the right triangle is given.

For example, for (3, 4, 5) the area is 6 and for (3/2, 20/3, 41/6) the area is 5. It can be shown that there are no right triangle with rational sides and area 1, 2, 3, or 4. Thus the problem is not as elementary as it would appear at first glance. We will see that it reduces to the question of rational points on certain cubic curves.

Observe that if A is the area of the right rational triangle with sides (a, b, c), then $m^2 A$ is the area of the rational right triangle (ma, mb, mc). Hence the question reduces to the case of right rational triangles with square-free integer area A. Further, we can order the triple so that $a < b < c$.

(3.4) Proposition. *For a square-free natural number A there is a bijective correspondence between the following three sets:*

(1) *Triples of strictly positive rational numbers (a, b, c) with $a^2 + b^2 = c^2$, $a < b < c$, and $A = (1/2)ab$.*
(2) *Rational numbers x such that x, $x + A$, and $x - A$ are squares.*
(3) *Rational points (x, y) on the cubic curve $y^2 = x^3 - A^2x$ such that x is a square of a rational number, the denominator of x is even, and $y > 0$.*

The sets (1) and (2) are related by observing that for $x = c^2/4$, we have

$$[(a+b)/2]^2 = x + A \quad \text{and} \quad [(a-b)/2]^2 = x - A.$$

Hence $x, x + A$, and $x - A$ are all squares. Conversely, for x as in (2) we define $c = 2\sqrt{x}$ and a and b with $a < b$ by the requirement that $[(a \pm b)/2]^2 = c^2/4 \pm A$. Then (a, b, c) is a Pythagorean triple with $A = 1/2\,ab$.

The sets (2) and (3) are related by assuming first that $x, x + A$, and $x - A$ are squares. Then $x = u^2$ and the product $(x + A)(x - A) = x^2 - A^2 = u^4 - A^2$ is a square denoted v^2. Hence $(uv)^2 = u^6 - A^2u^2$. Setting $y = uv$ and using $u^2 = x$, we obtain $y^2 = x^3 - A^2x$, i.e., (x, y) is a point on the cubic curve given by the equation $y^2 = x^3 - A^2x$. From $x = c^2/4$ we see that x is a square with denominator divisible by 2.

Conversely, if $x = u^2 = (c/2)^2$, i.e., x is a square with denominator divisible by 2, and if $x^3 - A^2x$ is a square y^2, then $v^2 = (y/u)^2 = y^2/x = x^2 - A^2 = (x + A)(x - A)$, and we have a Pythagorean triple $v^2 + A^2 = x^2$. The denominators of x^2 and v^2 are the same t^4 and t is even by assumption. Thus the Pythagorean triple of integers $(t^2v)^2 + (t^2A)^2 = (t^2x)^2$ is primitive, and, hence, it is of the form $t^2v = M^2 - N^2$, $t^2A = 2MN$, and $t^2x = M^2 + N^2$. By (3.1) this in turn yields a Pythagorean triple

$$\left(\frac{2N}{t}\right)^2 + \left(\frac{2M}{t}\right)^2 = 4x = (2u)^2$$

determining a right triangle of area $2MN/4t^2 = t^2A/t^2 = A$. This establishes the equivalence between the various sets and proves the proposition.

§4. Rational Cubics and Mordell's Theorem

Cubics have come up in two places in the previous section. Firstly, there is the Fermat cubic $x^3 + y^3 = 1$ which Euler showed had only two rational points, (1,0) and (0,1). Secondly, there is the cubic $y^2 = x^3 - A^2x$ whose rational points tell us about the existence of right rational triangles of area A. These are special cases of the general cubic which has the following form in projective coordinates $w : x : y$:

$$\begin{aligned} 0 = {} & c_1w^3 + c_2x^3 + c_3y^3 + c_4w^2x + c_5wx^2 \\ & + c_6x^2y + c_7xy^2 + c_8w^2y + c_9wy^2 + c_{10}wxy. \end{aligned}$$

The coefficients are determined only up to a nonzero constant multiple, and, hence, the cubic is given by $c_1 : c_2 : c_3 : c_4 : c_5 : c_6 : c_7 : c_8 : c_9 : c_{10}$, a point in a nine-dimensional projective space. This line of ideas is followed further in Chapter 2.

As in the case of conics, our main interest is to describe the rational points on a rational cuic relative to a given rational point O on the cubic. Again we use a geometric principle concerning the intersection of a line and a cubic. The difference in this case is that we do not compare the cubic with another curve as we did for the conic with a line, but, instead, we move between rational points within the cubic to give the cubic an algebraic structure. This is called the chord-tangent law of composition.

The intersection result needed for a line and a cubic, which is related to (2.2), is the following.

(4.1) Remark. If two of the three intersection points of a rational cubic with a rational line are rational points, then the third point is rational.

To see this, we use the equation $aw + bx + cy = 0$ of the line to eliminate one variable in the third-order equation $F(w, x, y) = 0$ of the cubic. For intersections off the line at infinity, given by $w = 0$, one comes up with a cubic equation in the x-coordinate or in the y-coordinate of the intersection points. Thus the intersection points will be rational if and only if the roots of the cubic equation are rational. Thus (4.1) reduces to the algebraic statement: if a cubic polynomial with rational coefficients has two rational roots, then the third root is rational.

(4.2) Definition. An irreducible cubic is one which cannot be factored over the complex numbers. A point O on a irreducible cubic C is called a singular point provided each line through O intersects C at, at most, one other point. An irreducible cubic without a singular point is called a nonsingular cubic curve, and one with singular points is called a singular cubic.

The description of rational points on rational cubics, which are either reducible or singular cubics follows very much the ideas used for conics. First we consider a cubic with a singular rational point O. A typical example is given by $y^2 = x^2(x+a)$ and $O = (0, 0)$, the origin.

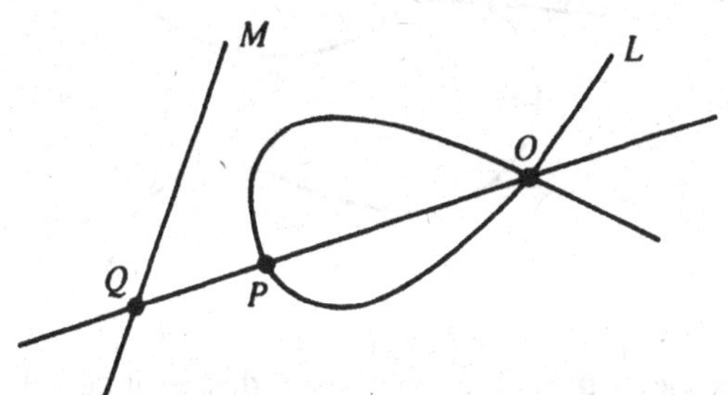

Since O is a singular point, each rational line L through O cuts the cubic at a second point P, and P is rational because its x-coordinate is the solution of a cubic equation in x or in y with a double rational root correspondingto the x- or y-coordinate of O. Thus, as with conics, we can project the singular cubic onto any fixed rational line M in such a way that rational points on the cubic correspond to rational points on the line M.

Next we consider a nonsingular cubic. A line meets these cubics in three points in general, and if we have one rational point, one cannot project the cubic in the naïve manner onto a line to obtain a description of the rational points. Under projection two points on the cubic correspond to one point on the line, and one rational point on the line does not necessarily correspond to a pair of rational points on the cubic.

This leads to a new approach to the description of the rational points. Observe that given two rational points on a rational nonsingular cubic C, we can construct a third one. Namely, you draw the line connecting the two points P and Q. This is a rational line since P and Q are rational, and this line meets the cubic at one more point, denoted PQ, which must be rational by (4.1). The formation of PQ from P and Q is some kind of law of composition for the rational points on a cubic.

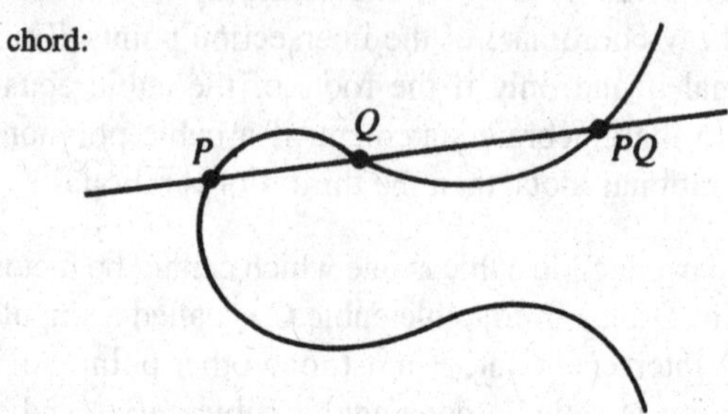

Even if you have only one rational point P, you can still find another, in general, because you draw a tangent to that point, i.e., you join the point to itself.

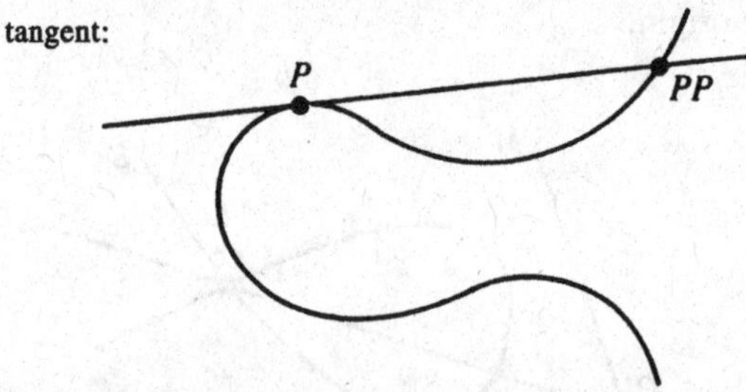

The tangent line meets the cubic twice at P, that is, it corresponds to a double root in the equation of the x-coordinate. By the above argument the third intersection point is rational. Thus, from a few rational points, one can, by forming compositions successively, generate lots of other rational points. The function which associates to a pair P and Q the point PQ is called the chord-tangent composition law.

(4.3) Primitive Form of Mordell's Theorem. *On a nonsingular rational cubic curve there exists a finite set of rational points such that all rational points on the curve are generated from these using iterates of the chord-tangent law of composition.*

In other words there is a finite set X of rational points on the nonsingular rational cubic such that every rational point P can be decomposed in the form,

$$P = (\ldots((P_1 P_2) P_3) \ldots P_r),$$

where $P_1, \ldots, P_r$ are elements of the finite set X with repetitions allowed.

The chord-tangent law of composition is not a group law, because, for example, there is no identity element, i.e., an element 1 with $1P = P = P1$ for all P. However it does satisfy a commutative law property $PQ = QP$.

(4.4) Remark. There are infinitely many rational points on a rational line, and there are either no rational points or infinitely many on a rational conic. The Mordell theorem points to a new phenomenon arising with curves in degree 3, namely the possibility of the set of rational points being finite but nonempty. This would be the case when only a finite number of chord-tangent compositions give all natural points. This theorem introduces the whole idea of finiteness of number of rational points on a rational plane curve. This fits with the Fermat conjecture where $x^n + y^n = 1$ has two points, (1,0) and (0,1), for n odd and four points, $(1, 0)$, $(-1, 0)$, $(0, 1)$, and $(0, -1)$, for n even where $n > 2$.

Finally, there is the question of the existence of any rational points on a rational cubic curve. For conics one could determine by Legendre's theorem (2.4) in a finite number of steps, whether a rational conic had a rational point on it or not. For cubics, there is no known method for determining, in a finite number of steps, whether there is a rational point. This very important question is still open, and it seems like a very difficult problem. The idea of looking at the cubic equation over the p-adic numbers for each prime p is not sufficient in this case, for, in the 1950s, Selmer gave the example

$$3x^3 + 4y^3 + 5z^3 = 0.$$

This is a cubic with a p-adic solution for each p, but with no nontrivial rational solution. The proof that there is no rational solution is quite a feat.

For the early considerations in this book we will leave aside the problem of the existence of a rational point and always assume that the cubics we consider have a given rational point O. Later, in 8 on Galois cohomology, the question of the existence of a rational point on an auxiliary curve plays a role in estimating the number of rational points on a given curve with a fixed rational point.

§5. The Group Law on Cubic Curves and Elliptic Curves

It was Jacobi [1835] in *Du usu Theoriae Integralium Ellipticorum et Integralium Abelianorum in Analysi Diophantea* who first suggested the use of a group law on a projective cubic curve. As we have already remarked the chord-tangent law of composition is not a group law, but with a choice of a rational point O as zero element and the chord-tangent composition PQ we can define the group law $P + Q$ by the relation

$$P + Q = O(PQ).$$

This means that $P+Q$ is the third intersection point on the line through O and PQ.

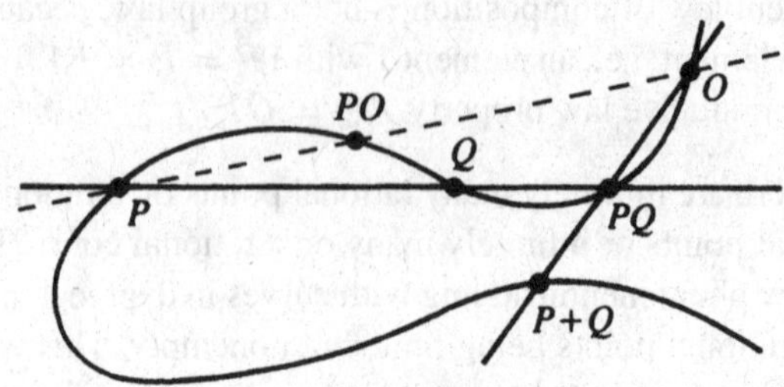

Clearly we have the commutative law $P+Q=Q+P$ since $PQ=QP$. From the fact that $O,\ PO$, and P are the three intersection points of the cubic with the line through O and P, we see that $P=PO=P+O$, and thus O is the zero element. To find $-P$ given P, we use the tangent line to the cubic at O and its third intersection point OO. Then we join P to $O(OO)=0$ with a line and $-P$ is the third intersection point.

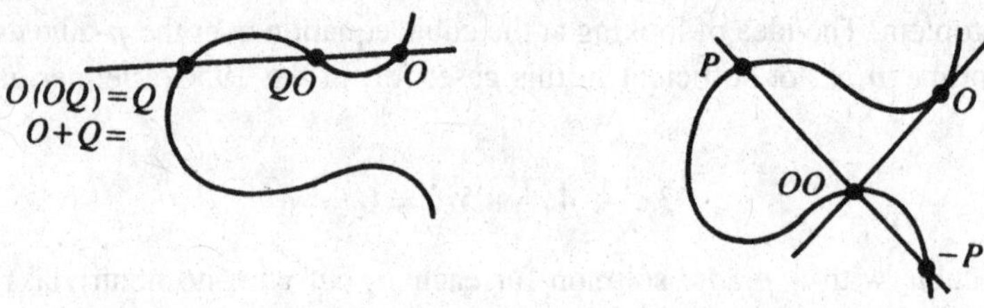

Note that $P+(-P)=O(OO)$ which is O in the above figure. The associative law is more complicated and is taken up in 2. It results from intersection theory for plane curves. Observe that if a line intersects the cubic in three rational points P, Q, and R, then we have $P+Q+R=OO$. We will be primarily interested in cubics where $O=OO$, i.e., the tangent to the cubic O has a triple intersection point. These points are called flexes of the cubic and are considered in 2.

In the next definition we formulate the notion of an elliptic curve over any field k, but, in keeping with the ideas of the introduction, we have in mind the rational field $\mathbb{Q}$, the real field $\mathbb{R}$, or the complex field $\mathbb{C}$.

(5.1) Definition. An elliptic curve E over a field k is a nonsingular cubic curve E over k together with a given point $O\in E(k)$. The group law on $E(k)$ is defined as above by O and the chord-tangent law of composition PQ with the relation $P+Q=O(PQ)$.

In all cases the first question is what can be said about the group $E(k)$ where E is an elliptic curve over k. The first chapter of the book is devoted to looking at examples of groups $E(k)$. Now we can restate Mordell's theorem in a more natural form.

(5.2) Theorem (Mordell 1921). *Let E be a rational elliptic curve. The group of rational points $E(\mathbb{Q})$ is a finitely generated abelian group.*

A rational elliptic curve is an elliptic curve defined over the rational numbers. The proof of this theorem will be given in Chapter 6 and is one of the main results in this book. The result was, at least implicitly, conjectured by Poincaré [1901] in *Sur les Properiétes Arithemétiques des Courbes Algébriques*, where he defined the rank of an elliptic curve over the rationals as the rank of the abelian group $E(\mathbb{Q})$. He studied the properties of the rank in terms of which elements are of the form $3P$. Mordell in his proof looked at the rank in terms of which elements are of the form $2P$ and then substracted off from a given element R elements of the form $2P$ to arrive at a finite set of generators. This is a descent procedure which goes back to Fermat.

In order to perform calculations with specific elliptic curves, it is convenient to put the cubic equation in a standard form. In 2 we show how, by changes of variable, we can eliminate three terms, y^3, xy^2, and wx^2, from the ten-term general cubic equation given at the beginning of §4 and further normalize the coefficients of x^3 and wy^2 to be one. The resulting equation is called an equation in normal form (or generalized Weierstrass equation)

$$wy^2 + a_1wxy + a_3w^2y = x^3 + a_2wx^2 + a_4w^2x + a_6w^3.$$

It has only one point of intersection with the line at infinity namely (0,0,1). In the x, y-plane the equation takes the form

$$y^2 + a_1xy + a_3y = x^3 + a_2x^2 + a_4x + a_6,$$

and it is this equation which is used for an elliptic curve throughout this book. If x has degree 2 and y has degree 3 in the graded polynomial, then the equation has weight 6 when a_i has weight i. The point at infinity (0,0,1) is the zero of the group and the lines through this zero in the x, y-plane are exactly the vertical lines. This zero has the property that $OO = O$ in terms of the chord-tangent composition so that three points add to zero in the elliptic curve if and only if they lie on a line in the plane of the cubic curve. In Chapter 1 we use this group law to calculate with an extensive number of examples.

For an elliptic curve E over $\mathbb{Q}$ we can apply the structure theorem for finitely generated abelian groups to $E(\mathbb{Q})$ to obtain a decomposition $E(\mathbb{Q}) = \mathbb{Z}^g \oplus \mathrm{Tors}\,E(\mathbb{Q})$, where g is an integer called the rank of E and Tors $E(\mathbb{Q})$ is a finite abelian group consisting of all the elements of finite order in $E(\mathbb{Q})$.

In 5 we study the torsion subgroup Tors $E(\mathbb{Q})$ and see that it is effectively computable. From elementary consideratons related to the implicit function theorem one can see that the group of real points $E(\mathbb{R})$ is either a circle group or the circle group direct sum with the group of order 2. Since Tors $E(\mathbb{Q})$ embeds into $E(\mathbb{R})$ as a finite subgroup, we have from this that Tors $E(\mathbb{Q})$ is either finite cyclic or the direct sum of a finite cyclic group with the group of order 2.

The question of a uniform bound on Tors $E(\mathbb{Q})$ as E varies over all curves E defined over $\mathbb{Q}$ was studied from the point of view of modular curves by G. Shimura, A. Ogg, and others, see Chapter 11, §3. In 1976 Barry Mazur proved the following deep result which had been conjectured by Ogg.

(5.3) Theorem (Mazur). *For an elliptic curve E defined over $\mathbb{Q}$ the group* Tors $E(\mathbb{Q})$ *of torsion points is isomorphic to either*

$$\mathbb{Z}/m\mathbb{Z} \qquad \textit{for } m = 1, 2, 3, \ldots, 10, 12$$

or

$$\mathbb{Z}/m\mathbb{Z} \oplus \mathbb{Z}/2\mathbb{Z} \qquad \textit{for } m = 2, 4, 6, \textit{ or } 8.$$

In particular there is no element of order 11, 13, or 14 in the group of rational points on an elliptic curve over $\mathbb{Q}$. There are examples which show that all above cases can occur.

This leaves the question of the rank g. There are examples of curves known with rank up to at least 24. It is unknown whether or not the rank is bounded as E varies over curves defined over $\mathbb{Q}$. Such a bound is generally considered to be unlikely. With our present understanding of elliptic curves the rank g is very mysterious and difficult to calculate in a particular case. See also Rubin and Silverberg [2002].

(5.4) Remark. Let E be an elliptic curve defined over $\mathbb{Q}$ by the equation $y^2 = x^3 + ax + b$. In fact, after a change of variable every elliptic curve over the rational numbers has this form. There is no known effective way to determine the rank of E from these two coefficients, a and b. In fact, there is no known effective way of determining whether or not $E(\mathbb{Q})$ is finite. Of course $E(\mathbb{Q})$ is finite if and only if the rank $g = 0$.

This is one of the basic problems in arithmetic algebraic geometry or diophantine geometry. In 16 we will associate an L-function $L_E(s)$ to E. Conjecturally it has an analytic continuation to the complex plane. This L-function was first introduced by Hasse and was studied further by A. Weil.

(5.5) Birch, Swinnerton–Dyer Conjecture. *The rank g of an elliptic curve E defined over the rational numbers is equal to the order of the zero of $L_E(s)$ at $s = 1$.*

Birch and Swinnerton–Dyer gathered a vast amount of supporting evidence for this conjecture. Coates and Wiles in 1977 made the first real progress on this conjecture for curves with complex multiplication and recently R. Greenberg has shown that the converse to some of their statements also holds. This subject has exploded in the last twenty years and we will not treat any of these developments. The reader should consult the book by K. Rubin, *Euler Systems, Annals of Math Studies*. The final part of the book is devoted to an elementary elaboration of this conjecture.

A refinement of their conjecture explains the number $\lim_{s\to 1}(s-1)^{-g}L_E(s)$. The final part of the book is devoted to an elementary elaboration of this conjecture.

§6. Rational Points on Rational Curves. Faltings and the Mordell Conjecture

The cases of rational points on curves of degrees 1, 2, and 3 have been considered, and we were led naturally into the study of elliptic curves by our simple geometric approach to these diophantine equations. Before going into elliptic curves, we mention some things about curves of degree strictly greater than 3.

(6.1) Mordell Conjecture (For Plane Curves). *Let C be a smooth rational plane curve of degree strictly greater than 3. Then the set $C(\mathbb{Q})$ of rational points on C is finite.*

This conjecture was proved by Faltings in 1983 and is a major achievement in diophantine geometry to which many mathematicians have contributed. Some of the ideas in the proof simplify known results for elliptic curves and we will come back to the subject later.

For curves other than lines, conics, and cubics, it is often necessary to consider models of the curve in higher dimensions and with more than one equation. This leads one directly into algebraic geometry and general notions of algebraic varieties. The topics in elliptic curves treated in detail in this book are exactly those which use only a minimum of algebraic geometry, namely the theory of plane curves given in 2.

From a descriptive point of view the complex points $X(\mathbb{C})$ of an algebraic curve defined over the complex numbers $\mathbb{C}$ have a local structure since $X(\mathbb{C})$ is homeomorphic to an open disc in the complex plane with change of variable given by analytic functions. Topologically $X(\mathbb{C})$ is a closed oriented surface with some number of g holes.

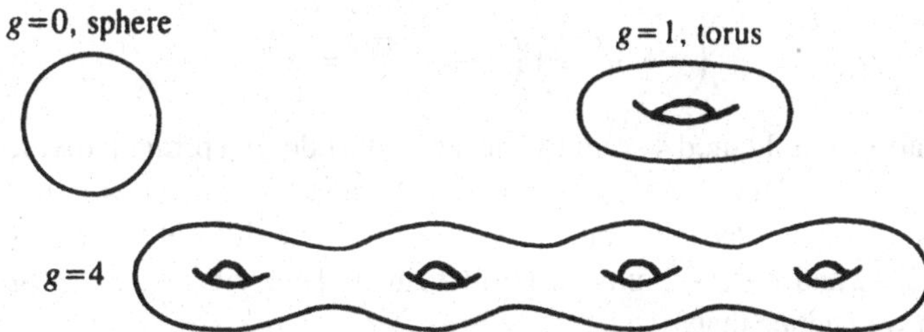

(6.2) Definition. The invariant g is called the genus of the curve.

There are algebraic formulations of the notion of genus, and it is a well-defined quantity associated with any algebraic curve. Lines and conics have genus $g = 0$, singular cubics have genus $g = 0$, and nonsingular cubics have genus $g = 1$.

(6.3) Assertion. A nonsingular plane curve of degree d has genus

$$g = \frac{(d-2)(d-1)}{2}.$$

(6.4) Remark. The Mordell conjecture and Faltings' proof of it are really for curves of genus strictly greater than 1. The curve must be nonsingular but not necessarily a plane curve.

Finally, a closely related subject is the study of integral points on rational curves. In terms of an equation $f(x, y) = 0$ we can ask if there are finitely many (x, y) with $f(x, y) = 0$ and x, y rational integers. With the solution of the Mordell conjecture this problem has less interest, but historically it was Siegel who established the finiteness result.

(6.5) Theorem of Siegel. *The number of integral points on a nonsingular curve of genus strictly greater than 0 and defined over the rational numbers is finite.*

In particular this applies to nonsingular cubic curves, but not to the singular cubic $y^2 = x^3$ which has infinitely many integral points of the form (n^2, n^3), where n is any integer.

(6.6) Remark. For certain explicit elliptic curves there are bounds on the size of the integral points. For exmaple, for $y^2 = x^3 - k$ one has:

$$\max(|x|, |y|) \le \exp\left(2^{7\cdot 2^4} k^{10^9\cdot 2^3}\right).$$

(6.7) Example. The only integral solutions of $y^2 + k = x^3$ for $k = 2$ occur when $y = \pm 5$ and for $k = 4$ occur when $y = \pm 2, \pm 11$. This question goes back to Diophantus and was taken up by Bachet in 1621. For the case $k = 2$ we will give an argument based on properties of the ring $\mathbb{Z}[\sqrt{-2}]$. We factor

$$\left(y + \sqrt{-2}\right)\left(y - \sqrt{-2}\right) = x^3.$$

For the equation to hold mod 4, x and y must both be odd. If a prime p divides x, then p^3 divides $(y + \sqrt{-2})(y - \sqrt{-2})$. If p divides both factors, then it would divide the sum $2y$, and this is impossible since $y^2 + 2 = x^3$. Since this holds for each p, both factors $y + \sqrt{-2}$ and $y - \sqrt{-2}$ must be perfect cubes. Thus $y + \sqrt{-2} = (a + b\sqrt{-2})^3$, from which we deduce that

$$y = a^3 - 6ab^2 = a\left(a^2 - 6b^2\right), \quad 1 = b\left(3a^2 - 2b^2\right).$$

The last equation gives $b = +1$ and $a = \pm 1$, and hence $y = \pm(-5)$ as was asserted. We have used that p has to be a prime of the unique factorization domain $\mathbb{Z}[\sqrt{-2}]$.

Finally we return to Remark (1.6) concerning the definition of a rational curve. A rational curve in the sense of geometry is a curve of genus 0. This definition makes sense over any field and has nothing to do with the rational numbers. The concept of genus is also extended to singular curves where it is called the arithmetic genus. The singular cubic $y^2 = x^3$ is a curve of arithmetic genus $= 0$.

§7. Real and Complex Points on Elliptic Curves

Let E be an elliptic curve over the real or complex numbers. The structure of the groups $E(\mathbb{R})$ and $E(\mathbb{C})$ as continuous groups or Lie groups is completely understood with a little background on the subject of Lie groups. Using E for either the real or the complex points of E, we point out several properties of these groups which allow us to determine their structure from a general result.

(1) There is a topology (or notion of convergence) on E such that E is locally Euclidean of dimension 1 in the real case and dimension 2 in the complex case. This locally Euclidean property comes from the implicit function theorem since E is nonsingular.
(2) The group operations are continuous, in fact, they are algebraic.
(3) The group E is a closed subspace of the projective plane and, since the projective plane is compact, the group E is compact (every sequence has a convergent subsequence).

Lie groups, which can be taken as locally Euclidean groups, have the following structure under suitable assumptions.

(7.1) Assertion. An abelian, compact, and connected Lie group is isomorphic to a product of circles. The number of factors is equal to the dimension of the locally Euclidean space.

To check whether or not this applies to E, we graph E to check its connectivity. Consider the elliptic curve given by the equation in normal form $y^2 + a_1xy + a_3y = f(x)$, where $f(x) = x^3 + \cdots$ is a cubic poynomial. Completing the square

$$\left(y + \frac{a_1x + a_3}{2}\right)^2 = f^*(x),$$

where $f^*(x) = x^3 + \cdots$ is also a cubic polynomial. Hence the graph of this equation for real coefficients is symmetric around the line $2y + a_1x + a_3 = 0$, so that is has one of the following two forms:

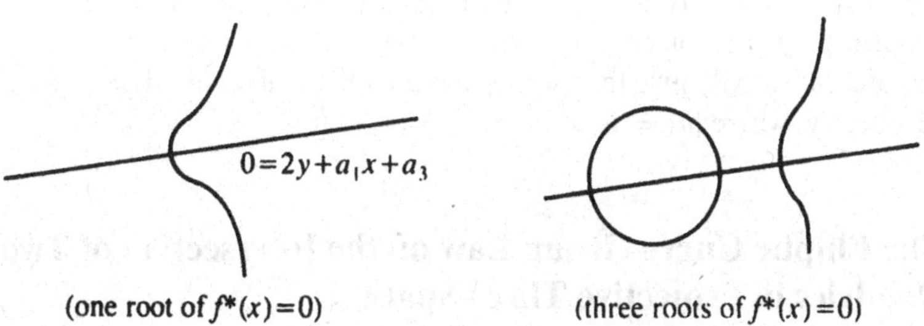

(one root of $f^*(x)=0$) (three roots of $f^*(x)=0$)

In the case of one real root, the group $E(\mathbb{R})$ has one connected component, and in the case of three real roots, the group $E(\mathbb{R})$ has two connected components. From this observation and (7.1), we deduce the following result.

(7.2) Proposition. *Let E be an elliptic curve defined by an equation in the form $(y + ax + b)^2 = g(x)$, where $g(x)$ is a cubic polynomial over the real numbers. If $g(x)$ has only one real root, then $E(\mathbb{R})$ is isomorphic as a Lie group to a circle, and if $g(x)$ has three real roots, then $E(\mathbb{R})$ is isomorphic to a circle direct sum with $\mathbb{Z}/2\mathbb{Z}$.*

Over the complex numbers every elliptic curve is connected so that the corresponding situation is easier to describe.

(7.3) Proposition. *Let E be an elliptic curve defined over the complex numbers. Then $E(\mathbb{C})$ is isomorphic as a Lie group to the product of two circles, hence, it is a torus.*

In the chapter on elliptic functions we will give a proof of this result using complex analysis and an explicit mapping using elliptic functions. Moreover, $E(\mathbb{C})$ will appear as $\mathbb{C}/L$, where L is a lattice in the plane having a basis of two elements. In other words $E(\mathbb{C})$ is isomorphic to $\mathbb{R}/\mathbb{Z} \times \mathbb{R}/\mathbb{Z}$, as asserted above, and $L = \mathbb{Z}\omega_1 + \mathbb{Z}\omega_2$ with $\mathbf{Im}\,(\omega_2/\omega_1) \neq 0$.

(7.4) Remark. From this we see that the kernel of multiplication by n is isomorphic to $\mathbb{Z}/n\mathbb{Z} \times \mathbb{Z}/n\mathbb{Z}$. By contrast the finite subgroups of $E(\mathbb{R})$ are of the form a cyclic group or a cyclic group direct sum with the group of order 2, i.e., of the form $\mathbb{Z}/n\mathbb{Z}$ or $\mathbb{Z}/n\mathbb{Z} \times \mathbb{Z}/2\mathbb{Z}$ up to isomorphism. Since for an elliptic curve E over the rational numbers $E(\mathbb{Q}) \subset E(\mathbb{R})$, the same holds for finite subgroups of $E(\mathbb{Q})$, and in particular for the torsion subgroup Tors $E(\mathbb{Q})$ of $E(\mathbb{Q})$ as remarked in (5.3).

Finally there is the question of why projective space over the real numbers or over the complex numbers is compact. This follows because they are separated quotient spaces of spheres.

(7.5) Remark. Each point in the real or complex projective plane has homogeneous coordinates $w : x : y$ where $|w|^2 + |x|^2 + |y|^2 = 1$. The real projective plane $\mathbb{P}_2(\mathbb{R})$ is a quotient of the 2-sphere S^2 in $\mathbb{R}^3$ where (w, x, y) and (w', x', y') give the same point in $\mathbb{P}_2(\mathbb{R})$ if and only if $w' = uw, x' = ux$, and $y' = uy$, where $u = \pm 1$. The complex projective plane $\mathbb{P}_2(\mathbb{C})$ is a quotient of the 5-sphere S^5 in $\mathbb{C}^3$ where (w, x, y) and (w', x', y') give the same point in $\mathbb{P}_2(\mathbb{C})$ if and only if $w' = uw,\ x' = ux$, and $y' = uy$, where $|u| = 1$.

§8. The Elliptic Curve Group Law on the Intersection of Two Quadrics in Projective Three Space

The content of this section is only sketched and not used in the rest of the book. Supplying the details is a serious exercise.

In Sections 4 and 5 we introduced elliptic curves as certain cubic curves in the projective plane, and using the intersection properties of lines and cubics, we defined

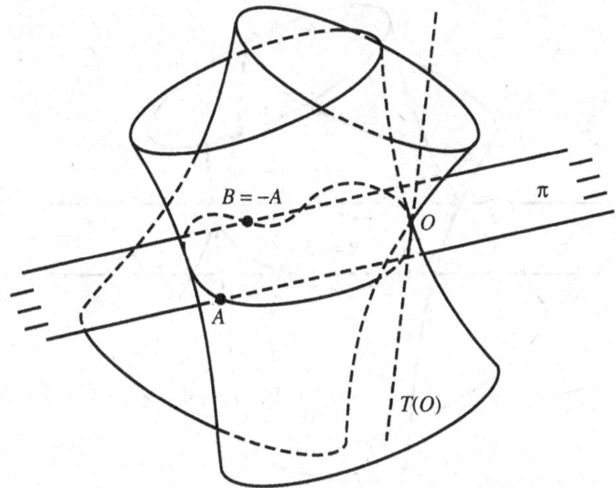

a group law. The zero point corresponded to a flex point of which the tangent line has a triple intersection point.

There is another classical picture of an elliptic curve as the smooth intersection of two quadric hypersurfaces in projective three space. The geometric construction of the group law, which was explained to me by Gizatulin, is outlined here.

(8.1) Zero Point and Negative of an Element. The intersection curve Γ of the two quadric hypersurfaces H' and H'' can be shown to have a hyperflex point, denoted $0 \in \Gamma$. If $T(0)$ denotes the tangent line to Γ at 0, then every plane π containing $T(0)$ intersects Γ in just two other points, i.e., $\pi \cap \Gamma = \{0, A, B\}$. Or in terms of cycles we have $\pi \cap \Gamma = 2.0 + A + B$. With this choice of zero we will make a group law with $B = -A$.

(8.2) Sum of Three Points Equal to Zero. Given $0 \in \Gamma = H' \cap H''$. We define the group law by starting with $P, Q \in \Gamma$, forming the plane $\pi(0, P, Q)$ containing the three points $0, P, Q$. Then there are four points of intersection $\pi(0, P, Q) \cap \Gamma = \{0, P, Q, R\}$. Since 0 is a hyperflex point, we have for cycles $0 + P + Q + R = 4.0$ so that $P + Q + R = 3.0$. If $\pi(P)$ is the plane through P containig $T(0)$, then the group law is given by the following cycle intersections:

$$\pi(0, P, Q) \cap \Gamma = 0 + P + Q + (-(P + Q))$$

and

$$\pi(P) \cap \Gamma = 2.0 + P + (-P).$$

These constructions should be compared with the intersection geometry of a plane cubic curve C where

$$L(P, Q) \cap C = P + Q + (-(P + Q)) \quad \text{and} \quad L(P) = P + (-P) + 0$$

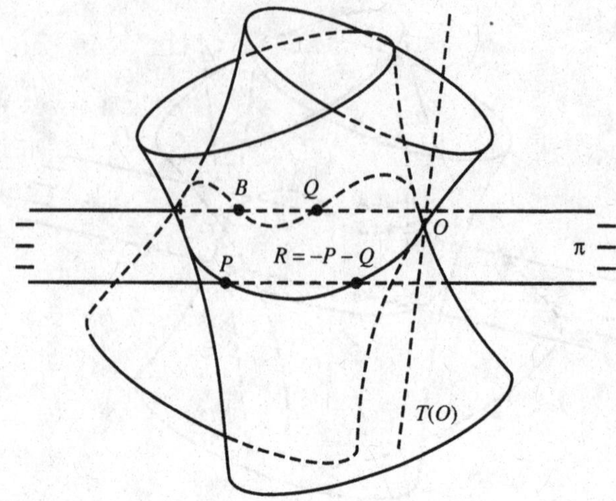

for the line $L(P, Q)$ through $P, Q \in C$ and the tangent line $L(P)$ to C through $P \in C$.

We return in 10(3.5) to elliptic curves as intersections of two quadrics in projective three space, and as quartic curves in 4(3.2).

1

Elementary Properties of the Chord-Tangent Group Law on a Cubic Curve

In this chapter we illustrate how, by using simple analytic geometry, a large number of numerical calculations are possible with the group law on a cubic curve. The cubic curves in x and y will be in normal form, that is, without x^2y, xy^2, or y^3 terms. In this form the entire curve lies in the affine x, y-plane with the exception of 0:0:1 which is to be zero in the group law. The lines through O are exactly the vertical lines in the x, y-plane, and all other lines used are of the form $y = \lambda x + \beta$. We use the definition of $P + Q$ as given in §5 of the Introduction.

We will postpone several questions related to the group law until Chapter 3 where they are dealt with using results from Chapter 2 on the general theory of algebraic curves. These include the associativity of the group law and a detailed discriminant criterion for a curve to be nonsingular. The procedure for transforming a general cubic into one in normal form will be worked out there too.

In Theorem (4.1) of this chapter there is a condition for an element (x', y') on an elliptic curve $E(k)$ to be of the form $2(x, y) = (x', y')$ in the group $E(k)$. This plays an important role in the proof of the Mordell theorem in Chapter 6.

In this chapter k will always denote a general field. For the beginning reader this can for most considerations be viewed as the rational numbers $\mathbb{Q}$.

§1. Chord-Tangent Computational Methods on a Normal Cubic Curve

A cubic equation in normal form, or generalized Weierstrass form, is an expression

$$y^2 + a_1xy + a_3y = x^3 + a_2x^2 + a_4x + a_6,$$

where the coefficients a_i are in the field k. Since there is no term of the form y^3 in the equation, a vertical line $x = x_0$ intersects the locus of the normal cubic at two points (x_0, y_1) and (x_0, y_2), where y_1 and y_2 are the roots of the quadratic equation $y^2 + (a_1x_0 + a_3)y - (x_0^3 + a_2x_0^2 + a_4x_0 + a_6) = 0$. In the completed plane, that is, the projective plane, we see that the cubic in normal form has one additional solution at infinity which we call O, and this O is the third point of intersection of the vertical line with the locus of the cubic equation in normal form in the projective plane.

(1.1) Definition. The elliptic curve E corresponding to the cubic equation in normal form is the locus of all solutions $(x, y) \in k^2$ of the equation

$$y^2 + a_1xy + a_3y = x^3 + a_2x^2 + a_4x + a_6$$

together with the point O which is on every vertical line.

When we wish to emphasize that we are looking at solutions (x, y) with x and y in k, we write $E(k)$ for E. Usually we use the term elliptic curve only for nonsingular cubics. We return to criteria for nonsingularity in the next chapter. The choice of O on cubic makes the nonsingular curve into an elliptic curve.

In the context of the normal form of a cubic equation for an elliptic curve $E(k)$ we give rules for the group law. Then O is the zero element in $E(k)$ and addition is carried out as in §5 of the Introduction.

(1.2) Assertion. Let E be an elliptic curve defined by an equation in normal form. If $P = (x, y)$ is a point on $E(k)$, then the negative $-P$ is (x, y^*), where $y + y^* = -a_1x - a_3$ or, in other words, $-(x, y) = (x, -y - a_1x - a_3)$.

Observe that O, (x, y), and (x, y^*) are the points of intersection of the vertical line through $(x, 0)$ with $E(k)$. As seen above, y and y^* are two roots of a quadratic equation over k where the sum of the roots is $-(a_1x + a_3)$ in k and so, if y is in k, then y^* is also in k.

The operation $P \mapsto -P$ defines a map $E(k) \to E(k)$ which is an involution of the curve onto itself, i.e., $-(-P) = P$. Also it shows that the curve has a vertical reflection symmetry with respect to the line

$$y = -\frac{a_1x + a_3}{2}$$

in the plane. For this we require $2 \neq 0$ in k, that is, the characteristic of $k \neq 2$.

(1.3) Example. For E given by the equation

$$y^2 + y - xy = x^3$$

we have $-(x, y) = (x, -y - 1 + x)$ and the curve is vertically symmetric about the line $y = (1/2)x - 1/2$.

In the diagram we have included for future reference two tangent lines to the curve T at $(1, 1)$ and T' at $(1, -1)$. The slopes of tangent lines are computed by implicit differentiation of the equation of the curve

$$(2y + 1 - x)y' = 3x^2 + y.$$

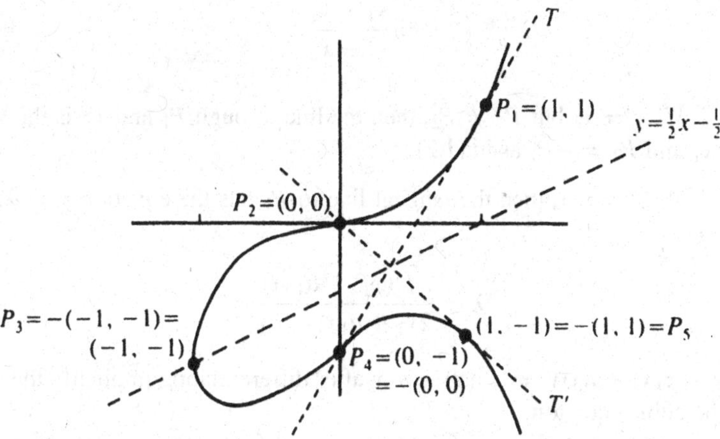

(1.4) Addition of Two Points. Let E be an elliptic curve defined by the equation in normal form

$$y^2 + a_1xy + a_3y = f(x) = x^3 + a_2x^2 + a_4x + a_6.$$

In order to add two points $P_1 = (x_1, y_1)$ and $P_2 = (x_2, y_2)$ we first form the line through P_1 and P_2 or the tangent line at P_1 when $P_1 = P_2$. Consider the third point of intersection denoted by $P_1P_2 = (x_3, y_3)$, so that

$$P_1 + P_2 = -P_1P_2.$$

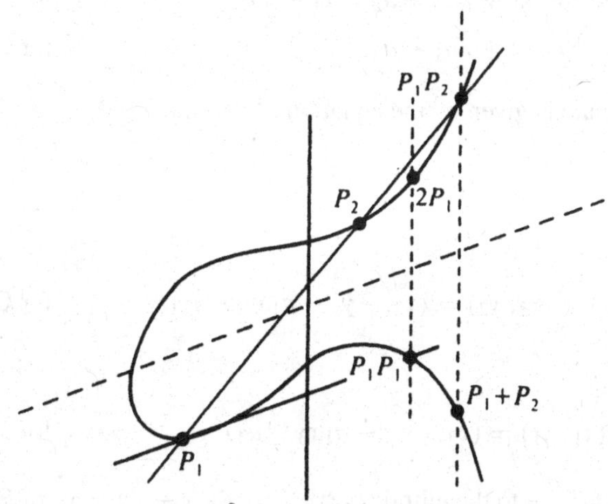

Case 1. If $x_1 \neq x_2$ so that $P_1 \neq P_2$, then the line through P_1 and P_2 has an equation $y = \lambda x + \beta$, where

$$\lambda = \frac{y_1 - y_2}{x_1 - x_2}.$$

Case 2. If $x_1 = x_2$ but $P_1 \neq P_2$, then the line through P_1 and P_2 is the vertical line $x = x_1$ and $P_2 = -P_1$ as in (1.2).

Case 3. If $P_1 = P_2$, then the tangent line to P_1 has the equation $y = \lambda x + \beta$, where

$$\lambda = \frac{f'(x_1) - a_1 y_1}{2y_1 + a_1 x_1 + a_3}$$

since $(2y + a_1 x_1 + a_3)y' = f'(x) - a_1 y$ after differentiating implicitly the normal form of the cubic equation.

Now substituting $y = \lambda x + \beta$ into the normal form of the cubic equation, and collecting all terms on one side, we have the following relations:

$$(\lambda x + \beta)^2 + a_1 x(\lambda x + \beta) + a_3(\lambda x + \beta) = x^3 + a_2 x^2 + a_4 x + a_6$$

and

$$0 = x^3 + (a_2 - \lambda^2 - \lambda a_1)x^2 + (a_4 - 2\lambda\beta - a_1\beta - \lambda a_3)x + (a_6 - \beta^2 - a_3\beta).$$

The three roots of this cubic equation are x_1, x_2, and x_3, the x-coordinates of the three intersection points, either P_1, P_2, and P_1P_2 in Case 1 or P_1, P_1, and P_1P_1 in Case 3. Since the sum of the roots is the negative of the coefficient of x^2 in the cubic equation for x, we have the following formula for x_3:

$$\begin{aligned} x_3 &= \lambda^2 + \lambda a_1 - a_2 - x_1 - x_2 && \text{for Case 1, Case 2} \\ &= \lambda^2 + \lambda a_1 - a_2 - 2x_1 && \text{for Case 3,} \end{aligned}$$

and the y-coordinate is given by the equation of the line

$$y_3 = \lambda x_3 + \beta.$$

Finally,

$$(x_1, y_1) + (x_2, y_2) = (x_3, -y_3 - a_1 x_3 - a_3) \qquad \text{for Case 1, Case 2}$$

and

$$2(x_1, y_1) = (x_3, -y_3 - a_1 x_3 - a_3) \qquad \text{for Case 3.}$$

(1.5) Example. Return to the elliptic curve E: $y^2 + y - xy = x^3$ of Example (1.3). Denote by P the point $(1, 1)$ on E, and observe that the tangent line T to P cuts the cubic at $(0, -1) = -(0, 0)$. Thus $2P = (0, 0)$ by the procedure in (1.4). Next observe that $y = x$ is the line through $P = (1, 1)$ and $2P = (0, 0)$, and the third point of intersection is $(-1, -1) = -(-1, -1)$. Hence $3P = (-1, -1)$. Further

$0 = 2(-1, -1) = 2 \cdot 3P = 6P$ since the tangent at $(-1, -1)$ is vertical, and we deduce that P is a point of order 6. In particular

$$4P = -2P = -(0, 0) = (0, -1)$$

and

$$5P = -P = -(1, 1) = (1, -1).$$

Thus $P_m = mP$, and O together with the five points $P, 2P, 3P, 4P$, and $5P$ shown on the cubic E form a cyclic subgroup of order 6 in $E(\mathbb{Q})$ in (1.3).

If we study the question of when the product of three consecutive numbers $y(y+1)$ is the product of three consecutive numbers $(x-1)x(x+1) = x^3 - x$, we are led to the following example of an elliptic curve.

(1.6) Example. The elliptic curve E defined by the normal cubic equation $y^2 + y = x^3 - x$ has six obvious points on it $(0,0)$, $(1,0)$, $(-1,0)$, $(0,-1)$, $(1,-1)$, and $(-1,-1)$. If $P = (0,0)$, then these points are all in the subgroup generated by P as with $P = (1,1)$ in (1.5), but in this case P generates an infinite cyclic group.

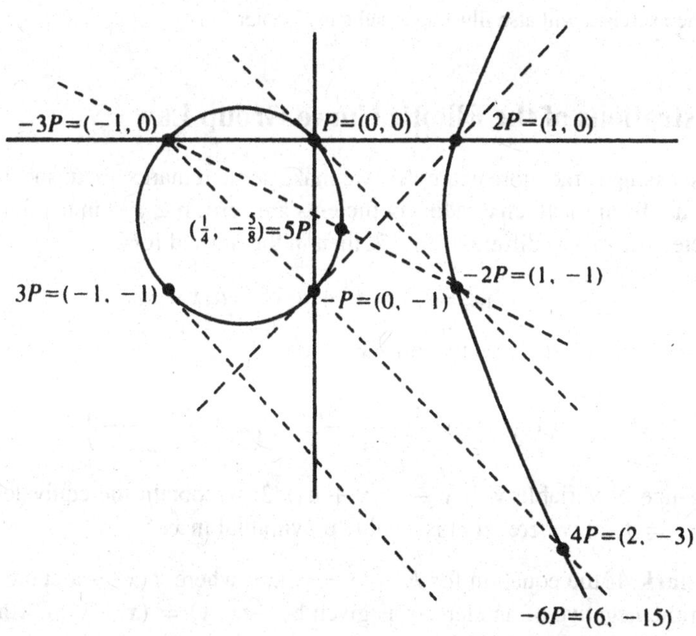

In general odd multiples of P are on the closed component of the curve which contains $P = (0,0)$ and the even multiples of P are on the open component which is closed up to a circle by adding O at infinity. We have the following values for a few multiples of P: $9P = (-20/7^2, -435/7^3)$, $16P = (a/65^2, b/65^3)$ for some integers a and b, and $51P = (a'/N^2, b'/N^2)$, where the natural number N has 32 digits and a', b' are integers.

Exercises

1. Find $7P, -7P, 8P, -8P, 9P, -9P, 10P$, and $-10P$ on $y^2 + y = x^3 - x$ in (1.6). Using Mazur's theorem in §5 of the Introduction, argue that P has infinite order where $P = (0, 0)$.
2. Show that $P = (0, 2)$ is a point of order 3 on $y^2 = x^3 + 4$.
3. Show that $P = (2, 4)$ is a point of order 4 on $y^2 = x^3 + 4x$.
4. Show that $P = (2, 3)$ is a point of order 6 on $y^2 = x^3 + 1$.
5. Show that $P = (-12, 108)$ is a point of order 5 on
$$y^2 = x^3 - 16 \cdot 27x + 19 \cdot 16 \cdot 27.$$
6. Let E denote the elliptic curve defined by the cubic equation
$$y^2 = x(x-1)(x+9).$$
Find a subgroup of order four in $E(\mathbb{Q})$, show that $(-1, 4) = P$ is on $E(\mathbb{Q})$ and not in this subgroup, and calculate nP for n between -7 and $+7$. Using Mazur's theorem in §5 of the Introduction, argue that P has infinte order.

Remark. The exercises will also illustrate results in Chapter 5.

§2. Illustrations of the Elliptic Curve Group Law

Before discussing some more examples, we make some remarks about special forms of the normal form in which certain coefficients are zero. If $2 \neq 0$ in the field k, i.e., the characteristic of k is different from 2, then in the normal form

$$y^2 + y(a_1x + a_3) = x^3 + a_2x^2 + a_4x + a_6,$$

we can complete the square on the left-hand side

$$y^2 + y(a_1x + a_3) + \frac{(a_1x + a_3)^2}{4} = \left(y + \frac{a_1x + a_3}{2}\right)^2.$$

With a change of variable y to $y - (a_1x + a_3)/2$, we obtain the equivalent cubic equation $y^2 = f(x)$, where $f(x)$ is a cubic polynomial in x.

(2.1) Remark. If the equation for E is $y^2 = f(x)$, where $f(x)$ is a cubic polynomial, then the negative of an element is given by $-(x, y) = (x, -y)$. Furthermore, the cubic will be nonsingular if and only if $f(x)$ has no repeated roots.

The reason that we might consider normal forms with terms a_1xy and a_3y is that the cubic might have a particularly simple form as in the case (1.3). These terms are always necessary potentially for a theory in characteristic 2, e.g., over the field $\mathbb{F}_2$ of two elements.

(2.2) Remark. The point $(0, 0)$ is on the curve with $y^2 = f(x)$ if and only if the equation has the form $y^2 = x^3 + ax^2 + bx$. If r is a root of $f(x)$, then $y^2 = f(x+r)$ has this form, and we will use the equation of the elliptic curve in this form frequently. If $3 \neq 0$ in the field, i.e., the characteristic of k is different from 3, then in the special normal form $y^2 = f(x)$ we can complete the cube in the right-hand side and after translation of x by a constant we have the Weierstrass form of the cubic

$$y^2 = x^3 + ax + b.$$

Now we consider some examples of subgroups of points on $E(\mathbb{Q})$ for elliptic curves which arise frequently.

(2.3) Example. For the curve E defined by $y^2 + y = x^3 - x^2$ we have four obvious points, $(1, 0)$, $(0, 0)$, $(0, -1) = -(0, 0)$ and $(1, -1) = -(1, 0)$, on the curve. The tangent line through $(1, 0)$ intersects E at $(0, -1)$ from which we deduce $2(1, 1) = (0, 0)$ and thus also $2(1, -1) = (0, -1)$. The tangent line to $(0, 0)$ intersects E at $(1, 0)$ from which we deduce $2(0, 0) = (1, -1)$.

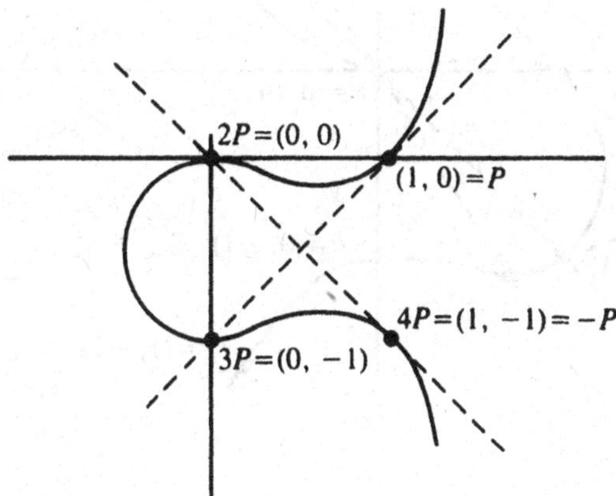

From $2(1, 1) = (0, 0)$ and $2(0, 0) = (1, -1) = -(1, 1)$, we obtain $4(1, 1) = (1, -1) = -(1, 1)$ or $5(1, 1) = O$. Hence the set

$$\{0, (1, 0), (0, 0), (0, -1), (1, -1)\}$$

is a cyclic subgroup of $E(\mathbb{Q})$ of order 5. In fact we will see later that $E(\mathbb{Q})$ is exactly this cyclic group of order 5.

The next example differs from the previous one by a single change in sign of the coefficient of x^2. The group of rational points is now infinite instead of finite as in (2.3).

(2.4) Example. For the curve E defined by $y^2 + y = x^3 + x^2$ we have four obvious points $(0, 0)$, $(-1, 0)$ $(0, -1) = -(0, 0)$ and $(-1, -1) = -(-1, 0)$, on the curve.

The point $P = (0, 0)$ generates an infinite cyclic subgroup of $E(\mathbb{Q})$. For example one can calculate $2P = (-1, -1)$, $-3P = (1, 1)$, $3P = (1, -2)$, $4P = (2, 3)$, and $5P = (-3/4, 1/8)$. This example is similar to that of (1.6).

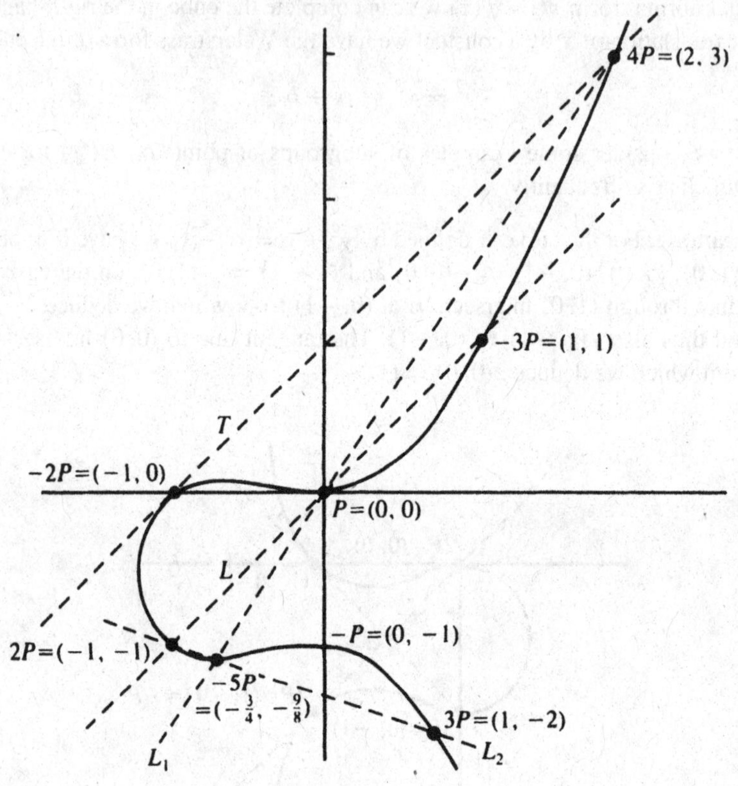

The tangent line T at $-2P$ goes through the cubic at $4P$, the line L through $2P$ and P goes through the cubic at $-3P$, and $-5P$ is calculated either by L_1 going through P and $4P$ or by L_2 going through $2P$ and $3P$.

(2.5) Example. This example is related to the Fermat equation $w^3 = u^3 + v^3$. It is known that the only rational solutions of this equation are $(u, v, w) = (0, 0, 0)$, $(1, 0, 1)$, $(-1, 0, -1)$, $(0, 1, 1)$, and $(0, -1, -1)$. This cubic equation is not in normal form, but under the transformation

$$x = \frac{3w}{u+v} \quad \text{and} \quad y = \frac{9}{2}\left(\frac{u-v}{u+v}\right) + \frac{1}{2},$$

we obtain the cubic curve E with equation in normal form

$$y^2 - y = x^3 - 7.$$

The calculation is an exercise; it is helpful to note that $y^2 - y = z^2 - 1/4$ when $y = z + 1/2$.

The transformation is not defined at the solution $(0, 0, 0)$ of the Fermat equation, but the solutions $(1, 0, 1)$ and $(-1, 0, -1)$ map to $(x, y) = (3, 5)$ on $E(\mathbb{Q})$ while $(0, 1, 1)$ and $(0, -1, -1)$ map to $(x, y) = (3, -4)$ on $E(\mathbb{Q})$. Thus $E(\mathbb{Q})$ is the cyclic group of three elements $\{0, (3, 5), (3, -4)\}$, for if $E(\mathbb{Q})$ contained other rational points (x, y) there would be corresponding (u, v, w) solutions to the Fermat equation $u^3 + v^3 = w^3$.

Now let us check the $3P = 0$ for $P = (3, 5)$ by calculating $-2P$ from the tangent line to E at P. From the derivative $(2y - 1)y' = 3x^2$ we calculate the slope of the tangent line to E at $(3, 5)$ to be $3 \cdot 3^2/(2 \cdot 5 - 1) = 27/9 = 3$. The tangent line is $y = 5 + 3(x - 3) = 3x - 4$, and this line intersects the cubic $y^2 - y = x^3 - 7$ at points whose x coordinates are roots of

$$x^3 - 7 = (3x - 4)(3x - 5) = 9x^2 - 27x + 20$$

or

$$\begin{aligned} 0 &= x^3 - 9x^2 + 27x - 27 \\ &= (x - 3)^3. \end{aligned}$$

Since the roots are the x-coordinates of the intersection of the tangent line to E at $(3, 5)$ with E at least two roots are equal to 3, and the third root, which is again 3, is the x-coordinate of $-2P$. Hence we must have $-2P = P$, so that $3P = 0$. The graph of $y^2 - y = x^3 - 7$ has just two rational points $(3, 5)$ and $(3, -4)$.

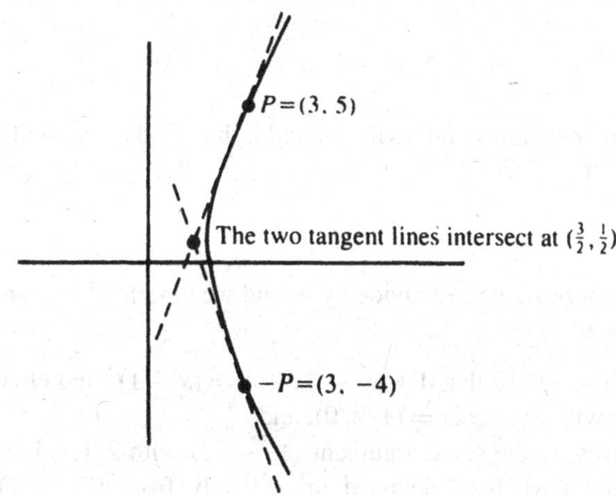

(2.6) Example. The example is related to the fourth-power Fermat equation $w^4 = u^4 + v^4$. Again it is known that the only rational solutions of this equation are

$(u, v, w) = (0,0,0), (1,0,1), (-1,0,-1), (0,1,1)$, and $(0,-1,-1)$. This is not even a cubic equation, but under the transformation

$$u^2 = \frac{2yw^2}{x^2} \quad \text{and} \quad v = w\left(1 - \frac{1}{x}\right),$$

we obtain

$$w^4 = \frac{4w^4y^2}{x^4} + w^4\left(1 - \frac{1}{x}\right)^4.$$

Dividing by w^4 and multiplying by x^4, we derive the cubic equation

$$y^2 = x^3 - \frac{3}{2}x^2 + x - \frac{1}{4}.$$

The equation factors giving one point $(1/2, 0)$ of order 2

$$y^2 = \left(x - \frac{1}{2}\right)\left(x^2 - x + \frac{1}{2}\right).$$

Consider the lines $y = \lambda(x - 1/2)$ through $(1/2, 0)$ and their other intersection points with the cubic. They have x-coordinates satisfying

$$\lambda^2\left(1 - \frac{1}{2}\right) = x^2 - x + \frac{1}{2}$$

or equivalently

$$0 = x^2 - (1 + \lambda^2)x + \frac{1}{2}(1 + \lambda^2).$$

The two other intersection points will coincide, that is, the line will be tangent to curve if and only if

$$(1 + \lambda^2)^2 = 2(1 + \lambda^2).$$

Since $\lambda^2 + 1$ is nonzero, we can divide by it, and we obtain $\lambda^2 = 1$ or $\lambda = +1, -1$. Substituting back for

1. $\lambda = 1$, $y = x - 1/2$ so that $0 = x^2 - 2x + 1 = (x-1)^2$ and $(1, 1/2)$ is a point on the curve with $2(1, 1/2) = (1/2, 0)$, and
2. $\lambda = -1$, gives by the same argument $(1, -1/2)$ with $2(1, -1/2) = (1/2, 0)$, but this could have been deduced immediately from (1) by $2(1, -1/2) = -2(1, 1/2) = -(1/2, 0) = (1/2, 0)$. It is an application of the group structure.

For this curve E with equation in normal form

$$y^2 = x^3 - \frac{3}{2}x^2 + x - \frac{1}{4},$$

the group $E(\mathbb{Q})$ is the cyclic group of four elements

$$\left\{0, \left(1, \frac{1}{2}\right), \left(\frac{1}{2}, 0\right), \left(1, -\frac{1}{2}\right)\right\},$$

for if $E(\mathbb{Q})$ contained other rational points (x, y) there would be corresponding (u, v, w) solutions to the Fermat equation $u^4 + v^4 = w^4$, see Exercise 2. This equation for the curve E takes a simplier form under the substitutions $x/4$ for $x - 1/2$ and $y/8$ for y. The new equations are

$$\left(\frac{y}{8}\right)^2 = \frac{x}{4}\left(\frac{x^2}{16} + \frac{1}{4}\right)$$

or

$$y^2 = x^3 + 4x.$$

This is the curve considered in Exercise 3 to §1 where the question was to show that a certain point was of order 4; in fact, from the relation to the Fermat curve we deduce that $E(\mathbb{Q})$ is cyclic of order 4 for the curve defined by $y^2 = x^3 + 4x$.

Exercises

1. Find $+6P, -6P, +7P, -7P, 8P$, and $-8P$ on $y^2 + y = x^3 + x^2$ in (2.4). Using Mazur's theorem in §5 of the Introduction, argue that P has infinite order.
2. In (2.5) and in (2.6) write u and v as functions of (w, x, y).
3. For $u^3 + v^3 = c$ consider the change of variable

$$u + v = \frac{12c}{x} \quad \text{and} \quad u - v = \frac{y}{3x}.$$

Show that

$$x = \frac{12c}{u + v} \quad \text{and} \quad y = 36c\frac{u - v}{u + v},$$

and that the curve is transformed into the cubic in normal form

$$y^2 = x^3 - 432c^2.$$

Determine the group of rational points on this elliptic curve for $c = 1$, for $c = 2$.

4. Show that $P = (3, 12)$ is a point of order 8 on

$$y^2 = x^3 - 14x^2 + 81x.$$

5. Show that $P = (1, 0)$ is a point of order 7 on

$$y^2 + xy + y = x^3 - x^2 - 3x + 3.$$

6. Calculate all multiples nP of $P = (3, 8)$ on

$$y^2 = x^3 - 43x + 166.$$

Find the order of $(3, 8)$ and of $(11, 32)$ on this curve.

7. The Fermat curve $u^3 + v^3 = 1$ is mapped to $y^2 + y = x^3$ by the functions $y = -u^3$ and $x = -uv$ (or $v = -x/u$). Determine $E(\mathbb{Q})$, where E is the elliptic curve defined by $y^2 + y = x^3$.
8. Let E by the elliptic curve given in normal form by

$$y^2 + a_1xy + a_3y = x^3 + a_2x^2 + a_4x + a_6.$$

A point $P = (x, y)$ satisfies $2P = 0$, or equivalently $P = -P$, if and only if x is a solution of

$$0 = x^3 + \left(a_2 + \frac{a_1^2}{4}\right)x^2 + \left(a_4 + \frac{a_1a_3}{2}\right)x + \left(a_6 + \frac{a_3^2}{4}\right)$$

and $y = -(a_1x + a_3)/2$. Show that the subgroup of all P with $2P = 0$ is isomorphic to either 0, $\mathbb{Z}/2\mathbb{Z}$, or $\mathbb{Z}/2\mathbb{Z} \oplus \mathbb{Z}/2\mathbb{Z}$.

§3. The Curves with Equations $y^2 = x^3 + ax$ and $y^2 = x^3 + a$

We study two sets of curves each with a single nonzero parameter a. In order that the curve be nonsingular, we need $a \neq 0$ in both cases, otherwise the curve has a cusp at the origin and is given by the equation $y^2 = x^3$. In Chapter 3 we will see that these are exactly the two families of curves where each curve has more symmetries, or automorphisms, than the obvious one $(x, y) \mapsto -(x, y) = (x, -y)$. If ζ_4 and ζ_3 are fourth and third roots of unity in k (unequal to -1 or $+1$), then $(x, y) \mapsto (-x, \zeta_4 y)$ is an automorphism of the curve defined by $y^2 = x^3 + ax$ and $(x, y) \mapsto (\zeta_3 x, y)$ is an automorphism of the curve defined by $y^2 = x^3 + a$. This topic is discussed in detail in Chapters 3 and 4.

(3.1) Remark. If we substitute u^2x for x and u^3y for y, we obtain $u^6y^2 = (u^3y)^2 = (u^2x)^3 + a(u^2x) = u^6(x^3 + (a/u^4)x)$ in the first equation. This means that we can assume that a is a nonzero integer which is free of any fourth-power factor. In the second equation we obtain $u^6y^2 = (u^2y)^2 + a = u^6(x^3 + a/u^6)$. This means that we can assume that a is a nonzero integer which is free of any sixth-power factor.

Now we study points of finite order on the elliptic curves defined by $y^2 = x^3 + ax$ and $y^2 = x^3 + a$. The significance of rational points on these curves in two special cases of $y^2 = x^3 + 4x$ and $y^2 = x^3 - 432$ was considered in its relation to the Fermat curves $u^4 + v^4 = w^4$ and $u^3 + v^3 = w^3$, respectively, in the previous sections (2.6) and (2.5) respectively.

(3.2) Theorem. *Let E be the elliptic curve defined by the equation $y^2 = x^3 + ax$, where a is a fourth-power free integer. The torsion subgroup of $E(\mathbb{Q})$ is*

$$\operatorname{Tors} E(\mathbb{Q}) = \begin{cases} \mathbb{Z}/2\mathbb{Z} \oplus \mathbb{Z}/2\mathbb{Z} & \text{if } -a \text{ is a square,} \\ \mathbb{Z}/4\mathbb{Z} & \text{if } a = 4, \\ \mathbb{Z}/2\mathbb{Z} & \text{if } -a \text{ is not a square, or } -4. \end{cases}$$

Proof. In all cases $(0, 0)$ is a point of order 2 because any point of order 2 has the form $(x, 0)$ where x is a root of the cubic equation $0 = x^3 + ax$. In particular, there are three points of order 2 if and only if $-a$ is a square.

Next consider the equation $2(x, y) = (0, 0)$ on $E(\mathbb{Q})$. For such a point there would be a line L: $y = \lambda x$ through $(0, 0)$ tangent to E at (x, y). Thus

$$(\lambda x)^2 = x^3 + ax \quad \text{or} \quad 0 = x(x^2 - \lambda^2 x + a).$$

Since L is tangent to E at (x, y) the quadratic equation

$$0 = x^2 - \lambda^2 x + a$$

would have a double root at R, and this condition is equivalent to the discriminant being zero, or, $0 = \lambda^4 - 4a$. Because a has no fourth-power factor, this has a rational solution λ if and only if $a = 4$ and $\lambda = +2$ or -2. In this case the points (x, y) satisfying $2(x, y) = 0$ are $(2, 4)$ and $(2, -4)$. This discussion shows that the 2-power torsion in $E(\mathbb{Q})$ has the above form, and we are left to show that there is no odd torsion.

Next we will show that there is no 3-torsion, that is, no points $P = 0$ on $E(\mathbb{Q})$ with $2P = -P$. If there were such a point P, then the tamgent line $y = \lambda x + \beta$ to E at P when substituted into

$$(\lambda x + \beta)^2 = x^3 + ax$$

or

$$0 = x^3 - \lambda^2 x^2 + (a - 2\beta\lambda)x - \beta^2$$

would be a perfect cube $0 = (x - r)^3$ with r the x-coordinate of P. This would mean that $3r = \lambda^2$ and $r^3 = \beta^2$ given the relation $\beta^2 = \lambda^6/27$. Finally, the third relation between coefficients $3r^2 = a - 2\beta\lambda$ gives the following formula

$$3\left(\frac{\lambda^4}{9}\right) = a - 2\left(\frac{\lambda^4}{3\sqrt{3}}\right)$$

which is impossible for a and λ rational numbers since $\sqrt{3}$ is irrational.

In this theorem and the next the nonexistence of p torsion for $p > 3$ will be taken up later.

(3.3) Theorem. *Let E be the elliptic curve defined by the equation $y^2 = x^3 + a$, where a is a sixth-power free integer. The torsion subgroup of $E(\mathbb{Q})$ is*

$$\operatorname{Tors} E(\mathbb{Q}) = \begin{cases} \mathbb{Z}/6\mathbb{Z} & \textit{if } a = 1, \\ \mathbb{Z}/3\mathbb{Z} & \textit{if } a \textit{ is a square different from } 1, \textit{ or } a = -432 = -2^4 3^3, \\ \mathbb{Z}/2\mathbb{Z} & \textit{if } a \textit{ is a cube different from } 1, \\ 0 & \textit{if } a \textit{ is not a cube, a square, or } -432. \end{cases}$$

Proof. A point of order 2 has the form $(x, 0)$, where x is a root of $x^3 + a$. This exists on $E(\mathbb{Q})$ if and only if a is a cube c^3 of an integer c and then $(-c, 0)$ is the point of order 2. We consider the equation $2(x, y) = -(c, 0)$ to see if there is a point (x, y) of order 4. If this were the case, the line $y = \lambda(x + c)$ through $(-c, 0)$ when substituted into the equation of the curve yields

$$\lambda^2(x+c)^2 = x^3 + c^3 \quad \text{or} \quad \lambda^2(x+c) = x^2 - cx + c^2.$$

The line through $(-c, 0)$ is tangent at another point (x, y) on E if and only if the following quadratic equation has a double root

$$0 = x^2 - (\lambda^2 + c)x + c(c - \lambda^2),$$

that is, if the discriminant $\lambda^4 + 2\lambda^2 c + c^2 - 4c(c - \lambda^2) = 0$. After completing the square in this equation, we obtain the following equation for λ^2:

$$(\lambda^2 + 3c)^2 = 12c^2.$$

There are no rational solutions to this equation because 12 is not a square, and thus, there are no points of order 4 on the curve. This proves that the 2-power torsion is what is stated in the theorem.

A point (x, y) is of order 3, that is, $2(x, y) = -(x, y)$ if and only if there is a line $y = \lambda x + \beta$ through (x, y) such that

$$(\lambda x + \beta)^2 = x^3 + a$$

is a perfect cube. In other words

$$(x - r)^3 = x^3 - \lambda^2 x^2 - 2\lambda\beta x + (a - \beta^2).$$

From the relations $-3r = -\lambda^2$ and $3r^2 = -2\beta\lambda$ we deduce the relation $\lambda^4 = -6\beta\lambda$.

For $\lambda = 0$ we see that a is a square b^2 and then $(0, +b)$ and $(0, -b)$ are the two points of order 3.

For $\lambda \neq 0$ we divide by λ to obtain $\lambda^3 = -6\beta$. From $3r^2 = -2\beta\lambda$ we derive the relation $27r^6 = -8\beta^3(-6\beta)$ or $3^2 r^6 = 2^4\beta^4$, and this relation is satisfied only in one case $\beta = 2^2 3^2 m^3$ and $r = 2^3 3m$. In this case we calculate

$$a = \beta^2 - r^3 = 2^4 3^4 m^6 - 2^6 3^3 m^6 = (3 - 4)2^4 3^3 m^6 = -2^4 3^3 m^6 = -432m^6,$$

where $m = 1$ since a is sixth-power free. This proves that the 3 torsion is what is stated above.

Again the nonexistence of p torsion for $p > 3$ will be taken up later.

While the torsion for the curves defined by the equations $y^2 = x^3 + ax$ and $y^2 = x^3 + a$ is completely understood, the rank g of the finitely generated abelian group $E(\mathbb{Q})$ is another matter. There is considerable numerical information which led Birch and Swinnerton-Dyer [1963] to formulate their conjectures which are taken up in Chapter 17. We quote some values of g for small a for each of the types of curves from their paper. These values a played a role in a two-part discussion leading to the conjectures relating the rank g with the zero of the L function at $s = 1$.

Included in Tables 1 and 2 are the cubics $y^2 = x^3 + 4x$ and $y^2 = x^3 - 432$ with only a finite number of rational points from their relation to the Fermat curve theorem. This is equivalent to $g = 0$.

Table 1. Values of the rank g for E given by $y^2 = x^3 + ax$.

$g = 0$:	$a =$	1	2	4	6	7	10	11	12	22
		−1	−3	−4	−8	−9	−11	−13	−18	−19
$g = 1$:	$a =$	3	5	8	9	13	15	18	19	20
		−2	−5	−6	−7	−10	−12	−14	−15	−20
$g = 2$:	$a =$	14	33	34	39	46	−17	−56	−65	−77
$g = 3$:	$a =$	82								

Table 2. Values of the rank g for E given by $y^2 = x^3 + a$.

$g = 0$:	$a =$	1	4	6	7	13	14	16	20	21
		−1	−3	−5	−6	−8	−9	−10	−14	−432
$g = 1$:	$a =$	2	3	5	8	9	10	11	12	18
		−2	−4	−7	−13	−15	−18	−19	−20	−21
$g = 2$:	$a =$	15	17	24	37	43	−11	−26	−39	−47
$g = 3$:	$a =$	113	141	316	346	359	−174	−307	−362	

Exercises

1. Determine the five points in the (x, y)-plane which together with 0 form the group of six torsion points on $y^2 = x^3 + 1$. Note that from (3.5) this is all of the group $E(\mathbb{Q})$. Which of these points are generators of this cyclic group of order 6?
2. Let E be the curve defined by $y^2 = x^3 + ax^2 + bx = x(x^2 + ax + b)$ over a field k of characteristic $\neq 2$, e.g., $k = Q$, the rational numbers, with $b \neq 0$ and the discriminant $a^2 - 4b \neq 0$.
   (a) Show that $(\mathbb{Z}/2\mathbb{Z})^2 \subset E(k)$ if and only if $a^2 - 4b$ is a square.
   (b) When $a^2 - 4b$ is not a square, show that $\mathbb{Z}/4\mathbb{Z} \subset E(k)$ if and only if $b = c^2$ is a square and either $a + 2c$ or $a - 2c$ is a nonzero square.
   (c) Show that $(\mathbb{Z}/4\mathbb{Z}) \times (\mathbb{Z}/2\mathbb{Z}) \subset E(k)$ with $(0, 0) \in 2E(k)$ if and only if $b = c^2$ is a square and $a + 2c$ or $a - 2c$ are squares. Note that if two of the following are nonzero squares, $a + 2c, a - 2c$, and $a^2 - 4b(b = c^2)$, then the third one is also.
3. Find a point of infinite order on the elliptic curves with equations $y^2 = x^3 + 3x$, $y^2 = x^3 - 2x$, $y^2 = x^3 + 2$, and $y^2 = x^3 - 2$.

The next two problems are very difficult and the beginner in the subject should not expect to carry them out completely.

4. Find two points P and Q of infinite order on the elliptic curve over $\mathbb{Q}$ defined by the equation $y^2 = x^3 - 11$ such that $nP \neq mQ$ for all integers m and n.
5. On the curve $y^2 + y = x^3 - 7x + 6$ show that an integral combination $a(0, 2) + b(1, 0) + c(2, 0) = 0$ only for $a = b = c = 0$.

§4. Multiplication by 2 on an Elliptic Curve

Let E be an elliptic curve given by an equation $y^2 = f(x)$, where $f(x)$ is a cubic polynomial without repeated roots. We assume through this section that the field of coefficients k has characteristic different from 2.

There is the homomorphism $E(k) \xrightarrow{2} E(k)$ given by multiplication by 2 in the abelian group $E(k)$. We have already studied the kernel of this homomorphism; it consists of 0 together with all $(x, 0)$, where x is a root of $f(x)$.

We wish to characterize the image $2E(k)$ of multiplication by 2. In Chapter 6 we will show that $E(k)$ is a finitely generated abelian group when k is the rational number field $\mathbb{Q}$ or more generally any number field. One step in the proof is the demonstration that the index $(E(k) : 2E(k))$ is finite. It is clearly a necessary condition for $E(k)$ to be finitely generated. In the next theorem we consider an elementary criterion for a point Q in $E(k)$ to be of the form $Q = 2P$ for P in $E(k)$, i.e., for Q to be in $2E(k)$. This theorem is proved using only the elementary methods of this chapter drawn from the classical theory of algebraic equations applied to analytic geometry.

(4.1) Theorem. *Let E be an elliptic curve defined over a field k by the equation*

$$y^2 = (x-\alpha)(x-\beta)(x-\gamma) = x^3 + ax^2 + bx + c, \qquad \text{with } \alpha, \beta, \gamma \in k.$$

For $(x', y') \in E(k)$ there exists $(x, y) \in E(k)$ with $2(x, y) = (x', y')$ if and only if $x' - \alpha$, $x' - \beta$, and $x' - \gamma$ are squares.

Proof. The equation $2(x, y) = (x', y')$ has a solution on $E(k)$ if and only if the related equation $2(x, y) = (0, y')$ has a solution on the curve defined by the normal cubic

$$y^2 = (x + x' - \alpha)(x + x' - \beta)(x + x' - \gamma).$$

Hence we are reduced to proving the assertion of the theorem for the special point $(0, y')$. In this case $y'^2 = -\alpha\beta\gamma$.

If $2(x, y) = (0, y')$, then substitute the equation of the tangent line $y = \lambda x + \delta$ to E at (x, y) into the cubic equation for E to obtain

$$(\lambda x + \delta)^2 = x^3 + ax^2 + bx + c.$$

Since $\delta = y'$ and $\delta^2 = y'^2 = -\alpha\beta\gamma - c$, this is equivalent to

$$0 = x(x^2 + (a - \lambda^2)x + (b - 2\lambda\delta)).$$

As $y = \lambda x + \delta$ is tangent to E at (x, y) and the root giving the point of tangency is a double root, the discriminant of the quadratic factor in the cubic in x must be zero. Thus we have

$$(\lambda^2 - a)^2 = 4(b - 2y'\lambda)$$

which is a quadratic equation for λ.

This quadratic equation in λ is solved by arranging the two sides into perfect squares

$$(\lambda^2 - a + u)^2 = 2u\lambda^2 - 8\lambda y' + (u^2 + 4b - 2ua).$$

The right-hand side is a perfect square if and only if again the discriminant is zero, that is,

$$0 = 8^2c - 4 \cdot 2u(u^2 + 4b - 2ua)$$

or

$$0 = u^3 - 2au^2 + 4bu - 8c.$$

To solve this cubic for u, we substitute $u = -2v$, and observe that the equation becomes

$$0 = (-8)(v^3 + av^2 + bv + c).$$

This is the cubic term in the equation of the curve, and, hence, the roots are $v = \alpha$, β, γ, so that $u = -2\alpha, -2\beta, -2\gamma$.

Now substitute $u = -2\alpha$ say (or $-2\beta, -2\gamma$) into the quadratic equation for λ, and use the relations

$$-a = \alpha + \beta + \gamma, \quad b = \alpha\beta + \beta\gamma + \gamma\alpha, \quad \text{and} \quad c = -\alpha\beta\gamma.$$

Thus the equations for λ become

$$\begin{aligned}(\lambda^2 &+ \alpha + \beta + \gamma - 2\alpha)^2 \\ &= -4\alpha\lambda^2 - 8y'\lambda + (4a^2 + 4[\alpha\beta + \beta\gamma + \gamma\alpha] - 4\alpha[\alpha + \beta + \gamma]),\end{aligned}$$

or

$$\begin{aligned}(\lambda^2 - \alpha + \beta + \gamma)^2 &= -4\alpha\lambda^2 - 8y'\lambda + 4\beta\gamma \\ &= 4(\alpha'\lambda - \beta'\gamma')^2,\end{aligned}$$

where $\alpha'^2 = -\alpha$, $\beta'^2 = -\beta$, and $\gamma'^2 = -\gamma$. Now we take the square root of both sides to obtain two quadratic equations for

$$\lambda^2 + \beta + \gamma - \alpha = \pm 2(\alpha'\lambda - \beta'\gamma').$$

In these equations we complete the square to get

$$(\lambda^2 \mp 2\alpha'\lambda - \alpha) = -\beta \mp 2\beta'\gamma' - \gamma$$

or

$$(\lambda \mp \alpha')^2 = (\beta' \mp \gamma')^2.$$

Taking the square root of both sides of the last equation, we find four solutions for λ proving the existence of λ in k and hence also of the point (x, y) since

$$x = \frac{1}{2}(\lambda^2 + \alpha + \beta + \gamma) \quad \text{and} \quad y = \lambda x + \delta.$$

This proves the theorem.

Observe that the values of λ are given by the following:

$$\lambda_1 = \alpha' + \beta' - \gamma', \qquad \lambda_3 = -\alpha' + \beta' + \gamma',$$
$$\lambda_2 = \alpha' - \beta' + \gamma', \qquad \lambda_4 = -\alpha' - \beta' - \gamma',$$

for $\alpha'^2 = -\alpha$, $\beta'^2 = -\beta$, and $\gamma'^2 = -\gamma$.

(4.2) Corollary. *For an elliptic curve E defined over an algebraically closed field the group homomorphism*

$$E(k) \xrightarrow{2} E(k)$$

is surjective, that is, the group $E(k)$ is 2 divisible.

The corollary gives rise to an exact sequence

$$0 \to {}_2E(k) = (\mathbb{Z}/2\mathbb{Z})^2 \to E(k) \xrightarrow{2} E(k) \to 0$$

for an algebraically closed field k, where ${}_2E(k)$ is the kernel of multiplication by 2. The points of order 2 form the noncyclic group of order 4, The "Viergruppe" since the cubic equation for the x-coordinate of the nonzero P with $2P = 0$ has three roots, see Exercise 8, §2.

(4.3) Remark. For any n prime to the characteristic of an algebraically closed field k and for any elliptic curve E over k we have an exact sequence

$$0 \to {}_nE(k) = (\mathbb{Z}/n\mathbb{Z})^2 \to E(k) \xrightarrow{n} E(k) \to 0,$$

where ${}_nE(k)$ is the kernel of multiplication by n. this is discussed in further detail in 12(1.4) for $k = \mathbb{C}$ and in 12(3.6) over any algebraically closed k.

Exercises

1. Let A be an abelian group, and for an integer n, let ${}_nA$ denote the kernel of $A \xrightarrow{n} A$ given by multiplication by n. If P is one solution to the equation $nP = Q$ for given Q, then show that any other solution P' to $nP' = Q$ is of the form $P' = P + K$, where K is in ${}_nA$.
2. Show that the condition in Exercise 2(c) to §3 is equivalent to the one in Theorem (4.1) for a solution (x, y) to $2(x, y) = (0, 0)$ on the curve defined by $y^2 = x(x - \alpha)(x - \beta)$.
3. Let E be an elliptic curve a field k of characteristic $\neq 2$. In terms of the three possible groups for ${}_2E(k)$ give all the possible groups ${}_4E(k)$ of points P with $4P = 0$. In each possible group give the number of points P of order 4, i.e., $4P = 0$ and $2P \neq 0$.
4. Show that an elliptic curve E defined over the rational numbers cannot have a subgroup $(\mathbb{Z}/4\mathbb{Z})^2$ contained in $E(\mathbb{Q})$, its group of rational points.

§5. Remarks on the Group Law on Singular Cubics

The two basic examples of singular points on cubic curves are:

(1) A double point $(0, 0)$ on $y^2 = x^2(x + a)$.
(2) A cusp $(0, 0)$ on $y^2 = x^3$.

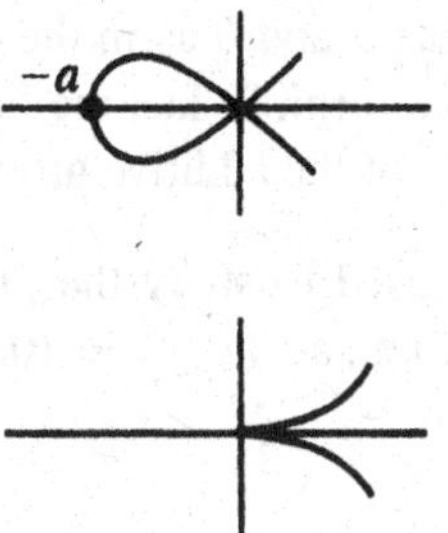

For a cubic in normal form

$$y^2 + a_1xy + a_3y = x^3 + a_2x^2 + a_4x + a_6,$$

the derivative y' satisfies

$$(2y + a_1x + a_3)y' = 3x^2 - a_1y + 2a_2x + a_4,$$

and at $(0, 0)$ we have

$$a_3y' = a_4.$$

The value of y' is indeterminate here, and these are singular points. Thus, the curve has a singularity at $(0, 0)$ if and only if $a_3 = a_4 = 0$.

Before discussing the group law for points on singular cubics, we consider the curve A of all (x, y) satisfying the cubic equation $y = x^3 + ax + b$. This is not in normal form since there is no y^2 term.

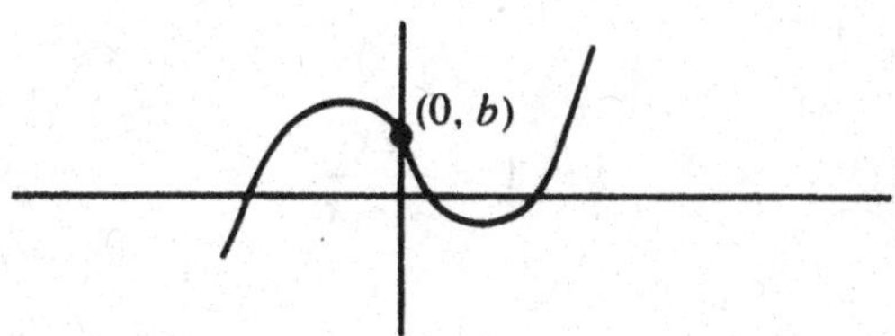

If (x_1, y_1), (x_2, y_2), and (x_3, y_3) are three points on the cubic A and on a line $y = \lambda x + \beta$, then

$$0 = x^3 + (a - \lambda)x + (b - \beta)$$
$$= (x - x_1)(x - x_2)(x - x_3)$$

and $x_1 + x_2 + x_3 = 0$. Hence the set of points of A defined over k, denoted $A(k)$, has the structure of a group where $(0, b) = 0$ and $-(x, y) = -(x, x^3 + ax + b) = (-x, (-x)^3 - ax + b)$. Again three points add to zero if and only if they lie on a line.

(5.1) Remark. This example does not give rise to a new group with the group structure depending on the coefficients a and b as in the case of $y^2 = x^3 + ax + b$, or, more generally, a cubic in normal form. In fact the function $f(t) = (t, t^3 + at + b)$ is an isomorphism $f : k \to A(k)$ of the additive group of the line k onto $A(k)$.

The above example can be used to study the group law on $E_{\text{ns}}(k)$, the set of nonsingular points of $y^2 = x^3$, i.e., all $(x, y) \neq (0, 0)$ with $y^2 = x^3$. Making the change of variable

$$x = \frac{u}{v}, \qquad y = \frac{1}{v}, \qquad \text{or} \quad u = \frac{x}{y}, \qquad v = \frac{1}{y},$$

the equation $y^2 = x^3$ is transformed into $(u/v)^3 = 1/v^2$ or $v = u^3$. A line $ax + by + c = 0$ is transformed into the line with equation $au + b + cv = 0$, and the point 0 at infinity in the x, y-plane is transformed to $(0, 0)$ in the u, v-plane. Thus we have:

(5.2) Proposition. *The function $g(t) = (1/t^2, 1/t^3)$ is an isomorphism $g : k \to E_{\text{ns}}(k)$ of the additive group k onto the group of nonsingular points of the cuspidal cubic curve $y^2 = x^3$.*

(5.3) Remark. The origin $(0, 0)$ cannot be included in the set with the chord-tangent group law on $E(k)$ for E defined by $y^2 = x^3$ since any line $y = \lambda x$ through $(0, 0)$ intersects the cubic at only *one* other point $(x, y) = (\lambda^2, \lambda^3)$. The chord-tangent group law is defined with all lines not passing through $(0, 0)$.

Consider the cubic $y^2 = x^2(x + 1) = x^3 + x^2$ with a double point (or node) at the origin $(0, 0)$. Using the substitution

$$u = \frac{y + x}{y - x} \quad \text{and} \quad v = \frac{1}{y - x},$$

and the calculations

$$u - 1 = \frac{2x}{y - x}$$

and

$$(y - x)^3(u - 1)^3 = 8x^3 = 8(y^2 - x^2) = 8uv(y - x)^3,$$

we obtain the equation $(u - 1)^3 = 8uv$, which is similar to the cubic in (5.1). Lines in the x, y-plane are transformed into lines in the u, v-plane with some exceptions, and 0 at infinity in the extended x, y-plane transforms to 1. If (u_1, v_1), (u_2, v_2), and (u_3, v_3) are three points on a line $v = \lambda u + \delta$, then

$$0 = (u-1)^3 - 8u(\lambda u + \delta) = (u - u_1)(u - u_2)(u - u_3),$$

and thus we have $u_1 u_2 u_3 = 1$. This means that $(x, y) \mapsto (y+x)/(y-x)$ is a group homomorphism of the nonsingular points of the cubic curve E given by $y^2 = x^3 + x^2$ into the multiplicative group $k^* = k - \{0\}$.

Let E be the cubic curve $y^2 = x^2(x+1)$ together with 0 at infinity, and let $E_{\mathrm{ns}}(k)$ denote $E(k) - \{0\}$. Given a nonzero u we define v by the equation $8uv = (u-1)^3$. From

$$y - x = \frac{1}{v} \quad \text{and} \quad y + x = \frac{u}{v},$$

we have the definition of

$$x = \frac{u-1}{2v} \quad \text{and} \quad y = \frac{u+1}{2v}.$$

These formulas require that 2 is nonzero in k, and with the above discussion yield the following proposition.

(5.4) Proposition. *With the above notations for E over a field k of characteristic $\neq 2$ the function $f : E_{\mathrm{ns}}(k) \to k^*$ given by $f(0) = 1$ and $f(x, y) = (y-x)(y+x)$ is a group isomorphism of the k-valued points on E_{ns} onto the multiplicative group of k.*

Exercise

1. Show that $(0, 0)$ is a singular point on

$$y^2 + a_1 xy = x^3 + a_2 x^2$$

by change of coordinates. Determine conditions on a_1 and a_2 when it is a double point and when it is a cusp.

2

Plane Algebraic Curves

In the Introduction and Chapter 1 we considered several ways of studying properties of rational points on curves using intersection theory, and we saw how modifications had to be made for singular points. In this chapter we develop background material on projective spaces, plane curves, and to a limited extent on hypersurfaces, and in the next chapter we apply these results to cubic curves.

Elementary intersection theory and the theory of singular points, as given here, is based on the resultant of two polynomials. This is a very classical and elementary approach. In the Appendix to this chapter the theory of the resultant is worked out and other background in algebra is supplied. In particular many aspects related to foundational questions in algebraic geometry are not made explicit. Frequently the same symbol is used for coordinates and for variables.

This chapter is used to give some details left open concerning the group law and singular points on cubics. It will be used also in Chapter 5 for the reduction modulo p of curves defined over the rational numbers.

§1. Projective Spaces

In the Introduction we considered the projective plane in order to have a satisfactory intersection theory of lines in the plane. The result of the basic geometric assertion:

(P) Two distinct points determine, i.e., lie on, a unique line, and two distinct lines determine, i.e., intersect at, a unique point.

The projective plane was modeled on the set of one-dimentional subspaces of the vector space k^3 over the field k. The lines in the projective plane were sets of all one-dimentional subspaces contained in a given two-dimentional subspace. In this context projective transformations are just induced by linear automorphisms by taking direct image.

It will be useful to have higher-dimentional projective spaces, as we did in the Introduction, in order to speak of the space of all cubic curves.

(1.1) Definition. The r-dimensional projective space $\mathbb{P}_r(k)$ over a field k consists of equivalence classes of $(r+1)$-tuples $y_0 : \cdots : y_r$, where the y_i are not all zero. The equivalence relation is defined by $y_0 : \cdots : y_r = y'_0 : \cdots : y'_r$ provided there is a nonzero constant a with $y'_i = ay_i$ for $i = 0, \ldots, r$.

If k is a subfield of K, then there is an obvious inclusion $\mathbb{P}_r(k) \subset \mathbb{P}_r(K)$.

(1.2) Definition. A hyperplane in $\mathbb{P}_r(k)$ is the set of all $y_0 : \cdots : y_r$ in $\mathbb{P}_r(k)$ satisfying an equation

$$a_0 y_0 + \cdots + a_r y_r = 0,$$

where not all a_i are zero.

Two equations $a_0 y_0 + \cdots + a_r y_r = 0$ and $a'_0 y_0 + \cdots + a'_r y_r = 0$ determine the same hyperplane if and only if there is a nonzero b with $a'_i = ba_i$ for all $i = 0, \ldots, r$. Hence the set of hyperplanes form a projective space where the point $a_0 : \cdots : a_r$ corresponds to the hyperplane given by the equation $a_0 y_0 + \cdots + a_r y_r = 0$.

(1.3) Remark. Let H_i denote the hyperplane $y_i = 0$ for $i = 0, \ldots, r$. The subset $\mathbb{P}_r(k) - H_i$ can be parametrized as k^r where $(x_1, \ldots, x_r)$ in k^r corresponds to the point $x_1 : \cdots : x_i : 1 : x_{i+1} : \cdots : x_r$ in $\mathbb{P}_r(k) - H_i$. If $y_i \neq 0$ in $y_0 : \cdots : y_r$ or $y_0 : \cdots : y_r \in \mathbb{P}_r(k) - H_i$, then it corresponds to $(y_0/y_i, \ldots, y_{i-1}/y_i, y_{i+1}/y_i \ldots, y_r/y_i)$ in k^n. Observe that

$$\mathbb{P}_r(k) = (\mathbb{P}_r(k) - H_0) \cup \cdots \cup (\mathbb{P}_r(i) - H_r).$$

We speak of H_0 as the hyperplane at infinity.

There is a coordinate-free version of projective space where we assign to any finite-dimensional vector space V of dimensional $r+1$ the r-dimensional projective space $\mathbb{P}(V)$ of all one-dimensional subspaces $P \subset V$. An s-dimensional linear subspace $M \subset \mathbb{P}(V)$ is determined by M^+, an $(s+1)$-dimensional subspace of V where for $P \in \mathbb{P}(V)$ we have $P \in M$ provided $P \subset M^+$.

A point $P \in \mathbb{P}(V)$ is represented $P = kv$ where $v \in P$ is nonzero vector in V defined over the field k. Properties of sets of nonzero vectors in V can be transferred to those of sets of points in $\mathbb{P}(V)$.

(1.4) Definition. Let Δ be a set of points in $\mathbb{P}(V)$. Let Δ' be the set of representatives of nonzero $v \in V$ with $kv \in \Delta$. The set Δ is a general position provided any subset $\Lambda' \subset \Delta'$ with $\#\Lambda' \leq \dim(V)$ is a linearly independent set.

Observe that for points $P_0 \ldots, P_m$ in $\mathbb{P}(V)$ with $m \leq r$, the following conditions are equivalent. We say they define the property of being in general position:

(1) No $s+1$ of the points $P_0, \ldots, P_m$ lie on an $(s-1)$-dimensional plane M in $\mathbb{P}(V)$ for each $s \leq m$.
(2) For nonzero vectors $v_i \in P_i$ the set of vectors $v_0 \ldots, v_m$ in V is linearly independent.

Exercises

1. For the finite field k of q elements determine the cardinality of the projective plane $\mathbb{P}_2(k)$ and more generally of the projective space $\mathbb{P}_r(k)$. How many points are there on a line and how many lines are there in $\mathbb{P}_2(k)$? Determine the number of s-dimensional subspaces in $\mathbb{P}_r(k)$.
2. If M_i is an s_i-dimensional subspace of $\mathbb{P}_r(k)$, where k is any field, then show that the intersection $M_1 \cap M_2$ is a subspace and determine its possible dimensions.
3. Show that a projective space of dimensional $n-r-1$ parametrizes the $(r+1)$-dimensional subspaces in $\mathbb{P}_n(k)$ containing a fixed r-dimensional subspace M_0.

§2. Irreducible Plane Algebraic Curves and Hypersurfaces

In (1.5) of the Introduction we considered a definition of a plane curve in the context of the complex numbers. Now we define and study plane curves over any field k using homogeneous polynomials $f(w, x, y) \in k[w, x, y]$. The polynomial has a degree d and the fact that it is homogeneous of degree d can be expressed by the relation

$$f(tw, tx, ty) = t^d f(w, x, y).$$

Again we use the notation $C_f(K)$ for the locus of all $w : x : y$ in $\mathbb{P}_2(K)$ with $f(w, x, y) = 0$, where K is any extension field of k.

Since $f(w, x, y) = 0$ if and only if $f(w, x, y)^2 = 0$, the set of points $C_f(K)$ does not determine the equation for any K. When f factors as $f = f_1 \dots f_r$, the algebraic curve can be represented as a union $C_f = C_{f_1} \cup \dots \cup C_{f_r}$. Similarly if f divides g, then the inclusion $C_f \subset C_g$ holds. Since the question of whether or not f factors depends on the field k where the coefficients of f are taken from, we will have to speak of curves over a given field k.

(2.1) Definition. An irreducible plane algebraic curve C_f of degree d defined over a field k is given by an irreducible homogeneous polynomial $f(w, x, y) \in k[w, x, y]$ of degree d. The points $C_f(K)$ of C_f in an extension field K of k consists of all $w : x : y$ in $\mathbb{P}_2(K)$ such that $f(w, x, y) = 0$.

A curve of degree 1 is called a line, 2 a conic, 3 a cubic, 4 a quartic, 5 a quintic, and 6 a sextic. For two extensions $K \subset K'$ of k, the inclusions $\mathbb{P}_2(K) \subset \mathbb{P}_2(K')$ and $C_f(K) \subset C_f(K')$ hold. For the reader with a background in categories and functors, it is now clear that C_f is a subfunctor of the functor $\mathbb{P}_2$ defined on the category of fields over k and k-morphisms to the category of sets.

If $u : k^3 \to k^3$ is a nonsingular linear transformation with inverse v, and if $u, v : \mathbb{P}_2(K) \to \mathbb{P}_2(K)$ are the associated projective transformations defined by direct image on lines, then for each homogeneous $f(w, x, y) \in k[w, x, y]$ of degree d the composite fv is homogeneous of degree d and $u(C_f(K)) = C_{fv}(K)$. This follows since $(fv)(u(w, x, y)) = 0$ if and only if $f(w, x, y) = 0$. Thus projective transformations carry algebraic curves to algebraic curves and preserve degree and the property of irreducibility. In this way we can frequently choose a convenient coordinate system for the discussion of properties of a curve.

(2.2) Remark. The above discussion extends to hypersurfaces H_f in $\mathbb{P}_n$ defined by homogeneous polynomials $f(y_0 \dots, y_n)$ of degree d in $k[y_0, \dots, y_n]$. For an extension field K of k the set $H_f(K)$ consists of all $y_0 : \dots : y_n$ in $\mathbb{P}_n(K)$ such that $f(y_0, \dots, y_n) = 0$.

Observe that the hyperplane H_0 given by $y_0 = 0$, called the hyperplane at infinity, is contained in H_f if and only if y_0 divides f. This is equivalent to $f(0, y_1, \dots, y_n) = 0$ or to the relation $\deg(g) < \deg(f)$ where $g(x_1, \dots, x_n) = f(1, x_1, \dots, x_n)$. The two polynomials f and g have the same factorization into irreducible factors except for the factor y_0^a, where $a = \deg(f) - \deg(g)$. The affine hypersurface $H_g(K)^{\text{aff}}$ is the set of all $(x_1, \dots, x_n)$ in K^n with $g(x_1, \dots, x_n) = 0$. Observe that $H_g(K)^{\text{aff}} = H_f(K) \cap K^n$. For some questions, as in the next theorem, it is more convenient to use the affine picture.

Note that if f divides f' in $k[y_0, \dots, y_n]$, then $H_f(K) \subset H_{f'}(K)$ and $H_g(K)^{\text{aff}} \subset H_{g'}(K)^{\text{aff}}$ with the above notations for the relation between f and g and between f' and g'.

(2.3) Theorem. *Let H_f and $H_{f'}$ be two hypersurfaces defined over k in $\mathbb{P}_n$, where f is irreducible. If for some algebraically closed extension field L of K we have the inclusion $H_f(L) \subset H_{f'}(L)$, then f divides f', and so $H_f(K) \subset H_{f'}(K)$ for all extension fields K of k.*

Proof. We can consider the inclusion of hypersurfaces in some affine space where $H_g(L)^{\text{aff}} \subset H_{g'}(L)^{\text{aff}}$. In affine coordinates we can write

$$g(x_1, \dots, x_{n-1}, x_n) = a_0(x_1, \dots, x_{n-1}) + \dots + a_d(x_1, \dots, x_{n-1})x_n^d,$$

where $d \geq 0$ and $a_i(x_1, \dots, x_{n-1}) \in k[x_1, \dots, x_{n-1}]$.

If $f' = 0$, then clearly f divides f'. Now consider the case where $g' \in k[x_1, \dots, x_{n-1}]$ with g' nonzero. Since L is infinite, there exists a point $(x_1, \dots, n_{n-1})$ in L^{n-1} with $g'(x)a_0(x)$ nonzero for $x = (x_1, \dots, x_{n-1})$. Since L is algebraically closed, the polynomial equation $g(x_1, \dots, x_{n-1}, t)$ has a root $t = x_n$ in L, and thus the point $(x_1, \dots, x_{n-1}, x_n)$ lies in $H_g(L)^{\text{aff}} - H_{f'}(L)^{\text{aff}}$, which is a contradiction. Therefore, if g' is in $k[x_1, \dots, x_{n-1}]$, then f' is zero.

Now suppose that $g = b_0 + b_1 x_n + \dots + b_e x_n^e$, where $e > 0$ and $b_i \in k[x_1, \dots, x_{n-1}]$. By (4.2) in the Appendix, there is a relation $R(x_1, \dots, x_{n-1}) = ug + vg'$ in $k[x_1, \dots, x_{n-1}][x_n]$. If $g(x_1, \dots, x_n) = 0$ for $(x_1, \dots, x_n) \in L^n$, then

$$g'(x_1, \dots, x_n) = 0$$

by hypothesis and thus by the resultant formula, $R(x_1, \dots, x_{n-1}) = 0$. In other words, as affine hypersurfaces in n-dimensional space $H_f(L)^{\text{aff}} \subset H_R(L)^{\text{aff}}$. Since R is a polynomial in $x_1, \dots, x_{n-1}$, it follows that $R = 0$ by the special case treated in the previous paragraph. By (4.2) in the Appendix again, g and g' must have a common prime factor which must be g itself since g is irreducible. Thus g divides g' and, hence, f divides f'. This proves the theorem.

(2.4) Corollary. *Let f and f' be two nonzero irreducible homogeneous polynomials in $k[y_0, \dots, y_n]$. If for some algebraically closed field L the sets $H_f(L) = H_{f'}(L)$, then $f' = cf$ for some nonzero c in k, and $H_f(K) = H_{f'}(K)$ for all extension fields K of k.*

For k the rational numbers and $f(w, x, y,) = w^2 + x^2 + y^2$ and $f'(w, x, y) = w^2 + 2x^2 + y^2$ the sets $H_f(\mathbb{R}) = H_{f'}(\mathbb{R})$ are empty, but clearly f' is not a constant multiple of f. So the hypothesis that L is algebraically closed is essential in (2.3) and (2.4).

Let $k[y_0, \dots, y_n]_d$ denote the vector space of homogeneous polynomials of degree d. Any nonzero element defines a hypersurface of degree d, and two nonzero elements define the same hypersurface if and only if they are in the same one-dimensional space. Thus the projective space $\mathbb{P}(k[y_0, \dots, y_n]_d)$ parametrizes the hypersurfaces over k of degree d in $\mathbb{P}_n$.

(2.5) Notation. Let $\mathbb{H}_n^d(k)$ denote the projective space $\mathbb{P}(k[y_0, \dots, y_n]_d)$ of hypersurfaces of degree d in $\mathbb{P}_n$ defined over k. The coordinates of this projective space are just the coefficients of the corresponding equation.

For k algebraically closed, the space $\mathbb{H}_1^d(k)$ is just the d-dimensional projective space of d points, repetitions allowed, on the projective line $\mathbb{P}_1(k)$.

(2.6) Proposition. *The dimension of $k[w, x, y]_d$ over k is $[(d + 1)(d + 2)]/2$, and the dimension of the projective space $\mathbb{H}_2^d$ is*

$$\frac{d(d+3)}{2} = \frac{(d+1)(d+2)}{2} - 1.$$

Proof. We have to count the number of monomials $w^a x^b y^c$, where $a + b + c = d$ since they form a basis of $k[w, x, y]_d$. For a fixed index a the number of monomials $w^a x^b y^c$ is $d - a + 1$. To obtain the dimension of the vector space in question, we must sum a from 0 to d. The dimension of the projective space is one less than the dimension of the vector space. We leave the computation to the reader.

We have the following table of values for these dimensions:

	1	2	3	4	5	6	7	8
dim $\mathbb{H}_2^d(k)$	2	5	9	14	20	27	35	44

(2.7) Remark. We can generalize the assertion that two points determine a line. Since the requirement that a plane algebraic curve goes through a point in $\mathbb{P}_2(k)$ is a hyperplane in $\mathbb{H}_2^d(k)$, and since the intersection of m hyperplanes in $\mathbb{H}_2^d(k)$ for $m \le [d(d+3)]/2$ is nonempty, it follows that there exists a curve of degree d through m given points if $m \le [d(d + 3)]/2$.

Exercises

1. Show that the dimension of $\mathbb{H}_n^d$ is the binomial coefficient $\binom{n+d}{n} - 1$.
2. If P_1, P_2, and P_3 are three points in $\mathbb{P}_2$ not on a line, then show that the set of conics through P_1, P_2, and P_3 form a two-dimensional subspace of the five-dimensional space $\mathbb{H}_2^2$ of conics in $\mathbb{P}_2$. Show that the set S of conics through $1:0:0, 0:1:0$, and $0:0:1$ can be parametrized by the coefficiens $a:b:c$ of the equations $axy + bwy + cwx = 0$. Describe the subfamilies of S consisting of conics which also go through $w':x':y'$ and which also go through two distinct points $w':x':y'$ and $w'':x'':y''$.

§3. Elements of Intersection Theory for Plane Curves

The following result gives some indication of the number of intersection points between two plane curves.

(3.1) Proposition. *Let C_f and C_g be two plane algebraic curves of degrees m and n, respectively, defined over k. If for some extension K of k the set $C_f(K) \cap C_g(K)$ has strictly more than mn points, then C_f and C_g have an entire curve in common.*

Proof. Suppose $C_f(K) \cap C_g(K)$ contains $mn + 1$ points. Join these points by lines and with a projective tranformation move the two curves so that the point $(0, 0, 1)$ is not on any of these lines. Decompose the polynomials f and g with respect to the variable y:

$$f(w, x, y) = a_0 y^m + a_1 y^{m-1} + \cdots + a_m,$$

$$g(w, x, y) = b_0 y^m + b_1 y^{n-1} + \cdots + b_n,$$

where $a_i(w, x)$ and $b_j(w, x)$ are homogeneous of degrees i and j, respectively. By (4.3) of the Appendix the resultant $R(f, g)(w, x)$ is homogeneous of degree mn. Moreover, $R(f, g)(w, x) = 0$ for $w, x \in K$ if and only if there exists y in K such that $w : x : y \in C_f(C) \cap C_g(K)$. Since there exists $mn + 1$ points (w_i, x_i, y_i) in $C_f(K) \cap C_g(K)$ for $i = 0, \ldots, mn$, it follows that the polynomial

$$\prod_{0 \le i \le mn} (x_i w - w_i x)$$

of degree $mn + 1$ divides the polynomial $R(f, g)(w, x)$ of degree mn. From this we deduce that $R(f, g) = 0$, and, hence, f and g have a common factor h. Then $C_h \subset C_f \cap C_g$, and this proves the theorem.

The above theorem is a corollary of Bezout's theorem which says that over an algebraically closed field the intersection of a plane curve of degree m with a plane curve of degree n will have exactly mn points in common when the intersection multiplicity is assigned to each intersection point, or the two curves will have a common subcurve. This is a more difficult result and requires both an analysis of singular points and a good definition of intersection multiplicity.

In (2.5) we described the space $\mathbb{H}_n^d(k)$ of hypersurfaces of degree d over k in $\mathbb{P}_n$. For two plane curves C_f and $C_{f'}$ of degree d over k we can speak of the line determined by these two points in $\mathbb{H}_2^d(k)$ and describe it with projective coordinates $a : a' \in \mathbb{P}_1(k)$ as the set of curves of the form $aC_f + a'C_{f'} = C_{af+a'f'}$. This one-dimensional family is called a pencil of curves of degree d over k. Classically two-dimensional families are called nets and three-dimensional families are called webs.

For example, two distinct lines L and L' have a point P in common and the pencil $aL + a'L'$ where $a : a' \in \mathbb{P}_1(k)$ is the pencil of all lines through P and defined over k. Every line in the dual projective space $\mathbb{P}'_2 = \mathbb{H}_2^1$ is a pencil, and it is determined by the unique point in $\mathbb{P}_2$ contained in all members of the pencil.

As for intersection properties of conics, we consider two conics C and C' with three intersection points on the line L. By (3.1) the conics have the line L in common and $C = L \cup M, C' = L \cup M'$, where M and M' are lines. The pencil $aC + a'C'$ becomes $L \cup (aM + a'M')$ which is effectively a pencil of lines.

(3.2) Proposition. *Let C and C' be two conics with exactly four distinct points, P_1, P_2, P_3, and P_4 in common, all defined over an infinite field k. Then any other conic C'' through P_1, P_2, P_3, and P_4 is the form $aC + a'C'$.*

Proof. Observe that no three of P_1, P_2, P_3, and P_4 lie on a line by the previous argument. Since k is infinite, there is a fifty point P on C'' distinct from the P_1 which can be taken to the intersection point if C'' is the union of two lines. Choose $a : a'$ such that P is on the conic $aC + a'C'$. Thus C'' and $aC + a'C'$ have five points in common. By (3.1) the equation of C'', even if it is reducible, divides the equation of $aC + a'C'$ so that $C = aC + a'C'$. This proves the proposition.

This previous proposition has a version for cubics which is basic for the proof that the group law on the nonsingular cubic satisfies the associative law.

(3.3) Theorem. *Let D and D' be two cubic curves intersecting at exactly nine points in $\mathbb{P}_2(k)$ all defined over an infinite field k. If D'' is a plane cubic curve through eight of the intersection points, then it goes through the ninth and has the form $D'' = aD + a'D'$.*

Proof. First, observe that no four of the nine intersection points lie on a line, for otherwise the line would be a common component of D and D' by (3.1). No seven of the nine intersection points lie on a conic, for otherwise a component of the conic would be common to both D and D' by (3.1). In either case the existence of such a common component would contradict the fact that there are exactly nine points of $D \cap D'$.

If D'' is not of the form $aD + a'D'$, then $aD + a'D' + a''D''$ is a two-dimensional family of cubics, and for any pair of distinct points, P' and P'' in the projective plane, we can find $a : a' : a''$ in $\mathbb{P}_2(k)$ such that $aD + a'D' + a''D''$ goes through these points. Now we will refine the statements in the previous paragraph.

Suppose that P_1, P_2, and P_3 are intersection points which lie on a line L. Choose P' on L, and choose P'' off L and off the conic C through P_4, P_5, P_6, P_7, and P_8. Then the cubic $aD + a'D' + a''D''$ going through P', P'', and the eight points $P_1, \ldots, P_8$ has L and C as components by (3.1). This contradicts the choice of P'' off of L and C. Hence no three intersection points lie on a line.

Suppose that $P_1, \dots, P_6$ are six intersection points which lie on a conic C. Choose P' on C and choose P'' off of C and the line L through P_7 and P_8. Then the cubic $aD+a'D'+a''D''$ going through P', P'', and the eight points $P_1, \dots, P_8$ has C and thus L as components by (3.1). This contradicts the choice of P'' off of L and C. Hence no six intersection points lie on a conic.

Now choose P' and P'' both on the line L through P_1 and P_2 but not on the conic C through P_3, P_4, P_5, P_6, and P_7. Then the cubic $aD+a'D'+a''D''$, going through P', P'', and the eight points $P_1, \dots, P_8$, has L and thus C as components by (3.1). This contradicts the fact that P_8 is not on L or C by the analysis in the previous two paragraphs. Thus we must have $D'' = aD+a'D'$ for some point $a : a' \in \mathbb{P}_1(k)$, and hence D will go through the ninth intersection point. This proves the theorem.

Exercises

1. If two curves of degree m intersect at exactly m^2 points, and if nm of these points lie on a curve of order n which is irreducible, then show that the remaining $(m-n)m$ points lie on a curve of degree $m-n$.
2. (Pascal's theorem.) The pairs of opposite sides of a hexagon inscribed in an irreducible conic meet in three points. Show these three points lie on a line.
3. (Pappus' theorem.) Let P_1, P_2, and P_3 be three points on a line L, and let Q_1, Q_2, and Q_3 be three points on a line M with none of these points on $L \cap M$. Let L_{ij} be a line through P_i and Q_j, and let R_k be the intersection point of $L_{ij} \cap L_{ji}$ for $\{i, j, k\} = \{1, 2, 3\}$. Show that R_1, R_2, and R_3 lie on a line.

§4. Multiple or Singular Points

In Theorem (3.1) we showed that two curves of degrees m and n, respectively, without a common component have at most mn intersection points by observing that the resultant between their equations was a homogeneous form of degree mn. Since the linear factors of this resultant form of degree mn correspond to intersection points, we must give a geometric interpretation of the repeated factors. Then we will be in a position to state the basic intersection theorem of Bezout, namely that the number of intersections, counted with the appropriate multiplicities, is exactly mn. A preliminary step is to consider the order of a hypersurface at a point.

(4.1) Definition. Let H_f be a hypersurface of degree d, and let P be a point on $H_f(k)$. We can choose affine coordinates such that P is the origin and the equation of the hypersurface in affine coordinates becomes

$$0 = f(x_1, \dots, x_n) = f_r(x_1, \dots, x_n) + \cdots + f_d(x_1, \dots, x_n),$$

where $f_i(x_1, \dots, x_n)$ is a homogeneous polynomial of degree i and the forms where $r \le d$ and f_r and f_d are nonzero. Then P is called a point of order r on H_f, and $f_r(x_1, \dots, x_n)$ is called a leading form at P.

The leading form is well defined up to a projective change of variables coming from the subgroup of the projective linear leaving one point fixed.

(4.2) Definition. A point on a hypersurface is called simple or nonsingular provided its order is 1 and is called multiple or singular provided the order is strictly greater than 1.

For example all points on a multiple component, defined by a power of an irreducible polynomial, are singular with order equal to an integral multiple of the multiplicity of the multiple component. If H is a hypersurface of degree d consisting of the union of d hyperplanes $H_1 \cup \cdots \cup H_d$, then P has order r on H if and only if P is on exactly r of the hyperplanes H_i for $i = 1, \ldots, d$. Thus every point on $H_i \cap H_j$ for $i \neq j$ is singular.

(4.3) Definition. Let P be a point of order r on a plane algebraic curve $C_f(k)$ of degree m where k is algebraically closed. The tangent cone to C_f at P is the union of the lines whose equations are the linear factors of the leading form of the point P. If P is a nonsingular point, then the tangent cone reduces to a single line called the tangent line.

If P is transformed by a projective tranformation to $1 : 0 : 0$, then the leading form is $f_r(x, y) = \prod_{1 \leq i \leq r}(a_i x + b_i y)$ where the equation of the curve is

$$f(w, x, y) = w^{m-r} f_r(x, y) + \cdots + f_m(x, y).$$

For $f(w, x, y)$ a homogeneous polynomial of degree m, the following formula, called Euler's formula, holds:

$$mf(w, x, y) = w\frac{\partial f}{\partial w} + x\frac{\partial f}{\partial x} + y\frac{\partial f}{\partial y}.$$

The reader can easily check that for a simple point (w_0, x_0, y_0) on $C_f(k)$ the tangent line is of the form

$$\frac{\partial f}{\partial w}(w_0, x_0, y_0)w + \frac{\partial f}{\partial x}(w_0, x_0, y_0)x + \frac{\partial f}{\partial y}(w_0, x_0, y_0)y = 0.$$

Now we relate the order of a point to a special case of intersection multiplicities. The line L determined by two distinct points $y_0 : \cdots : y_n$ and $y'_0 : \cdots : y'_n$ in $\mathbb{P}_n(k)$, can be parametrized by the function which assigns to each $s : t \in \mathbb{P}_1(k)$ the point $(sy_0 + ty'_0) : \cdots : (sy_n + ty'_n)$ in $\mathbb{P}_n(k)$. The function which assigns to each $t \in k$ the point $(y_0 + ty'_0) : \cdots : (y_n + ty'_n)$ parametrizes the affine line $L - \{y'_0 : \cdots : y'_n\} = L^{\text{aff}}$.

If H_f is a hypersurface of degree d, then the intersection set of $L^{\text{aff}} \cap H_f$ consists of points P_t, where t is a root of the polynomial equation $0 = \varphi(t) = f(y_0 + ty'_0, \ldots, y_n + ty'_n)$. Observe that $y_0 : \cdots : y_n \in H_f(k)$ if and only if $\varphi(0) = 0$, and $L \not\subset H_f$ if and only if φ is not identically zero.

(4.4) Definition. With these notations, the intersection multiplicity of the line L with the hypersurface H_f at $P = y_0 : \cdots : y_n$, denoted $i(P; L, H_f)$, is the order of the zero of $\varphi(t)$ at $t = 0$. It is defined only when $L \not\subset H_f$.

Now we can relate intersection multiplicity at P with the order of P on a hypersurface H_f.

(4.5) Proposition. *Let k be an infinite field. A point $P = y_0 : \cdot : y_n$ is of order r on $H_f(k)$ if and only if $i(P; L, H_f) \geq r$ for all lines L through P not contained in H_f and $i(P; L, H_f) = r$ for some line L through P.*

Proof. After a projective change of coordinates, we can assume that $y_0 : \cdots : y_n = 1:0:\cdots:0$, and, without changing the line L, we can assume that the point $y'_0 : \cdots : y'_n$ has the property that $y'_0 = 0$, that is, the point lies on the hyperplane at infinity. By definition of the order of $1:0:\cdots:0$ on H_f, the polynomial $f(1, x_1, \ldots, x_n)$ has the form

$$f(1, x_1, \ldots, x_n) = f_r(x_1, \ldots, x_n) + \cdots + f_d(x_1, \ldots, x_n),$$

where $f_1(x_1, \ldots, x_n)$ is homogeneous of degree i and $f_r \neq 0$. The polynomial $\partial(t)$ associated with f and the two points has the form

$$\begin{aligned}\varphi(t) &= f_r(ty'_1, \ldots, ty'_n) + \cdots + f_d(ty'_1, \ldots, ty'_n)\\ &= t^r f_r(y'_1, \ldots, y'_n) + \cdots + t^d f_d(y'_1, \ldots, y'_n)\end{aligned}$$

and $\varphi(t) = 0$ has $t = 0$ as a root of order $\geq r$. For $(y'_1, \ldots, y'_n)$ with $f_r(y'_1, \ldots, y'_n) \neq 0$, such a point exists since k is infinite, the root $t = 0$ of $\varphi(t)$ has multiplicity equal to r. This proves the proposition.

For the case $n = 2$, that of a place curve $f(w, x, y) = 0$, the above relations for L at the points $1 : 0 : 0$ and $0 : a : b$ take the form

$$f(1, x, y) = f_r(x, y) + \cdots + f_d(x, y)$$

and

$$\begin{aligned}\varphi(t) &= f_r(ta, tb) + \cdots + f_d(ta, tb)\\ &= t^r f_r(a, b) + \cdots + t^d f_d(a, b).\end{aligned}$$

The condition $f_r(a, b) = 0$ is equivalent to the condition that $bx - ay$ divides $f_r(x, y)$, that is, the line L through $1 : 0 : 0$ and $0 : a : b$ given by the equation $bx - ay = 0$ is part of the tangent cone. This leads to the following result extending (4.5).

(4.6) Proposition. *Let k be algebraically closed. The point $P = w : x : y$ is of order r on the plane curve C_f if and only if $i(P; L, C_j) \geq r$ for all lines L through P with $L \not\subset C_f$, and $i(P; L, C_f) = r$ for some line through P. The line L through P is part of the tangent cone if and only if $L \subset C_f$ or $i(P; L, C_f) > r$.*

In the special case of a nonsingular point $w : x : y$ on $C_f(k)$, the polynomial $\varphi(t)$, given above, will have a simple root for all lines L except the tangent line (at, bt) where $bx - ay$ is the leading form of $f(1, x, y)$. In other words,

$$f(1, x, y) = (bx - ay) + f_2(x, y) + \cdots + f_d(x, y).$$

Then in that case the order of the root $t = 0$ of $\varphi(t)$ is ≥ 1. This leads to the following definition which is used in the study of cubic curves.

(4.7) Definition. A point of inflection (or flex) is a nonsingular point $P = w : x : y$ on a curve C_f such that $i(P; L, C_f) \geq 3$ for the tangent line L at P.

In other words, if $1 : 0 : 0$ is a flex of the curve C_f with a tangent line given by the equation $f_1(x, y) = bx - ay = 0$, then the polynomial $f(1, x, y)$ has the following form:

$$f(1, x, y) = f_1(x, y) + \cdots + f_d(x, y),$$

where $f_i(x, y)$ is homogeneous of degree i with $f_1(x, y)$ dividing $f_2(x, y)$. Since any flex, by projective transformation, can be moved to the origin $1 : 0 : 0$, we see that a nonsingular conic will have no flexes.

For a curve C_f of degree d with $f(w, x, y) \in k[w, x, y]_d$ we consider the condition of f that $1 : r : s$ is a flex on C_f. We expand

$$f(1, x + r, y + s) = f_0(r, s) + f_1(r, s)(x, y) + \cdots + f_d(r, s)(x, y),$$

where $f_i(r, s)(x, y) \in k[r, s][x, y]_i$. Setting $x = y = 0$, we see that $f(1, r, s) = f_0(r, s)$ and hence $1 : r : s \in C_f$ if and only if $f_0(r, s) = 0$.

Now assume that $1 : r : s \in C_f$. Then $1 : r : s$ is a nonsingular point if and only if $f_1(r, s)(x, y) = b(r, s)x - a(r, s)y$ is nonzero in $k[r, s][x, y]_1$. A smooth point $1 : r : s$ on C_f is a flex if and only if $f_1(r, s)(x, y)$ divides $f_2(r, s)(x, y)$ in $k[x, y]$. In terms of the resultant $R(r, s) = R(f_1(r, s), f_2(r, s)) \in k[r, s]$ we see that $1 : r : s$ is a flex of C_f if and only if $f(1, r, s) = 0$ and $R(r, s) = 0$. This resultant can be calculated as a 3 by 3 determinant for $f_2(r, s)(x, y) = A(r, s)x^2 + B(r, s)xy + C(r, s)y^2$ as follows:

$$R(r, s) = A(r, s)a(r, s)^2 + B(r, s)a(r, s)b(r, s) + C(r, s)b(r, s)^2.$$

Since for degree of f strictly bigger that 2 the curve C_f given by $f(1, x, y) = 0$ and the resultant $R(x, y) = 0$ have an intersection point $1 : r : s$ we deduce the following result.

(4.8) Proposition. *Let C_f be an algebraic plane curve without singularities of degree $d \geq 3$ over an algebraically closed field k. Then C_f has at least one flex.*

Exercises

1. For k infinite prove that P is a point of order $\leq r$ on a hypersurface H_f of degree n if and only if there exists a line L through P intersecting H_f at $n-r$ additional points.
2. In the $[m(m+3)/2]$-dimensional family $\mathbb{H}_2^m$ of all curves of degree m, calculate the dimension of the subfamily of all curves passing through a point P and having order $\geq r$ at that point. Give a lower bound for the dimension of the subfamily of all curves in $\mathbb{H}_2^m$ passing through the points P_i and having order $\geq r_i$ at P_i for $i = 1, \ldots, t$.
3. Let C_f and C_g be two plane algebraic curves of degrees m and n, respectively, where $f(w, x, y) = a_0 y^m + \cdots + a_m$, $g(w, x, y) = b_0 y^n + \cdots + b_n$ with $a_i = a_i(w, x)$ and $b_j = b_j(w, x)$ homogeneous polynomials of degree $i.j$. If $w : x : y$ is a point on $C_f \cap C_g$ with order r on C_f and order s on C_g, then show that $R(f, g)(W, X)$ has a factor of the form $(xW - wX)^{sr}$ in its factorization as a product of linear forms.
4. Let C_f and C_g be two plane algebraic curves of degrees m and n, respectively, without common factors. If $P_1, \ldots, P_t$ are the intersection points where r_i is the order of C_f at P_i and s_i is the order of C_g at P_i, then show that the following relation holds

$$\sum_{1 \leq i \leq t} r_i s_i \leq mn.$$

5. Let C_f be a curve of degree m with no multiple components and with orders r_i of the singular points P_i. Then show

$$m(m-1) \geq \sum r_i(r_i - 1).$$

   Also, show that this is the best possible result of this kind by examining the case of n lines through one point. Note that this shows that the number of singular points is finite. *Hint:* Compare f with a suitable derivative of f which will have degree $m-1$.
6. Let C_f be a curve of degree m which is irreducible and with orders r_i of the singular points P_i. Then show

$$(m-1)(m-2) \geq \sum r_i(r_i - 1).$$

   Hint: Compare a suitable derivative of f having order $r_i - 1$ at P_i with the $[(m-1)(m+2)/2]$-dimensional family of all curves of degree $m-1$ and the subfamily of curves having degree $r_i - 1$ at P_i. Also, show that this is the best possible result of this kind by examining the curve given by the equation $X^n + WY^{n-1} = 0$.
7. Prove that in characteristic zero a conic C_f is reducible if and only if the 3 by 3 matrix of second partial derivatives of f has a determinant equal to zero. When does this assertion hold in characteristic p?
8. For a plane curve C_f and $w_0 : x_0 : y_0$ on C_f derive the equation of the tangent line in terms of the first partial derivatives of f at $w_0 : x_0 : y_0$. In characteristic zero prove that $w_0 : x_0 : y_0$ is a flex if and only if the linear form of first partial derivatives divides the quadratic form of second partial derivatives. When does this assertion hold in characteristic p?
9. Prove that in characteristic zero the flexes of C_f are the intersections between C_f and the curve whose equation is the determinant of the matrix of second partial derivatives of f. When does this assertion hold in characteristic p?
10. Prove that every nonsingular curve of degree 3 or more has at least one flex over an algebraically closed field.

Appendix to Chapter 2
Factorial Rings and Elimination Theory

Factorial rings, also called unique factorization domains, have most of the strong divisibility properties of the ring of integers $\mathbb{Z}$. These properties are used in Chapter 5 for the study of the reduction of a plane curve modulo a prime number. The concept is also useful for the understanding of divisibility properties of polynomials over a field. Included in our brief introduction to the theory of these rings is a discussion of the resultant of two polynomials which was used in Chapter 2, §§2 and 3.

All rings considered in this appendix are commutative.

§1. Divisibility Properties of Factorial Rings

A *unit* in a ring R is an element $u \in R$ such that there exists an element $v \in R$ with $uv = 1$. The units in R form a group under multiplication which we denote by R^*. Note that R is a field if and only if $R^* = R - \{0\}$.

(1.1) Definition. For a, b in R the element a divides b, denoted $a|b$, provided there exists x in R with $b = ax$. In terms of ideals this condition can be written $Ra \supset Rb$.

Observe that $u|a$ for any unit u and $a|0$ for all a in R. Moreover, $Ra = Rb$ if and only if $b = ua$, where u is a unit.

(1.2) Definition. A nonzero element p in R is an irreducible provided for each factorization $p = ab$ either a or b is a unit but not both.

Since each factorization of a unit is by units, an irreducible is not a unit. In a field there are no irreducibles.

(1.3) Definition. A factorial ring R is an integral domain such that for any nonzero a in R we can decompose a as

$$a = up_1, \ldots, p_r,$$

where u is a unit and $p_1 \ldots, p_r$ are irreducibles, and also this factorization is unique in the following sense: for a second decomposition $a = vq_i, \ldots, q_s$ by irreducibles we have $r = s$ and after permutation of the q_i's each $p_i = u_i q_i$, where u_i is a unit in R.

(1.4) Alternative Formulations. Let R be a ring. The following remarks give conditions under which unique factorization is possible.

(a) Every nonzero element of a ring R can be factored as a product of irreducible elements if and only if every sequence of principal ideals

$$Ra_1 \subset Ra_2 \subset \cdots$$

is stationary, that is, there exists m with $Ra_m = Ra_{m+1} = \cdots$.

(b) For an integral domain R such that every nonzero element is a product of irreducible elements the following are equivalent:

(1) R is factorial.

(2) If p is an irreducible dividing ab, then p divides either a or b.

(3) If p is an irreducible, then Rp is a prime ideal.

(1.5) Example. Every principal ring is a factorial ring for example the integers $\mathbb{Z}$, the Gaussian integers $\mathbb{Z}[i]$, the Jacobi integers $\mathbb{Z}[\rho]$, where $\rho = \exp(2\pi i/3)$, and $k[X]$, where k is a field.

For R a factorial ring with field of fractions F and an irreducible p in R, each nonzero x in F can be represented as $x = p^r(a/b)$, where a, b are in R and not divisible by p and r is an integer. Moreover, r and a/b are unique. We define the order function $\mathrm{ord}_p : F^* \to \mathbb{Z}$ at the irreducible p by the relation $\mathrm{ord}_p(x) = r$. Observe that it has the following properties:

$$\mathrm{ord}_p(xy) = \mathrm{ord}_p(x) + \mathrm{ord}_p(y),$$
$$\mathrm{ord}_p(x + y) \geq \min\{\mathrm{ord}_p(x), \mathrm{ord}_p(y)\}.$$

Moreover, any nonzero a in F is in R if and only if $\mathrm{ord}_p(a) \geq 0$ for all irreducibles p in R. This leads to the following definition.

(1.6) Definition. A (discrete) valuation on a field F is a function $v : F' \to \mathbb{Z}$ such that

$$v(xy) = v(x) + v(y) \quad \text{and} \quad v(x + y) \geq \min\{v(x), v(y)\},$$

where x, y are in F. By convention we set $v(0) = +\infty$ so that the case $x + y = 0$ is covered in the second relation. The value group of v is defined to be the image of v in $\mathbb{Z}$.

As for some elementary properties of valuations, observe that

$$v\left(\frac{x}{y}\right) = v(x) - v(y), \quad v(1) = v(-1) = 0, \quad v(x^n) = nv(x),$$

and if $v(x) < v(y)$, then it follows that $v(x) = v(x + y)$. To see this, we have $v(x) \leq v(x + y)$ from the definition, and from $x = (x + y) + (-y)$, it follows that $v(x + y) \leq v(x)$ too since $v(y) = v(-y)$.

For a prime number p in $\mathbb{Z}$ the associated valuation ord_p on $\mathbb{Q}$ is called the p-adic valuation, and up to an integral multiple every valuation on the field of rational numbers is a p-adic valuation for some p.

To each valuation we can associate a principal, hence factorial, ring with exactly one irreducible up to multiplication by units.

(1.7) Definition. Let v be a valuation on a field F. The valuation ring R_v is the set of all x in F with $v(x) \geq 0$.

Observe that the units R_v^* in R_v is the group of all x with $v(x) = 0$ and the nonunits form the unique maximal ideal M_v which equals the set of all x in F with $v(x) > 0$. The residue class field $k(v)$ is the quotient ring R_v/M_v. If the valuation is trivial, that is, if $v(F^*) = 0$, then $R_v = F$ and $M_v = 0$. Otherwise $R_v \neq F$ and $M_v \neq 0$. If the value group is $v(F^*) = m\mathbb{Z}$, then there exists an element t in R_v with $v(t) = m$. Then it follows that $M_v = R_v t$ and every x in F^* can be represented as $x = t^r u$, where u is in R_v^* and r is the integer where $v(x) = rm$. Note that v is $m \cdot \mathrm{ord}_t$.

If k is a subfield of F with $v(k^*) = 0$, then $k^* \subset R_v^*$ and the restriction of the residue class morphism $R_v \to k(v)$ to $k \to k(v)$ is a monomorphism of fields. In this case we say that the valuation v is trivial on k.

§2. Factorial Properties of Polynomial Rings

Valuations are useful for discussing factorization properties of polynomial rings. Let F be a field with a valuation v and define v^+ on $F[x_1, \ldots, x_n]$ by the relation

$$v^+\left(\sum a_{i,\ldots i_n} x_1^{i_1} \ldots x_n^{i_n}\right) = \min\left\{v\left(a_{i_1\ldots i_n}\right)\right\}.$$

(2.1) Proposition (Gauss's Lemma). *Let F be a field with a valuation v. For f, g in $F[x_1, \ldots, x_n]$ we have*

$$v^+(fg) = v^+(f) + v^+(g).$$

Proof. Consider the case of one variable x. Since $v^+(cf) = v(c) + v^+(f)$, we can assume that $v^+(f) = v^+(g) = 0$ so that f, g are in $R_v[x]$. If $f(x) = a_m x^m + \cdots + a_0$ and $g(x) = b_n x^n + \cdots + b_0$, then there exists indices i and j where a_i and b_j are units and a_p and b_q are not units for $p < i$ and $q < j$. From this we see that the coefficient of x^{i+j} in $f(x)g(x)$ is a unit since $v(a) = v(a+b)$ if $v(a) < v(b)$. Thus we have $v^+(fg) = 0$, and this proves the formula in the case of one variable.

For n variables choose $d > \deg(fg)$. Then substitute $x_i = x^{d^{i-1}}$ for all i. This maps

$$f(x_1, \ldots, x_n) \mapsto f^*(x) = f\left(x, x^d, \ldots, x^{d^{n-1}}\right)$$

such that $v^+(f) = v^+(f^*)$. In this way the n variable case is reduced to the 1 variable result which proves the proposition.

(2.2) Theorem. *If R is a factorial ring, then $R[x_1, \ldots, x_n]$ is a factorial ring.*

Proof. Since $R[x_1, \ldots, x_n] = R[x_1, \ldots, x_{n-1}][x_n]$, it suffices, by induction on n, to prove the theorem for $n = 1$ and $x_1 = x$. If F is the field of fractions of R, then $F[x]$ is a factorial ring since $F[x]$ is a principal ring. Thus every f in $R[x] \subset F[x]$ is a product $f = f_1 \ldots f_r$, where the f_i are irreducibles in $F[x]$. For each irreducible p in R, we have

$$0 \leq \mathrm{ord}_p^+(f) = \mathrm{ord}_p^+(f_1) + \cdots + \mathrm{ord}_p^+(f_r).$$

We can assume that $\mathrm{ord}_p^+(f_i) \geq 0$ for all i and that p is irreducible in R. This means that each f_i is an irreducible in $R[x]$.

Let f be an irreducible in $R[x]$ with $f(x)$ dividing $a(x)b(x)$ in $R[x]$. If $\deg(f) = 0$, then f is an irreducible in R and

$$0 < \mathrm{ord}_f^+(a(x)b(x)) = \mathrm{ord}_f^+(a(x)) + \mathrm{ord}_f^+(b(x)).$$

Hence f divides $a(x)$ when $\mathrm{ord}_f^+(a(x)) > 0$ or $b(x)$ when $\mathrm{ord}_f^+(b(x)) > 0$. If $\deg(f) > 0$, then $\mathrm{ord}_p^+(f) = 0$ for every irreducible $p \in R$ and f is irreducible in $F[x]$ by the argument in the previous paragraph. Since $F[x]$ is factorial, $f(x)$ divides $a(x)$ or $b(x)$ in $F[x]$ and so $a(x) = f(x)q(x)$ for example. Then $0 \leq \mathrm{ord}_p^+(a(x)) = \mathrm{ord}_p^+(f(x)) + \mathrm{ord}_p^+(q(x)) = \mathrm{ord}_p^+(q(x))$ and $q(x) \in R[x]$. Hence $f(x)$ divides $a(x)$ in $R[x]$ with quotient $q(x)$. This proves the theorem.

§3. Remarks on Valuations and Algebraic Curves

Let k be an algebraically closed field, and let V denote the set of valuations up to equivalence on the field $k(x)$ of rational functions in one variable over k which are trivial on k. We map $\mathbb{P}_1(k) \mapsto V$ by

$$\begin{aligned} (1, a) \text{ in } \mathbb{P}_1(k) &\mapsto v_{(1,a)} = \mathrm{ord}_{x-a} && \text{for } k[x] \subset k(x), \\ \infty = (0, 1) \text{ in } \mathbb{P}_1(k) &\mapsto v_{(0,1)} = \mathrm{ord}_{1/x} && \text{for } k[1/x] \subset k(x). \end{aligned}$$

Then the function $\mathbb{P}_1(k) \to V$ given by $P \mapsto v_P$, is a bijection. For it is clearly an injection, and to show that each valuation v is of the form v_P, consider the ideal $M_v \cap k[x]$ in $k[x]$ if $x \in R_v$ or the ideal $M_v \cap k[1/x]$ if $x \in R_v$. The remainder of the argument is left to the reader.

(3.1) Remark. Let C_f be an irreducible plane curve defined over k. Then the integral domains

$$\frac{k[x, y]}{f(1, x, y)}, \qquad \frac{k[w, y]}{f(w, 1, y)}, \qquad \frac{k[w, x]}{f(w, x, 1)}$$

all have naturally isomorphic fields of fractions, denoted $k(C_f)$, and called the function field of the curve. The subrings $k[w]$, $k[x]$, and $k[y]$ all inject into $k(C_f)$ except when f is w, x, or y, respectively. Thus the field $k(C_f)$ is a finite extension of $k(w)$, $k(x)$, or $k(y)$. The main assertion is: there is a natural map from $C_f(k)$ to valuations on $k(C_f)$ trivial over k up to equivalence which is a bijection when C_f is nonsingular and k is algebraically closed. This map extends $\mathbb{P}_1(k) \to V$ considered above in the sense that for $k(\mathbb{P}_1) = k(t) \subset k(C_f)$ the points of C_f mapping onto $P \in \mathbb{P}_1$ under $t : C_f \to \mathbb{P}_1$ correspond to the extensions to $k(C_f)$ of the valuation v_P on $k(\mathbb{P}_1)$.

§4. Resultant of Two Polynomials

The considerations in the section were used in Chapter 2, §2 to determine to what extent the sets $H_f(K)$ determine the form f.

(4.1) Definition. For polynomials f, g in $R[x]$ given by

$$f(x) = a_0 + a_1 x + \cdots + a_m x^m \qquad \text{and} \qquad g(x) = b_0 + b_1 x + \cdots + b_n x^n$$

the resultant $R(f, g)$ of f and g is the element of R given by the following $(m+n) \times (m+n)$ determinant.

$$R(f,g) = \begin{vmatrix} a_0 & a_1 & \cdots & a_m & & 0 & \cdots & 0 \\ 0 & a_0 & \cdots & a_{m-1} & a_m & & \cdots & 0 \\ \cdots & \cdots & \cdots & \cdots & \cdots & \cdots & \cdots & \cdots \\ 0 & 0 & \cdots & & a_0 & a_1 & \cdots\cdots & a_m \\ b_0 & b_1 & \cdots & b_{n-1} & b_n & 0 & \cdots & 0 \\ 0 & b_0 & \cdots & & b_{n-1} & b_n & \cdots & 0 \\ \cdots & \cdots & \cdots & \cdots & \cdots & \cdots & \cdots & \cdots \\ 0 & 0 & \cdots & b_0 & b_1 & & \cdots\cdots & b_n \end{vmatrix} \begin{matrix} \left.\vphantom{\begin{matrix}a\\a\\a\\a\end{matrix}}\right\} n \text{ rows} \\ \left.\vphantom{\begin{matrix}a\\a\\a\\a\end{matrix}}\right\} m \text{ rows} \end{matrix}$$

We denote the corresponding resultant matrix by $[R(f, g)]$.

(4.2) Theorem. *For a factorial ring R and polynomials $f, g \in R[x]$, the following statements are equivalent:*

(1) *The polynomials f and g have a common factor of strictly positive degree in $R[x]$.*
(2) *There exists nonzero polynomials $a(x), b(x)$ in $R[x]$ such that* $\deg a < n = \deg g, \deg b < m = \deg f$, *and* $af + bg = 0$.
(3) $R(f, g) = 0$.

When $R(f, g) \neq 0$ there exists polynomials $a(x), b(x)$ in $R[x]$ such that $\deg a < n$, $\deg b < m$, and $R(f, g) = af + bg$.

Proof. If $u(x)$ is a common factor of strictly positive degree, then we can write $f(x) = b(x)u(x)$ and $g(x) = -a(x)u(x)$ where a and b have the desired properties, so that (1) implies (2). The converse (2) implies (1) follows from factoring of f and g in $R[x]$ since $R[x]$ is factorial by (4.1).

For any polynomials of the form

$$a(x) = \alpha_0 + \alpha_1 x + \cdots + \alpha_{n-1} x^{n-1}$$

and

$$b(x) = \beta_0 + \beta_1 x + \cdots + \beta_{m-1} x^{m-1},$$

we see from the formulas for matrix multiplication that

$$(c_0, c_1, \ldots, c_{m+n-1}) = (\alpha_0, \ldots, \alpha_{n-1}, \beta_0, \ldots, \beta_{m-1})[R(f,g)],$$

where $a(x)f(x)+b(x)g(x) = c(x) = c_0+c_1x+\cdots+c_{m+n-1}x^{m+n-1}$. Now (2) and (3) are equivalent because there exists a nonzero $(\alpha_0, \ldots, \alpha_{n-1}, \beta_0, \ldots, \beta_{m-1})$ for $c_0, \ldots, c_{m+n-1}) = (0, \ldots, 0)$ if and only if $R(f, g) = \det[R(f, g)] = 0$. Moreover, the second statement holds by choosing $(c_0, \ldots, c_{m+n-1}) = (R(f,g), 0, \ldots, 0)$, and using the cofactor formulas for the inverse of the matrix $[R(f, g)]$. This proves the theorem.

(4.3) Proposition. *Let $f = a_m + a_{m-1}x + \cdots + a_0x^m$ and $g = b_n + b_{n-1}x + \cdots + b_0x^n$ be polynomials over the ring $R[y_0, \ldots, y_r]$, where R is factorial. If a_k and b_k are homogeneous of degree k, then $R(f, g) \in R[y_0, \ldots, y_r]$ is homogeneous of degree mn.*

Proof. We calculate $R(f, g)(ty)$ using the homogeneous character of $a_1(y_0, \ldots, y_r)$ and $b_j(y_0, \ldots, y_r)$ and at the time multiplying each row by a suitable power of t so that each column contains a constant power of t.

$$t^u R(f,g)(ty) = \begin{vmatrix} t^nt^m a_m & t^nt^{m-1}a_{m-1} & \cdots & t^n a_0 & 0 & \cdots & 0 \\ 0 & t^{n-1}t^m a_m & \cdots & t^{n-1}ta_1 & t^{n-1}a_0 & \cdots & 0 \\ \cdots & \cdots & \cdots & \cdots & \cdots & \cdots & \cdots \\ 0 & 0 & tt^m a_m & tt^{m-1}a_{m-1} & \cdots & & ta_0 \\ t^mt^n b_n & t^mt^{n-1}b_{n-1} & \cdots & t^m b_0 & 0 & \cdots & 0 \\ 0 & t^{m-1}t^n b_n & \cdots & t^{m-1}tb_1 & t^{m-1}b_0 & \cdots & 0 \\ \cdots & \cdots & \cdots & \cdots & \cdots & \cdots & \cdots \\ 0 & 0 & \cdots & tt^n b_m & tt^{n-1}b & \cdots & tb_0 \end{vmatrix}$$

Since each column contains a fixed power of t, this determinant equals $t^v R(f, g)(y)$ where

$$u = \frac{m(m+1)}{2} + \frac{n(n+1)}{2}$$

and

$$v = \frac{(m+n)(m+n+1)}{2}.$$

Since $v - u = mn$, we deduce that $R(f, g)(ty) = t^{mn} R(f, g)(y)$. Thus $R(f, g)(y)$ is homogeneous of degree mn and this proves the proposition.

Exercises to Appendix

1. Verify the elementary properties of valuations given in definition (1.6), and verify that ord_p is a valuation on the field F of fractions of R where p is an irreducible in a factorial ring R.
2. In §3 carry out the details to show $P \mapsto v_P$ defining $\mathbb{P}_1(k) \to V$ is a bijection.

3. Let $k[[x]]$ be the ring of formal series over the field k, that is, expressions of the form $a(x) = \sum_{0 \leq n} a_n x^n$. Let $O(a(x)) = n$, where $a_n \neq 0$ and $a_i = 0$ for $i < n$. Show that

$$O(ab) = O(a) + O(b), \quad O(a+b) \geq \min\{O(a), O(b)\}.$$

Show that the field of fractions $k((x))$ of $k[[x]]$ is equal to $k[[x]][1/x]$, and that $v(a/b) = O(a) - O(b)$ is a valuation on $k((x))$. Moreover, show that the elements c of $k((x))$ can be written in the form $c(x) = \sum_{m \leq i} c_i x^i$, and if $c_m \neq 0$, then $v(c) = m$.
4. The discriminant $D(f)$ of a polynomial $f(x)$ is defined to be the resultant $R(f, f')$, where f' is the derivative of f. Prove that f in $k[x]$ has a repeated root in an extension field of k if and only if $D(f) = 0$.
5. Calculate the discriminant $D(f)$ of the polynomials $ax^2 + bx + c$ and $x^3 + px + q$.

3

Elliptic Curves and Their Isomorphisms

Using the results of the previous chapter, we complete the unfinished business concerning elliptic curves described as cubic curves, namely the associative law, the transformation into normal form, and the discriminant criterion for nonsingularity or smoothness. At the same time admissible changes of variables are introduced; these are equivalent to isomorphisms given by a change of variable in the first equation from an elliptic curve defined by one cubic onto another.

The special cases of characteristic 2 and 3 are considered in detail. All elliptic curves over $\mathbb{F}_2$, the field of two elements, are described and their isomorphism relations over extension fields are given. Finally, we return to the subject of singular cubics and their group of nonsingular points.

§1. The Group Law on a Nonsingular Cubic

In the Introduction we described the chord-tangent group law on a nonsingular cubic curve, and in Chapter 1 we made extensive calculations with this group law. Now using the intersection theory of Chapter 2, we show that the group law satisfies the associative law and point out how the intersection multiplicity $i(P, L, C)$ enters into the definition of the group law.

For the chord-tangent composition PQ of P and Q on a nonsingular cubic the following assertions based on 2(3.1), 2(4.5), and 2(4.6) are used.

(1.1) Remarks. Let L be a line and C a cubic curve both defined over a field k. Let k' be an algebraically closed extension of k. One of the following situations hold for the intersection $L(k') \cap C(k')$:

(a) $L(k') \cap C(k') = \{P_1, P_2, P_3\}$, three points where $i(P_i, L, C) = 1$ for $i = 1, 2, 3$. The composition is given by $P_i P_j = P_k$, and if P_i and P_j are rational over k, then so is P_k for $\{i, j, k\} = \{1, 2, 3\}$.

(b) $L(k') \cap C(k') = \{P, P'\}$, two points where $i(P; L, C) = 2$ and $i(P'; L, C) = 1$. Either L is tangent to C at P or P is a singular point of C, and the compositions are given by $PP = P'$ and $PP' = P$. If P is rational over k, then so is P'.

(c) $L(k') \cap C(k') = \{P\}$, one point where $i(P; L, C) = 3$ and in this case $PP = P$ and P is either a point of inflection or a singular point.

In particular, the chord-tangent composition is a function $C(k) \times C(k) \to C(k)$.

(1.2) Theorem. *Let C be a nonsingular cubic curve defined over a field k, and let O be a point on $C(k)$. Then the law of composition defined by $P + Q = O(PQ)$ on $C(k)$ makes $C(k)$ into an abelian group with O as zero element and $-P = P(OO)$. Moreover, O is a point of inflection if and only if $P + Q + R = 0$ whenever P, Q, and R are the three intersection points of C with a line. In this case we also have $-P = PO$ and $OO = O$.*

Proof. The above group law was introduced in §5 of the Introduction. All of the group axioms were considered except for the associative law, and, in fact, they are immediate from the definition. The statement about O being an inflection point was also considered in the Introduction.

The associativity relation would follow if we knew that $P(Q + R) = (P + Q)R$ since composing this with O yields our composition law.

We begin with the case where P, Q, and R are distinct. To form $P(Q + R)$, we find first QR, join that to O, and take the third intersection point $Q + R$. Now join $Q + R$ to P, which gives the point $P(Q + R)$, and we need to show that it is the same as $(P + Q)R$. In the figure each of the points O, P, Q, R, PQ, $P + Q$, QR, and $Q + R$ lies on one dotted line and one solid line.

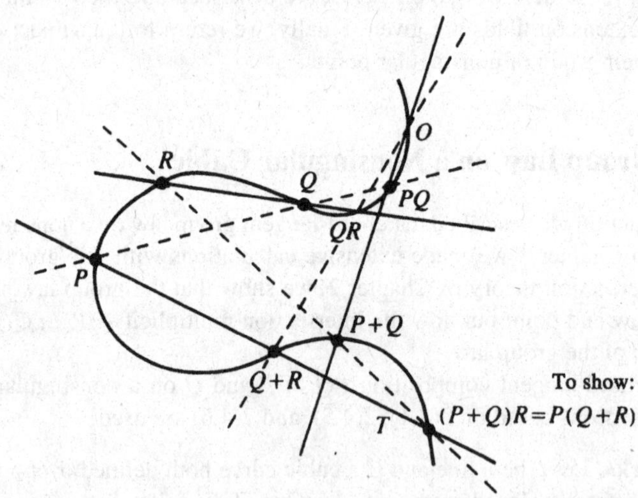

We have nine points, O, P, Q, R, PQ, $P + Q$, QR, $Q + R$, and the intersection T of the line joining P to $Q + R$ and the line joining $P + Q$ to R. We must show that T is on the cubic.

For this, observe that there are two degenerate cubics which go through the nine points, namely the union C_1 of the three dotted lines and the union C_2 of the three solid lines. By construction these three cubics go through the nine points. Since the given cubic C goes through the first eight points, by Theorem 2(3.3) we deduce that the ninth point T is also on C and thus $(P+Q)R = P(Q+R)$.

When P, Q, and R are not all distinct, using the commutative law, we see that one has only to check the case $P+(P+Q) = (P+P)+Q$. This comes from a limiting position of the above diagram or can be proved using a special direct argument. Hence $P+Q$ is an abelian group law, and this proves the theorem.

With this theorem we complete the formulation of the notion of an elliptic curve given in (5.1) of the Introduction as a nonsingular cubic plane curve, with a chosen point, and the group law given by the chord-tangent composition. Observe that any projective transformation preserving the zero point will also be a group homomorphism since it preserves the chord-tangent construction.

Let E be an elliptic curve over a field k. For each field extension K of k the points $E(K)$ form an abelian group, and if $K \to K'$ is a k-morphism of fields, then the induced function $E(K) \to E(K')$ is a group morphism. Thus E is a functor from the category of fields over k to the category of abelian groups.

Exercises

1. Verify completely the assertions in (1.1).
2. Carry out the details of the proof of the associative law for the ases $P+(P+Q) = (P+P)+Q$.

§2. Normal Forms for Cubic Curves

In the examples of Chapter 1 we always chose a cubic equation in normal form and mentioned that any elliptic curve could be transformed into normal form. This can be done in two ways. Transforming a flex, which exists by 2(4.8), to infinity with a tangent line, the line at infinity gives the normal form after suitable rescaling of the variables. For the reader with more background we sketch how it can be derived from the Riemann–Roch theorem. Here the background reference is Hartshorne [1977], Chapter 4.

(2.1) Remark. Any flex on a nonsingular cubic curve can be transformed to 0:0:1 in such a way that the tangent line to the transformed curve is $w = 0$, the line at infinity. In this case the cubic equation has the form in the w, x-plane

$$f(w,x,1) = w + f_2(w,x) + f_3(w,x) = w + a_1wx + a_3w^2 + f_3(w,x)$$

since w must divide $f_2(w,x)$ from the remarks following 2(4.7). Now normalize the coefficient of x^3 in $f_3(w,x)$ to -1, and we have the normal form $f(w,x,y) = 0$, namely

$$0 = wy^2 + a_1wxy + a_3w^2y - x^3 - a_2wx^2 - a_4w^2x - a_6w^3.$$

Away from the origin in the elliptic curve, we set $w = 1$ and consider the differential

$$0 = df(1,x,y) = f_x(1,x,y)dx + f_y(1,x,y)\,dy.$$

(2.2) Invariant Differential. For an elliptic curve E given by the equation $0 = f(1, x, y)$ in normal form the invariant differential is given by the following two expressions:

$$\omega = \frac{dx}{2y + a_1x + a_3} = \frac{dx}{f_y} = -\frac{dy}{f_x} = \frac{dy}{3x^2 + 2a_2x + a_4 - a_1y}.$$

Henceforth when we speak of an elliptic curve we will always mean a cubic curve E with O at $0 : 0 : 1$ and the equation in normal form. We still need to study to what extend these forms are unique. This is related to transformations of an elliptic curve, or from another point of view, changes of variable which preserves the normal form. We make this precise with the next definition.

(2.3) Definition. An admissible change of variables in the equation of an elliptic curve is one of the form

$$x = u^2\bar{x} + r \quad \text{and} \quad y = u^3\bar{y} + su^2\bar{x} + t,$$

where u, r, s, and t in k with u invertible, i.e., nonzero.

(2.4) Remark. Substitution by an admissible change of variables into the equation $0 = f(w, x, y)$ in normal form yields a new form of the equation in terms of the variables $\bar{x}$ and $\bar{y}$:

$$\bar{y}^2 + \bar{a}_1\overline{xy} + \bar{a}_3\bar{y} = \bar{x}^3 + \bar{a}_2\bar{x}^2 + \bar{a}_4\bar{x} + \bar{a}_6.$$

The invariant differential ω is changed to $\bar{\omega}$ and $\bar{\omega} = u\omega$. Also,

$$\begin{aligned}
u\bar{a}_1 &= a_1 + 2s,\\
u^2\bar{a}_2 &= a_2 - sa_1 + 3r - s^2,\\
u^3\bar{a}_3 &= a_3 + ra_1 + 2t = f_y(r, t),\\
u^4\bar{a}_4 &= a_4 - sa_3 + 2ra_2 - (t + rs)a_1 + 3r^2 - 2st = -f_x(r, t) - sf_y(r, t),\\
u^6\bar{a}_6 &= a_6 + ra_4 + r^2a_2 + r^3 - ta_3 - rta_1 - t^2 = -f(r, t).
\end{aligned}$$

These relations are left to the reader who should keep in mind that $a_6 = -f(0, 0)$, $a_4 = -f_x(0, 0)$, and $a_3 = f_y(0, 0)$.

(2.5) Remark. Consider two elliptic curves $\bar{E}$ and E defined by the equations in $\bar{a}_i$ and a_i, respectively, in normal form. Then $\phi : \bar{E} \to E$ is an isomorphism such that the functions x, y on E composed with ϕ are related to the functions $\bar{x}, \bar{y}$ on $\bar{E}$ by

$$x\phi = u^2\bar{x} + r \quad \text{and} \quad y\phi = u^3\bar{y} + su^2\bar{x} + t$$

as in an admissible change of variable. An easy calculation shows that the composition of two admissible changes of variable is again one and that the inverse of an admissible change of variables is again one.

For an admissible change of variables with $r = s = t = 0$ we have $a_i = u^i\bar{a}_i$, $x = u^2\bar{x}$, $y = u^3\bar{y}$, and $\omega = u^{-1}\bar{\omega}$. Thus there is a homogeneity where x has weight 2, y weight 3, a_i weight i, and ω weight -1. The polynomial $f(x, y)$ has weight 6 where $f(x, y) = 0$ is the equation in normal form.

Now for the second derivation of the normal form of the cubic equation for an elliptic curve we use Hartshorne [1977], Chapter 4 as background reference. The reader unfamiliar with this general curve theory can skip to the next section.

For the remainder of this section a curve C is a complete, nonsingular curve over an algebraically closed field k and 0 is a point of $C(k)$.

(2.6) Riemann–Roch for Curves of Genus 1. Let $\mathcal{O}_C$ be the structure sheaf on C of germs of regular functions. Let $\mathcal{O}_C(m \cdot 0)$ be the sheaf of germs of functions having at most an mth order pole at 0. For the vector space of sections $\Gamma(\mathcal{O}_C(m \cdot 0))$ the Riemann–Roch theorem gives the following formula when C is of genus 1:

$$\dim_k \Gamma(\mathcal{O}_C(m \cdot 0)) = \begin{cases} m & \text{for } m > 0, \\ 1 & \text{for } m = 0. \end{cases}$$

We can choose a basis for these spaces $\Gamma(\mathcal{O}_C(m \cdot 0))$ for small m as follows where we make use of the inclusions $\Gamma(\mathcal{O}_C(m \cdot 0)) \subset \Gamma(\mathcal{O}_C(m' \cdot 0))$ for $m \le m'$:

$$\begin{aligned}
&\Gamma(\mathcal{O}_C(1 \cdot 0)) = \Gamma(\mathcal{O}_C(0 \cdot 0)) = k \cdot 1, \\
&\Gamma(\mathcal{O}_C(2 \cdot 0)) = k \cdot 1 \oplus k \cdot x, \\
&\Gamma(\mathcal{O}_C(3 \cdot 0)) = k \cdot 1 \oplus k \cdot x \oplus k \cdot y, \\
&\Gamma(\mathcal{O}_C(4 \cdot 0)) = k \cdot 1 \oplus k \cdot x \oplus k \cdot y \oplus k \cdot x^2, \\
&\Gamma(\mathcal{O}_C(5 \cdot 0)) = k \cdot 1 \oplus k \cdot x \oplus k \cdot y \oplus k \cdot x^2 \oplus k \cdot xy.
\end{aligned}$$

Here x has a pole of order 2 and y of order 3 at 0. In $\Gamma(\mathcal{O}_C(6 \cdot 0))$ there are seven natural basis elements,

$$1, x, y, x^2, xy, x^3, \text{ and } y^2,$$

but the space is six dimensional. Hence there is a linear relation which is the defining equation of the image under

$$(1 : x : y) : C \to \mathbb{P}_2(k).$$

Further, given a local uniformizing parameter z at 0 generating the maximal idea of the local ring $\mathcal{O}_C$ of C at 0, we can specify that the formal analytic expansion of x and y be of the form

$$x = \frac{1}{z^2} + \cdots \quad \text{and} \quad y = -\frac{1}{z^3} + \cdots.$$

Note that x is unique up to a constant and y up to a linear combination of x and a constant. With this normalization the linear relation satisfied by $1, x, y, x^2, xy, x^3$, and y^3 is exactly the normal form of the cubic.

(2.7) Remark. The origin of admissible changes of variables can be seen from the point of view of changing z to a new local uniformizing parameter $uz = \bar{z}$ by multiplying z by a nonzero element of the ground field. Then x and y are changed to new functions $\bar{x}$ and $\bar{y}$ satisfying $\bar{x} = (\bar{z})^{-2} + \cdots$ and $\bar{y} = -(\bar{z})^{-3} + \cdots$. Since $x = u^2(\bar{z})^{-2} + \cdots$, it follows that $x = u^2\bar{x}$ up to a constant, i.e., $x = u^2\bar{x} + r$ for r in k. Since $y = -u^3(\bar{z})^{-3} + \cdots$, it follows that $y = u^3\bar{y}$ up to a linear combination of x and a constant, i.e., $y = u^3\bar{y} + su^2\bar{x} + t$.

We summarize (2.6) and (2.7) in the next theorem.

(2.8) Theorem. *Any isomorphism between two elliptic curves is given by an admissible change of variables relative to two given equations in normal form.*

Exercises

1. Verify the formulas for $u^i\bar{a}_i$, where $i = 1, 2, 3, 4$, and 6 in (2.4).
2. Show that the inverse and the composite of two admissible changes of variable are again admissible changes of variable.

§3. The Discriminant and the Invariant j

We associate to a cubic equation in normal form

$$y^2 + a_1xy + a_3y = x^3 + a_2x^2 + a_4x + a_6, \tag{N_1}$$

two new sets of coefficients b_i for $i = 2, 4, 6$, and 8 and c_j for $j = 4$ and 6 which arise first from completing the square and second from completing the cube.

(3.1) Notations. Associated to the coefficients a_1 in the cubic (N_1) are the coefficients

$$\begin{aligned} b_2 &= a_1^2 + 4a_2, \\ b_4 &= a_1a_3 + 2a_4, \\ b_6 &= a_3^2 + 4a_6, \\ b_8 &= a_1^2a_6 - a_1a_3a_4 + 4a_2a_6 + a_2a_3^2 - a_4^2. \end{aligned}$$

These quantities are related by $4b_8 = b_2b_6 - b_4^2$. We also introduce the discriminant in terms of the b_i's:

$$\Delta = -b_2^2b_8 - 8b_4^3 - 27b_6^2 + 9b_2b_4b_6.$$

With the discriminant we can decide whether or not the cubic is nonsingular. It is nonsingular if and only if $\Delta \neq 0$. Firstly we study the b_i and introduce the c_j coefficients.

(3.2) Remarks. Under an admissible change of variable as in (2.3) and (2.4) we have, for the corresponding $\bar{b}_i$ and $\bar{\Delta}$, the relations

$$u^2\bar{b}_2 = b_2 + 12r,$$

$$u^4\bar{b}_4 = b_4 + rb_2 + 6r^2,$$
$$u^6\bar{b}_6 = b_6 + 2rb_4 + r^2b_2 + 4r^3,$$
$$u^8\bar{b}_8 = b_8 + 3rb_6 + 3r^2b_4 + 3r_4,$$

and $u^{12}\bar{\Delta} = \Delta$. Moreover, if k is a field of characteristic different from 2, then for $y' = y + (a_1x + a_3)/2$ and $x' = x$ the equation in normal form becomes

$$(y')^2 = (x')^3 + \frac{b_2}{4}(x')^2 + \frac{b_4}{2}x' + \frac{b_6}{4}. \tag{N$_2$}$$

(3.3) Notations. Associated to the coefficients a_i in (N_1) and b_j in (3.1) are the coefficients

$$c_4 = b_2^2 - 24b_4, \qquad c_6 = -b_2^3 + 36b_2b_4 - 216b_6.$$

For Δ invertible, we introduce the quantity

$$j(E) = j = \frac{c_4^3}{\Delta}.$$

We have the following relation $12^3\Delta = c_4^3 - c_6^2$, and, therefore,

$$j = 12^3\frac{c_4^3}{c_4^3 - c_6^2}.$$

(3.4) Remarks. Under an admissible change of variable as in (2.3) and (2.4) we have for the corresponding $\bar{c}_j$ and $\bar{j}$ the relations

$$u^4\bar{c}_4 = c_4 \quad \text{and} \quad u^6\bar{c}_6 = c_6 \quad \text{and finally } \bar{j} = j.$$

For the j-invariant we have $j = \bar{j}$ which means that $j(E)$ is an invariant of an elliptic curve E up to isomorphism. If E and $\bar{E}$ are isomorphic, as in (2.5) and (2.8), then we have $j(E) = j(\bar{E})$. Moreover, if k is a field of characteristic different from 2 and 3, then for $y'' = y'$ and $x'' = x' + b_2/12$ the equation in normal form becomes

$$(y'') = (x'')^3 - \frac{c_4}{48}x'' - \frac{c_6}{864} \tag{N$_3$}$$

and $\omega = dx''/2y''$.

Now we take up the question of when the normal form defines a nonsingular curve. First consider a cubic polynomial $f(x) = x^3 + px + q$. The discriminant $D(f)$ is the resultant $R(f, f')$ where $f'(x) = 3x^2 + p$ is the derivative of $f(x)$. We compute

$$D(f) = \begin{vmatrix} q & p & 0 & 1 & 0 \\ 0 & q & p & 0 & 1 \\ p & 0 & 3 & 0 & 0 \\ 0 & p & 0 & 3 & 0 \\ 0 & 0 & p & 0 & 3 \end{vmatrix} = 27q^2 + 4p^3.$$

(3.5) Remarks. The polynomial $f(x) = x^3 + px + q \in k[x]$ has a repeated root in some extension field of k if and only if $D(f) = 0$. For a field k of characteristic unequal to 2 the cubic curve given by the equation $y^2 = f(x)$ is nonsingular if and only if $D(f) \neq 0$. This was already remarked in 1(2.1), and it is related to the calculation $2yy' = f'(x)$ showing that there is a well-defined tangent line if and only if $f(x)$ and $f'(x)$ do not have a common solution See also (8.2).

Return now to

$$(\mathrm{N}_2) \qquad y^2 = x^3 - \frac{c_4}{48}x - \frac{c_6}{864} = f(x),$$

so that $p = -c_4/48$ and $q = -c_6/864$. Using $864 = 2^5 \cdot 3^3$ and $48 = 2^4 \cdot 3$, we see that

$$-2^4 D(f) = \frac{c_4^3 - c_6^2}{12^3} = \Delta.$$

In conclusion we have the following result in characteristic different from 2 and 3.

(3.6) Proposition. *Over a field k of characteristic different from* 2 *or* 3, *the cubic equation*

$$y^2 = x^3 - \frac{c^4}{48}x - \frac{c_6}{864}$$

represents an elliptic curve if and only if $\Delta \neq 0$. *Also* $\omega - dx/2y$.

(3.7) Remark. For $j \neq 0$ or 12^3 the following cubic

$$y^2 + xy = x^3 - \frac{36}{j - 1728}x - \frac{1}{j - 1728}$$

defines an elliptic curve with j-invariant equal to j over any field k. This is a straightforward calculation which is left to the reader to verify. The elliptic curve with equation $y^2 = x^3 + a$ has $j = 0$, and the elliptic curve with equation $y^2 = x^3 + ax$ has $j = 12^3 = 1728$.

This topic is also taken up in §8.

Exercises

1. Derive the formula for the discriminant of the cubic polynomial $f(x) = x^3 + ax^2 + bx + c$.
2. Derive the formula for the discriminant of the cubic polynomial

$$f(x) = (x - \alpha_1)(x - \alpha_2)(x - \alpha_3).$$

3. Derive the formula for the discriminant of the quartic polynomial

$$f(x) = x^4 + ax^2 + bx + c.$$

4. Calculate the discriminant for the following elliptic curves. Calculate the value of j.

(a) $y^2 + y = x^3 - x^2$. (b) $y^2 + y = x^3 + x^2$.

(c) $y^2 + y = x^3 - x$. (d) $y^2 + y = x^3 + x$.

(e) $y^2 - xy + y = x^3$. (f) $y^2 - y = x^3 - 7$.

5. Give the general formula for the discriminant for $x^3 + ax^2 + bx$ and compare with the discriminant of the cubic curve with equation $y^2 = x^3 + ax^2 + bx$. Determine the discriminant of the following elliptic curves.

(a) $y^2 = x^3 + x^2 - x$. (b) $y^2 = x^3 - x^2 + x$.

(c) $y^2 = x^3 - 2x - x$. (d) $y^2 = x^3 - 2x^2 - 15x$.

§4. Isomorphism Classification in Characteristics $\neq 2, 3$

In §4, §5, §6 we have a common approach to the isomorphism classification, and in §8 there is an alternative approach to §4.

Let k be a field of characteristic $\neq 2, 3$ in this section. For an elliptic curve E over k we can choose coordinates x and y giving the Weierstrass model in the following form:

$$y^2 = x^3 + a_4x + a_6 \quad \text{and} \quad \omega = \frac{dx}{2y}.$$

By (3.4) we see that

$$c_4 = -48a_4, \quad c_6 = -864a_6, \quad \text{and} \quad \Delta = -16(4a_4^3 + 27a_6^2).$$

(4.1) Conditions for Smoothness. For $f = y^2 - x^3 - a_4x - a_6$ the curve E is smooth or nonsingular if and only if f, f_x, and f_y have no common zero. By (3.6) the curve E is smooth if and only if $\Delta \neq 0$.

Further, $j = j(E)$ is given by

$$j = \frac{c_4^3}{\Delta} = 12^3\frac{c_4^3}{c_4^3 - c_6^2} = \frac{-4^3 12^3 (a_4)^3}{-16(4a_4^3 + 27a_6^2)} = 12^3\frac{4a_4^3}{4a_4^3 + 27a_6^2}.$$

(4.2) Isomorphisms Between Two Curves with the Same j-Invariant. Suppose E and $\bar{E}$ are two elliptic curves defined over k with equations $y^2 = x^3 + a_4x + a_6$ and $y^2 = x^3 + \bar{a}_4x + \bar{a}_6$ such that $j = j(E) = j(\bar{E})$. If $\phi : \bar{E} \to E$ is an isomorphism, or equivalently admissible change of variables, then

$$x\phi = u^2\bar{x}, \quad y\phi = u^3\bar{y}, \quad a_4 = u^4\bar{a}_4, \quad \text{and} \quad a_6 = u^6\bar{a}_6.$$

These relations are now studied in three separate cases. Observe that $12^3 \neq 0$ in the characteristics under consideration.

Case 1. $j \neq 0$ or 12^3, or equivalently $a_4a_6 \neq 0$. Then we see that E and $\bar{E}$ are isomorphic only if the quotient $a_4\bar{a}_6/\bar{a}_4a_6$ is a square u^{-2}. Hence E and $\bar{E}$ are isomorphic over any field extension of k containing the square root of the quotient $a_4\bar{a}_6/\bar{a}_4a_6$. Further, specializing to $E = \bar{E}$, we have that the automorphism group $\mathrm{Aut}(E) = \{+1, -1\}$, the group of square roots of 1.

Case 2. $j = 12^3$, or equivalently $a_6 = 0$. The basic example is the curve $y^2 = x^3 - x$. Then E and $\bar{E}$ are isomorphic if and only if the quotient $a_4/\bar{a}_4$ is a fourth power u^4. Hence E and $\bar{E}$ are isomorphic over any field extension of k containing a fourth root of the quotient $a_4/\bar{a}_4$. Further, specializing to $E = \bar{E}$, we have that the automorphism group $\mathrm{Aut}(E) = \{+1, -1, +i, -i\}$, the group of fourth roots of unity.

Case 3. $j = 0$, or equivalently $a_4 = 0$. The basic example is the curve $y^2 = x^3 - 1$. Then E and $\bar{E}$ are isomorphic if and only if the quotient $a_6/\bar{a}_6$ is a sixth power u^6. Hence E and $\bar{E}$ are isomorphic over any field extension of k containing a sixth root of the quotient $a_6/\bar{a}_6$. Further, specializing to $E = \bar{E}$, we have that the automorphism group $\mathrm{Aut}(E) = \{+1, -1, +\rho, -\rho, +\rho^2, -\rho^2\}$, the group of sixth roots of unity where $\rho^2 + \rho + 1 = 0$.

It is natural to ask (1) $j(E) = j(\bar{E})$ implies E and $\bar{E}$ are isomorphic for k algebraically closed, and (2) whether all values in k, besides 0 and 12^3, are j values of some elliptic curve. When $a_4a_6 \neq 0$ and k is algebraically closed, we can rescale x and y so that the Weierstrass equation has the form $y^2 = 4x^3 - cx - c$. In terms of c we calculate with (3.4)

$$j = 12^3 \frac{c^3}{c^3 - 27c^2} = 12^3 \frac{c}{c - 27} = 12^3 J \qquad (12^3 = 1728).$$

From the relation $J = c/(c - 27)$ we can solve for c in terms of j as

$$c = 27\frac{J}{J - 1} = 27\frac{j}{j - 1728}.$$

Thus we have the proposition.

(4.3) Proposition. *Two elliptic curves E and $\bar{E}$ over k are isomorphic over $\bar{k}$ if and only if $j(E) = j(\bar{E})$. The curve with classical Weierstrass equation*

$$y^2 = 4x^3 - 27\frac{j}{j - 1728}x - 27\frac{j}{j - 1728}$$

has j-invariant equal to the parameter j in the formula for the coefficients.

This is another version of the result (3.7) where for all j values unequal to 0 and 12^3 an elliptic curve E with given $j = j(E)$. These curves are in a family of elliptic curves over the twice punctured plane with fibre over j equal to an elliptic curve with j value equal to the given j.

§5. Isomorphism Classification in Characteristic 3

Let k be a field of characteristic 3 in this section. For an elliptic curve E over k we can choose coordinates x and y giving the Weierstrass model in the form

$$y^2 = x^3 + a_2x^2 + a_4x + a_6 \quad \text{and} \quad \omega = -\frac{dx}{y}.$$

By (3.1) and (3.3) and using $3 = 0$, we see that

$$b_2 = a_2, \qquad b_4 = -a_4, \qquad b_6 = a_6, \qquad b_8 = -a_4^2 + a_2a_6$$

and

$$c_4 = a_2^2, \qquad c_6 = -a_2^3, \qquad \Delta = a_2^2a_4^2 - a_2^3a_6 - a_4^3.$$

(5.1) Conditions for Smoothness. As in (4.1) the curve E is smooth if and only if the cubic polynomial $g(x) = x^3 + a_2x^2 + a_4x + a_6$ and its derivative $g'(x)$ have no common zero. Since Δ and the discriminant of $g(x)$ are equal up to a nonzero constant, we deduce again that E is smooth if and only if $\Delta \neq 0$.

Further, $j = j(E)$ is given by

$$j = \frac{c_4^3}{\Delta} = \frac{a_2^6}{a_2^2a_4^2 - a_2^3a_6 - a_4^3}.$$

(5.2) Isomorphisms Between Two Curves with the Same *j*-Invariant. Suppose E and $\bar{E}$ are two elliptic curves defined over k with Weierstrass equations $y^2 = x^3 + a_2x^2 + a_4x + a_6$ and $y^2 = x^3 + \bar{a}_2x^2 + \bar{a}_4x + \bar{a}_6$ such that $j = j(E) = j(\bar{E})$. If $f : \bar{E} \to E$ is an isomorphism, then its form is determined by whether $j \neq 0$ or $j = 0 = 12^3$.

Case 1. $j \neq 0$ or equivalently $a_2 \neq 0$. By completing the square in both Weierstrass equations, we can assume that $a_4 = \bar{a}_4 = 0$. Then $j(E) = -a_2^3/a_6 = -\bar{a}_2^3/\bar{a}_6 = j(\bar{E})$, and the following hold:

$$xf = u^2\bar{x}, \qquad yf = u^3\bar{y}, \quad \text{and} \quad a_2 = u^2\bar{a}_2.$$

Thus E and $\bar{E}$ are isomorphic if and only if the quotient $a_2/\bar{a}_2$ is a square u^2. Hence E and $\bar{E}$ are isomorphic over any field extension of k containing the square root of the quotient $a_2/\bar{a}_2$. Further, specializing to $E = \bar{E}$, we have that the automorphism group $\mathrm{Aut}(E) = \{+1, -1\}$, the group of square roots of 1.

Case 2. $j = 0$, or equivalently $a_2 = 0$. Then $\Delta = a_4$ and $\omega = dy/a_4$, and the following hold for the isomorphism f:

$$xf = u^2\bar{x} + r, \qquad yf = u^3\bar{y}, \qquad a_4 = u^4\bar{a}_4,$$

and

$$u^6\bar{a}_6 = a_6 + ra_4 + r^3.$$

Then E and $\bar{E}$ are isomorphic if and only if the quotient $a_4/\bar{a}_4$ is a fourth power u^4 and $u^6\bar{a}_6 - a_6$ is of the form $r^3 + ra_4$. Hence E and $\bar{E}$ are isomorphic over any field where $a_4/\bar{a}_4$ is a fourth power and there is a solution for the cubic equation for r. This can always be realized by going to a separable extension of degree dividing 12.

Further, specializing to $E = \bar{E}$, we have that the automorphism group is a semi-direct product $\mathrm{Aut}(E) = (\mathbb{Z}/4) \times (\mathbb{Z}/3)$ where the cyclic group of order 4 acts nontrivially on the normal subgroup of order 3. The only nontrivial action is for a generator of $\mathbb{Z}/4$ to carry every element to its inverse. The automorphisms of E are parametrized by pairs (u, r), where u is in $+1, -1, +i, -i$ and r satisfies

(i) $\quad r^3 + a_4r = 0 \quad$ if $u = +1, -1$,

(ii) $\quad r^3 + a_4r + 2a_6 = 0 \quad$ if $u = +i, -i$.

For the curve $y^2 = x^3 - x$ the automorphisms are given by (u, r) with $u^4 = 1$ and $r^3 - r = 0$, i.e., r in $\mathbb{F}_3$. The action of (u, r) on the curve E is given by $(u, r)(x, y) = (x + r, uy)$ in terms of points on the curve.

The problem of realizing all j values, different from $0 = 12^3$, has been solved in (3.7). This is the case $a_2 \neq 0$ and when k is algebraically closed, we can rescale x and y so that the cubic equation has the form $y^2 = x^3 + x^2 + a_6$ or in characteristic 3 the form $y^2 + xy = x^3 + a_6$. In both cases $j(E) = -1/a_6$.

(5.3) Proposition. *The curves with normal cubic equations*

$$y^2 = x^3 + x^2 - \frac{1}{j} \quad \text{or} \quad y^2 + xy = x^3 - \frac{1}{j}$$

have j-invariant equal to the parameter j appearing in the formula for the coefficient for all nonzero j.

Exercises

1. Find the relation between Δ for the curve defined by the cubic equation $y^2 = g(x) = x^3 + a_2x^2 + a_4x + a_6$, and the discriminant of the cubic polynomial $g(x)$.
2. Find an elliptic curve in characteristic 3 with j value 0.

§6. Isomorphism Classification in Characteristic 2

Let k be a field of characteristic 2 in this section. For an elliptic curve E over k the invariant differential has the form $\omega = dx/(a_1x + a_3)$ so that either a_1 or $a_3 \neq 0$. Also by (3.4) using $2 = 0$, we see that $b_2 = a_1^2$ and $c_4 = b_2^2$, and therefore

$$c_4 = a_1^4 \quad \text{and} \quad j = a_1^{12}/\Delta.$$

In particular, $a_1 = 0$ if and only if $j = 0 = 12^3$.

(6.1) Conditions for Smoothness. Unlike characteristics different from 2 we have to treat the two cases $j = 0$ and $j \neq 0$ separately.

Case 1. $j \neq 0$ or equivalently $a_1 \neq 0$. Under a change x to $x + c$, the term $y^2 + a_1xy + a_3$ becomes $y^2 + a_1xy + (a_1c + a_3)y$, and for a_1 nonzero we can choose $a_3 = 0$. Changing x to a_1^3x and y to a_1^3y allows us to normalize $a_1 = 1$, and a linear change of the variable allows us to choose $a_4 = 0$. The normal form becomes

$$y^2 + xy = x^3 + a_2x^2 + a_6 \quad \text{and} \quad \omega = \frac{dx}{x}.$$

Then $b_2 = 1,\ b_4 = b_6 = 0$, and $b_8 = a_6$, and, moreover, $c_4 = 1$ and $\Delta = a_6 = 1/j$. The partial derivatives $f_x = y + x^2$ and $f_y = x$ have only $x = y = 0$ as a common zero, and this lies on the curve if and only if $a_6 = \Delta = 0$. Hence E is smooth if and only if $\Delta \neq 0$.

Case 2. $j = 0$ or equivalently $a_1 = 0$. By completing the cube, we can choose the normal form of the cubic to be

$$y^2 + a_3y = x^3 + a_4x + a_6 \quad \text{and} \quad \omega = \frac{dx}{a_3}.$$

Then $b_2 = b_4 = 0,\ b_6 = a_3^2$, and $b_8 = a_4^2$, and, moreover, $\Delta = a_3^4$ and $j = 0$. Since the partial derivative $f_x = x^2 + a_4$ and $f_y = a_3$, it follows that the curve is smooth if and only if $a_3 \neq 0$ or equivalently $\Delta \neq 0$.

(6.2) Isomorphisms Between Two Curves with the Same j-Invariant. Suppose E and $\bar{E}$ are two elliptic curves defined over k such that $j = j(E) = j(\bar{E})$. If $f : \bar{E} \to E$ is an isomorphism, then its form is determined for $j \neq 0$ or $j = 0 = 12^3$.

Case 1. $j \neq 0$ or equivalently $a_1 \neq 0$. Using the form of the Weierstrass equations in (6.1), Case 1 for E and $\bar{E}$, we have

$$xf = \bar{x}, \qquad yf = \bar{y} + s\bar{x},$$

and for the coefficients $\bar{a}_2 = a_2 + s^2 + s$ and $\bar{a}_6 = a_6$. Then E and $\bar{E}$ are isomorphic if and only if the difference $\bar{a}_2 - a_2$ is of the form $s_2 + s$. Hence E and $\bar{E}$ are isomorphic over any field extension of k containing a solution to the quadratic equation

$$s^2 + s = \bar{a}_2 - a_2.$$

Further, specializing to $E = \bar{E}$, we have that the automorphism group $\mathrm{Aut}(E) = \{0, 1\}$ under addition.

Case 2. $j = 0$ or equivalently $a_1 = 0$. Using the form of the Weierstrass equations in (6.1), Case 2 for E and $\bar{E}$, we have

$$xf = u^2\bar{x}, \qquad yf = u^3y + su^2\bar{x} + t,$$

and for the coefficients

$$u^3\bar{a}_3 = a_3, \qquad u^4\bar{a}_4 = a_4 + sa_3 + s^4,$$
$$u^6\bar{a}_6 = a_6 + s^2a_4 + ta_3 + s^6 + t^2.$$

Then E and $\bar{E}$ are isomorphic if and only if:

(1) the quotient $a_3/\bar{a}_3$ is a cube u^3,
(2) the separable quartic equation $s^4 + a_3t + a_4 + u^4\bar{a}_4 = 0$ in s has a solution, and
(3) the quadratic equation $t^2 + a_3t + (s^6 + s^2a_4 + a_6 + u^6\bar{a}_6) = 0$ in t has a solution.

Hence E and $\bar{E}$ are isomorphic over any field extension of k containing the roots to the equations (1), (2), and (3). Further, specializing to $E = \bar{E}$, we have that the automorphism group $\mathrm{Aut}(E)$ is a certain group of order 24 provided that k contains the roots of the above equations.

(6.3) Remark. The group $\mathrm{Aut}(E)$ can be described two ways either as $\mathrm{SL}_2(\mathbb{F}_3)$ or as the units in the integral quaternions $\pm 1, \pm i, \pm j, \pm k, (\pm 1 \pm i \pm j \pm k)/2$.

Now we consider the special case of $k = \mathbb{F}_2$, the field of two elements and write down all the elliptic curves. There are five up to isomorphism (over $\mathbb{F}_2$) two with $j = 1$ and three with $j = 0$.

Case 1. $j = 1$. Since $y^2 + xy = f(x)$ with $f(x) = x^3$ and $f(x) = x^3 + x^2$ have a singularity at $(0, 0)$, we have four equations related by replacing y by $y + 1$ pairwise:

$$E_1 : y^2 + xy = x^3 + x^2 + 1 \cong E_1' : y^2 + xy = x^3 + x^2 + x,$$
$$E_2 : y^2 + xy = x^3 + 1 \qquad \cong E_2' : y^2 + xy = x^3 + x.$$

We can "graph" the elliptic curves E_1 and E_2 over $\mathbb{F}_2$.

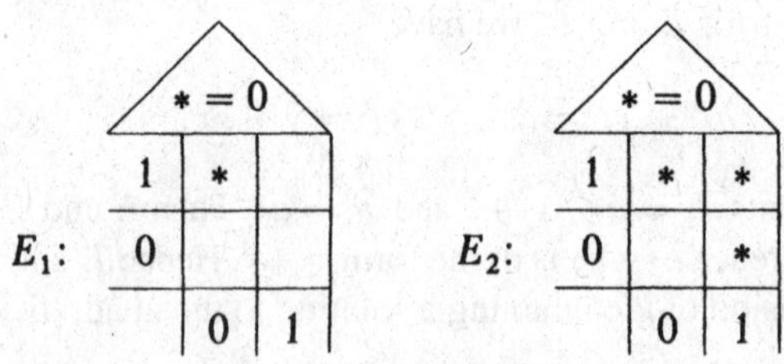

The groups $E_1(\mathbb{F}_2) = \mathbb{Z}/2\mathbb{Z} = \{0, (0, 1)\}$ and $E_2(\mathbb{F}_2) = \mathbb{Z}/4\mathbb{Z} = \{0, (0, 1), (1, 0), (1, 1)\}$.

Case 2. $j = 0$. Since the general form of such a curve is $y^2 + y = x^3 + a_4x + a_6$, there are four possibilities of which two are isomorphic. Also we can replace y by $y + x$ or x by $x + 1$ to obtain forms up to isomorphism of the same elliptic curve.

$$E_3 : y^2 + y = x^3 + x \cong E_3' : y^2 + y = x^3 + x^2,$$

E_3: 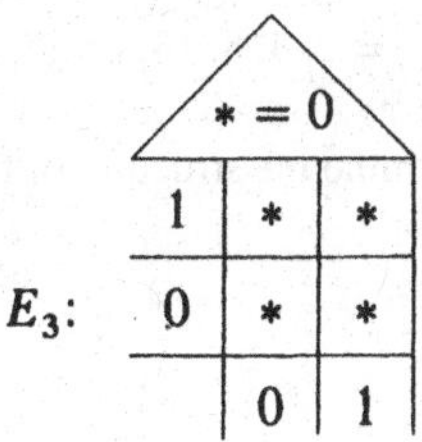

E_4: 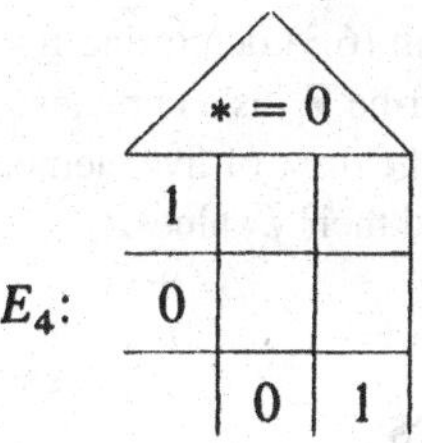

E_5: 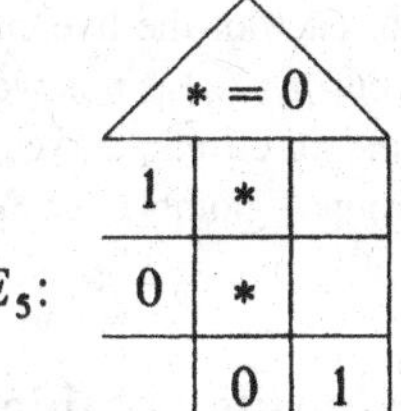

$$\begin{aligned} E_4 : y^2 + y = x^3 + x + 1 &\cong E_4' : y^2 + y = x^3 + x^2 + 1, \\ E_5 : y^2 + y = x^3 &\cong E_5' : y^2 + y = x^3 + 1, \\ &\cong E_5'' : y^2 + y = x^3 + x^2 + x, \\ &\cong E_5''' : y^2 + y = x^3 + x^2 + x + 1. \end{aligned}$$

We can "graph" the elliptic curves E_3, E_4, and E_5 over $\mathbb{F}_2$. The groups $E_3(\mathbb{F}_2) = \mathbb{Z}/5\mathbb{Z} = \{0, (0, 0), (1, 0), (0, 1), (1, 1)\}$, $E_4(\mathbb{F}_2) = 0 = \{0\}$, and $E_5(\mathbb{F}_2) = \mathbb{Z}/3\mathbb{Z} = \{0, (0, 0), (0, 1)\}$. Observe that the five elliptic curves could not be isomorphic over $\mathbb{F}_2$ in any sense preserving their structure as algebraic curves because all five have different numbers of points over the field $\mathbb{F}_2$ of two elements. In summary we have:

(6.4) Proposition. *Up to isomorphism over $\mathbb{F}_2$ there are five elliptic curves defined over $\mathbb{F}_2$. Two, namely E_1 and E_2 above, have $j = 1$ and three, namely E_3, E_4 and E_5 above, have $j = 0$.*

Exercises

1. Show that the field of four elements $\mathbb{F}_4$ will be of the form $0, 1, u, u'$ where $u^2 = u'$, $uu' = 1$, $1 + u + u^2 = 0$, and $1 + u' + u'^2 = 1$. "Graph" the curves $E_1, \ldots, E_5$ of (6.4) over the field of four elements. Show that E_1 and E_2 are isomorphic and E_3 and E_4 are isomorphic over $\mathbb{F}_4$. Show that E_3 and E_5 are not isomorphic over $\mathbb{F}_4$.
2. Show that there is a field F_{16} of 16 elements which is additively of the form $\mathbb{F}_{16} = \mathbb{F}_4 + \mathbb{F}_4 v$ where $v^2 + v = u$. Show that every nonzero element is a square. Show that there is a field $\mathbb{F}_{256}$ of 256 elements which is additively of the form $\mathbb{F}_{256} = \mathbb{F}_{16} + \mathbb{F}_{16} w$ where $w^2 + w = v^3$. Observe that we have the inclusions $\mathbb{F}_2 \subset \mathbb{F}_4 \subset \mathbb{F}_{16} \subset \mathbb{F}_{256}$ and $\mathbb{F}_{q^2}$ is a two-dimensional vector space over $\mathbb{F}_q$.
3. Show that E_3 and E_5 are not isomorphic over $\mathbb{F}_{16}$, but that they are isomorphic over $\mathbb{F}_{256}$.
4. Determine the automorphism groups $\mathrm{Aut}_k(E_i)$ with $i = 1, \ldots, 5$ as in (6.3) and $k = \mathbb{F}_q$ for $q = 2, 4, 16, 256$.
5. Find all elliptic curves over $\mathbb{F}_4$ up to isomorphism over $\mathbb{F}_4$. Which curves become isomorphic over $\mathbb{F}_{16}$? Find their j-invariants.
6. Find all elliptic curves over $\mathbb{F}_3$ up to isomorphism over $\mathbb{F}_3$. Show there are four with $j = 1$ or -1 and four with $j = 0$. Determine their groups of points over $\mathbb{F}_9$ and which ones are isomorphic over $\mathbb{F}_9$.
7. For an elliptic curve E over $\mathbb{F}_3$ determine $\mathrm{Aut}_k(E)$, where $k = \mathbb{F}_3$ and $\mathbb{F}_9$.

8. For each of the five curves E_i in (6.3) determine for which $q = 2, 4, 8, 16$, or 256 the group $E(\mathbb{F}_q)$ is noncyclic. Describe $\mathbb{F}_8$ as a cubic extension of $\mathbb{F}_2$.
9. Find all elliptic curves over the field $\mathbb{F}_5$ of five elements, determine the structure of their group of points over $\mathbb{F}_5$, and find their j values.

§7. Singular Cubic Curves

Singular cubic curves will arise naturally in Chapter 5 when an elliptic curve is reduced modulo a prime. For this reason and as a rounding off of our discussion of cubics in the plane, we study cubics over a field k wih irreducible equation

$$F(x, y) = y^2 + a_1xy + a_3y - x^3 - a_2x^2 - a_4x - a_6 = 0$$

having a singular point which is rational.

The singular point can be transformed to the affine origin $(x, y) = (0, 0)$ by an admissible change of variables as discussed in (2.4). Observe that $(0, 0)$ is on the curve, i.e., $F(0, 0) = 0$ if and only if $a_6 = 0$. To determine whether $(0, 0)$ is a singular point, we substitute into the partial derivatives

$$F_y = 2y + a_1x + a_3 \quad \text{and} \quad F_x = a_1y - 3x^2 - 2a_2x - a_4.$$

(7.1) Remark. The point $(0, 0)$ is a singular point on the Weierstrass cubic with equation $F(x, y) = 0$ if and only if

$$a_3 = a_4 = a_6 = 0.$$

This follows by just substituting $F_y(0, 0) = a_3 = 0$ and $F_x(0, 0) = -a_4 = 0$. With reference to the formulas (3.1) we see that $(0, 0)$ a singular point on the cubic curve implies that

$$b_4 = b_6 = b_8 = 0 \quad \text{and} \quad \Delta = 0.$$

The Weierstrass form becomes

$$y^2 = a_1xy - a_2x^2 = x^3,$$

and the discriminant of the quadratic form on the left is the coefficient $b_2 = a_1^2 + 4a_2$.

As in the Introduction, we go to the (t, s)-plane with $t = -x/y$ and $s = -1/y$. The singular Weierstrass equation becomes $s = t^3 + a_1ts + a_2t^2s$, and thus s is a rational function of t, namely

$$s = \frac{t^3}{1 - a_1t - a_2t^2}.$$

This means that a singular cubic C is a rational curve in the geometric sense, and the set of nonsingular points C_{ns} consists of all (t, s) with $s = t^3/(1 - a_1t - a_2t^2)$ and

$1 - a_1 t - a_2 t^2 \neq 0$. The discriminant of this quadratic is $b_2 = a_1^2 + 4a_2$. In fact, we can go further and put a group structure on C_{ns} using the chord-tangent construction considered in Chapter 1. Unlike the nonsingular cubics which give entirely new objects of study, singular cubics are isomorphic to the familiar multiplicative $\mathbb{G}_m$ and additive $\mathbb{G}_a$ groups. This relation is carried out explicitly in the next theorem.

(7.2) Theorem. *Let E be a cubic curve over k with equation $y^2 + a_1 xy = x^3 + a_2 x^2$ which we factor as $(y - \alpha x)(y - \beta x) = x^3$ over the field $k_1 = k(\alpha) = k(\beta)$.*

(1) *Multiplicative case, $\alpha \neq \beta$: The function $(x, y) \to (y - \beta x)/(y - \alpha x)$ defines a homomorphism $E_{ns} \to \mathbb{G}_m$ over k_1.*
   (a) *If $k = k_1$, that is, α and β are in k, then the map $E_{ns}(k) \to \mathbb{G}_m(k) = k^*$ is an isomorphism onto the multiplicative group of k.*
   (b) *If k_1 is a quadratic extension of k, that is, α and β are not in k, then the map defines an isomorphism $E_{ns}(k) \to \ker(N_{k_1/k})$, where $N_{k_1/k} : k_1^* \to k^*$ is the norm map and $\ker(N_{k_1/k})$ is the subgroup elements in k_1^* with norm 1.*

(2) *Additive case, $\alpha = \beta$: The function $(x, y) \to x/(y - \alpha x)$ defines a homomorphism $E_{ns} \to \mathbb{G}_a$ over k_1. The map $E_{ns}(k_1) \to \mathbb{G}_a(k_1)$ is an isomorphism onto the additive group of k_1. Observe that $k = k(\alpha)$ except possibly in characteristic 2.*

Proof. In the multiplicative case we introduce the new variables

$$u = \frac{y - \beta x}{y - \alpha x} \quad \text{and} \quad v = \frac{1}{y - \alpha x}.$$

Using the relation $(y - \alpha x)^3 (u - 1)^3 = (\alpha - \beta)^3 x^3 = (\alpha - \beta)^3 (y - \alpha x)(y - \beta x)$, we obtain the equation for E_{ns} in (u, v)-coordinates as $(\alpha - \beta)^3 uv = (u - 1)^3$. Moreover, lines in x, y with equations $Ax + By + C = 0$ are transformed into lines in u, v with equations $A'u + B'v + C' = 0$. If (u_1, v_1), (u_2, v_2), and (u_3, v_3) are three points on the cubic E_{ns} which lie on a line $v = \lambda u + \delta$, then we have the factorization

$$0 = (u - 1)^3 - (\alpha - \beta)^3 u(\lambda u + \delta) = (u - u_1)(u - u_2)(u - u_3),$$

and hence the relation $u_1 u_2 u_3 = 1$ in the multiplicative group. This means that the function $(x, y) \to u$ carries the group law on E_{ns} into the multiplicative group law on k_1^*.

Finally the elements $u = (y - \beta x)/(y - \alpha x)$ have norm one

$$N_{k_1/k}(u) = uu' = \frac{y - \beta x}{y - \alpha x} \cdot \frac{y - \alpha x}{y - \beta x} = 1,$$

where $\alpha' = \beta$ and $\beta' = \alpha$ are conjugates of each other in k_1 over k. Conversely, if z in k_1 has norm 1, then for some c in k_1 the element $w = c + zc' \neq 0$ and from $w' = c' + z'c$ we deduce that

$$zw' = zc' + zz'c = c + zc' = w.$$

Hence $z = w/w' = (y - \beta x)/(y - \alpha x)$ for x, y in k.

In the additive case we introduce the new variables

$$u = \frac{x}{y - \alpha x} \quad \text{and} \quad v = \frac{1}{y - \alpha x}.$$

Using the relation $(y - \alpha x)^2 = x^3 = (y - \alpha x)^3 u^3$, we obtain the equation for E_{ns} in (u, v)-coordinates as $v = u^3$. Moreover, lines in x, y are transformed into lines in u, v. If (u_1, v_1), (u_2, v_2), and (u_3, v_3) are three points on the cubic E_{ns} which lie of a line $v = \lambda u + \delta$, then we have the factorization

$$0 = u^3 - (\lambda u + \delta) = (u - u_1)(u - u_2)(u - u_3),$$

and hence the relation $u_1 + u_2 + u_3 = 0$ in the additive group. This means that the function $(x, y) \to u$ carries the group law on E_{ns} into the additive group law on k_1, and $E_{ns}(k_1) \to k_1^+$ is an isomorphism of groups. This proves the theorem.

(7.3) Remark. At the singular point (0, 0) on the cubic

$$\begin{aligned} x^3 &= y^2 + a_1 xy - a_2 x^2 \\ &= (y - \alpha x)(y - \beta x) \end{aligned}$$

the tangent lines are given by $y = \alpha x$ and $y = \beta x$. The discriminant of the quadratic form factoring into the equations of the tangent lines is $D = a_1^2 + 4a_2 = b_2$. The two cases considered in the previous theorem correspond to two kinds of singularities and significantly for further questions to two kinds of j values:

(1) The singularity (0, 0) is a node (for simple double point) if and only if $D = b_2 \neq 0$, i.e., $\alpha \neq \beta$. Observe that $\beta_2 \neq 0$, $c_4 \neq 0$, and $c_6 \neq 0$ are all equivalent in this case and

$$j = \frac{c_4^3}{\Delta} = \infty, \qquad \text{where } \Delta = 0.$$

The tangents are rational over k if and only if b_2 is a square in k, i.e., $b_2 \in (k^*)^2$.

(2) The singularity (0, 0) is a cusp if and only if $D = b_2 = 0$, i.e., $\alpha = \beta$. Observe that $\beta_2 = 0$, $c_4 = 0$, and $c_6 = 0$ are all equivalent in this case and $j = 0/0$ is indeterminate.

§8. Parameterization of Curves in Characteristic Unequal to 2 or 3

In this section K is a field of characteristic $\neq 2, 3$.

(8.1) Notation. For $\alpha, \beta \in K$ we denote the elliptic curve with cubic equation

$$y^2 = x^3 - 3\alpha x + 2\beta$$

by $E\langle \alpha, \beta \rangle$.

(8.2) Nonsingular Curves. $E\langle\alpha, \beta\rangle$. We study the question of when $E\langle\alpha, \beta\rangle$ is nonsingular using $(d/dx)(x^3 - 3\alpha x + 2\beta) = 3x^2 - 3\alpha$. Multiple roots of the cubic are solutions of the two equations

$$0 = x^3 - 3\alpha x + 2\beta \quad \text{and} \quad 0 = 3x^3 - 3\alpha x.$$

Eliminating the x^3 term, we obtain $-6\alpha x + 6\beta = 0$ or $x = \beta/\alpha$ as the only possible multiple root. Substituting this back into the multiple $\alpha^2(3x^2 - 3\alpha)$ of the derivative, we obtain $3(\beta^2 - \alpha^3)$, and hence $E\langle\alpha, \beta\rangle$ is nonsingular if and only if $\Delta(\alpha, \beta) = \Delta = \alpha^3 - \beta^2 \neq 0$.

(8.3) Isomorphism Classification. We introduce the invariant

$$J(\alpha, \beta) = \frac{\alpha^3}{\alpha^3 - \beta^2} = \frac{\alpha^3}{\Delta(\alpha, \beta)}$$

for $E\langle\alpha, \beta\rangle$. Then the isomorphism classification given in (4.3) takes the form of the following three equivalent statements:

(1) the elliptic curves $E\langle\alpha, \beta\rangle$ and $E\langle\alpha', \beta'\rangle$ are isomorphic,
(2) there exists $\lambda \in K^*$ with $\lambda^4\alpha = \alpha'$ and $\lambda^6\beta = \beta'$, and
(3) the J-invariants are equal $J(\alpha, \beta) = J(\alpha', \beta')$.

To construct a space with one point representing each isomorphism class of an elliptic curve over K we use the following action.

(8.4) A Filtration of K^2 Stable under the Action of the Multiplicative Group. The multiplicative group K^* of K acts on K^2 by the formula $\lambda\cdot(\alpha, \beta) = (\lambda^4\alpha, \lambda^6\beta)$, and J is equivariant by the above (8.3) (2) and (3), that is, $J(\lambda\cdot(\alpha, \beta)) = J(\lambda^4\alpha, \lambda^6\beta) = J(\alpha, \beta)$. We exhibit a K^*-equivariant filtration of K^2 which maps by J to a filtration of the projective line $\mathbb{P}_1(K)$ under J as follows

$$\begin{array}{ccccccc}
K^2 & \supset & K^2 - \{(0,0)\} & \supset & K^2 - \{\alpha^3 = \beta^2\} & \supset & (K^*)^2 - \{\alpha^3 = \beta^2\} \\
 & & \downarrow J & & \downarrow J & & \downarrow J \\
 & & \mathbb{P}_1(K) & \supset & K & \supset & K - \{0, 1\}.
\end{array}$$

Now (8.3) can be summarized in terms of this filtration as follows:

(1) $E\langle 0, 0\rangle$ is the cuspidal singular cubic.
(2) $J(\alpha, \beta) = \infty$ if and only if $\alpha \neq 0$ and $\alpha^3 = \beta^2$, in which case $E\langle\alpha, \beta\rangle$ is singular cubic with a double point at the origin.
(3) $J(\alpha, \beta) = 0$ if and only if the curve is $E\langle 0, \beta\rangle : y^2 = x^3 + 2\beta,\ \beta \neq 0$, with an automorphism group of order 6.
(4) $J(\alpha, \beta) = 1$ if and only if the curve is $E\langle\alpha, 0\rangle : y^2 = x^3 - 3\alpha x,\ \alpha \neq 0$, with automorphism group cyclic of order 4.

Now we come to the construction of a family of elliptic curves, where each isomorphism class of curves is represented exactly once in the family. It begins with a pair of coefficients. Although the concept of the moduli space of elliptic curves is more involved in that automorphisms enter the picture, these next examples are crude moduli spaces which are useful for explicit computations.

(8.5) Notation. In terms of equivalence classes of coefficients we introduce the following sets:

(1) isomorphism classes of elliptic curves $E\langle\alpha, \beta\rangle$:

$$\mathcal{E}\ell\ell(K) = (K^2 - \{\alpha^3 = \beta^2\})/K^* \quad \text{and}$$

(2) isomorphism classes of possible singular elliptic curves with at most a double point $E\langle\alpha, \beta\rangle : \mathcal{E}\ell\ell'(K) = (K^2 - \{(0, 0)\})/K^*$.

In both cases the action of K^* on K^2 is $\lambda \cdot (\alpha, \beta) = (\lambda^4\alpha, \lambda^6\beta)$. The J-function is a bijection $J : \mathcal{E}\ell\ell(K) \to K$ which extends to $J : \mathcal{E}\ell\ell'(K) \to \mathbb{P}_1(K) = K \cup \{\infty\}$. The value $J = \infty$ corresponds to the curve with a double point $E\langle\lambda^2, \lambda^3\rangle$ at $(\lambda, 0)$ and third root giving the point $(-2\lambda, 0)$. Hence sets of isomorphisms classes of elliptic curves are parametrized by quotients of subsets of K^2.

(8.6) A Family on the Diagonal Subset of K^2. On the diagonal of all $(\alpha, \alpha) \in (K^2 - \{(\alpha, \beta) : \alpha^3 = \beta^2\})$ the J-function has the form

$$J = J(\alpha, \alpha) = \frac{\alpha}{\alpha - 1} \quad \text{and solving for } \alpha \text{ it is } \alpha = \frac{J}{J - 1}.$$

This means that the equation of the curve has the following form, but with $J = 1$ excluded, that is, $J \in K - \{0, 1\}$

$$E\langle\alpha, \alpha\rangle : y^2 = x^3 - \frac{3J}{J - 1}x + \frac{2J}{J - 1} \quad \text{where } \alpha = \beta = \frac{J}{J - 1}.$$

The family of curves $\mathcal{E}$ is a subset of $(K - \{0, 1\}) \times \mathbb{P}_2(K)$ where $(J : w : x : y) \in \mathcal{E}$ if and only if it satisfies the equation

$$E\left\langle\frac{J}{J - 1}, \frac{J}{J - 1}\right\rangle$$

in homogeneous form

$$wy^2 = x^3 - \frac{3J}{J - 1}w^2x + \frac{2J}{J - 1}w^3.$$

The family has a projection $\pi : \mathcal{E} \to (K - \{0, 1\})$ defined by the restriction of the product projection on the first factor, that is, $\pi(J; w : x : y) = J$. The fiber $\pi^{-1}(J)$ is the elliptic curve

$$E\left\langle\frac{J}{J - 1}, \frac{J}{J - 1}\right\rangle,$$

and it is the unique curve in the family with this J-value for $J \in (K - \{0, 1\})$.

4

Families of Elliptic Curves and Geometric Properties of Torsion Points

In this chapter we consider families of elliptic curves by studying cubics in normal form with coefficients depending on a parameter. The most important example is the Legendre family $E : y^2 = x(x-1)(x-\lambda)$ over $k(\lambda)$, where k is a field of characteristic unequal to 2. The points $\{(0,0), (1,0), (\lambda, 0)\}$ are the three 2-torsion points on E_λ for each value of $\lambda \in k - \{0, 1\}$, and they are specified with a given ordering.

There are families for points of order 3 and points of order 4. Then various families for higher-order points are considered. For example, there is the Tate normal form of the cubic

$$E(b, c) : y^2 + (1-c)xy - by = x^3 - bx^2,$$

where with polynomial conditions on b and c the point $(0, 0)$ has a given order n.

We close the chapter with an explicit isogeny, that is, a homomorphism of elliptic curves given by algebraic change of coordinates. The curve in question is $E[a, b] : y^2 = x^3 + ax^2 + bx$ which in characteristic unequal to 2 is nonsingular for b and $a^2 - 4b$ different from 0. The isogeny is a morphism $E[a, b] \to E[-2a, a^2 - 4b]$ and has kernel of order 2 containing $(0, 0)$. This isogeny when composed with its dual is multiplication by 2, hence is called a 2-isogeny.

§1. The Legendre Family

Consider a normal cubic equation with $a_i(t) \in k[t]$

$$y^2 + a_1(t)xy + a_3(t)y = x^3 + a_2(t)x^2 + a_4(t)x + a_6(t)$$

giving an elliptic curve E over $k(t)$, then we can substitute in any value for $t \in T$, the parameter space, and obtain a normal cubic equation, and, hence, an elliptic curve E_t over k at all points T where $\Delta(E_t) \neq 0$. Each point $P(t) = (x(t), y(t)) \in E(k(t))$ can be viewed as a mapping $t \to P(t) \in E(k(t)) = E_t(k)$ by substitution of specific values of t, defining a map $T \to E$. Such a map is called a cross-section.

(1.1) Remark. The group $E(k(t))$ of points of E over $k(t)$ is the group of rational cross-sections of the algebraic family $E \to T$ of elliptic curves E_t over k. One such cross-section of the family is always the zero cross-section.

(1.2) Definition. For a field k of characteristic $\neq 2$, the Legendre family of elliptic curves is $E_\lambda : y^2 = x(x-1)(x-\lambda)$.

The curve E_λ is nonsingular for $\lambda \neq 0, 1$, so that over $k - \{0, 1\}$, it is a family of nonsingular elliptic curves. There are four basic cross-sections:

$$0(\lambda) = 0, \qquad e_1(\lambda) = (0, 0), \qquad e_2(\lambda) = (1, 0), \qquad e_3(\lambda) = (\lambda, 0).$$

The values in $E_\lambda(k)$ of these four cross-sections give the group ${}_2E_\lambda(k)$ of 2-division points in E_λ, and the four cross-sections as points in $E(k(\lambda))$ give the group ${}_2E(k(\lambda))$ of 2-division in E. The Legendre family E_λ of cubics over the λ-line has two singular fibres which are nodel cubics $E_0 : y^2 = x^2(x-1)$ and $E_1 : y^2 = x(x-1)^2$. At $\lambda = 0$ the cross-sections e_1 and e_3 take the same value equal to the double point $(0, 0)$ and at $\lambda = 1$ the cross-sections e_2 and e_3 take the same value equal to the double point $(1, 0)$.

There are six possible orderings for the 2-division points on an elliptic curve E, or equivalently, six possible bases (e_1, e_2) for the subgroup ${}_2E$. Now we consider the effect of some of these permutations on the Legendre form where the basis is implicitly given from the form of the cubic.

EXAMPLE 1. The effect of the transposition of $(0, 0)$ and $(1, 0)$ in this three-element set in ${}_2E - \{0\}$ can be viewed as a mapping $E_\lambda \to E_{\lambda'}$, such that $(0, 0)$ corresponds to $(1, 0)$, $(1, 0)$ to $(0, 0)$, and $(\lambda, 0)$ to $(\lambda', 0)$ on $E_{\lambda'}$. In this case $\lambda' = 1 - \lambda$, and the mapping is given by $(x, y) \mapsto (1 - x, iy)$ in terms of a change of variable, since $(iy)^2 = (1-x)(1-x-1)(1-x-\lambda)$, becomes

$$y^2 = x(x-1)\,(x - (1-\lambda))\,.$$

EXAMPLE 2. The effect of the transposition of $(1, 0)$ and $(\lambda, 0)$ can be viewed as a mapping $E_\lambda \to E_{\lambda'}$ such that $(0, 0)$ corresponds to $(0, 0)$, $(\lambda, 0)$ to $(1, 0)$, and $(1, 0)$ to $(\lambda', 0)$ on $E_{\lambda'}$. In this case $\lambda' = 1/\lambda$, and the mapping is given by $(x, y) \mapsto (\lambda x, \lambda^{3/2} y)$ in terms of a change of variable, since $(\lambda^{3/2} y)^2 = (\lambda x)(\lambda x - 1)(\lambda x - \lambda)$, becomes

$$y^2 = x(x-1)\left(x - \frac{1}{\lambda}\right).$$

These changes of variable generate the group G of order 6 acting on $\mathbb{P}_1$, which is nonabelian and hence is isomorphic to the symmetric group on three letters. The changes of variable are defined over the field $k(\lambda^{1/2}, i)$, and this field is an extension field of $k(\lambda)$, the field of definition of the Legendre curve.

(1.3) Proposition. *The orbit of λ under G acting on $\mathbb{P}_1 - \{0, 1, \infty\}$ is*

$$\lambda, 1-\lambda, \frac{1}{\lambda}, \frac{1}{1-\lambda}, \frac{\lambda-1}{\lambda}, \frac{\lambda}{\lambda-1}.$$

If s is an element in G, then $s(\lambda)$ equals one of these expressions. The curves E_λ and $E_{\lambda'}$ are isomorphic, in the sense that their equations differ by a linear change of variable which preserves the group structure, if and only if there exists $s \in G$ with $s(\lambda) = \lambda'$.

Proof. When E_λ and $E_{\lambda'}$ are isomorphic in this sense, the three points of order 2 are preserved, and, hence, the change of variable must be a composite of the two given explicitly above. Remaining computations are left to the reader.

(1.4) Remark. The j-invariant $j(\lambda)$ or $j(E_\lambda)$ of $E_\lambda : y^2 = x(x-1)(x-\lambda)$ is the value.

$$j(\lambda) = 2^8 \frac{(\lambda^2 - \lambda + 1)^3}{\lambda^2(\lambda-1)^2}.$$

This j-invariant is a special case of the j-invariant of any cubic in normal form, which was considered in detail in 3, §3, and the normalization factor 2^8 arises naturally in the general context.

(1.5) Proposition. *The j-invariant has the property that $j(\lambda) = j(\lambda')$ if and only if E_λ and $E_{\lambda'}$ are isomorphic under changes of variable preserving the group structure.*

Proof. This follows from (1.3) and the observation that $j(\lambda) = j(\lambda')$ if and only if $\lambda' = s(\lambda)$ for some $s \in G$. This is a direct calculation coming from $j(1-\lambda) = j(\lambda)$ and $j(1/\lambda) = j(\lambda)$.

(1.6) Remark. The orbit of $\lambda \in \mathbb{P}_1$ under G has six distinct elements, $\lambda, 1/\lambda, 1-\lambda, 1/(1-\lambda), \lambda/(\lambda-1)$, and $(\lambda-1)/\lambda$, except in these cases:

(a) $j(\lambda) = \infty$ where the orbit is $\{0, 1, \infty\}$.
(b) $j(\lambda) = 0$ where the orbit is $\{-\rho, -\rho^2\}$ for $\rho^2 + \rho + 1 = 0$, i.e., ρ is a primitive third root of unity.
(c) $j(\lambda) = 12^3 = 1728$ where the orbit is $\{1/2, -1, 2\}$.

The mapping from λ to $j(\lambda)$ is shown in the following diagram with the ramification behavior.

Now we point out a few things about the two exceptional values $j = 12^3$ and $j = 0$.

1. For $j = 12^3$ take $\lambda = -1$. Then the curve is the familiar

$$y^2 = x(x-1)(x+1) = x^3 - x$$

which is one of the family $y^2 = x^3 + ax$ studied in (3.2). It is closely related to the Fermat curve $u^4 + v^4 = w^4$.

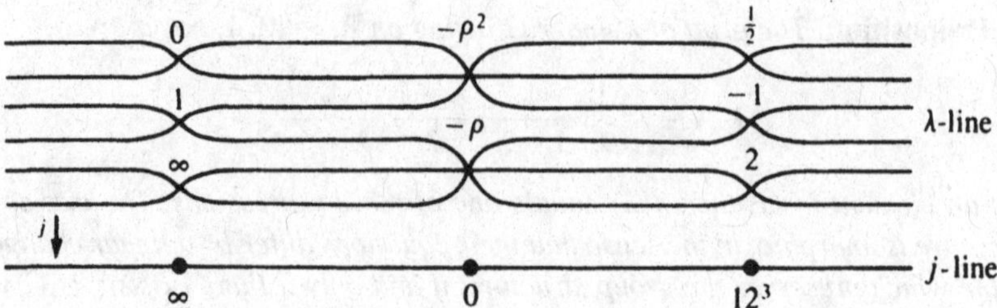

2. For $j = 0$ take $\lambda = -\rho$. Then the curve has equation

$$y^2 = x(x-1)(x+\rho),$$

and make a change of variable $x + (1-\rho)/3$ for x. This gives the equation

$$y^2 = \left(x + \frac{1-\rho}{3}\right)\left(x + \frac{-2-\rho}{3}\right)\left(x + \frac{1+2\rho}{3}\right).$$

Observe that this is just

$$y^2 = x^3 - \frac{i}{3\sqrt{3}}$$

which is one of the family $y^2 = x^3 + a$ studied in (3.3). It is closely related to the Fermat curve $u^3 + v^3 = w^3$.

The discriminant for the family E_λ is given by $\Delta_\lambda = 2^4\lambda^2(\lambda-1)^2$. This is the special case of the discriminant of an elliptic curve considered in (3.1).

Exercises

1. Verify the expression $2^4\lambda^2(\lambda-1)^2$ for the discriminant of E_λ.
2. Verify that $j(\lambda)$ is the j-invariant of E_λ, see (1.4).

§2. Families of Curves with Points of Order 3: The Hessian Family

Returning to the general normal form of the cubic defining an elliptic curve, we observe that $(0,0)$ is a point on the curve if and only if $a_6 = 0$. The family of these cubics reduces to

$$E_0 : y^2 + a_1xy + a_3y = x^3 + a_2x^2 + a_4x.$$

From the calculation of the derivative y' in the relation

$$(2y + a_1x + a_3)\,y' = 3x^2a_2x + a_4 - a_1y$$

we see that the slope of the tangent line at $(0,0)$ is a_4/a_3 on E_0.

(2.1) Remark. For the cubic E_0 in normal form with $(0, 0)$ on the curve, the point $(0, 0)$ is singular if and only if $a_3 = a_4 = 0$. The point $(0, 0)$ is nonsingular and of order 2 in the group E if and only if $a_3 = 0$ and $a_4 \neq 0$, the case of a vertical tangent at $(0, 0)$. The family of these cubics reduces to

$$E_{00} : y^2 + a_1xy = x^3 + a_2x^2 + a_4x.$$

Now we assume that $(0, 0)$ is a nonsingular point which is not of order 2. By a change of variable of the form

$$y = y' + \left(\frac{a_2}{a_3}\right)x' \quad \text{and} \quad x = x'.$$

the equation for E_0 takes the form

$$E' : y^2 + a_1xy + a_3y = x^3 + a_2x^2,$$

where the tangent line to the curve at $(0, 0)$ has slope equal to 0.

(2.2) Remark. The point $(0, 0)$ on E' has order 3 if and only if $a_2 = 0$ and $a_3 \neq 0$. This is the condition for the curve E' to have a third-order intersection with the tangent line $y = 0$ at $(0, 0)$. The family reduces to

$$E\,(a_1, a_3) : y^2 + a_1xy + a_3y = x^3.$$

For these curves some of the basic invariants defined in 3(3.1) and 3(3.3) are the following: $b_2 = a_1^2$, $b_4 = a_1a_3$, $b_6 = a_3^2$, and $b_8 = 0$. Further we have $\Delta = a_1^3a_3^3 - 27a_3^4$ and $c_4 = a_1(a_1^3 - 24a_3)$ with $j = c_4^3/\Delta$.

Since $a_3 \neq 0$ in the curve $E(a_1, a_3)$, we can normalize $a_3 = 1$, and we obtain an important family of elliptic curves with given point of order 3.

(2.3) Definition. The Hessian family of elliptic curves $E_\alpha : y^2 + \alpha xy + y = x^3$ is defined for any field of characteristic different from 3 with j-invariant $j(\alpha) = j(E_\alpha)$ of E_α given by

$$j(\alpha) = \frac{\alpha^3\,(\alpha^3 - 24)^3}{\alpha^3 - 27}.$$

The curve E_α is nonsingular for $\alpha^3 \neq 27$, that is, if α is not in $3\mu_3$, where μ_3 is the group of third roots of unity. Over the line k minus 3 points, $k - 3\mu_3$, the family E_α consists of elliptic curves with a constant section $(0, 0)$ of order 3 where $2(0, 0) = (0, -1)$. The Hessian family E_α has three singular fibres which are nodal cubics at the points of $3\mu_3 = \{3, 3\rho, 3\rho^2\}$, where $\rho^2 + \rho + 1 = 0$.

Now we consider conditions on a_1 and a_3 in the cubic equation $y^2 + a_1xy + a_3y = x^3$ such that both $y = 0$ and $y = x + u$ intersect the cubic with points generating distinct subgroups of order 3. The line $y = x + u$ has a triple intersection point $(v, v + u)$ with the cubic if and only if

$$x^3 - (x+u)^2 - (a_1x + a_3)(x+u) = (x-v)^3.$$

Comparing coefficients of x^2, x^1, and x^0 yields the relations

$$3v = 1 + a_1,$$
$$-3v^2 = 2u + a_1u + a_3,$$
$$v^3 = u^2 + a_3u.$$

Multiply the second relation by u, we obtain

$$-3uv^2 = 2u^2 + a_1u^2 + a_3u,$$

and subtracting it from the third relation yields

$$v^3 + 3v^2u = -(1+a_1)u^2 = -3vu^2 \quad \text{or just} \quad (v+u)^3 = u^3.$$

This means that the second point of order 3 has the form $(v, v+u)$ where $(v+u)^3 = u^3$. Since $v \neq 0$, we must have $v+u = \rho u$ where $\rho^2+\rho+1 = 0$. Thus $v = (\rho = 1)u$, and $u = (\rho-1)^{-1}v$. Since $T^2+T+1 = (T-\rho)(T-\rho^2)$, so that $3 = (1-\rho)(1-\rho^2)$, we have

$$u = (\rho-1)^{-1}v = \frac{1}{3}\left(\rho^2 - 1\right)v = -\frac{1}{3}(\rho+2)v$$

and

$$u + v = \frac{1}{3}(1-\rho)v.$$

In terms of the parameter v with $v \neq 0$ we have a basis

$$(0,0), \qquad \left(v, \frac{1}{3}(1-\rho)v\right)$$

of the group of points of order 3 on the curve with equation $y^2 + a_1xy + a_3y = x^3$. Now solving for a_1 and a_3 in terms of the parameter which is the x-coordinate of the second point of order 3, we obtain a one-parameter family of curves with a basis for the 3-division points, i.e., the points of order 3 on the curve.

(2.4) Assertion. The family of cubic curves

$$E_\gamma : y^2 + a_1(\gamma)xy + a_3(\gamma)y = x^3,$$

where $a_1(\gamma) = 3\gamma - 1$ and $a_3(\gamma) = \gamma(\rho-1)(\gamma - 1/3(\rho+1))$ defines for $\Delta(\gamma) = (a_1(\gamma)^3 - 27a_3(\gamma))a_3(\gamma) \neq 0$ a family of elliptic curves with a basis $(0,0)$, $(\gamma, 1/3(1-\rho)\gamma)$ for the subgroup of points of order 3 on E_γ.

This assertion follows from the above calculations. There are other versions of the Hessian family which in homogeneous coordinates take the form

$$H_\mu : u^3 + v^3 + w^3 = 3\mu uvw,$$

or in affine coordinates with $w = -1$, it has the form

$$u^3 + v^3 = 1 - 3\mu uv.$$

If we set $y = -v^3$ and $x = -uv$, we obtain $x^3/y - y = 1 + 3\mu x$, or

$$E_{3\mu} : y^2 + 3\mu xy + y = x^3.$$

This change of variable defines what is called a 3-isogeny of H_μ onto $E_{3\mu}$. There are nine cross-sections of the family H_μ given by

$$\begin{array}{lll} (0,-1,1), & (0,-\rho,1), & (0,-\rho^2,1), \\ (1,0,-1), & (\rho,0,-\rho^2), & (\rho^2,0,-\rho), \\ (-1,1,0), & (-1,\rho^2,0), & (1-,\rho,0). \end{array}$$

Again ρ is a primitive third root of unity satisfying $\rho^2+\rho+1 = 0$. The family H_μ is nonsingular over the line minus μ_3, and choosing, for example, $0 = (-1, 1, 0)$, one can show that these nine points form the subgroup of 3-division points of the family H_μ.

Exercise

1. Show that $E_t : y^2 - 2txy + ty = x^3 + (1 - 2t)x^2 + tx$ is a family of cubics with $s_1(t) = (0, 0)$ and $s_2(t) = (0, -t)$ two cross-sections of order 3 inverse to each other. Show that E_0 is the nodal cubic $y^2 = x^2(x + 1)$.

§3. The Jacobi Family

Finally we consider the Jacobi family, which along with the Legendre family and the Hessian family, give the three basic classical families of elliptic curves. The Jacobi family is given by a quartic equation and we begin by explaining how to transform a quartic equation to a cubic equation.

(3.1) Remark. Let $v^2 = f_4(u) = a_0u^4 + a_1u^3 + a_2u^2 + a_3u + a_4$ be a quartic equation. Let

$$u = \frac{ax+b}{cx+d} \quad \text{and} \quad v = \frac{ad-bc}{(cx+d)^2}y.$$

The idea is that v behaves like the derivative of u, and y like the derivative of x. Substituting into the quartic equation we obtain

$$v^2 = \frac{(ad-bc)^2}{(cx+d)^4} y^2 = f_4\left(\frac{ax+b}{cx+d}\right)$$

or

$$(ad-bc)^2 y^2 = f^4\left(\frac{ax+b}{cx+d}\right)(cx+d)^4 = \sum_{i=0}^{4} a_i (ax+b)^{4-i}(cx+d)^i$$
$$= c^4 f^4\left(\frac{a}{c}\right)x^4 + f_3(x),$$

where $f_3(x)$ is a cubic polynomial whose coefficient of x^3 is $c^3 f_4'(a/c)$. For a/c a simple root of f_4 and $ad - bc = 1$, we reduce to the equation $y^2 = f_3(x)$ a cubic in x.

(3.2) Definition. The Jacobi family of quartic curves is given by

$$J_\sigma : v^2 = \left(1 - \sigma^2 u^2\right)\left(1 - \frac{u^2}{\sigma^2}\right) = 1 - 2\rho u^2 + u^4$$

over any field of characteristic different from 2. Here $\rho = (1/2)(\sigma^2 + 1/\sigma^2)$ so that $\rho + 1 = (1/2)(\sigma + 1/\sigma)^2$.

We map $J_\sigma \to E_\lambda$ where $E_\lambda : y^2 = x(x-1)(x-\lambda)$ is a Legendre family with $\lambda = (1/4)(\sigma + 1/\sigma)^2$ by the following change of variable

$$x = \frac{\sigma^2+1}{2\sigma^2}\left(\frac{u-\sigma}{u-1/\sigma}\right) \quad \text{and} \quad y = \frac{\sigma^4-1}{4\sigma^3}\frac{v}{(u-1/\sigma)^2}.$$

The points on J_σ with u-coordinates

$$0, \infty, \pm\sigma, \pm\frac{1}{\sigma}, \pm 1, \pm i$$

map to the $16 = 4^2$ points of order 4 on the elliptic curve E_λ.

The detailed calculations relating the Jacobi family to the Legendre family are left as an exercise to the reader.

§4. Tate's Normal Form for a Cubic with a Torsion Point

Now we return to the normal form of the cubic E'. Assume that $(0,0)$ is on the curve with tangent line of slope 0.

$$E' : y^2 + a_1 xy + a_3 y = x^3 + a_2 x^2.$$

In particular, $(0, 0)$ is not a point of order 2. By (2.2) the point $(0, 0)$ has order 3 if and only if $a_2 = 0$ and $a_3 \neq 0$.

Assume that $(0, 0)$ is not of order 2 or 3, that is , both a_2 and a_3 are each unequal to 0. By changing x to u^2x and y to u^3y, we can make $a_3 = a_2 = -b$, and we obtain the following form of the equation of the curve which now depends on two parameters. Observe that all of the above discussion is carried out in a given field k, and the changes of variable did not require field extensions.

(4.1) Definition. The Tate normal form of an elliptic curve E with point $P = (0, 0)$ is

$$E = E(b, c) : y^2 + (1 - c)xy - by = x^3 - bx^2.$$

Further a calculation, using the relations in 3(3.3), yields the following formula for the discriminant $\Delta = \Delta(b, c)$ of $E(b, c)$:

$$\Delta(b, c) = (1 - c)^4b^3 - (1 - c)^3b^3 - 8(1 - c)^2b^4 + 36(1 - c)b^4 - 27b^4 + 16b^5.$$

(4.2) Remark. The Tate normal form gives a description in terms of equations for the set of pairs (E, P) consisting of an elliptic curve E together with a point P on E where $P, 2P$, and $3P$ are all unequal to zero. These pairs correspond to pairs (b, c) with both $b \neq 0$ and $\Delta(b, c) \neq 0$. The curve in the family corresponding to (b, c) is $E(b, c)$ and the point is $P = (0, 0)$. In the two-parameter Tate family $E(b, c)$ there will be many cases where curves in different fibres $E(b, c)$ are isomorphic, for example, $E(b, 1)$ and $E(b, -1)$ are isomorphic curves.

If we require, further, that $nP = 0$ for some integer $n > 3$, then there is a polynomial equation $f_n(b, c) = 0$ over $\mathbb{Z}$ which b and c must satisfy. The result is that the relations

$$T_n : f_n(b, c) = 0, \qquad b \neq 0, \qquad \Delta(b, c) \neq 0$$

define an open algebraic curve with a family $E(b, c)$ of elliptic curves over it together with a given n division point P, i.e., a cross-section P in the family of order n. We will determine f_n explicitly in several cases, see (4.6).

(4.3) Remarks. Since this family contains, up to isomorphism, all elliptic curves E with torsion point P of order n, the curve T_n maps onto the open curve $Y_1(n)$ which is the parameter space for isomorphism classes of pairs (E, P) of elliptic curves E together with a point P of order n. The curve $Y_1(n)$ has a completion $X_1(n)$ which is nonsingular where the completing points, called cusps, correspond to degenerate elliptic curves. We return to this later.

(4.4) Remark. There is an elliptic curve E over a field k with torsion point P of order n over the field k if and only if the open algebraic curve T_n has k rational points, or in other words, the set $T_n(k)$ is nonempty. This is equivalent to the statement that $Y_1(n)(k)$ is nonempty. This is also equivalent to $X_1(n)(k)$ having noncuspidal k rational points. For further discussion of $Y_1(n)$ and $X_1(n)$ see Chapter 11, §2 and §3.

In order to derive the equations $f_n(b, c) = 0$ in special cases, we make use of the following formulas.

(4.5) Calculations. On the curve $E(b, c)$ we have the following:

$$P = (0, 0),\quad 2P = (b, bc),\quad 3P = (c, b - c),\quad 4P = \big(d(-1), d^2(c - d + 1)\big),$$
$$-P = (0, b),\ -2P = (b, 0),\quad -3P = (c, c^2),\quad -4P = \big(d(d - 1), d(d - 1)^2\big),$$

where $d = b/c$ in the formulas for $4P$ and $-4P$. Finally, introducing $e = c/(d - 1)$, we have

$$-5P = \Big(de(e - 1), d^2e(e - 1)^2\Big) \quad \text{and} \quad 5P = \Big(de(e - 1), de^2(d - e)\Big).$$

(4.6) Examples. We have the following formulas for $f_n(b, c)$ arising from the condition that $nP = 0$.

(a) $4P = 0$ is equivalent to $2P = -2P$ which by (4.5) reduces to the relation $c = 0$. Thus $f_4(b, c) = c$ is the equation of a projective line. Moreover, the equation for the family becomes

$$E(b, 0) : y^2 + xy - by = x^3 - bx^2$$

with discriminant $\Delta = b^4(1 + 16b) \neq 0$. For a given x, the y-coordinate of a point (x, y) on $E(b, 0)$ satisfies the quadratic equation $y^2 + (x - b)y + (bx^2 - x^3) = 0$. The point (x, y) has order 2 if and only if this equation in y has a double root, or, in other words, the discriminant of the quadratic equation in zero.

$$(x - b)^2 - 4x^2(b - x) = 0.$$

One solution is $2P = (b, bc) = (b, 0) = -2P$ for $c = 0$. The other solutions are points whose x-coordinates satisfy

$$0 = 4x^2 + x - b = 4x^2 + x + \frac{1}{16} - \left(b + \frac{1}{16}\right).$$

There are two other 2-division points on $E(b, 0)$ over a field other than coming from $2P = -2P$ if and only if $b + 1/16$ is a square v^2. Moreover, v can take on any value except 0, 1/4, and $-1/4$ since b must be unequal to 0 and $-1/16$. The x values for these two points are

$$x = -\frac{1}{8} \pm \frac{v}{2}.$$

(b) $5P = 0$ if equivalent to $3P = -2P$ which by (4.5) reduces to the relation $b = c$. Thus $f_5(b, c) = b - c$ is also the equation of a projective line. Moreover, the equation for the family becomes

$$E(b, b) : y^2 + (1 - b)xy - by = x^3 - bx^2$$

with discriminant $\Delta = b^5(b^2 - 11b - 1) \neq 0$.

(c) $6P = 0$ is equivalent to $3P = -3P$ which by (4.5) reduces to the relation $c^2 + c = b$. Hence $f_6(b, c) = 0$ is a conic parametrized rationally by c. Also the discriminant is

$$\Delta(b, c) = c^6(c + 1)^3(9c + 1) \neq 0.$$

(d) $7P = 0$ is equivalent to $4P = -3P$ which by (4.5) reduces to the relations $c = d(d-1) = d^2 - d$ and $c^2 = d^2(c - d + 1)$. The second relation is a consequence of the first since $c - d + 1 = d^2 - 2d + 1$. Then $b = d^3 - d^2, c = d^2 - d$ is the rational parametrization of the cubic curve

$$c^3 = b(b - c)$$

with a double point at the origin. In this case the discriminant is

$$\Delta(b, c) = d^7(d - 1)^7(d^3 - 8d^2 + 5d + 1) \neq 0.$$

(e) $8P = 0$ is equivalent to $4P = -4P$ which by (4.5) reduces to the relations $d^2(c - d + 1) = d(d - 1)^2$. Since $b \neq 0$ implies $d \neq 0$, we can divide by d to obtain $d(c - d + 1) = (d - 1)^2$. Solving for c, we obtain

$$c = \frac{(d - 1)(2d - 1)}{d} \quad \text{or} \quad b = cd = (d - 1)(2d - 1).$$

(f) $9P = 0$ is equivalent to $5P = -4P$ which by (4.5) reduces to the relations $de(e - 1) = d(d - 1)$ or $e(e - 1) = d - 1$ since $d = b/c \neq 0$. Thus $d = e^2 - e + 1$ is the equation which for (b, c) becomes

$$c = de - e = e^3 - e^2,$$
$$b = cd = \left(e^3 - e^2\right)\left(e^2 - e + 1\right) = e^5 - 2e^{4'} + 2e^3 - e^2.$$

Then $b = e^5 - 2e^4 + 2e^3 - e^2, c = e^3 - e^2$ is the rational parametrization of the fifth-order curve with singularity at the origin.

We return to this subject when we study elliptic curves over the complex numbers and construct the curves $Y_1(n)$ and $X_1(n)$ along with related families.

§5. An Explicit 2-Isogeny

We consider an example of a 2-isogeny φ which is explicitly given by equations and which we will use later to illustrate other ideas in the global theory. The dual isogeny $\hat{\varphi}$ is also given explicitly, and it is shown that $\hat{\varphi}\varphi$ is multiplication by 2. These formulas work over any field k of characteristic $\neq 2$. The curve for this isogeny φ is

$$E = E[a, b] : y^2 = x^3 + ax^2 + bx,$$

and the kernel of φ is normalized to be $\{0, (0, 0)\}$.

Using the formulas of 6(2.1) for $a_2 = a, a_4 = b$, and $a_1 = a_3 = a_6 = 0$, we have $b_2 = 4a, b_4 = 2b, b_6 = 0$, and $b_8 = -b^2$. This leads to the following relations.

(5.1) Invariants of the Curve $E[a, b]$. For the curve $E[a, b]$ defined by

$$y^2 = x^3 + ax^2 + bx$$

the following hold:

$$c_4 = 16\left(a^2 - 3b\right), \qquad c_6 = 2^5\left(9ab - 2a^3\right),$$

$$\Delta = 2^4 b^2\left(a^2 - 4b\right), \qquad j = \frac{c_4^3}{\Delta} = 2^8 \frac{\left(a^2 - 3b\right)^3}{b^2\left(a^2 - 4b\right)}.$$

The two basic special cases are:

(1) $j = 12^3$ if and only if $a = 0$ where we abbreviate the notation $E[0, b] = E[b]$: $y^2 = x^3 + bx$.

(2) $j = 0$ if and only if $3b = a^2 = (3c)^2$ for characteristic unequal to 3. This is the curve $E[3c, 3c^2]$ where $y^2 = x^3 + 3cx^2 + 3cx^2x$, or $E[3c, 3c^2]$: $y^2 = (x + c)^3 - c^3$. It has the form $y^2 = x^3 - c^3$ after translation of x by c.

Before giving the formulas for the isogeny, we observe that the function $h(a, b) = (-2a, a^2 - 4b)$ when iterated twice is $h^2(a, b) = (4a, 16b)$ since $(-2a)^2 - 4(a^2 - 4b) = 16b$. This means $h : k^2 \to k^2$ is a bijection for $\operatorname{char}(k) \neq 2$, and it is its own inverse up to the division of the coordinates by a power of 2, namely

$$h^{-1}(a, b) = (-a/2, (a^2 + 4b)/16).$$

Finally, note that $a^2 - 4b$ is the discriminant of the quadratic $x^2 + ax + b$.

(5.2) Formulas for the 2-Isogeny. The 2-isogeny with kernel $\{0, (0, 0)\}$ is given by $\varphi : E[a, b] \to E[-2a, a^2 - 4b]$ where

$$\varphi(x, y) = \left(\frac{y^2}{x^2}, \frac{y\left(x^2 - b\right)}{x^2}\right) = \left(x + a + \frac{b}{x}, y\left(1 - \frac{b}{x^2}\right)\right).$$

The formula for the dual isogeny $\hat{\varphi} : E[-2a, a^2 - 4b] \to E[a, b]$ is given by

$$\hat{\varphi}(x, y) = \left(\frac{y^2}{4x^2}, \frac{y}{8x^2}\left(x^2 - \left(a^2 - 4b\right)\right)\right)$$

$$= \left(\frac{1}{4}\left(x - 2a + \frac{a^2 - 4b}{x}\right), \frac{y}{8}\left(1 - \frac{a^2 - 4b}{x^2}\right)\right).$$

Note that $\hat{\varphi}$ is given by the same formulas as φ up to powers of 2 in the coordinates which reflects the fact that the dual $\hat{\varphi}$ maps $E[-2a, a^2 - 4b]$ to $E[a, b]$ while φ maps $E[-2a, a^2 - 4b]$ to $E[4a, 16b]$. Multiplying the equation for $E[a, b]$ by 2^6, we see that $(x, y) \to (4x, 8y)$ is an isomorphism of $E[a, b] \to E[4a, 16b]$. Hence any property that holds for φ will also hold for the dual example an easy calculation shows that $\varphi(x, y) = (x', y')$ satisfies

$$(y')^2 = y^2\left(1 - \frac{b}{x^2}\right)^2 = \frac{y^2}{x^4}\left(x^4 - 2x^2b + b^2\right),$$

$$\frac{x^6}{y^2}(y')^2 = x^6 - 2bx^4 + b^2x^2,$$

and

$$f(x') = \frac{y^6}{x^6} - 2a\frac{y^4}{x^4} + \left(a^2 - 4b\right)\frac{y^2}{x^2},$$

$$\frac{x^6}{y^2}f(x') = y^4 - 2ay^2x^2 + \left(a^2 - 4b\right)x^4,$$

which boils down to $(x^6/y^2)(y')^2$ using $y^2 = x^3 + ax^2 + bx$.

(5.3) Remark. Next we calculate $\hat{\varphi}\varphi$ as follows:

$$\begin{aligned}\hat{\varphi}(\varphi(x, y)) &= \hat{\varphi}\left(\frac{y^2}{x^2}, \frac{y(x^2 - b)}{x^2}\right)\\ &= \left(\frac{y^2(x^2 - b)^2}{4x^4} \cdot \frac{x^4}{y^4}, \frac{y(x^2 - b)}{8x^2} \cdot \frac{x^4}{y^4}\left(\frac{y^4}{x^4} - (a^2 - 4b)\right)\right)\\ &= \left(\frac{(x^2 - b)^2}{4y^2}, \frac{x^2 - b}{8x^2y^3} \cdot \left(y^4 - \left(a^2 - 4b\right)x^4\right)\right).\end{aligned}$$

In order to show that $\hat{\varphi}(\varphi(x, y)) = 2(x, y)$, we consider the tangent line to a point (x, y) on the curve. To find the slope of the tangent line, we differentiate the equation for $E[a, b]$ and we obtain

$$\begin{aligned}2yy' = 3x^2 + 2ax + b &= 2(x^2 + ax + b) + (x^2 - b)\\ &= \frac{2y^2}{x} + (x^2 - b),\end{aligned}$$

or, in other words,

$$\frac{dy}{dx} = \frac{y}{x} + \frac{x^2 - b}{2y}.$$

If $2(x_1, y_1) = (x_2, y_2)$ on $E[a, b]$, then the tangent line $y = \sigma(x - x_1) + y_1$ to $E[a, b]$ at (x_1, y_1) must intersect $E[a, b]$ at $(x_2, -y_2)$. In particular, x_2 is a root and x_1 is a double root of the cubic equation

$$x^3 + ax^2 + bx = (\sigma(x - x_1) + y_1)^2$$

or

$$0 = x^3 + \left(a - \sigma^2\right)x^2 + \cdots$$

from which we deduce that

$$2x_1 + x_2 = \sigma^2 - a.$$

We just calculated σ above so that

$$\begin{aligned} x_2 &= \sigma^2 - a - 2x_1 \\ &= \left[\frac{y_2}{x_1} + \frac{s_1^2 - b}{2y_1}\right]^2 - a - 2x_1. \end{aligned}$$

Dropping the subscript 1 on x_1 and y_1, we have

$$\begin{aligned} x_2 &= \frac{\left(x^2-b\right)^2}{4y^2} + \frac{x^3 - bx + y^2 - ax^2 - 2x^3}{x^2} \\ &= \frac{\left(x^2-b\right)^2}{4y^2} \end{aligned}$$

using $y^2 = x^3 + ax^2 + bx$. This is just the x-coordinate in the above calculation of $\hat{\varphi}(\varphi(x, y))$. For the y-coordinate we use $y = \sigma(x - x_1) + y_1$ and hence $y = (y_1/x_1 + (x_1^2 - b)/2y_1)(x_2 - x_1) + y_1$. Again dropping the subscript 1 on x_1 and y_1 and substituting the above expression for x_2, we obtain

$$\begin{aligned} y_2 &= \left(\frac{y}{x} + \frac{x^2-b}{2y}\right)\left(\frac{\left(x^2-b\right)^2}{4y^2} - x\right) + y \\ &= \frac{x\left(x^2-b\right)^3 + 2y^2\left(x^2-b\right)^2 - 4x^2y^2\left(x^2-b\right)}{8xy^3} \\ &= \left(x^2-b\right)\frac{x\left(x^2-b\right)^2 + 2x^2y^2 - 2by^2 - 4x^2y^2}{8xy^3} \\ &= \left(x^2-b\right)\frac{x^2\left(x^2-b\right)^2 - 2y^2\left(y^2-ax^2\right)}{8x^2y^3}. \end{aligned}$$

Now using $x^3 - bx = y^2 - ax^2 - 2bx$, we have

$$\begin{aligned} &\left(x^3-bx\right)^2 - 2y^2\left(y^2-ax^2\right) \\ &\quad = y^4 - 2y^2\left(ax^2+2bx\right) + \left(a^2x^4 + 4abx^3 + 4b^2x^2\right) - 2y^4 + 2ax^2y^2 \\ &\quad = -y^4 - 4bx\left(x^3 + ax^2 + bx\right) + \left(a^2x^4 + 4abx^3 + 4b^2x^2\right) \\ &\quad = -\left[y^4 - \left(a^2-4b\right)x^4\right]. \end{aligned}$$

Hence we deduce that $2(x_1, y_1) = \hat{\varphi}\varphi(x_1, y_1)$.

Although we have not checked directly that φ is a group homomorphism, the previous relation $\hat{\varphi}(\varphi(x, y)) = 2(x, y)$ and the following useful results show that the composite

$$E \xrightarrow{\varphi} E' \to \frac{E'}{\{0, (0, 0)\}}$$

is a homomorphism of groups.

(5.4) Proposition. *On the curve $E[a, b]$ we have*

$$(0, 0) + (x, y) = \left(\frac{b}{x}, -\frac{by}{x^2}\right).$$

Proof. For $(x_1, y_1) + (0, 0) = (x_2, y_2)$ consider the line $y = (x_1/x_1)x$ through $(0, 0)$ and (x_1, y_1). For the third point of intersection we compute

$$\left(\frac{y_1^2}{x_1^2}\right) x^2 = x^3 + ax^2 + bx$$

$$0 = x^3 - \left(\frac{y_1^2}{x_1^2} - a\right) x^2 + bx.$$

This means that

$$x_2 = \frac{y_1^2}{x_1^2} - a - x_1 = \frac{y_1^2 - ax_1^2 - x_1^3}{x_1^2} = \frac{bx_1}{x_1^2} = \frac{b}{x_1}.$$

Moreover, $y_2 = -(y_1/x_1)x_2 = -by_1/x_1^2$ which proves the proposition.

(5.5) Remark. We can show that $\varphi((x, y) + (0, 0)) = \varphi(x, y)$ by the direct calculation

$$\frac{(-by/x^2)^2}{(b/x)^2} = \left(\frac{b^2y^2}{x^4}\right)\left(\frac{x^2}{b^2}\right) = \frac{y^2}{x^2}$$

and

$$\frac{(-by/x^2)\,(b^2/x^2 - b)}{(b/x)^2} = \frac{x^2}{b^2} \cdot \frac{by}{x^2} \left(\frac{bx^2 - b^2}{x^2}\right) = \frac{y}{x^2}\left(x^2 - b\right).$$

In order to study the image of φ, we introduce the following homomorphism from the curve to the multiplicative group of the field modulo squares.

(5.6) Proposition. *The function $\alpha : E[a, b] \to k^*/(k^*)^2$ defined by $\alpha(0) = 1$, $\alpha((0, 0)) = b \bmod (k^*)^2$, and $\alpha((x, y)) = x \bmod (k^*)^2$ is a group homomorphism.*

Proof. For three points (x_1, y_1), (x_2, y_2), and (x_3, y_3) on a line $y = \lambda x + \nu$ and on the curve $E[a, b]$ the roots of the cubic equation

$$(\lambda x + \nu)^2 = x^3 + ax^2 + bx$$

are x_1, x_2, and x_3. The relation $x_1 x_2, x_3 = v^2$, a square, shows that $\alpha(x_1, y_1)\alpha(x_2, y_2)\cdot\alpha(x_3, y_3) = 1$. This and the calculation

$$\begin{aligned}\alpha\left((0,0)+(x,y)\right) &= \alpha\left(\frac{b}{x}, \frac{-bx}{x^2}\right) = \frac{b}{x} \mod (k^*)^2 = bx \mod (k^*)^2\\ &= \alpha(0,0)\alpha(x,y)\end{aligned}$$

shows that α is a group homomorphism.

(5.7) Proposition. *The sequence*

$$E[a,b] \xrightarrow{\varphi} E[-2a, a^2-4b] \xrightarrow{\alpha} \frac{k^*}{(k^*)^2}$$

is exact.

Proof. First $\alpha(\varphi(x, y)) = \alpha(y^2/x^2, *) = (y.x)^2 \mod (k^*)^2 = 1$ so that the composite is trivial. Next, if $\alpha(x, y) = 1$, i.e., if $\alpha^2 = t$, then we choose two points (x_+, y_+) and (x_-, y_-), where $x_\pm = 1/2(t^2 - a \pm y/t)$ and $y_\pm = x_\pm t$. We wish to show that $(x_\pm, y_\pm)$ is on $E[a, b]$ and $\varphi(x_\pm, y_\pm) = (x, y)$, where (x, y) is on $E[-2a, a^2 - 4b]$.

First we see that $x_+ x_- = b$ by the direct calculation

$$x_+x_- = \frac{1}{4}\left[(x-a)^2 - \frac{y^2}{x}\right] = \frac{x^3 - 2ax^2 + a^2x - y^2}{4x} = b$$

since $y^2 = x^3 - 2ax + (a^2 - 4b)x$.

Now a point $(x_\pm, y_\pm)$ is on $E[a, b]$ if and only if

$$\left(\frac{y_\pm}{x_\pm}\right)^2 = x_\pm + a + \frac{b}{x_\pm} = x_\pm + a + x_\mp .$$

This is just the relation $t^2 = x_+ + a + x_-$, where $x_+ + x_- = t^2 - a$ immediately from the definition of $x_\pm$.

Finally we must show that $\varphi(x_\pm, y_\pm) = (x, y)$. For this, we calculate

$$\begin{aligned}\varphi(x_\pm, y_\pm) &= \left(\left(\frac{y_\pm}{y_\pm}\right)^2, y_\pm\left(1 - \frac{b}{x_\pm^2}\right)\right) = \left(t^2, x_\pm t\left(1 - \frac{b}{x_\pm^2}\right)\right)\\ &= \left(x, t\left(x_\pm - \frac{b}{x_\pm}\right)\right) = (x, t\,(x_\pm - x_\mp))\\ &= \left(x, t\left(\pm\frac{y}{t}\right)\right) = (x, \pm y).\end{aligned}$$

This proves the proposition.

In 8, §2, see 8(2.3), we give a second interpretation of the homomorphism α in terms of Galois cohomology.

§6. Examples of Noncyclic Subgroups of Torsion Points

Under the curve $E[a, b]$ studied in the previous section, we look for examples of subgroups of $E(k)$ of the form $\mathbb{Z}/n\mathbb{Z} \oplus \mathbb{Z}/2\mathbb{Z}$. We start with the cases $n = 2, 4$, and 8 always assuming that k is of characteristic different from 2.

(6.1) Example. For $n = 2$ we have $\mathbb{Z}/2\mathbb{Z} \oplus \mathbb{Z}/2\mathbb{Z} \subset E(k)$ for $E = E[a, b]$ if and only if the quadratic polynomial $x^2 + ax + b$ factors, that is, $y^2 = x^3 + ax^2 + bx = x(x - r_1)(x - r_2)$ and $\{0, (0, 0), (r_1, 0), (r_2, 0)\}$ is the subgroup of 2-division points.

Now assuming that $x^2 + ax + b$ factors, we go on to $n = 4$ by applying 1(4.1) to see when there exists a point $P \in E(k)$ with $2P = (0, 0)$.

(6.2) Example. For $n = 4$ we have $\mathbb{Z}/4\mathbb{Z} \oplus \mathbb{Z}/2\mathbb{Z} \subset E(k)$ for $E = E[a, b] = E[-r_1 - r_2, r_1 r_2]$ with some P satisfying $2P = (0, 0)$ if and only if $y^2 = x^3 + ax^2 + bx = x(x + r^2)(x + s^2)$, that is, $-r_1$ and $-r_2$ are squares in k. The four solutions to the equation $2P = (0, 0)$ are $(rs, \pm rs(r - s))$ and $(-rs, \pm rs(r + s))$.

Assuming that $2P = (0, 0)$ has a solution as in (6.2), then $2P = (-r^2, 0)$ has a solution by (4.1) if and only if the elements $-r^2 - 0 = -r^2, -r^2 - (-r^2) = 0$, and $-r^2 - (-s^2) = s^2 - r^2$ are all squares in k. This cannot happen in $\mathbb{Q}$ or even $\mathbb{R}$, as we know, but over $\mathbb{Q}(i)$, the Gaussian numbers, we need only choose s and r such that $s^2 = r^2 + t^2$, and this takes us back to Pythagorean triples again.

Continuing with the curve given in (6.2), we ask when is it true that

$$(rs, \pm rs(r - s)) \in 2E(k) \quad \text{or} \quad (-rs, \pm rs(r + s)) \in 2E(k)?$$

Again we apply 1(4.1) to see that $2P = (rs, rs(r - s))$ has a solution if and only if $rs, rs + r^2 = r(r + s)$, and $rs + s^2 = s(r + s)$ are all three squares in k. This is the case when $r = u^2, s = v^2$, and $r + s = t^2$ for $t^2 = u^2 + v^2$.

(6.3) Example. For $n = 8$ we have $\mathbb{Z}/8\mathbb{Z} \oplus \mathbb{Z}/2\mathbb{Z} \subset E(k)$ for $E = E[r^2 + s^2, r^2 s^2]$ when $r = u^2, s = v^2$, and $u^2 + v^2 = t^2$ in k. This is equivalent to E is a curve of the form $E[u^4 + v^4, u^4 v^4]$. In particular, over the rational numbers such curves correspond to Pythagorean triples (u, v, t).

Applying 1(4.1) to the equation $2P = (-rs, rs(r + s))$, we obtain almost the same result, that is, $-rs, -rs + r^2$, and $-rs + s^2$ must be squares. This can be realized for $r = u^2, -s = v^2$, and $r - s = t^2 = u^2 + v^2$.

For the question of a point of order 3 on $E[a, b]$ we recall from (5.3) that if $(x_2, y_2) = 2(x_1, y_1)$, then $x_2 = (x_1^2 - b)^2/4y_1^2$. The case $2(x_1, y_1) = (x_1, -y_1)$ occurs for $x = x_1 = x_2$, that is, $4xy^2 = (x^2 - b)^2$. For $x = t^2$, $y = st^2$ we see that $b = t^3(t - 2s)$ and from $ax^2 = y^2 - x^3 - bx$, we obtain

$$a = (s - t)^2 - 3t^2 \quad \text{and} \quad b = t^3(t - 2s).$$

(6.4) Example. The point (t^2, st^2) is the order 3 on $E = E[(s-t)^2-3t^2, t^3(t-2s)]$ and the point

$$\left(3t^2 + 2st - 2s^2, 3t^3 + 2st^2 - 2s^2t\right)$$

is of order 6. We do the calculations $(s^2, s^2t) + (0, 0) = (x, tx)$ where

$$t^2x^2 = x^3 + \left((s-t)^2 - 3t^2\right)x^2 + t^3(t-2s)$$

and

$$0 = x^3 - \left(4t^2 - (s-t)^2\right)x^2 + \cdots .$$

The three roots which add up to $4t^2-(s-t)^2$ are 0, s^2, and x. Thus $x = 3t^2+2st-2s^2$ holds as asserted.

Some of the above examples were brought to our attention by Hans Peter Kraft and were worked out in a seminar in Basel, October 1984 by him, Friedrich Knopp, Gisela Menzel, and Erhar Senn. The reader can find further examples by combining the above results with the statements in 1(3, Ex. 2).

5

Reduction mod p and Torsion Points

The reduction modulo p morphism $\mathbb{Z} \to \mathbb{Z}/p\mathbb{Z} = \mathbb{F}_p$ is a fundamental construction for studying equations in arithmetic. A basic advantage of projective space over affine space is that the entire rational projective space can be reduced modulo p, yielding a map $\mathbb{P}_n(\mathbb{Q}) \to \mathbb{P}_n(\mathbb{F}_p)$, in such a way that rational curves (curves defined over $\mathbb{Q}$) and their intersection points also reduce modulo p. The first task is to study when the reduced curve is again smooth and when intersection multiplicities are preserved. This is an extension of the ideas in Chapter 2 to arithmetic.

Turning to the special case of an elliptic curve E, we need to choose a "minimal" cubic equation for E. These have the best possible reduction properties, and reduction is a group homomorphism when the curve E reduces to a nonsingular E_p modulo p. We say E has a good reduction at p in this case, and we look for criterions for good reduction.

Finally we turn to the global arithmetic of elliptic curves, and prove the Nagell–Lutz theorem which says that for the reduction at p homomorphism $E(\mathbb{Q}) \to E_p(\mathbb{F}_p)$ the restriction to

$$\mathrm{Tors}(E(\mathbb{Q})) \to E_p(\mathbb{F}_p)$$

is injective for p odd and has kernel at most of order 2 at 2. Combining this arithmetic study with the structure of the group of real points $E(\mathbb{R})$, we deduce that the torsion subgroup $\mathrm{Tors}(E(\mathbb{Q}))$ of an elliptic curve E over the rational numbers is either finite cyclic or finite cyclic direct sum with the group of order 2. Moreover, it is effectively computable.

§1. Reduction mod p of Projective Space and Curves

(1.1) Notations. We will use the following notation in the next three sections. Let R be a factorial ring with field of fractions k. For each irreducible p in R we form the quotient ring $R/p = R/Rp$ and denote its field of fractions by $k(p)$. Each element a in k can be decomposed as a quotient

$$a = p^n \cdot \frac{u}{v},$$

where p does not divide either u or v and n is an integer uniquely determined by a. Let $\mathrm{ord}_p(a) = n$ denote the order function associated with p. Let $r_p(a) = \bar{a}$ denote the canonical reduction mod p defined $R \to k(p)$. If $R_{(p)}$ denotes the subring of all a in k with $\mathrm{ord}_p(a) \geq 0$, then the mod p reduction function is well defined on $R_{(p)} \to k(p)$.

The order function satisfies the following properties:

(V1) $$\mathrm{ord}_p(ab) = \mathrm{ord}_p(a) + \mathrm{ord}_p(b)$$

and

(V2) $$\mathrm{ord}_p(a+b) \geq \min\{\mathrm{ord}_p(a), \mathrm{ord}_p(b)\}.$$

The second property can be refined in the case where $\mathrm{ord}_p(a) < \mathrm{ord}_p(b)$ in which case $\mathrm{ord}_p(a) = \mathrm{ord}_p(a+b)$. This refinement can be deduced directly from the second property of $\mathrm{ord}_p(a)$ using an elementary argument. The function is a special case of a discrete valuation (of rank 1) and $R_{(p)}$ is the valuation ring associated to ord_p. It is a principal ring with one irreducible p and one maximal ideal $R_{(p)}p$ where r_p induces an isomorphism $R_{(p)}/R_{(p)}p \to k(p)$.

There are some cases, especially over a number field, where it is confusing to use p for the irreducible or local uniformizing element of R, in those cases we will tend to use t or π.

The reduction mod p function $r_p : R_{(p)} \to k(p)$ can be defined on affine space by taking products

$$k^n \supset R^n_{(p)} \to k(p)^n,$$

but it is only defined on the points $x = (x_1, \ldots, x_n)$ all of whose coordinates x_j have positive ord_p. Then the formula is $r_p(x_1, \ldots, x_n) = (\bar{x}_1, \ldots, \bar{x}_n)$. On the other hand, reduction mod p is defined on the entire projective space.

(1.2) Definition. The reduction mod p function $r_p : \mathbb{P}_n(k) \to \mathbb{P}_n(k(p))$ is defined by the relation

$$r_p(y_0 : \cdots : y_n) = (\bar{y}_0 : \cdots : \bar{y}_n),$$

where $(y_0 : \cdots : y_n)$ is the homogeneous coordinates of a point in $\mathbb{P}_n(k)$ with all y_i in R and without a common irreducible factor. Such a representative of a point is called reduced.

Observe that $r_p(y_0 : \cdots : y_n) = (\bar{y}_0 : \cdots : \bar{y}_n)$ is defined when each $\mathrm{ord}_p(y_i) \geq 0$ and some $\mathrm{ord}_p(y_j) = 0$ so that $\bar{y}_j \neq 0$. Such a representative $(y_0 : \cdots : y_n)$ of a point in $\mathbb{P}_n(k)$ is called p-reduced. The reduced representatives are unique up to multiplication by a unit in R, and the p-reduced representatives are unique up to multiplication by a unit in $R_{(p)}$.

We saw in Chapter 2 that a good intersection theory had to be formulated in projective space. The above construction of mod p reduction shows another advantage of projective space over affine space which is arithmetic instead of geometric.

(1.3) Remark. Let $F(y_0, \dots, y_n) \in k[y_0, \dots, y_n]$ and multiply the polynomial by an appropriate nonzero element of k such that the coefficients of the new polynomial, also denoted f, are all in R and have no common irreducible factor. Then we denote by $\bar{f}(y_0, \dots, y_n)$ the polynomial over $k(p)$ which is f with all its coefficients reduced modulo p. Observe that $\bar{f}$ is nonzero and homogeneous of the same degree as f.

(1.4) Definition. Let C be an algebraic curve of degree d in $\mathbb{P}_2$ defined over k. Choose an equation $f = 0$ for C of degree d over R with coefficients not having a common irreducible factor. The reduction mod p of $C = C_f$ is the plane curve $C_{\bar{f}}$ of degree d in $\mathbb{P}_2$ defined over $k(p)$.

The modulo p reduction function $r_p : \mathbb{P}_2(k) \to \mathbb{P}_2(k(p))$ restricts to a function $r_p : C_f(k) \to C_{\bar{f}}(k(p))$, for if $(w, x, y) \in C_f(k)$, then $f(w, x, y) = 0$, and we can apply r_p to obtain

$$0 = r_p(f(w, x, y)) = \bar{f}(\bar{w}, \bar{x}, \bar{y}) = \bar{f}(r_p(w, x, y)).$$

Observe that the same construction and definition holds for hypersurfaces in n variables over k.

(1.5) Examples. The nonsingular conic defined by $wx + py^2 = 0$ reduces to the singular conic equal to the union of two lines defined by $wx = 0$, and the conic defined by $pwx + y^2 = 0$ reduces to $y^2 = 0$ which is a double line. This example shows that there are some subtleties related to reduction modulo p concerning whether $y^2 = 0$ defines a line or a conic. These questions are taken care of in the context of schemes, but we do not have to get into this now since a cubic in normal form reduces to another cubic in normal form.

In order to see what happens with the group law of a cubic under reduction recall from 2(4.4) that the intersection multiplicity $i(P; L, C_f)$ of P on L and C_f is defined by forming the polynomial

$$\varphi(t) = f(w + tw', x + tx', y + ty'),$$

where $P = (w, x, y)$ and $(w', x', y') \in L - C_f$. The intersection points $L \cap C_f$ are of the form $(w + tw', x + tx', y + ty')$ where $\varphi(t) = 0$, and the order of the zero is the intersection multiplicity. Further the order of P on C_f is always $\le i(P; L, C_f)$.

Now we reduce all of the above constructions including the extra polynomial $\varphi(t)$ mod p from k to $k(p)$. We obtain

$$\bar{\varphi}(t) = \bar{f}(\bar{w} + t\bar{w}', \bar{x} + t\bar{x}', \bar{y} + t\bar{y}'),$$

where now we must use further care in choosing (w', x', y') such that $(\bar{w}', \bar{x}', \bar{y}') \in \bar{L} - C_{\bar{f}}$. This is possible provided $\bar{L} \not\subset C_{\bar{f}}$.

(1.6) Remark. With the above notations we have the following inequalities:

$$i(P; L, C_f) \le i(\bar{P}; \bar{L}, C_f);$$
$$\text{order of } P \text{ on } C_f \le \text{order of } \bar{P} \text{ on } C_{\bar{f}}.$$

For $P = (1,0,0)$ and $(w',x',y') = (0,a,b)$ the polynomial $\varphi(t)$ takes the form $\varphi(t) = f_r(ta,tb) + \cdots + f_d(ta,tb)$, where r is the order of P on the curve C_f. Recall from 2(4.6) that L is part of the tangent cone to C_f at P if and only if $r < i(P; L, C_f)$. For P' on L where $P' = (at_0, bt_0)$ we have $P' \in C_f$ if and only if $\varphi(t_0) = 0$.

(1.7) Proposition. *With the above notations let $P, P' \in L \cap C_f$ where $P \neq P'$ and $\bar{P} = r_p(P) = r_p(P')$. If the order of P on C_f equals the order of $\bar{P}$ on $C_{\bar{f}}$, then the reduced line $\bar{L}$ is part of the tangent cone of $C_{\bar{f}}$. If $\bar{P}$ has order 1 on $C_{\bar{f}}$, then $\bar{L}$ is the tangent line to $C_{\bar{f}}$ at $\bar{P}$.*

Proof. Since P is on $L \cap C_f$ of order $\geq r$, the polynomial t^r divides $\varphi(t)$. Since P' is on $L \cap C_f$, the polynomial $t - t_0$ divides $\varphi(t)$. Thus the product $t^r(t - t_0)$ divides $\varphi(t)$. Since $r_p(P) = r_p(P')$, we have $\bar{t}_0$ mod p and therefore t^{r+1} divides $\bar{\varphi}(t)$. Now the proposition follows from the criterion in 2(4.6).

In Exercise 1 we give an example which shows that the hypothesis that the order of P on C_f equals the order of $\bar{R}$ on $C_{\bar{f}}$ is necessary in the proposition.

Exercise

1. Let C be the conic defined by $wx - p^3y^2 = 0$ and show that C is nonsingular over $\mathbb{Q}$. The reduction mod p is $wx = 0$, and show that the reduction $\bar{C}$ has a singular point at $(0,0,1)$. Find two distinct points P and P' on C whose reduction is in both cases $(0,0,1)$ and such that the reduction $\bar{L}$ of the line L through P and P' is different from either the line defined by $w = 0$ or by $x = 0$.

§2. Minimal Normal Forms for an Elliptic Curve

(2.1) Proposition. *Let k be the field of fractions for an integral domain R, and let E be an elliptic curve over k. Then there is a cubic equation for E in normal form with all $a_i \in R$.*

Proof. Choose any normal form for E over k with coefficients $\bar{a}_i$ in variables $\bar{x}$ and $\bar{y}$. Let u be a common denominator for all $\bar{a}_i$, in particular all $u\bar{a}_i \in R$. and make the change of variable $x = u^2\bar{x}$ and $y = u^3\bar{y}$. Then the coefficients $a_i = u^i\bar{a}_i$ is in R for all i. This proves the proposition.

In the previous proposition, observe that once all a_j are in R, all the related constants b_j, c_j, and Δ, defined in 3(3.1), are also in R. Assume now that R is a discrete valuation ring, that is, R is principal with one nonzero prime ideal Rp and the order function ord_p is a valuation denoted by v. Then R is the set of all a in K, the field of fractions of fractions of R, such that $v(a) \geq 0$.

(2.2) Definition. Let K be a field with a discrete valuation v, and let E be an elliptic curve over K. A minimal normal form for E is a normal form with all a_j in the valuation ring R of K such that $v(\Delta)$ is minimal among all such equations with coefficients a_j in R.

Since the valuation v takes positive integer values on the discriminant of a given equation in normal form over R, it is clear that every elliptic curve has a minimal model. In the next proposition we see in what sense they are unique. Following the literature, we use the terminology minimal model, minimal normal form, and minimal Weierstrass model interchangeably.

(2.3) Proposition. *Let E and E' be two elliptic curves over K with minimal models having coefficients a_j and a'_j, respectively. Let $f : E' \to E$ be an isomorphism with $xf = u^2x' + r$ and $yf = u^3y' + su^2x' + t$. Then $v(\Delta) = v(\Delta')$, u is in R^*, and r, s, and t are in R. The differential ω is unique up to a unit in R.*

Proof. The equality $v(\Delta) = v(\Delta')$ follows from the definition of minimal, and hence $v(u) = 0$. So u is a unit in R from $u^{12}\Delta' = \Delta$. The relation $u^8b'_8 = b_8 + \cdots$ in R implies that $3r$ is in R, and the relation $u^6b'_6 = b_6 + \cdots$ in R implies that $4r$ is in R. Hence the difference r is in R. The relation $u^2a'_2 = a_2 + \cdots$ in R implies that s is in R and the relation $u^6a'_6 = a_6 + \cdots$ implies that t is in R. The last assertion follows from the formula $\omega f = u^{-1}\omega'$.

Now we look for estimates on $v(\Delta) \geq 0$ which hold for minimal models and which further might characterize those equations over R corresponding to minimal models. With $f : E' \to E$ as above $xf = u^2\bar{x} + \cdots$ and $yf = u^3\bar{u} + \cdots$, we can change the valuation of the discriminant by

$$v(\Delta) = v(u^{12}\bar{\Delta}) = 12 \cdot v(u) + v(\bar{\Delta}).$$

Therefore, we see that the following assertion holds.

(2.4) Proposition. *If all a_j are in R, and if $0 \leq v(\Delta) < 12$, then the model is minimal.*

Observe that if $\pi^4 \mid A$ and $\pi^6 \mid B$ in R, then the equation $y^2 = x^3 + Ax + B$ is not minimal. Since all elements of K equal $j(E)$ for some E over K, we will not have $j(E)$ in R for all E, that is, $v(j(E))$ may be arbitrary, even with coefficients a_i in R. In terms of $v(j)$ we can obtain estimates on $v(\Delta)$ for a minimal model.

(2.5) Proposition. *Let E be an elliptic curve over K, and assume that the characteristic of K is not 2 or 3. For a minimal model the valuation of the discriminant satisfies*

$$v(\Delta) + \min\{v(j), 0\} < 12 + 12v(2) + 6v(3).$$

In addition, assuming that the residue class characteristic is different from 2 and 3, it follows that a model over R is minimal if and only if $v(\Delta) + \min\{v(j), 0\} < 12$.

Proof. Since $c_4^3 = \Delta j$ and $c_6^2 = \Delta(j - 12^3)$, we have the relations $v(\Delta) + v(j) = 3v(c_4)$ and $v(\Delta) + V(j - 12^3) = 2v(c_6)$. By 3(3.4) the equation of the cubic can be transformed into the form

$$y^2 = x^3 - \left(\frac{c_4}{48}\right) - \frac{c_6}{864}.$$

By the above remark, if $48p^4 \mid c_4$ and $864p^6 \mid c_6$, then the equation is not minimal. Since the equation is minimal and since $48 = 2^4 \cdot 3$ and $864 = 2^5 \cdot 3^3$, it follows that

$$v(\Delta) + v(j) = 3v(c_4) < 12 + 3v(48) = 12 + 12v(2) + 3v(3),$$

or

$$v(\Delta) + v(j - 12^3) = 6v(c_6) < 12 + 2v(864) = 12 + 10v(2) + 6v(3).$$

Since $v(\Delta) + \min\{v(j), 0\} \leq v(\Delta) + v(j)$ or $v(\Delta) + v(j - 12^3)$, we obtain the first inequality.

For the second statement observe that for $v(2) = v(3) = 0$, the minimal model satisfies $v(\Delta) + \min\{v(j), 0\} < 12$. The converse holds by Remark (2.4) since $0 < v(\Delta) + \min\{v(j), 0\}$ and the relation between two valuations of the discriminants given above (2.3). This proves the proposition.

Now we return to the notations of (1.1) where R is a factorial ring with field of fractions k. Two normal forms of equations for E with coefficients a_j in R are related by an admissible change of variables 3(2.3) where $x = u^2\bar{x} + r$ and $y = u^3\bar{y} + su^2\bar{x} + t$. The discriminants are related by $u^{12}\bar{\Delta} = \Delta$ by 3(3.2). For an irreducible element p in R we have

$$\mathrm{ord}_p(\Delta) = 12\,\mathrm{ord}_p(u) + \mathrm{ord}_p(\bar{\Delta}).$$

This leads to the global version of (2.2) since by a change of variable we can always choose an equation where $\mathrm{ord}_p(\Delta)$ is minimal for all irreducibles p in R.

(2.6) Definition. Let k be the field of fractions of a factorial ring R, and let E be an elliptic curve over k. A minimal normal form for E is a normal form with all a_j in R such that all $\mathrm{ord}_p(\Delta)$ is minimal among all equations in normal form with coefficients a_j in R.

By the above discussion a minimal normal form for an elliptic curve E always exists.

Unfortunately we are interested in rings R, namely the ring of integers in a number field k, which are not always factorial. In the case there is only a minimal model of an elliptic curve E locally at valuations v of k with valuation ring $R_{(v)}$ associated with prime ideals of R. For questions of reduction mod v it is good enough to work locally with coefficients in $R_{(v)}$. We return later to the more general case of a Dedekind ring R to define the conductor of an elliptic curve. In any case the present theory includes curves defined over $\mathbb{Q}$.

Exercise

1. Over the rational numbers $\mathbb{Q}$ under what conditions on the constant a are the normal forms $y^2 = x^3 + ax$ and $y^2 = x^3 + a$ minimal.

§3. Good Reduction of Elliptic Curves

Continuing with the notations (1.1), we have for an irreducible p a canonical reduction homomorphism $r_p : R_{(p)} \rightarrow k(p)$ denoted by $r_p(a) = \bar{a}$.

(3.1) Definition. Let E be an elliptic curve over k with minimal normal form $y^2 + a_1xy + a_3y = x^3 + a_2x^2 + a_4x + a_6$. The reduction $\bar{E}$ of E modulo p is given by

$$y^2 + \bar{a}_1xy + \bar{a}_3y = x^3 + \bar{a}_2x^2 + \bar{a}_4x + \bar{a}_6.$$

It is a plane cubic curve over $k(p)$. The curve $\bar{E}$ is also denoted $E_{(p)}$.

Note that the normal form of the equation for E only has to be minimal at p for this definition. Observe that an admissible change of variable between two minimal normal forms of E at p given by $x = u^2x' + r$ and $y = u^3y' + su^2x' + t$ over $R_{(p)}$ reduces to $x = \bar{u}^2x' + \bar{r}$ and $y = \bar{u}^3y' + \bar{s}\bar{u}^2x' + \bar{t}$ an admissible change of variable for $\bar{E}$ over $k(p)$. Hence the reduction $\bar{E}$ is well defined up to isomorphism.

(3.2) Remarks. With the above notations the discriminant of the reduced curve $\bar{E}$ is $\bar{\Delta}$, the reduction mod p of the discriminant Δ of E. Clearly $\bar{E}$ is nonsingular if and only if $\bar{\Delta} \neq 0$, or equivalently, if and only if $\text{ord}_p(\Delta) = 0$.

(3.3) Definition. An elliptic curve E defined over k has a good reduction at p provided $\bar{E}$, the reduced curve at p, is nonsingular. When $\bar{E}$ is singular, we say E has bad reduction at p.

In general the reduction function $r_p : \mathbb{P}_2(k) \rightarrow \mathbb{P}_2(k(p))$ restricts to $r_p : E(k) \rightarrow \bar{E}(k(p))$, and in the case of good reduction we have the following result.

(3.4) Proposition. *Let E be an elliptic curve over k with good reduction at p. Then the reduction function $r_p : E(k) \rightarrow \bar{E}(k(p))$ is a group morphism.*

Proof. Clearly $r_p(0 : 0 : 1) = 0 : 0 : 1$ so that zero is preserved. For $P, Q \in E(k)$ let L be the line through P and Q when $P \neq Q$ and the tangent line to E at P when $P = Q$. Then L reduces to $\bar{L}$, the line through $r_p(P)$ and $r_p(Q)$. Again $\bar{L}$ is tangent to $\bar{E}$ at $r_p(P)$ when $r_p(P) = r_p(Q)$ by (1.7). If PQ denotes, as usual, the third intersection point of L with $\bar{E}$, then we have $r_p(PQ) = r_p(P)r_p(Q)$. Since the third intersection point of $\bar{L}$ with $\bar{E}$ for the x-coordinate is given by reduction mod p of the equation giving the x-coordinate of PQ. Thus we calculate

$$r_p(P + Q) = r_p((PQ)0) = (r_p(P)r_p(Q))r_p(0) = r_p(P) + r_p(Q),$$

and thus r_p is a group morphism. This proves the proposition.

(3.5) Remark. Since $0 = 0 : 0 : 1$ on both E and the reduced curve over $k(p)$, we see that the p-reduced $w : x : y$ on $E(k)$ is in $\ker(r_p)$ if and only if $\text{ord}_p(y) = 0$, $\text{ord}_p(w) > 0$, and $\text{ord}_p(x) > 0$. In fact, we can divide by y and assume that the point is of the form $w : x : 1$, where w and x have strictly positive ordinal at p.

(3.6) Example. If the minimal normal form of E over k is $y^2 = f(x)$, where $f(x)$ is a cubic polynomial, then E has bad reduction at all p where $k(p)$ has characteristic 2 and at all irreducibles p which divide the discriminant $D(f)$ of the cubic $f(x)$.

(3.7) Example. If the minimal normal form of E over k is

$$y^2 = (x-\alpha)(x-\beta)(x-\gamma),$$

then no p^2 divides all roots α, β, and γ for any irreducible p. The elliptic curve E has good reduction at $p > 2$ if and only if p does not divide any of the differences $\alpha - \beta, \beta - \gamma$, and $\gamma - \alpha$.

(3.8) Example. The curve E defined by $y^2 + y = x^3 - x^2$ has good reduction at 2 since over $\mathbb{F}_2$ it is the curve E_3' of (4.2). Denoting the curve mod p by $E_{(p)}$ and the reduction modulo p by $r_p : E(\mathbb{Q}) \to E_{(p)}(\mathbb{F}_p)$, we have in this case a cyclic subgroup of order 5 mapping isomorphically onto $E_{(2)}(\mathbb{F}_2)$. In order to determine whether $E_{(p)}$ has a singularity for odd primes p, we calculate the derivative y', that is, $(2y+1)y' = 3x^2 - 2x$, and look for points where it is indeterminate, i.e., $0 \cdot y' = 0$.

Mod 3. The tangent slope $(2y+1)y' = -2x$ is indeterminate for (0, 1), but the point (0, 1) is not on the curve. Now reduction modulo 3 is defined $r_3 : E(\mathbb{Q}) \to E_{(3)}(\mathbb{F}_3)$ and is an isomorphism of the cyclic group of order 5 generated by $(0,0)$ onto the group $E_{(3)}(\mathbb{F}_3)$.

Mod 11. The point $(-3, 5)$ is a point on the curve $E_{(11)}$ which is singular so that E has bad reduction at 11. Substituting $x - 3$ for x and $y + 5$ for y, we obtain the equation $y^2 = x^3 + x^2$ with a node at $(0,0)$ over $\mathbb{F}_{11}$. The reduction r_{11} is defined on the cyclic group of order 5 generated by $(0,0)$ mapping into $E_{(11)}(\mathbb{F}_{11})_{\text{ns}}$ which is a cyclic group of order 10.

Exercises

1. For which rational odd primes do the following curves have bad reduction?
   a) $y^2 = x^3 + ax$.
   b) $y^2 = x^3 + a$.
   c) $y^2 = x^3 - 43x + 166$.
   d) $y^2 = x^3 - 16 \cdot 27x + 19 \cdot 16 \cdot 27$.
2. For which rational odd primes do the following curves have bad reduction?
   a) $y^2 = x^3 + x^2 - x$.
   b) $y^2 = x^3 - x^2 + x$.
   c) $y^2 = x^3 - 2x^2 - x$.
   d) $y^2 = x^3 - 2x^2 - 15x$.
3. For which rational primes do the following curves have bad reduction? If the curve has good reduction at 2, then identify the curve reduced at 2 in the list 3(6.4), and if one has the list of curves over $\mathbb{F}_3$ as asked for in 3(6, Ex.6), then identify the curve reduced modulo 3 in the list when it has good reduction at 3.
   a) $y^2 + y = x^3 - x$.
   b) $y^2 + y = x^3 + x^2$.
   c) $y^2 + y = x^3 + x$.

d) $y^2 - y = x^3 - 7$.
e) $y^2 + xy + y = x^3 - x^2$.
f) $y^2 + xy = x^3 - x^2 - 5$.

4. Which of the following curves have good reduction at $p = 2, p = 3, p = 5$, and $p = 7$?
   a) $y^2 + xy + y = x^3 - x$.
   b) $y^2 + xy = x^3 - x^2 - 2x - 1$.
   c) $y^2 + xy + y = x^3 - x^2 - 3x + 3$.
   d) $y^2 + xy = x^3 + x^2 - 2x - 7$.
   e) $y^2 + xy + y = x^3 + x^2 - 4x + 5$.
5. The curves $y^2 + y = x^3 - x^2$ and $y^2 + y = x^3 + x^2$ have good reduction at 2 and the same reduced curve mod 2. Determine r_2 on the subgroup generated by (0, 0) with values in the points over $\mathbb{F}_2$ of the reduced curve in each case, see 1(2.3) and 1(2.4).
6. Find the discriminant of $y^2 + y = x^3 - 7x + 6$ and show that it is a prime.

§4. The Kernel of Reduction mod p and the p-Adic Filtration

Following (1.1), we use the notations of the previous section where R is a factorial ring, p is an irreducible, and k is its field of fractions. Recall from (3.5) that the inverse image of $(0, 0, 1)$ under the reduction mapping $r_p : E(k) \to \bar{E}(k(p))$ consists of all $(w, x, 1) \in E(k)$ such that $\mathrm{ord}_p(w), \mathrm{ord}_p(x) > 0$. These conditions are analyzed further in the next two propositions using the normal form of the cubic equation.

(4.1) Proposition. *Let E be an elliptic curve over k, and $(w, x, 1) \in E(k)$. If $\mathrm{ord}_p(w) > 0$, then we have $\mathrm{ord}_p(x) > 0$, and the relation $\mathrm{ord}_p(w) = 3\,\mathrm{ord}_p(x)$ holds.*

Proof. For $y = 1$, the projective normal form of the cubic equation for E has the form

$$w + a_1wx + a_3w^2 = x^3 + a_2wx^2 + a_4w^2x + a_6w^3.$$

We assume that $\mathrm{ord}_p(w) > 0$ and $\mathrm{ord}_p(x) \le 0$, and we derive a contradiction. If R is the right-hand side of the normal cubic, then

$$\mathrm{ord}_p(R) = \mathrm{ord}_p(x^3) = 3\,\mathrm{ord}_p(x) \le 0,$$

and if L is the left-hand side of the normal cubic, then

$$\mathrm{ord}_p(L) = \min\{\mathrm{ord}_p(w), \mathrm{ord}_p(x) + \mathrm{ord}_p(w) + \mathrm{ord}_p(a_1)\}.$$

Since $\mathrm{ord}_p(w) > 0$, we would derive the relation

$$3\,\mathrm{ord}_p(x) \ge \mathrm{ord}_p(x) + \mathrm{ord}_p(w) \quad \text{or} \quad 0 \ge 2\,\mathrm{ord}_p(x) \ge \mathrm{ord}_p(w).$$

This is a contradiction.

Next observe that $\mathrm{ord}_p(w) = \mathrm{ord}_p(w + a_1wx + a_3w^2)$ since $\mathrm{ord}_p(w) < \min\{\mathrm{ord}_p(a_1wx), \mathrm{ord}_p(a_3w^2)\}$. Thus we obtain the second assertion

$$\text{ord}_p(w) = \text{ord}_p\left(x^3 + a_2wx^2 + a_4w^2x + a_6w^3\right) = 3\,\text{ord}_p(x)$$

where we have checked the four possible minima in

$$\text{ord}_p(w) \geq \min\left\{\text{ord}_p(x^3), \text{ord}_p(a_2wx^2), \text{ord}_p(a_4w^2x), \text{ord}_p(a_6w^3)\right\}.$$

This proves the proposition.

The converse to (4.1) does not hold; it is possible to have $\text{ord}_p(x) > 0$ and $\text{ord}_p(w) \leq 0$. For example, if all $a_i = 0$ except a_6 in the normal form, then the cubic reduces to

$$w = x^3 + a_6w^3,$$

and we would have only the relation $0 = \text{ord}_p(a_6) + 2\,\text{ord}_p(w)$.

The next definition leads to a sequence of subgroups of $E(k)$ which are used to analyse $\ker(r_p : E(k) \to \bar{E}(k(p)))$.

(4.2) Definition. Let E be a elliptic curve over k defined by a cubic in normal form. The p-adic filtration on E is the sequence of subgroups $E^{(n)}(k)$ defined by the condition that $w : x : 1 \in E^{(n)}(k)$ provided $\text{ord}_p(w) > 0$ and $\text{ord}_p(x) \geq n$, or equivalently provided $\text{ord}_p(w) \geq 3n$ for $n \geq 1$.

The equivalence of the two conditions for a point to be in $E^{(n)}(k)$ follows from (4.1). Observe that $E^{(1)}(k) = r_p^{-1}(0)$, where $0 = 0 : 0 : 1$ in $\bar{E}^{(n)}(k(p))$. We extend the definition to one other term $E^{(0)}(k) = r_p^{-1}(\bar{E}(k(p))_{\text{ns}})$. The curve E has good reduction at p if and only if $E(k) = E^{(0)}(k)$. The subset $E^{(0)}(k)$ is a subgroup of $E(k)$ since a line L intersecting $E(k)$ at one point of $E(k) - E^{(0)}(k)$ must intersect it twice or at another point. We leave it to the reader to check in Exercise 3 that the restriction $r_p : E^{(0)}(k) \to \bar{E}(k)_{\text{ns}}$ is a group morphism.

(4.3) Proposition. *Let $P = w : x : 1$, $P' = w' : x' : 1$, and $P'' = w'' : x'' : 1$ be three points on the intersection $E \cap L$ over k, where E is an elliptic curve defined by a cubic in normal form and L is a line. If P, $P' \in E^{(n)}(k)$ for $n \geq 1$, then we have*

$$\text{ord}_p(x + x' + x'') \geq 2n \quad \textit{and} \quad \text{ord}_p(w'') = 3\,\text{ord}_p(w'') \geq 3n.$$

Proof. Let $w = cx + b$ be the equation of the line L through the three points P, P', and P''. Now we calculate c and $\text{ord}_p(c)$ using the equation of the cubic.

Case 1. $P \neq P'$. Then $c = (w - w')/(x - x')$ is the slope of the line. Subtracting the normal form of the cubic

$$w + a_1wx + a_3w^2 = x^3 + a_2wx^2 + a_4w^2x + a_6w^3$$

for P' from the equation for P on the curve, we get

$$(w - w') + \cdots = (x - x')(x^2 + xx' + x'^2) + \cdots.$$

Each term in the previous relation is of the form

$$w^a x^b - w'^a x'^b = (w^a - w'^a)x^b + w'^a(x^b - x'^b).$$

Hence the difference of the two normal forms has the form

$$(w - w')(1 + u) = (x - x')(x^2 + xx' + x'^2 + v), \qquad \text{for } u, v \in k,$$

where $\text{ord}_p(u) > 0$ so that $\text{ord}_p(1 + u) = 0$. Further, each term of v is divisible by some w or w' so that $\text{ord}_p(v) \geq 3n$. Since $\text{ord}_p(x)$ and $\text{ord}_p(x') \geq n$, we have $\text{ord}_p(x^2 + xx' + x'^2 + v) \geq 2n$. Thus we obtain the following inequality:

$$\begin{aligned}\text{ord}_p(c) = \text{ord}_p\left(\frac{w - w'}{x - x'}\right) &\geq \text{ord}_p(x^2 + xx' + x'^2 - v) - \text{ord}_p(1 + u)\\ &\geq 2n.\end{aligned}$$

Case 2. $P = P'$. Then $c = dw/dx$. Differentiate the normal form of the cubic implicitly

$$\begin{aligned}&\frac{dw}{dx} + a_1\left(w + x\frac{dw}{dx}\right) + 2a_3 w\frac{dw}{dx}\\ &\quad = 3x^2 + a_2\left(2wx + x^2\frac{dw}{dx}\right) + a_4\left(w^2 + 2wx\frac{dw}{dx}\right) + 3a_6 w^2\frac{dw}{dx},\end{aligned}$$

and collecting the terms we obtain

$$\begin{aligned}&\left(1 + a_1 x + 2a_3 w - a_2 x^2 - 2a_4 wx - 3a_6 w^2\right)\frac{dw}{dx}\\ &\quad = 3x^2 + 2a_2 xw + a_4 w^2 - a_1 w.\end{aligned}$$

The coefficient of dw/dx has the form $1 + u$ where $\text{ord}_p(u) > 0$ and this means that $\text{ord}_p(1 + u) = 0$. The right-hand side has the form $3x^2 + v$ where $\text{ord}_p(3x^2 + v) \geq \text{ord}_p(3x^2) \geq 2\text{ord}_p(x)$ since $\text{ord}_p(w) = 3n$, and thus we have the inequality

$$\text{ord}_p(c) = \text{ord}_p\left(\frac{dw}{dx}\right) = \text{ord}_p(3x^2 + v) \geq 2n.$$

Therefore in both cases $\text{ord}_p(c) \geq 2n$. As for the coefficient b in the equation of the line through P and P', we use $b = w - cx$ to obtain the inequality

$$\text{ord}_p(b) \geq \min\{\text{ord}_p(w), \text{ord}_p(c) + \text{ord}_p(x)\} \geq 3n.$$

In order to estimate $\text{ord}_p(x + x' + x'')$, we substitute the equation for the line $w = cx + b$ through P and P' in the normal form for the equation of the cubic to obtain an equation in x

$$(cx + b) + a_1(cx + b)x + a_3(cx + b)^2$$

$$= x^3 + a_2(cx+b)x^2 + a_4(cx+b)^2x + a_6(cx+b)^3.$$

Collecting coefficients of powers of x, we have the following first two terms:

$$0 = x^3\left(1 + a_2c + a_4c^2 + a_6c^3\right)$$
$$+ x^2\left(a_2b + 2a_4bc + 3a_6bc^2 - a_1c - a_3c^2\right) + \cdots.$$

The sum of the roots of this cubic equation in x is $x + x' + x''$ the x-coordinates of P, P', and P'' and is given by the following expression:

$$x + x' + x'' = -\frac{a_2b + 2a_4bc + 3a_6bc^2 - a_1c - a_3c^2}{1 + a_2c + a_4c^2 + a_6c^3}.$$

Since the denominator is of the form $1 + u$ where $\mathrm{ord}_p(u) > 0$, we see that

$$\mathrm{ord}_p(1 + a_2c + a_4c^2 + a_6c^3) = 0.$$

From the relations $\mathrm{ord}_p(b) \geq 3n$ and $\mathrm{ord}_p(c) \geq 2n$, we deduce that $\mathrm{ord}_p(x + x' + x'') \geq 2n$.

For the last statement of the proposition observe that

$$\mathrm{ord}_p(x'') \geq \min\left\{\mathrm{ord}_p(x + x' + x''), \mathrm{ord}_p(-x), \mathrm{ord}_p(-x')\right\} \geq n.$$

Since $w = cx + b$, or from (4.1), it is now clear that $\mathrm{ord}_p(w'') \geq 3n$ and the point $(w'', x'', 1)$ is in $E^{(n)}(k)$. This proves the proposition.

(4.4) Remark. At the end of the proof of the previous proposition, we see that $\mathrm{ord}_p(x + x' + x'') \geq 3n$ whenever $a_1 = 0$, so that the term a_1c is not in the numerator of the expression for $x + x' + x''$.

(4.5) Theorem. *Let E be an elliptic curve over k defined by a cubic in normal form with p-adic filtration $E^{(n)}(k)$ on $E(k)$. The subsets $E^{(n)}(k)$ are subgroups. The function $P = (w, x, 1) \mapsto x(P) = x$ defined $E^{(n)}(k) \to p^nR$ composed with the canonical quotient morphism $p^nR \to p^nR/p^{2n}R$ defines a group morphism $E^{(n)}(k) \to p^nR/p^{2n}R$ with kernel in $E^{(2n)}(k)$ and it induces a monomorphism $E^{(n)}(R)/E^{(2n)}(k) \to p^nR/p^{2n}R$ for $n \geq 1$.*

Proof. For the first statement observe that the condition on $P = (w, x, 1)$ where $\mathrm{ord}_p(w) \geq 3n$ and $\mathrm{ord}_p(x) \geq n$ is equivalent for the coordinates $P(1, x, y)$ to the condition $\mathrm{ord}_p(x) \leq -2n$ and $\mathrm{ord}_p(y) \leq -3n$. Since

$$-(1, x, y) = (1, x, -y - a_1x - a_3),$$

it follows that the sets $E^{(n)}(k)$ are stable under taking inverse of elements. This together with (4.3) implies that $E^{(n)}(k)$ is a subgroup for $n \geq 1$.

For the second statement, consider $PQ = T$. By (4.3) we have

$$x(P) + x(Q) + x(T) \equiv 0 \mod p^{2n} R,$$

and this means that

$$x(P + Q) + x(T) \equiv x(P) + x(Q) \mod p^{2n} R.$$

For the last statement note that $x(P) \in p^{2n} R$ if and only if $P \in E^{(2n)}(k)$. This proves the theorem.

(4.6) Remark. The above result can be modified when $a_1 = 0$ to give an injection

$$\frac{E^{(n)}(k)}{E^{(3n)}(k)} \to \frac{p^n R}{p^{3n} R}.$$

In the next section we will see one case where this modification is useful.

Exercises

1. Carry out the proofs of the modifications indicated in (4.4) and (4.6).
2. Using the notations in (2.1), show that R is a valuation ring with maximal ideal Rp if and only if $R_{(p)} = R$. Then show that Rp is the unique maximal ideal in R. Show that the following is a filtration by subgroups of the multiplicative group k^*:

$$k^* \supset R^* \supset 1 + Rp \supset 1 + Rp^2 \supset \cdots \supset 1 + Rp^n \supset \cdots.$$

In this case show that ord_p induces an isomorphism $\text{ord}_p : k^*/R^* \to \mathbb{Z}$, reduction mod p induces an isomorphism $R^*/(1 + Rp) \to k(p)^* = (R/Rp)^*$, and $1 + ap^n \mapsto a$ induces an isomorphism $(1 + Rp^n)/(1 + Rp^{n+1}) \to k(p) = R/Rp$ as additive groups.
3. Let $\bar{E}(k)_{\text{ns}}$ denote the subgroup of nonsingular points of $\bar{E}(k)$. Show that $E^{(0)}(k) = r_p^{-1}(\bar{E}(k)_{\text{ns}})$ is a subgroup of $E(k)$ with zero equal to $0 : 0 : 1$. Show that the restriction $r_p : E^{(0)}(k) \to \bar{E}(k(p))_{\text{ns}}$ is a group homomorphism.

§5. Torsion in Elliptic Curves over $\mathbb{Q}$: Nagell–Lutz Theorem

Let $\text{Tors}(A)$ or A_{tors} denote the torsion subgroup of an abelian group A. In this section we will see that $E(\mathbb{Q})_{\text{tors}}$ is a group that can be determined effectively from the equation of the curve E, and is in particular a finite group.

(5.1) Theorem (Nagell–Lutz). *Let E be an elliptic curve over the rational numbers $\mathbb{Q}$.*

(1) *The subgroup $E(\mathbb{Q})_{\text{tors}} \cap E^{(1)}(\mathbb{Q})$ is zero for each odd prime p and*

$$E(\mathbb{Q})_{\text{tors}} \cap E^{(2)}(\mathbb{Q})$$

is zero for $p = 2$.

(2) *The restriction of the reduction homomorphism $r_p|E(\mathbb{Q})_{\text{tors}} : E(\mathbb{Q})_{\text{tors}} \to E_{(p)}(\mathbb{F}_p)$ is injective for any odd prime p where E has good reduction and $r_2|E(\mathbb{Q})_{\text{tors}} : E(\mathbb{Q})_{\text{tors}} \to E_{(2)}(\mathbb{F}_2)$ has kernel at most $\mathbb{Z}/2\mathbb{Z}$ when E has good reduction at 2.*

Proof. The function $T \mapsto x(T)$ defines a monomorphism $E^{(n)}(\mathbb{Q})/E^{(2n)}(\mathbb{Q}) \to \mathbb{Z}p^n/\mathbb{Z}p^{2n} \cong \mathbb{Z}/\mathbb{Z}p$ by (4.5), and this implies that there is no torsion prime to p in $E^{(1)}(\mathbb{Q})$ prime to p. Assume that $pT = 0$ where $T \in E^{(r)}(\mathbb{Q}) - E^{(r+1)}(\mathbb{Q})$ and $r \geq 1$. If p is odd, then we can use (4.5) and (4.6) to show that

$$0 = x(pT) \equiv px(T) \qquad \text{mod } p^{3r}.$$

Hence $x(T) \in p^{3r-1}\mathbb{Z}$, and this means that $T \in E^{(3r-1)}(\mathbb{Q})$. This implies that $r \leq 3r - 1$ or $2r \leq 1$ so that $r = 0$. If $p = 2$, then we can only use (4.5) to show that

$$0 = x(2T) \equiv 2x(T) \qquad \text{mod } 2^{2r}.$$

Hence $x(T) \in 2^{2r-1}\mathbb{Z}$, and this means that $T \in E^{(2r-1)}(\mathbb{Q})$. This implies that $r = 2r - 1$ $r = 1$. Hence $E(\mathbb{Q})_{\text{tors}}$ has zero intersection with $E^{(1)}(\mathbb{Q})$ for p odd and with $E^{(2)}(\mathbb{Q})$ for $p = 2$.

For the second assertion recall that $\ker(r_p) = E^{(1)}(\mathbb{Q})$ at any prime p. The first assertion implies that for good reduction at p the restriction $r_p|E(\mathbb{Q})_{\text{tors}}$ has zero kernel for p odd and kernel isomorphic to $E(\mathbb{Q})_{\text{tors}} \cap E^{(1)}(\mathbb{Q})/E^{(2)}(\mathbb{Q})$ for $p = 2$. The group $E(\mathbb{Q})_{\text{tors}} \cap E^{(1)}(\mathbb{Q})/E^{(2)}(\mathbb{Q})$ injects into $2\mathbb{Z}/4\mathbb{Z} = \mathbb{Z}/2\mathbb{Z}$ by (4.5). This proves the second assertion and the theorem.

(5.2) Remark. If C is a cubic curve defined by an equation over $\mathbb{F}_q$ in normal form, then for each x in $\mathbb{F}_q$ we have at most two possible (x, y) on the curve $C(\mathbb{F}_q)$ and so the cardinality $\#C(\mathbb{F}_q) \leq 2q + 1$.

(5.3) Corollary. *Let E be an elliptic curve over $\mathbb{Q}$. If E has good reduction at an odd prime p, then the cardinality of the torsion subgroup satisfies $\#E(\mathbb{Q})_{\text{tors}} \leq 2p + 1$. If E has good reduction at 2, then $\#E(\mathbb{Q})_{\text{tors}} \leq 10$.*

(5.4) Corollary. *For an elliptic curve E over $\mathbb{Q}$, the torsion subgroup $E(\mathbb{Q})_{\text{tors}}$ of $E(\mathbb{Q})$ is finite and is either cyclic or cyclic direct sum with $\mathbb{Z}/2\mathbb{Z}$.*

Proof. Every elliptic curve has good reduction at some p giving the finiteness assertion. Since $E(\mathbb{Q})_{\text{tors}}$ is a finite subgroup of $E(\mathbb{R})$, we can apply (7.2) of the Introduction and the structure of such finite subgroups of the cicle or the circle direct sum with $\mathbb{Z}/2\mathbb{Z}$.

In the first few sections of Chapter 1, we introduced some elliptic curves in order to see how the group law works, and now we return to these curves to illustrate the use of Theorem (5.1) and some other general ideas.

(5.5) Example. The curve $E : y^2 + y - xy = x^3$ was considered in 1(1.6) where we saw that (1, 1) generated a subgroup of order 6 in $E(\mathbb{Q})_{\text{tors}}$. Reducing this curve mod 2, we apply the ideas in 3(6.4) to see that the curve is isomorphic to $y^2 + xy = x^3 + x^2$

over $\mathbb{F}_2$ which is singular. Hence it has bad reduction at 2. Modulo 3 the situation is better. From the derivative

$$(2y + 1 - x)y' = 3x^2 + y = y,$$

we see that $(1, 0)$ is the only indeterminate value for y', but $(1, 0)$ is not on $E(\mathbb{F}_3)$. We have the following "graph"

0			
+1			*
0		*	
−1	*	*	*
	−1	0	+1

and from this we see that $E_{(3)}(\mathbb{F}_3)$ is cyclic of order 6. Applying (5.1), we deduce that $E(\mathbb{Q})_{\text{tors}}$ is cyclic of order 6 generated by $(1, 1)$. The question of points of infinite order is still pending.

(5.6) Example. In, respectively, 1(1.6), 1(2.3), and 1(2.4), we introduced three curves

$$E : y^2 + y = x^3 - x, \qquad E' : y^2 + y = x^3 - x^2, \qquad E'' : y^2 + y = x^3 + x^2.$$

On E and E'' the point $(0, 0)$ has infinite order and on E' it has order 5. Mod 2 all three curves reduce to the same curve up to isomorphism, namely the curve called E_3 in 3(6.4), and $E_3(\mathbb{F}_2)$ is cyclic of order 5. Mod 3 we leave it to the reader to check that all three curves are nonsingular and they have the following graphs:

E :

0			
+1			
0	*	*	*
−1	*	*	*
	−1	0	+1

E' :

0			
+1			
0		*	*
−1		*	*
	−1	0	+1

E'' :

0			
+1			*
0	*	*	
−1	*	*	
	−1	0	+1

Hence $E(\mathbb{F}_3) = \mathbb{Z}/7\mathbb{Z}$, $E'(\mathbb{F}_3) = \mathbb{Z}/5\mathbb{Z}$, and $E''(\mathbb{F}_3) = \mathbb{Z}/6\mathbb{Z}$. Any torsion in $E(\mathbb{Q})$ would have to inject into both the cyclic group of order 5, mod 2, and the cyclic group of order 7, mod 3. Hence $E(\mathbb{Q})_{\text{tors}} = 0$, and by the same principle, $E''(\mathbb{Q})_{\text{tors}} = 0$, while $E'(\mathbb{Q})_{\text{tors}}$ is cyclic of order 5.

Exercises

1. For the elliptic curve $E: y^2 - y = x^3 - 7$, see 1(2.5), show that E is nonsingular mod 2. Show that $E(\mathbb{Q})$ has no point of order 2, and deduce that $E(\mathbb{Q})_{\text{tors}}$ is cyclic of order 3.
2. Study $E: y^2 + xy + y = x^3 - x^2$ both modulo 2 and modulo 3. Determine $E(\mathbb{Q})_{\text{tors}}$.
3. Study $E: y^2 + xy = x^3 - x^2 - 2x - 1$ both modulo 2 and modulo 3. Determine $E(\mathbb{Q})_{\text{tors}}$.
4. In 1(2. Ex.5) we noted that $(1, 0)$ has order 7 on $E: y^2 + xy + y = x^3 - x^2 - 3x + 3$. Show that E is nonsingular modulo 3 and determine $E(\mathbb{Q})_{\text{tors}}$.
5. Let $M_n(R)$ denote the algebra of n by n matrices over the ring R. Reduction modulo q of integral matrices defines a ring homomorphism $r_q : M_n(\mathbb{Z}) \to M_n(\mathbb{Z}/q\mathbb{Z})$, and show that it restricts to $r_q : \mathrm{GL}_n(\mathbb{Z}) \to \mathrm{GL}_n(\mathbb{Z}/q\mathbb{Z})$ on the invertible matrices. For $X \in M_n(\mathbb{Z})$ such that $r_p(X) = 0$ show that
$$(I_n + p^a X)^n \equiv I_n + np^a X \qquad \text{mod } p^{a+b+1},$$
where $b = \mathrm{ord}_p(n)$ and $p > 2$ or $a \geq 1$. Prove that for a finite subgroup G of $\mathrm{GL}_n(\mathbb{Z})$ that $G \cap \ker(r_p) = 1$ for $p > 2$ and $G \cap \ker(r_4) = 1$ at $p = 2$. Compare with (5.1).
6. Show that the curve $y^2 + y = x^3 - 7x + 6$ has no torsion in its group of rational points.

§6. Computability of Torsion Points on Elliptic Curves from Integrality and Divisibility Properties of Coordinates

The first theorem says that when looking for torsion points we have only to look at integral points.

(6.1) Theorem. *Let E be an elliptic curve defined over $\mathbb{Q}$ with an equation in normal form with integer coefficients. If $(x, y) \in E(\mathbb{Q})_{\text{tors}}$, then the coordinates x and y are integers.*

Proof. If $y = 0$, then x is a solution of the cubic equation
$$0 = x^3 + a_2x^2 + a_4x + a_6$$
with integer coefficients. Since x is rational, it is also an integer, for $x = m/n$ reduced to lowest terms satisfies
$$0 = m^3 + a_2m^2n + a_4mn^2 + a_6n^3,$$
and any prime dividing n must divide m. Thus $x = m$ is an integer.

If $y \neq 0$, then the point with homogeneous coordinates has the form $(w : x' : 1) = (1 : x : y)$, where $w = 1/y$ and $x' = x/y$. By Theorem (4.1), we have that $(w : x' : 1) \in r_p^{-1}(0)$ for p odd and $(w : x' : 1) \in E^{(2)}(\mathbb{Q})$ at 2. In other words we have $\mathrm{ord}_p(w) \leq 0$ for p odd and $\mathrm{ord}_2(w) \leq -1$ at 2. This condition becomes for y the relation $\mathrm{ord}_p(y) \geq 0$ for all odd p and $\mathrm{ord}_2(y) \geq -1$ at 2. Thus y has the form $y = h/2$ for an integer h. Again write $x = m/n$, and x satisfies a cubic equation with coefficient of x^3 one, coefficient of x^2 an integer, coefficient of x an integer over 2, and constant coefficent an integer over 4. A modification, using 2, of the above argument showing $x = m$ shows again that $x = m$ and h is even. This proves the theorem.

The last step in our analysis of torsion points, which we now know must have integer coordinates, is to obtain a divisibility requirement on the y-coordinate.

(6.2) Theorem. *Let E be an elliptic curve over $\mathbb{Q}$, and let $y^2 = f(x)$ be a Weierstrass equation for E where $f(x)$ has integer coefficients. If (x, y) is a torsion point on E, then the integer y is zero or y divides $D(f)$, the discriminant of the cubic polynomial $f(x)$.*

Proof. If $y = 0$, then $(x, 0)$ is of order 2 and 0 divides the discriminant. Otherwise, $2(x, y) = (\bar{x}, \bar{y})$ unequal to 0 on $E(\mathbb{Q})$. The tangent line to E at (x, y) has slope $f'(x)/2y$, and when its equation $y = \lambda x + \beta$ is substituted into the Weierstrass equation $y^2 = f(x) = x^3 + ax^2 + bx + c$, we obtain a cubic equation with x as double root and $\bar{x}$ as a single root. This equation has coefficient $a - (f'(x)/2y)^2$ of x^2, and hence the sum of the roots of the cubic in x is the negative of this coefficient, so

$$2x + \bar{x} = a - \left(\frac{f'(x)}{2y}\right)^2.$$

Since $x, \bar{x}$, and a are integers, it follows that $f'(x)/2y$ is an integer, and $2y$ divides $f'(x)$.

By (4.2) in the Appendix to Chapter 2, we can write the discriminant $D(f)$ of $f(x)$ as a linear combination $D(f) = u(x)f(x) + v(x)f'(x)$, where $u(x), v(x) \in \mathbb{Z}[x]$. Since $y = f(x)$ and y divides $f'(x)$ for the point (x, y) on E, we deduce that y divides $D(f)$. This proves the theorem.

(6.3) Remark. The effectively computable method for finding $E(\mathbb{Q})_{\text{tors}}$ is to take for E a Weierstrass equation $y^2 = f(x)$, where $f(x) = x^3 + ax^2 + bx + c$ and a, b, and c are integers. Consider the finite set of all divisors y_0 of $D(f)$, the discriminant of $f(x)$. Solve the cubic $f(x) = y_0^2$ for integer solutions x_0. Among these (x_0, y_0) are all points of $E(\mathbb{Q})_{\text{tors}}$ which are unequal to 0.

(6.4) Example. For a cubic in the form $x^3 + ax^2 + bx$ the discriminant is $D(x^3 + ax^2 + bx) = b^2(4b - a^2)$. In the case of E: $y^2 = x^3 + x^2 - x$ we have $D = -4 - 1 = -5$. The divisors of 5 are $+1, -1, +5$, and -5. For $y = 0$ we have the point $(0, 0)$, for $y = +1$ we have the points $(1, 1)$ and $(-1, 1)$, for $y = -1$ we have the points $(1, -1)$ and $(-1, -1)$, and for $y = 5$ we obtain the cubic equation $0 = x^3 + x^2 - x - 25$. It has a root strictly between 2 and 3 which is not integral, and, in fact, it is not rational. The same applies to $y = -5$. Thus we have six rational torsion points on $E(\mathbb{Q})$, and $E(\mathbb{Q})_{\text{tors}}$ is cyclic of order 6. Although not all points need be torsion points in general.

(6.5) Remark. If (x, y) is a point $E(\mathbb{Q})$ such that some multiple $n(x, y)$ has nonintegral coefficients, then (x, y) is not a torsion point. This applies to $(0, 0)$ on E and E'' in (5.6) by the calculations in 1(1.6) and 1(2.4).

Exercises

1. Find all torsion points on E: $y^2 = x^3 - x^2 + x$ by the method using (6.1) and (6.2). Show that the curve has good reduction modulo 5 and compare with $E(\mathbb{F}_5)$.
2. Find all torsion points on E: $y^2 = x^3 - 2x^2 - x$ by the method using (6.1) and (6.2). Show that the curve has good reduction both modulo 3 and modulo 5, and compare with $E(\mathbb{F}_3)$ and $E(\mathbb{F}_5)$.
3. Show that E: $y^2 = x^3 + x^2 - x$ has good reduction modulo 3 and compare $E(\mathbb{Q})_{\text{tors}}$ with $E(\mathbb{F}_3)$.
4. Let (x, y) be a point on $E(\mathbb{Q})$ defined by a normal cubic over the integers. If the slope of the tangent line of E at (x, y) is not an integer, then show that (x, y) is not a torsion point.

§7. Bad Reduction and Potentially Good Reduction

We continue with the notations of (1.1) and recall that an elliptic curve E over k with discriminant Δ has good (resp. bad) reduction at p if and only if $\mathrm{ord}_p(\Delta) = 0$ (resp. $\mathrm{ord}_p(\Delta) > 0$). Bad reduction divides into two cases using the description of singular cubic curves in Chapter 3, §7.

(7.1) Definition. An elliptic curve E over k has:

(1) multiplicative reduction (or semistable reduction) at p provided the reduction $E_{(p)}$ has a double point (or node), or
(2) additive reduction (or unstable reduction) at p provided the reduction $E_{(p)}$ has a cusp.

Multiplicative reduction is divided into split or nonsplit depending on whether or not $E_{(p)}(k(p))$ is isomorphic to the multiplicative group of $k(p)$ or to the elements of norm one in a quadratic extension of $k(p)$.

(7.2) Remark. Let E be an elliptic curve E over k with discriminant Δ and having bad reduction at p, that is, $\mathrm{ord}_p(\Delta) > 0$ or $\bar{\Delta} = 0$. The reduction is:

(1) multiplicative reduction if and only if $\mathrm{ord}_p(c_4) = 0$ or, equivalently, $\mathrm{ord}_p(b_2) = 0$, or
(2) additive reduction if and only if $\mathrm{ord}_p(c_4) > 0$ or, equivalently, $\mathrm{ord}_p(b_2) > 0$.

Observe that $\mathrm{ord}_p(j(E))$ is positive if E has good reduction and can be either positive or negative if E has bad reduction since it is a quotient of elements in R.

(7.3) Remark. Let E be an elliptic curve over k with good reduction at p. Then the reduction modulo p of $j(E)$ is given by $r_p(j(E)) = j(E_{(p)})$ and $\mathrm{ord}_p(j(E)) \geq 0$. We also have two congruences:

$$\mathrm{ord}_p(j(E)) \equiv 0 \pmod 3 \quad \text{and} \quad \mathrm{ord}_p(j(E) - 12^3) \equiv 0 \pmod 2$$

since $j(E) = c_4^3/\Delta$, $j(E) - 12^3 = c_6^2/\Delta$, and $12^3\Delta = c_4^3 - c_6^2$. Conversely, if $\mathrm{ord}_p(j - 12^3) = 0 = \mathrm{ord}_p(j)$, then the equation

$$y^2 + xy = x^3 - \frac{12^3}{j - 12^3}x - \frac{1}{j - 12^3}$$

shows that the curve E with $j(E) = j$ can be defined over k.

(7.4) Remark. If E has multiplicative reduction at p, then $\mathrm{ord}_p(j(E)) < 0$ since $j(E) = c_4^3/\Delta$ and $\mathrm{ord}_p(c_4) = 0$ for multiplicative reduction by (7.2(1)).

Now we consider an elliptic curve E over k with $j(E) \in R$, that is, with $\mathrm{ord}_p(j(E)) \geq 0$ for all p even at those irreducibles p where E has bad reduction.

Case 1. If k has characteristic unequal to 2, then E can be defined by an equation of the form $y^2 = f(x)$. If we extend the field k to k' including the roots of $f(x)$, then E becomes isomorphic to a curve with equation in Legendre form E_λ: $y^2 = x(x-1)(x-\lambda)$ where the j-invariant in this case is

$$j(E_\lambda) = j(\lambda) = 2^8\frac{(\lambda^2 - \lambda + 1)^3}{\lambda^2(\lambda - 1)^2}$$

by 4(1.4). Given a value $j = j(E)$, we can solve for one of the values of λ such that $j(\lambda) = j$. In general there are six possible λ. If $\mathrm{ord}_p(\lambda) < 0$, then

$$\mathrm{ord}_p(j(\lambda)) = 3\,\mathrm{ord}_p(\lambda^2) - 2\,\mathrm{ord}_p(\lambda) - 2\,\mathrm{ord}_p(\lambda) < 0,$$

and hence $\mathrm{ord}_p(j) \geq 0$ implies that $\mathrm{ord}_p(\lambda) \geq 0$. Furthermore, the curve E_λ has good reduction since $\bar{\lambda}$ is not 0 or 1 because it is a solution of $\bar{j} = j(\lambda)$ over the residue class field $k(p)$.

Case 2. If k has characteristic unequal to 3, then E can be transformed into an equation of the form E_α: $y^2 + \alpha xy + y = x^3$ where the j-invariant in this case is

$$j(E_a) = j(\alpha) = \frac{\alpha^3(\alpha^3 - 24)^3}{\alpha^3 - 27}$$

by 4(2.2) and 4(2.3). Given a value $j = j(E)$, we can solve for one of the values of α such that $j(\alpha) = j$. If $\mathrm{ord}_p(\alpha) < 0$, then

$$\mathrm{ord}_p(j(\alpha)) = 3\,\mathrm{ord}_p(\alpha) + 3\,\mathrm{ord}_p(\alpha^3) - 3\,\mathrm{ord}_p(\alpha^3) < 0,$$

and hence $\mathrm{ord}_p(j) \geq 0$ implies that $\mathrm{ord}_p(\alpha) \geq 0$. Furthermore, the curve E_α has good reduction since $\bar{\alpha}^3$ is unequal to 27 because it is a solution of $\bar{j} = j(\alpha)$ over the residue class field $k(p)$.

The above discussion is a proof of Deuring [1941] characterizing potentially good reduction. We formulate the concept and theorem for fields K with a discrete valuation in terms of finite extensions L of K and prolongations w of v to L. Recall that these extensions w of v always exist.

(7.5) Definition. An elliptic curve E over K has potential good reduction provided there exists a finite extension L and an extension w of v to L such that E over L has good reduction at the valuation w.

Be extending the ground field from K to L, the minimal equation of E can change to one with good reduction, or, in other words, E can become isomorphic over L to a curve E' with good reduction at w. In the above case E' was the Legendre or Hessian form of the equation. In general the discriminant Δ of E over K is different from the discriminant Δ' of E' over L and in this case of potential good reduction $v(\Delta) > 0$ and $w(\Delta') = 0$ while $j(E) = j(E')$. Since any curve E' with good reduction has $v(j(E')) \geq 0$, it follows immediately that if E has potential good reduction, then $j(E)$ is a local integer, i.e., $v(j(E)) \geq 0$. The above argument with Legendre or Hessian forms of the equations shows that the converse holds, and this is the theorem of Deuring.

(7.6) Theorem. *An elliptic curve E defined over K has potential good reduction if and only if $j(E)$ is a local integer, i.e., $v(j(E)) \geq 0$.*

§8. Tate's Theorem on Good Reduction over the Rational Numbers

Tate's theorem says that over the rational numbers there are no elliptic curves with good reduction everywhere. We give a proof using formulas for coefficients following Ogg [1966].

(8.1) Theorem. *Every elliptic curve E over the rational numbers $\mathbb{Q}$ has bad reduction at some prime, i.e., Δ cannot be equal to ± 1.*

Proof. Assume that $\Delta = \pm 1$. Recall from 3(3.1) that

$$\Delta = -b_2^2 b_8 - 8b_4^3 - 27b_6^2 - 9b_2 b_4 b_6$$

and

$$b_2 = a_1^2 + 4a_2, \qquad b_4 = a_1 a_3 + 2a_4.$$

If a_1 is even, then $4|b_2$ and $2|b_4$ so that b_6 will have to be odd, and in fact, $\pm 1 = \Delta \equiv 5b_6^2 \pmod 8$. Since any square modulo 8 is conguent to $0, 1, 4 \pmod 8$, this is impossible.

Now consider the case where a_1 is odd, and hence b_2 is also odd. Then the coefficient $c_4 = b_2^2 - 24b_4 \equiv 1 \pmod 8$. We write $c_4 = x \pm 12$ from the relation $c_4^3 - c_6^2 = 12^3 \Delta = \pm 12^3$, and thus

$$c_6^2 = x(x^2 \pm 36x + 3 \cdot 12^3) \equiv x^2(x+4) \pmod 8.$$

This means that $x \equiv 5 \pmod 8$. Now $3|x$, for otherwise any $p|x$ with $p > 3$, it would follow that $p^2|x$ and $\pm x$ would be a square. This would contradict $x \equiv 5 \pmod 8$.

Let $x = 3y$ so that $y \equiv 7 \pmod 8$, and hence $c_6 = 9c$. This gives the equation

$$3c^2 = y(y^2 \pm 12y + 4 \cdot 12^2) = y((y \pm 6)^2 + 540).$$

Now $y > 0$ since $y((y \pm 6)^2 + 540)$ is positive. If p is unequal to 3 and divides y, it does so to an even power. Also the relation for $3c^2$ shows that if 3 divides y, then 27 divides $3c^2$. In this case let $y = 3z$ and $c = 3d$ which leads to

$$d^2 = z(z^2 \pm 4z + 64)$$

from the relation for $3c^2$. If an odd prime p divides z, then $p^2|z$ and z is a square. But $y \equiv 7 \pmod 8$ implies $z \equiv 5 \pmod 8$ which contradicts the fact that z is a square. This proves the theorem.

(8.2) Example. The following curve

$$y^2 + xy + \varepsilon^2 y = x^3, \qquad \text{where } \varepsilon = \frac{5 + \sqrt{29}}{2}$$

over $K = \mathbb{Q}(\sqrt{29})$ was shown by Tate to have good reduction at all places of K, see Serre [1972, p. 320]. The norm of ε is -1, and the group of totally positive units in $R = \mathbb{Z}[\varepsilon]$ is generated by ε^2. The reader can check that $\Delta = -\varepsilon^{10}$ and verify the good reduction properties.

(8.3) Example (Unpublished example of R. Oort).

$$y^2 + xy = x^3 - \varepsilon x$$

where $\varepsilon = 32 + 5\sqrt{41}$ the fundamental unit in $\mathbb{Z}[(1/2) + (1/2)\sqrt{41}]$. Here $\Delta = \varepsilon^4$ and the curve has good reduction at all places of $\mathbb{Q}(\sqrt{41})$.

6

Proof of Mordell's Finite Generation Theorem

In this chapter we prove that the rational points $E(\mathbb{Q})$ on an elliptic curve E over the rational numbers $\mathbb{Q}$ form a finitely generated group. We will follow a line of argument which generalizes to prove A. Weil's extension to number fields: The Mordell–Weil group $E(k)$ of points over a number field k on an elliptic curve E is a finitely generated group.

There are two conditions on an abelian group A which together are equivalent to A being finitely generated. First the index $(A : nA)$ is finite for some $n > 1$, usually one shows that $(A : 2A)$ is finite, and second the existence of a norm on the group. This last metric property allows a descent procedure to take place similar to the descent arguments first introduced by Fermat.

The proof of the finiteness of $(E(k) : 2E(k))$ is done over any number field. The argument uses the equation of E in factored Weierstrass form

$$y^2 = (x - \alpha)(x - \beta)(x - \gamma).$$

A factored form of the equation is always possible after at most a ground field extension of degree 6. Thus we are led to number fields even if our interest is elliptic curves over $\mathbb{Q}$.

The norm is constructed from the canonical height function of projective space. The subject of heights comes up again in considerations related to the Birch and Swinnerton–Dyer conjectures in Chapter 17.

§1. A Condition for Finite Generation of an Abelian Group

Multiplication by a natural number m on an abelian group A is a homomorphism $A \overset{m}{\to} A$ with a kernel ${}_mA$ and a cokernel A/mA. The order of A/mA is also the index $(A : mA)$ of the subgroup mA in A. If A is finitely generated, then $(A : mA)$ is finite for all nonzero m.

(1.1) Definition. A norm function $|\ \ |$ on an abelian group A is a function $|\ \ | : A \to \mathbb{R}$ such that:

(1) $|P| \geq 0$ for all $P \in A$, and for each real number r the set of $P \in A$ with $|P| \leq r$ is finite.
(2) $|mP| = |m||P|$ for all $P \in A$ and integers m.
(3) $|P + Q| \leq |P| + |Q|$ for all $P, Q \in A$.

Observe that from these axioms P is a torsion point if and only if $|P| = 0$. By the first axiom the torsion subgroup $\mathrm{Tors}(A)$ of A is therefore finite.

(1.2) Examples. The restriction of the Euclidean norm on $\mathbb{R}^n$ to $\mathbb{Z}^n$ is a norm on $\mathbb{Z}^n$. The zero norm is a norm on any finite abelian group, and, in fact, it is the only norm. If A_i is an abelian group with norm $|\ \ |_i$ for $i = 1, \ldots, n$, then $A_1 \oplus \cdots \oplus A_n = A$ has a norm given by $|(P_1, \ldots, P_n)| = |P_1|_1 + \cdots + |P_n|_n$. In particular, any finitely generated abelian group A has a norm since A is isomorphic to $\mathbb{Z}^n \times \mathrm{Tors}(A)$. The converse is the next proposition.

(1.3) Proposition. *An abelian group A is finitely generated if and only if the index $(A : mA)$ is finite for some $m > 1$ and the group A has a norm function.*

Proof. The direct implication is contained in the above discussion. Conversely, let $R_1, \ldots, R_n$ be representatives in A for the cosets modulo mA, and let $c = \max_i |R_i| + 1$. Let X denote the finite set of all $P \in A$ with $|P| \leq c$, and let G be the subgroup of A generated by X. In particular, each $R_i \in X \subset G$.

If there exists $P \in A - G$, then there exists such a P with minimal norm by the first axiom. Since $P \equiv R_i \pmod{mA}$ for some coset representative of A/mA, we have $P = R_i + mQ$ or $mQ = P + (-R_i)$, and

$$m|Q| = |mQ| \leq |P| + |R_i| < |P| + c \leq m|P|.$$

From the minimal character of $P \in A - G$, we see that $Q \in G$, and hence the sum $P = R_i + mQ$ is also in G. This contradicts the assumption that $A - G$ is nonempty and, therefore, $A = G$, because G is finitely generated by X. This proves the proposition.

(1.4) Remark. The above argument has a constructive character and is frequently referred to as a descent procedure. For a given $P \in A$ we can choose a sequence

$$P = R_{i(0)} + mP_1, \quad P_1 = R_{i(1)} + mP_2, \ldots, \quad P_j = R_{i(j)} + mP_{j+1}, \ldots$$

with

$$|P| > |P_1| > |P_2| > \cdots > |P_j| > |P_{j+1}| > \cdots$$

and the descent stops when some $|P_j| \leq c$. Thus P is a sum of chosen coset representatives.

In the next section we show how descent was first used by Fermat to show the nonexistence of solutions to $x^4 + y^4 = 1$, where x and y are rational numbers both nonzero. The program for the remainder of this chapter is to check the two conditions of (1.3) for $A = E(k)$ the rational points on an elliptic curve over a number field.

(1.5) Remark. If the index $(A : pA) = p^r$ for a prime number p, then r is an upper bound for the rank of A. This principle is used frequently in obtaining information on the rank of the group of rational points on an elliptic curve especially for $p = 2$.

§2. Fermat Descent and $x^4 + y^4 = 1$

This section is a short detour away from the main purpose of the chapter and is not used in the rest of the book. We wish to explain the context in which the Fermat descent first arose.

We consider the equation $x^4 + y^4 = z^2$, where x, y, and z are integers without common factors.

Step 1. We apply (3.1) of the Introduction to the primitive Pythagorean triple x^2, y^2, z, where $z^2 = (x^2)^2 + (y^2)^2$ giving a representation

$$x^2 = m^2 - n^2, \quad y^2 = 2mn, \quad z = m^2 + n^2.$$

Step 2. Now look at the relation $x^2 + n^2 = m^2$ which says that n, x, m is a Pythagorean triple which is again primitive, for otherwise x^2, y^2, z would not be primitive. Applying (3.1) of the Introduction again, we can write

$$x = t^2 - s^2, \quad n = 2ts, \quad m = t^2 + s^2,$$

where s and t have no common prime factor.

Step 3. Substituting into y^2, we obtain $y^2 = 2mn = 4st(s^2 + t^2)$. Since s and t have no common prime factor, we deduce that both s and t are squares, that is, $s = a^2$ and $t = b^2$ and hence

$$c^2 = (a^2)^2 + (b^2)^2$$

for some c.

Step 4. We started with a primitive $z^2 = x^4 + y^4$ and we produced $c^2 = a^4 + b^4$ with the additional feature that $z > c$. To see that this inequality holds, we calculate

$$\begin{aligned} z = m^2 + n^2 &= (t^2 + s^2)^2 + 4t^2s^2 \\ &= (a^4 + b^4) + 4a^4b^4 > (a^4 + b^4)^2 = c^4. \end{aligned}$$

This establishes the inequality which is the core of the Fermat descent procedure.

(2.1) Theorem (Fermat, 1621). *The equation $x^4 + y^4 = t^2$ has no integral solutions with both x and y nonzero. The only rational points on the Fermat curve*

$$F_4 : x^4 + y^4 = 1$$

are the intersection points with the axes $(\pm 1, 0)$ *and* $(0, \pm 1)$.

Proof. If the first equation had an integral solution x, y, t, then we could choose x, y, and t strictly positive and consider a solution with t a minimum among all such solutions. By the Fermat descent procedure there would be a solution $a^4 + b^4 = c^2$ with c strictly smaller than t contradicting the minimal property of t. Hence there is no integral solution.

For the second assertion, consider a rational solution and clear the denominators of x and y to obtain a relation between integers of the form $u^4 + v^4 = t^2$. By the first result either $u = 0$ or $v = 0$. This proves the theorem.

§3. Finiteness of $(E(\mathbb{Q}) : 2E(\mathbb{Q}))$ for $E = E[a, b]$

Using the factorization of multiplication by 2 on the elliptic curve $E = E[a, b]$ considered in 4(5.3), we will show by relatively elementary means that $(E(\mathbb{Q}) : 2E(\mathbb{Q}))$ is finite for an elliptic curve of this form over the rational numbers. Note $E[a, b]$ is defined by $y^2 = x^3 + ax^2 + bx$ for $a, b \in k$.

Recall from 4(5.6) the function $\alpha : E[a, b] \to k^*/(k^*)^2$ defined by

$$\alpha(0) = 1,$$
$$\alpha((0, 0)) = b \mod (k^*)^2,$$
$$\alpha((x, y)) = x \mod (k^*)^2 \qquad \text{for } x \neq 0,$$

and having the property, see 4(5.7), that the sequence

$$E[a, b] \xrightarrow{\varphi} E[-2a, a^2 - 4b] \xrightarrow{\alpha} k^*/(k^*)^2$$

is exact. In the case $k = \mathbb{Q}$, the field of rational numbers, the quotient group $\mathbb{Q}^*/(\mathbb{Q}^*)^2$ is additively an $\mathbb{F}_2$-vector space with a basis consisting of -1 and all the positive prime numbers p.

(3.1) Proposition. *Let $E = E[a, b]$ be an elliptic curve over the rational numbers $\mathbb{Q}$. The homomorphism*

$$\alpha : E[a, b] \to \mathbb{Q}^*/(\mathbb{Q}^*)^2$$

has image $\mathrm{im}(\alpha)$ *contained in the $\mathbb{F}_2$-vector subspace generated by -1 and all primes p dividing b. If r distinct primes divide b, then the cardinality of* $\mathrm{im}(\alpha)$ *is less than 2^{r+1}.*

Proof. Let $x = m/e^2$ and $y = n/e^3$ be the coordinates of (x, y) on $E[a, b]$ with rational coefficients where the representations are reduced to lowest terms. Note we can always choose denominators in this form. The equation of the curve gives

$$n^2 = m^3 + am^2e^2 = bme^4$$
$$= m(m^2 + ame^2 + be^4).$$

If m and $m^2 + ame^2 + be^4$ are relatively prime, then it follows that m is a square up to sign and $\alpha(x, y) = \pm 1$.

More generally, let d be the greatest common divisor of m and $m^2 + ame^2 + be^4$. Then d divides be^4, and since m and e are relatively prime, the integer d also divides b. Moreover, $m = M^2 d$ up to sign, and $\alpha(x, y) = d \mod \mathbb{Q}^*/(\mathbb{Q}^*)^2$. This proves the proposition.

(3.2) Theorem. *Let a and b be two integers with $\Delta = 2^4 b^2(a^2 - 4b)$ unequal to zero, and let r be the number of distinct prime divisors of b and s the number of $a^2 - 4b$. Then for $E = E[a, b]$ we have*

$$(E(\mathbb{Q}) : 2E(\mathbb{Q})) \leq 2^{r+s+2}.$$

Proof. We factor multiplication by 2 as

$$E = E[a, b] \xrightarrow{\varphi} E' = E[-2a, a^2 - 4b] \xrightarrow{\varphi'} E.$$

By 4(5.7) two different maps α induce isomorphisms $E(\mathbb{Q})/\varphi' E'(\mathbb{Q}) \to \operatorname{im}(\alpha)$ and $E'(\mathbb{Q})/\varphi E(\mathbb{Q}) \to \operatorname{im}(\alpha)$. Also φ' induces an isomorphism $E'(\mathbb{Q})/\varphi E(\mathbb{Q}) \to \varphi E'(\mathbb{Q})/\varphi'\varphi E(\mathbb{Q})$. Since we have two-stage filtration $\varphi'\varphi E(\mathbb{Q}) = 2E(\mathbb{Q}) \subset \varphi' E'(\mathbb{Q}) \subset E(\mathbb{Q})$, it follows that the index $(E(\mathbb{Q}) : 2E(\mathbb{Q}))$ is the product of the orders of the two $\operatorname{im}(\alpha)$ where in one case α is defined on $E(\mathbb{Q})$ and on $E'(\mathbb{Q})$ in the other case. Now we apply the previous proposition, (3.1), to derive the theorem.

(3.3) Remark. With the notations of the previous theorem we see that $r + s + 2$ is an upper bound on the rank of $E(\mathbb{Q})$. Since "most" curves over $\mathbb{Q}$ have rank 0 or 1, this is not a very good upper bound. Using the methods of the next chapter we will derive improved bounds.

§4. Finiteness of the Index $(E(k) : 2E(k))$

Throughout this section, we use the notations of 5(1.1) where R is a factorial ring with field of fractions k. Observe that if c is a square in k, then $\operatorname{ord}_p(c)$ is an even number for all irreducibles p, and, conversely, if $\operatorname{ord}_p(c)$ is an even number for all irreducibles p, then c is a square times a unit in R.

Let (x, y) be a point on the elliptic curve $E(k)$ defined by a factored Weierstrass equation $y^2 = (x - r_1)(x - r_2)(x - r_3)$. By Theorem 1(4.1) we know that $(x, y) \in 2E(k)$ if and only if all $x - r_i$ are squares for $i = 1, 2, 3$. In particular, $\operatorname{ord}_p(x - r_i)$ is even for such points. For an arbitrary point on $E(k)$ we have the following 2-divisibility property.

(4.1) Proposition. *Let E be an elliptic curve defined by*

$$y^2 = (x - r_1)(x - r_2)(x - r_3),$$

where r_1, r_2, and r_3 are in R. If (x, y) is a point on $E(k)$, then $\operatorname{ord}_p(x - r_i)$ is even for all irreducibles p not dividing any elements $r_i - r_j$ for $i \neq j$.

Proof. Let p be an irreducible not dividing $r_i - r_j$, or, equivalently p such that $\operatorname{ord}_p(r_i - r_j) = 0$ for $i \neq j$. If $\operatorname{ord}_p(x - r_i) < 0$ for one i, then for all $j = 1, 2, 3$, we have $\operatorname{ord}_p(x) = \operatorname{ord}_p(x - r_i) = \operatorname{ord}_p(x - r_j)$ since each $\operatorname{ord}_p(r_j) \geq 0$. Thus it follows that

$$2 \operatorname{ord}_p(y) = \operatorname{ord}_p(y^2) = \operatorname{ord}_p((x - r_1)(x - r_2)(x - r_3)) = 3 \operatorname{ord}_p(x),$$

and hence $\operatorname{ord}_p(x) = \operatorname{ord}_p(x - r_j)$ is even for each j.

We cannot have both $\operatorname{ord}_p(x - r_i) > 0$ and $\operatorname{ord}_p(x - r_j) > 0$ for $i \neq j$ since $r_j - r_i = (x - r_i) - (x - r_j)$ and $\operatorname{ord}_p(r_i - r_j) = 0$. Hence, if $\operatorname{ord}_p(x - r_i) > 0$, for one root r_i, then we have the relation

$$2\,\mathrm{ord}_p(y) = \mathrm{ord}_p(y^2) = \mathrm{ord}_p(x - r_i),$$

and thus all $\mathrm{ord}_p(x - r_j)$ are even. This proves the proposition.

(4.2) Notations. Let E be an elliptic curve defined by the equation

$$y^2 = (x - r_i)(x - r_2)(x - r_3)$$

where each $r_i \in R$.

(a) Let $P(E)$ denote the set of all irreducibles p (defined up to units in R) such that p divides some $r_i - r_j$, where $i \neq j$. Then $P(E)$ is a finite set. Let $A(E)$ denote the subgroup of all cosets $a(k^*)^2$ in $k^*/(k^*)^2$ such that $\mathrm{ord}_p(a)$ is even for $p \notin P(E)$.

(b) Let $\theta_1, \theta_2, \theta_3 : E(k) \to A(E) \subset k^*/(k^*)^2$ be three functions given by the relations:

(1) $\theta_i(0) = 1$;
(2) $\theta_i((r_i, 0)) = (r_j - r_i)(r_k - r_i) \bmod (k^*)^2$ for $\{i, j, k\} = \{1, 2, 3\}$;
(3) $\theta_i((x, y)) = (x - r_i) \bmod (k^*)^2$ otherwise.

Observe that the set $P(E)$ is close to the set of irreducibles P where the curve E has bad reduction and where $k(p)$ has characteristic 2.

(4.3) Proposition. *With the notations of (4.2) the functions $\theta_i : E(k) \to A(E)$ are group homomorphisms and*

$$\ker(\theta_1) \cap \ker(\theta_2) \cap \ker(\theta_3) \subset 2E(k).$$

Proof. Consider three points $P_i = (x_i, y_i)$ on $E(k) \cap L$, where L is a line and show that $\theta_i(P_1), \theta_i(P_2), \theta_i(P_3) \in k^*/(k^*)^2$. The line is vertical if and only if some $P_j = 0$, and then by inspection $\theta_i(P_1)\theta_i(P_2)\theta_i(P_3) = 1$ in $k^*/(k^*)^2$. Otherwise the line is of the form $y = \lambda x + \beta$, and x_1, x_2, and x_3 are the roots of the equation $(\lambda x + \beta)^2 = f(x)$, where $f(x) = (x - r_1)(x - r_2)(x - r_3)$. Hence $x_1 - r_i, x_2 - r_i, x_3 - r_i$ are roots of the equation

$$(\lambda(x + r_i) + \beta)^2 = f(x + r_i) = x^3 + ax^2 + bx,$$

where $f(r_i) = 0$. Collecting terms, we obtain

$$0 = x^3 + (a - \lambda^2)x^2 + (b - 2\lambda(\lambda r_i + \beta))x - (\lambda r_i + \beta)^2,$$

and this leads to the following cases.

Case 1. All $P_j = (r_j, 0)$ for $j = 1, 2, 3$. Then we calculate

$$\begin{aligned}\theta_i(P_1)\theta_i(P_2)\theta_i(P_3) &= (x_1 - r_i)(x_2 - r_i)(x_3 - r_i) = -[-(r_i + \beta)^2] \\ &\equiv 1 \mod (k^*)^2.\end{aligned}$$

Case 2. Some $P_j = (r_i, 0)$, which we can take to be $P_i = (r_i, 0)$. Then 0, $x_2 - r_1$, $x_3 - r_1$ are the roots of the cubic which means $\beta = -\lambda r_i$, and the equation becomes $0 = x^3 + (a - \lambda^2)x^2 + bx$. Now we have

$$(x_2 - r_1)(x_3 - r_1) = b = (r_2 - r_1)(r_3 - r_1),$$

and, using this, we calculate for $\{i, j, k\} = \{1, 2, 3\}$

$$\begin{aligned}\theta_i(P_i)\theta_i(P_j)\theta_i(P_k) &= (r_j - r_i)(r_k - r_i)(x_j - r_i)(x_k - r_i)\\ &= (r_j - r_i)^2(r_k - r_i)^2\\ &\equiv 1 \quad \operatorname{mod}(k^*)^2.\end{aligned}$$

Hence each θ_i is a group morphism.

The last statement follows from 1(4.1), and this proves the proposition.

(4.4) Remark. The three morphisms of the previous proposition collect to define a group homomorphism

$$\theta = (\theta_1, \theta_2, \theta_3) : E(k) \to A(E)^3,$$

where $\ker(\theta) \subset 2E(k)$ by the previous proposition. Thus $E(k)/2E(k)$ is a subquotient of $A(E)^3$, and $(E(k) : 2E(k))$ is finite whenever $A(E)$ is finite.

(4.5) Remark. The group $A(E)$ is finite for any principal ring R where each $k(p)$ and $R^*/(R^*)^2$ are finite. For example, if $R = \mathbb{Z}$ and $k = \mathbb{Q}$, then the cardinality of $A(E)$ is 2^{m+1}, where m is the number of primes in $P(E)$. Hence we have the following assertion which we will generalize immediately.

(4.6) Assertion. Let $y^2 = (x - r_1)(x - r_2)(x - r_3)$ define an elliptic curve E over $\mathbb{Q}$ where each $r_i \in \mathbb{Z}$. Then the index $(E(\mathbb{Q}) : 2E(\mathbb{Q}))$ is finite.

To prove this assertion for $y^2 = f(x)$ where the cubic $f(x)$ does not necessarily factor over $\mathbb{Q}$, we will extend the ground field to k a number field, and factor the cubic in this field k. An extension of degree 6 will be sufficient. Thus the following more general result would be necessary even if our primary interest is elliptic curves over $\mathbb{Q}$.

(4.7) Theorem. *Let E be an elliptic curve over an algebraic number field k. Then the index $(E(k) : 2E(k))$ is finite.*

Proof. We can assume that E is defined by an equation $y^2 = f(x)$, where $f(x)$ is a cubic with three integral roots in k. We take for R in k the principal ideal ring equal to the ring of integers in k with a finite set of primes in k localized. By the finiteness of the ideal class group we could localize those primes which divide a finite set of representatives of the ideal class group. With a zero ideal class group the ring is principal.

The group of units R^* is finitely generated by the Dirichlet unit theorem, and thus, $R^*/(R^*)^2$ is finite. Now $A(E)$ is finite by (4.6) and we can apply (4.4) to obtain the proof of the theorem.

Exercises

1. Let $f(x)$ be a monic polynomial of degree n over k, and let (f) denote the ideal generated by f in $k[x]$. Show that $R_f = k[x]/(f)$ is an algebra of dimension n over k generated by one element. If $f(x)$ factors into distinct linear factors, then show that R_f is isomorphic to k^n as algebras over k. When $f(x)$ factors with repeated linear factors, describe a structure theorem for R_f as a direct sum of indecomposible algebras.
2. Let $y^2 = f(x) = (x - e_1)(x - e_2)(x - e_3)$ with distinct e_i, and let E be the elliptic curve over k defined by this equation. Show that we can define, using the notation of Exercise 1, a homomorphism $g : E(k) \to R_f^*/(R_f^*)^2$ by the relation for $2P = 0$,
$$g(P) = x(P) - e \mod (R_f^*)^2, \quad \text{where} \quad e \equiv x \mod (f).$$
Define $g(P)$ for $2P = 0$ in such a way that g is a group morphism. Show that $\ker(g) \subset 2E(k)$, and $\operatorname{im}(g) \subset R_{f,1}^*/(R_f^*)^2$, where $R_{f,1}^*$ consists of all $a \in R_f^*$ with $N(a) \subset (k^*)^2$. Finally relate these results with (4.3).

§5. Quasilinear and Quasiquadratic Maps

This is a preliminary section to our discussion of heights on projective space and on elliptic curves. From heights we derive the norm function on the group of rational points on an elliptic curve.

(5.1) Definition. For a set X a function $h : X \to \mathbb{R}$ is proper provided $h^{-1}([-c, +c])$ is finite for all $c \geq 0$.

In general a map between two locally compact spaces is proper if and only if the inverse image of compact subsets is compact. Thus a function $h : X \to \mathbb{R}$ is proper if and only if h is a proper map when X has the discrete topology and $\mathbb{R}$ the usual topology.

(5.2) Definition. Two functions $h, h' : X \to \mathbb{R}$ are equivalent, denoted $h \sim h'$, provided $h - h'$ is bounded, that is, there exists $a > 0$ such that $|h(x) - h'(x)| \leq a$ for all $x \in X$.

(5.3) Remark. Equivalence of real valued functions on a set is an equivalence relation. If two functions are equivalent, and if one function is proper, then the other function is proper.

Using this equivalence relation, we formulate quasilinearity, quasibilinearity, and quasiquadratic and then relate them to the usual algebraic concepts.

(5.4) Definition. Let A be an abelian group.

(1) A function $u : A \to \mathbb{R}$ is quasilinear provided $u(x + y)$ and $u(x) + u(y)$ are equivalent functions $A \times A \to \mathbb{R}$, i.e., $\Delta u(x, y) - u(x + y) - u(x) - u(y)$ is bounded on $A \times A$.

(2) A function $\beta : A \times A \to \mathbb{R}$ is quasibilinear provided the pairs of functions $\beta(x + x', y)$ and $\beta(x, y) + \beta(x', y)$ and $\beta(x, y + y')$ and $\beta(x, y) + \beta(x, y')$ are equivalent functions $A \times A \times A \to \mathbb{R}$.

(3) A function $q : A \to \mathbb{R}$ is quasiquadratic provided $\Delta q(x, y)$ is quasibilinear and $q(x) = q(-x)$. Moreover, q is positive provided $q(x) \geq 0$ for all $x \in A$. Again $\Delta q(x, y) = q(x + y) - q(x) - q(y)$.

We will construct a norm on a group from a quasiquadratic function. First we need a characterization of quasibilinearity.

(5.5) Lemma. *Let $f : A \to \mathbb{R}$ be a function on an abelian group A satisfying $f(x) = f(-x)$. Then $\Delta f : A \times A \to \mathbb{R}$ is quasibilinear (resp. bilinear) if and only if the weak (resp. ordinary) parallelogram law holds, i.e., for $(x, y) \in A \times A$*

$$f(x + y) + f(x - y) \sim 2f(x) + 2f(y)$$
$$(\textit{resp. } f(x + y) + f(x - y) = 2f(x) + 2f(y)).$$

Proof. We form the following real valued function on $A \times A \times A$:

$$g(x, y, z) = f(x + y + z) - f(x + y) - f(x + z) - f(y + z) + f(x) + f(y) + f(z).$$

The function Δf is quasibilinear (resp. bilinear) if and only if $g(x, y, z)$ is bounded (resp. zero). The parallelogram law is easily seen to be equivalent to the relation $g(x, y, -z) \sim -g(x, y, z)$ (resp. $g(x, y, -z) = -g(x, y, z)$). Since $f(x) = f(-x)$ implies that $g(-x, -y, -z) = g(x, y, z)$, and since $g(x, y, z)$ is symmetric in x, y, z, we see that the parallelogram law is equivalent to $2g(x, y, z) \sim 0$ (resp. $g(x, y, z) = 0$). This proves the lemma.

In connection with Definition (5.4)(3) we recall that a function $q : A \to \mathbb{R}$ is quadratic provided Δq is bilinear on $A \times A$.

(5.6) Proposition. *A function $q : A \to \mathbb{R}$ is quadratic if and only if $q(x) = q(-x)$, $q(2x) = 4q(x)$, and q is quasiquadratic.*

Proof. The direct implication is immediate. Conversely, suppose that $\Delta q(x, y)$ and $\Delta q(x, -y)$ are bounded. Then the sum of the two

$$s(x, y) = q(x + y) + q(x - y) - 2q(x) - 2q(y)$$

is bounded in the absolute value by a constant A, and since the hypothesis implies that $2^{-2n}s(2^n x, 2^n y) = s(x, y)$, we have zero in the limit as $n \to +\infty$, $|s(x, y)| = 2^{-2n}|s(2^n x, 2^n y)| \leq 2^{-2n}A \to 0$. Hence the function $s(x, y) = 0$ for all (x, y) and the parallelogram law holds. Thus by (5.5) the function $q(x)$ is quadratic, and this proves the proposition.

(5.7) Theorem. *If $q : A \to \mathbb{R}$ is a quadratic function satisfying $q(x) = q(-x)$, then $q^*(x) = \lim_{n\to\infty} 2^{-2n} q(2^n x)$ exists and the function $q^*(x)$ is quadratic.*

Proof. By (5.5) the function q is quasiquadratic if and only if the weak parallelogram law holds $q(x+y)+q(x-y) \sim 2q(x)+2q(y)$. Setting $x = y$ we have $q(2x) \sim 4q(x)$, that is, $|q(2x) - 4q(x)| \le A$ for positive constant A. Replacing x by $2^n x$, we obtain the inequality

$$|2^{-2(n+1)}q(2^{n+1}x) - 2^{-2n}q(2^n x)| = 2^{-2n}A$$

which leads to the following estimate for all n and p

$$|2^{-2(n+p)}q(2^{n+p}x) - 2^{-2n}q(2^n x)| = 2^{-2n} \cdot \frac{4A}{3}.$$

Thus the sequence defining $q^*(x)$ is Cauchy, and, therefore, it is convergent. The last assertion follows now from the previous proposition since the condition $q^*(2x) = 4q^*(x)$ follows from the defining limit and the conditions that $q^*(x) = q^*(-x)$ and $q^*(x)$ be quasiquadratic are preserved in the limit. This proves the theorem.

Another important situation where an equivalence can be modified to become an equality is contained in the next proposition.

(5.8) Proposition. *Let $f : X \to X$ be a function, let $d > 1$, and let $h : X \to \mathbb{R}$ be a function such that $hf \sim d \cdot h$. Then the limit $h_f(x) = \lim_{n\to\infty} d^{-n} \cdot h(f^n(x))$ exists uniformly on X and:*

(1) *$h_f(f(x)) = dh_f(x)$ and $|h_f(x) - h(x)| \le cd^2/(d-1)^2$ where*

$$|h(f(x)) - d \cdot h(x)| \le c$$

for all $x \in X$.

(2) *If h is proper, then h_f is proper.*

(3) *If $g : X \to X$ with $hg \sim d' \cdot h$ for a constant d', then $h_f g \sim d' \cdot h_f$ holds on X.*

Proof. From the inequality $|h(f(x) - d \cdot h(x)| \le c$, we have

$$|d^{-n-1}h(f^{n+1}(x) - d^{-n}h(f^n(x))| \le cd^{-n-1},$$

and, therefore, we have the estimate

$$|d^{-(n+p)}h(f^{n+p}(x)) - d^{-n}h(f^n(x))| \le \frac{c}{1-(1/d)}\left(\frac{1}{d}\right)^n.$$

Thus $d^{-n}h(f^n(x))$ is a Cauchy sequence, and it converges uniformly to h_f.

As for assertion (1), the relation $h_f(f(x)) = d \cdot h_f(x)$ follows since the two sides are rearrangements of the same limit. The second part results from the above estimate by letting $p \to \infty$ and taking $n = 0$. Assertion (2) follows from (1) and (5.3). As for (3), we calculate the difference

$$\begin{aligned} &h_f(g(x)) - d' \cdot h_f(x) \\ &\quad = [h_f(g(x)) - h(g(x))] + [h(g(x)) - d' \cdot h(x)] + [d' \cdot h(x) - h_f(x)], \end{aligned}$$

and see that each of the three terms is bounded. This proves the proposition.

§6. The General Notion of Height on Projective Space

A height on projective space is a proper, positive real valued function with a certain behavior when composed with algebraic maps of projective space onto itself.

(6.1) Definition. Let k be a field. A k-morphism $f : \mathbb{P}_m(k) \to \mathbb{P}_m(k)$ of degree d is a function of the form

$$f(y_0 : \cdots : y_m) = f_0(y_0, \ldots, y_m) : \cdots : f_m(y_0, \ldots, y_m),$$

where each $f_i(y_0, \ldots, y_m) \in k[y_o, \ldots, y_m]$ is homogeneous of degree d and not all are equal to zero at any $y_0 : \cdots : y_m \in \mathbb{P}_m(\bar{k})$. Here $\bar{k}$ denotes an algebraic closure of k.

(6.2) Definition. A height h on $\mathbb{P}_m(k)$ is a proper function $h : \mathbb{P}_m(k) \to \mathbb{R}$ such that for any k-morphism $f : \mathbb{P}_m(k) \to \mathbb{P}_m(k)$ of degree d the composite hf is equivalent to $d \cdot h$, that is, there is a constant c with $|h(f(y)) - d \cdot h(y)| \le c$ for all $y \in \mathbb{P}_m(k)$.

There is a canonical height function on projective space over a global field which is basic in many considerations in diophantine geometry. We will consider some special cases which are used to construct a norm on the rational points of an elliptic curve over the rational numbers.

(6.3) Notations. For a point in $\mathbb{P}_m(\mathbb{Q})$ we choose a $\mathbb{Z}$-reduced representative $y_0 : \cdots : y_m$ and denote by

$$H(y_0 : \cdots : y_m) = \max\{|y_0|, \ldots, |y_m|\} \quad \text{and} \quad h(P) = \log H(P),$$

where $P = y_0 : \cdots : y_m$. Recall that a $\mathbb{Z}$-reduced representative of P is integral and without common divisor, and so unique up to sign. This $h(P)$ is called the canonical height on $\mathbb{P}_m(\mathbb{Q})$.

In the one-dimensional case there is a bijection $u : \mathbb{Q} \cup \{+\infty\} \to \mathbb{P}_1(\mathbb{Q})$ defined by $u(m/n) = n : m$ and $u(\infty) = 0 : 1$. The composite hu restricted to $\mathbb{Q}$ is given by $h(m/n) = \log \max\{|m|, |n|\}$, where m/n is reduced to lowest terms.

(6.4) Remark. Since $h(y_0, \ldots, y_m) = \log(\max\{|y_0|, \ldots, |y_m|\})$ is a proper map of $\mathbb{R}^{m+1} - \{0\} \to \mathbb{R}$ and since $\mathbb{Z}^{m+1}$ is a discrete subset of $\mathbb{R}^{m+1}$, the map $h : \mathbb{P}_m(\mathbb{Q}) \to \mathbb{R}$ is proper where $\mathbb{P}_m(\mathbb{Q})$ has the discrete topology.

To obtain a norm on an elliptic curve from a height, we will make use of the following function $s : \mathbb{P}_1(k) \times \mathbb{P}_1(k) \to \mathbb{P}_2(k)$ given by $s(w : x, w' : x') = ww' : (xw' + x'w) : xx'$. It has the following elementary property related to heights.

(6.5) Proposition. *For* $s : \mathbb{P}_1(\mathbb{Q}) \times \mathbb{P}_1(\mathbb{Q}) \to \mathbb{P}_2(\mathbb{Q})$ *the difference* $h(s(w : x, w' : x')) - h(w : x) - h(w' : x')$ *has absolute value less than* $\log 2$ *on the product* $\mathbb{P}_1(\mathbb{Q}) \times \mathbb{P}_1(\mathbb{Q})$.

Proof. It suffices to show that $(1/2)M \cdot M' \le M'' \le 2M \cdot M'$, where $M'' = \max\{|ww'|, |wx' + xw'|, |xx'|\}$, $M = \max\{|w|, |x|\}$, and $M' = \max\{|w'|, |x'|\}$. If $|w| \le |x|, |w'| \le |x'|$, then $M \cdot M' = |x| \cdot |x'| \le M''$. If $|w| \le |x| = M$ and $|x'| \le |w'| = M'$, then for $M/2 \le |w|$ we have $MM'/2 \le |w||w'| \le M''$ and for $M'/2 \le |x'|$ we have $MM'/2 \le |x'||x| \le M''$. Finally, if $|x'| \le M'/2$ and $|w| \le M/2$, then $|wx'| \le (1/4)|w'x| = MM'/4$, and $(3/4)MM' \le |wx' + w'x| \le M''$. The other cases result from these by switching variables. The other inequality results from $|wx' + xw'| \le 2\max\{|wx'|, |xw'|\}$. The proposition itself follows now by applying the log to the above inequality.

Now we return to showing that the canonical height satisfies the basic property with respect to $\mathbb{Q}$-morphisms.

(6.6) Lemma. *Let φ be a form of degree d in $y_0, \ldots, y_m$. Then there exists a positive constant $c(\varphi)$ such for $\mathbb{Z}$-reduced $y \in \mathbb{P}_m(\mathbb{Q})$ we have $|\varphi(y)| \le c(\varphi)H(y)^d$.*

Proof. Decompose $\varphi(y) = \sum a_\alpha m_\alpha(y)$, where the index α counts off the monomials $m_\alpha(y)$ of degree d. Then clearly we have

$$|\varphi(y)| \le \sum_\alpha |a_\alpha| \cdot |m_\alpha(y)| \le \left(\sum_\alpha |a_\alpha|\right)(\max\{|y_0|, \ldots, |y_m|\})^d = c(\varphi)H(y)^d,$$

where $c(\varphi) = \sum_\alpha |a_\alpha|$. This proves the lemma.

This lemma will give an upper estimate for $H(f(y))$ or $(h(f(y))$, but for a lower estimate we need the following assertion which is a consequence of the Hilbert Nullstellensatz.

(6.7) Assertion. A sequence of forms $(f_0, \ldots, f_m)$ of degree d in $\mathbb{Z}[y_0, \ldots, y_m]$ defines a $\mathbb{Q}$-morphism, i.e., they have no common zero in $\mathbb{P}_m(\bar{\mathbb{Q}})$, if and only if there exists a positive integer s, and integer b, and polynomials $g_{ij}(y) \in \mathbb{Z}[y_0, \ldots, y_m]$ such that

$$\sum_{0 \le j \le m} g_{ij} f_j = b y_i^{s+d} \qquad \text{for all } i = 0, \ldots, m.$$

The Nullstellensatz says that $f_0, \ldots, f_m$ have no common zeros in $\mathbb{P}_m(\bar{\mathbb{Q}})$ if and only if the ideal I generated by $f_0, \ldots, f_m$ contains a power $(y_0, \ldots, y_m)^s$ of the ideal generated by $y_0, \ldots, y_m$, that is, if and only if there exist forms $g_{ij}(y)$ over $\mathbb{Q}$ such that

$$\sum_{0 \le j \le m} g_{ij} f_j = y_i^{s+d} \qquad \text{for all } i = 0, \ldots, m.$$

The integer b is a common denominator of the $g_{ij}(y)$.

For $m = 1$ the condition on $(f_0(y_0, y_1), f_1(y_0, y_1))$ to be a $\mathbb{Q}$-morphism is that f_0 and f_1 be relatively prime. In this case (6.7) can be verified directly without the Nullstellensatz.

(6.8) Theorem. *For the canonical height h on $\mathbb{P}_m(\mathbb{Q})$ and a $\mathbb{Q}$-morphism $f : \mathbb{P}_m(\mathbb{Q}) \to \mathbb{P}_m(\mathbb{Q})$ of degree d the difference*

$$h(f(y)) - d \cdot h(y)$$

is bounded on $\mathbb{P}_m(\mathbb{Q})$. In particular, h is a height on $\mathbb{P}_m(\mathbb{Q})$.

Proof. Using (6.6), we have an upper estimate for $H(f(y))$ where

$$H(f(y)) = \max_i |f_i(y)| \le \left(\max_i c(f_i)\right) H(y)^d = c_2 H(y)^d.$$

To obtain a lower estimate, we use (6.7)

$$\begin{aligned} |b| \cdot |y_i|^{s+d} &= \left(\max_{i,j} c(g_{ij})\right) H(y)^s \sum_j |f_j(y)| \\ &\le \left(\max_{i,j} c(g_{ij})\right) (m+1) H(y)^s \left(\max_j |f_j(y)|\right). \end{aligned}$$

Since any common factor among the $f_j(y)$ divides b, by (6.7), it follows that $\max_j |f_j(y)| = |b| H(f(y))$. Using this inequality and taking the maximum over i of the above inequality, we have

$$|b| \cdot H(y)^{s+d} \le \left(\max_{i,j} c(g_{ij})\right) (m+1) H(y)^s |b| H(f(y)).$$

After cancellation, we have for some $c_1 > 0$ the inequality

$$c_1 \cdot H(y)^d \le H(f(y)).$$

Thus for all y in $\mathbb{P}_m(\mathbb{Q})$ it follows that

$$c_1 \cdot H(y)^d \le H(f(y)) \le c_2 \cdot H(y)^d,$$

and after taking the logarithm of both sides, we see that $\log(H(f(y))/H(y)^d) = h(f(y)) - d \cdot h(y)$ is bounded on $\mathbb{P}_m(\mathbb{Q})$. This proves the theorem.

§7. The Canonical Height and Norm on an Elliptic Curve

Using the canonical height h on $\mathbb{P}_m(\mathbb{Q})$ for k a number field which is defined in §8 for $k \ne \mathbb{Q}$, we can define a height h_E on $E(k)$ the group of k-valued points on an elliptic curve E over k. First, we need another lemma about multiplication by 2 on an elliptic curve.

(7.1) Lemma. *Let E be an elliptic curve defined by $y^2 = f(x) = x^3 + ax^2 + bx + c$ over a field k. Define the function $q : E(k) \to \mathbb{P}_1(k)$ defined by $q(x, y) = (1, x)$ and*

$q(0) = (0, 1)$. *Then there is a k-morphism $g : \mathbb{P}_1(k) \to \mathbb{P}_1(k)$ of degree 4 such that the following diagram is commutative:*

$$\begin{array}{ccc} E(k) & \xrightarrow{2} & E(k) \\ q\downarrow & & \downarrow q \\ \mathbb{P}_1(k) & \xrightarrow{g} & \mathbb{P}_1(k). \end{array}$$

Proof. The relation between (x, y) and $2(x, y) = (x', y')$ is given by considering the tangent line $y = \lambda x + \beta$ to E at (x, y). This line goes through $(x', -y')$ and as in 1(1.4) we have

$$2x + x' = \lambda^2 - a \quad \text{and} \quad \lambda = \frac{f'(x)}{2y}.$$

Using the relation $y^2 = f(x)$, we obtain

$$\lambda^2 = \frac{(3x^2 + 2ax + b)^2}{4(x^3 + ax^2 + bx + c)},$$

and

$$x' = \lambda^2 - a - 2x = \frac{x^4 - 2bx^2 - 8cx + (b^2 - 4ac)}{4x^3 + 4ax^2 + 4bx + 4c}.$$

Thus $g(w, x) = (g_0(w, x), g_1(w, x))$ is given by the forms

$$g_0(w, x) = 4wx^3 + 4aw^2x^2 + 4bw^3x + 4cw^4,$$
$$g_1(w, x) = x^4 - 2bw^2x^2 - 8cw^3x + (b^2 - 4ac)w^4.$$

This proves the lemma.

Observe that for the map $q : E(k) \to \mathbb{P}_1(k)$ the inverse image $q^{-1}(1, x)$ is empty when $y^2 = f(x)$ has no solution in k and $q^{-1}(1, x) = \{(x, \pm y)\}$ when $\pm y$ are the solutions of $y^2 = f(x)$. Now we can easily describe the height function h_E on an elliptic curve over a number field in terms of the canonical height h on $\mathbb{P}_1(k)$.

(7.2) Theorem. *Let E be an elliptic curve over a number field k in Weierstrass form $y^2 = f(x) = x^3 + bx + c$. Then there is a unique function $h_E : E(k) \to \mathbb{R}$ such that:*

(1) $h_E(P) - (1/2)h(x(P))$ is bounded, where $x(P) = q(P)$ is the x-coordinate of P and h is the canonical height on $\mathbb{P}_1(k)$, and
(2) $h_E(2P) = 4h_E(P)$ and $h_E(P) = h_E(-P)$.

Moreover, h_E is proper, positive, and quadratic.

Proof. Since we must have $h_E(P) = h_E(-P)$, we consider the proper map $(1/2)hq = h' : E(k) \to \mathbb{R}$. Then $h'(P) = (1/2)h(q(P))$ satisfies $h'(-P) = h'(P)$. Now form the limit $h_E(P) = \lim_{n\to\infty} 2^{-2n}h'(2^n P)$ which exists by (5.8) and (7.1). Then by (7.1), it follows that h_E satisfies (1) and (2). Moreover, h_E is proper ince h' is proper and it is positive.

To prove that h_E is quadratic on $E(k)$, we have only to check the weak parallelogram law by (5.5) and (5.6) in view of (2). For this we use the following commutative diagram:

$$\begin{array}{ccccccc} & & E\times E & \xrightarrow{u} & E\times E & & \\ & \overset{q\times q}{\swarrow} & \Big\downarrow{\scriptstyle\theta} & & \Big\downarrow{\scriptstyle\theta} & \overset{q\times q}{\searrow} & \\ \mathbb{P}_1\times\mathbb{P}_1 & & & & & & \mathbb{P}_1\times\mathbb{P}_1 \\ & \underset{s}{\searrow} & & & & \underset{s}{\swarrow} & \\ & & \mathbb{P}_2 & \xrightarrow{f} & \mathbb{P}_2 & & \end{array}$$

where $s((w,x),(w',x')) = (ww', wx' + w'x, xx')$, $u(P,Q) \doteq (P+Q, P-Q)$, and $f(\alpha,\beta,\gamma) = (f_0(\alpha,\beta,\gamma), f_1(\alpha,\beta,\gamma), f_2(\alpha,\beta,\gamma))$ such that

$$f_0(\alpha,\beta,\gamma) = \beta^2 - 4\alpha\gamma, \quad f_1(\alpha,\beta,\gamma) = 2\beta(b\alpha+\gamma) + 4c\alpha^2,$$
$$f_2(\alpha,\beta,\gamma) = (\gamma - b\alpha)^2 - 4c\alpha\beta.$$

To check the commutativity of the above diagram we consider $P = 1 : x : y$ and $Q = 1 : x' : y'$ with x and x' unequal. Then we have

$$\theta(P,Q) = s(q\times q)(P,Q) = s(1 : x, 1 : x') = 1 : (x+x') : xx'.$$

From 1(1.4) we calculate

$$P + Q = (x-x')^2 : (y-y')^2 - (x+x')(x-x')^2 : *,$$
$$P - Q = (x-x')^2 : (y+y')^2 - (x+x')(x-x')^2 : *,$$

and hence the composite with the upper arrow in the diagram takes the following form:

$$\begin{aligned}\theta u(P,Q) &= (x-x')^4 : 2(x-x')^2(y^2+y'^2 - (x+x')(x-x')^2) : \\ &\quad (y^2-y'^2)^2 + (x-x')^4(x+x')^2 - 2(x-x')^2(x+x')(y^2+y'^2) \\ &= (x-x')^2 : 2(xx'^2 + x^2x' + b(x+x') + 2c) : \\ &\quad ((xx'-b)^2 - 4c(x+x')) \\ &= f_0(1, x+x', xx') : f_1(1, x+x', xx') : f_2(1, x+x', xx'),\end{aligned}$$

where as homogeneous forms of degree 2 in three variables $f_0(\alpha,\beta,\gamma) = \beta^2 - 4\alpha\gamma$, $f_1(\alpha,\beta,\gamma) = 2\beta\gamma + 2b\alpha\beta + 4c\alpha^2$, and $f_2(\alpha,\beta,\gamma) = (\gamma - b\alpha)^2 - 4c\alpha\beta$.

From this commutative diagram we calculate

$$h_E(P+Q)+h_E(P-Q)=h(\theta(P+Q,P-Q))=h(\theta(u(P,Q)))$$
$$=h(f(\theta(P,Q)))\sim 2h(\theta(p,Q)).$$

The last expression can be studied using (6.5), and we have

$$2h(\theta(P,Q))=2h(s(q(P),q(Q)))\sim 2h(q(P))+2h(q(Q))$$
$$=2h_E(P)+2h_E(Q).$$

Now we apply (5.5) to see that h_E is quasiquadratic and further (5.6) to see that h_E is a quadratic function.

Finally, the function h_E is unique since two possible functions would be quadratic by the above argument and equivalent to h', and hence they are equivalent to each other. If two quadratic functions differ by a bounded function, then they are equal. This proves the theorem.

(7.3) Corollary. *With the hypothesis and notations of the previous theorem, the function* $|P|=\sqrt{h_E(P)}$ *is a norm on* $E(k)$.

Now we assume Theorems (3.2) and (7.2) to deduce the following theorem of Mordell. This theorem was generalized to number fields by A. Weil and is one of the main results of this book.

(7.4) Theorem (Mordell–Weil). *Let* E *be an elliptic curve over a number field* k*. Then the group* $E(k)$ *is finitely generated.*

Proof of (7.4) *for* $k=\mathbb{Q}$ *where* $E=E[a,b]$. By (3.2) the index $(E(\mathbb{Q}):2E(\mathbb{Q}))$ is finite. By (7.2) and (7.3) the function $|P|=\sqrt{h_E(P)}$ is a norm on $E(\mathbb{Q})$. The criterion (1.3) applies to show that $(E(\mathbb{Q})$ is a finitely generated abeian group. This proves the theorem in this case.

Sketch of proof for k *any number field and any* E *over* k. In the general case we extend k to k' so that the equation of E has the form $y^2=(x-\alpha)(x-\beta)(x-\gamma)$. By (4.7) the index $E(k'):2E(k'))$ is finite. Again by (7.3) the function $|P|=\sqrt{h_E(P)}$ is a norm on $E(k')$; here we assume there is a height function on projective space over k, and (6.5) holds in order to carry out the proof of (7.2) for a number field. Finally, $E(k)$ is then a subgroup of a finitely generated group $E(k')$, and thus, it is finitely generated.

§8. The Canonical Height on Projective Spaces over Global Fields

In order to formulate the notion of height in the case of number fields and more generally global fields, we reformulate the definition over the rational numbers without using $\mathbb{Z}$-reduced coordinates of a point but using instead all the valuations or absolute values on $\mathbb{Q}$. Up to equivalence, each absolute value on $\mathbb{Q}$ is either of the form:

(1) (Archimedean case) $|x| = |x|_\infty$, the ordinary absolute value, or
(2) (non-Archimedean case) $|a/b|_p = (1/p)^{\mathrm{ord}_p(a)-\mathrm{ord}_p(b)}$, the p-adic absolute value where a and b are integers and p is a prime number in $\mathbb{Z}$.

Let $V(\mathbb{Q})$ denote the totality of these absolute values on $\mathbb{Q}$. The product formula for $\mathbb{Q}$ is

$$\prod_{v\in V} |x|_v = \begin{cases} 1 & \text{for} \quad x \neq 0, \\ 0 & \text{for} \quad x = 0, \end{cases}$$

or

$$\sum_{v\in V} \log |x|_v = 0 \quad \text{for} \quad x \neq 0.$$

(8.1) Definition. A number field k is a finite extension of $\mathbb{Q}$, a function field in one variable over a field F is a finite separable extension of $F(t)$, and a global field is either a number field or a function field in one variable over $\mathbb{F}_q$, a finite field.

A global field k is a number field if and only if it has characteristic zero and is a function field if and only if it has positive characteristic.

(8.2) Remark. Global fields have several things in common which can be used to axiomatically characterize them. The most important feature is a family $V(k)$ of absolute values $|\ |_v$ such that the product formula holds

$$\prod_{v\in V(k)} |x|_v = \begin{cases} 1 & \text{if} \quad x \neq 0, \\ 0 & \text{if} \quad x = 0. \end{cases}$$

It is this product formula generalizing the product formula for $\mathbb{Q}$ which leads to a height function on $\mathbb{P}_m(k)$ for a global field k.

(8.3) Notation. For $y = (y_0, \ldots, y_m) \in k^{m+1} - \{0\}$, we introduce

$$h^*(y_0, \ldots, y_m) = \sum_{v\in V(k)} \max\{\log |y_0|_v, \ldots, \log |y_m|_v\}$$

This is a finite sum.

(8.4) Lemma. *For a nonzero a in k and $(y_0, \ldots, y_m) \in k^{m+1} - 0$, we have $h^*(ay_0, \ldots, ay_m) = h^*(y_0, \ldots, y_m)$.*

Proof. We calculate

$$\begin{aligned}
&\sum_{v\in V} \max\{\log |ay_o|_v, \ldots, \log |ay_m|_v\} \\
&\quad = \sum_{v\in V} [\log |a|_v + \max\{\log |y_o|_v, \ldots, \log |y_m|_v\}] \\
&\quad = \sum_{v\in V} \max\{\log |y_o|_v, \ldots, \log |y_m|_v\} = h^*(y_0, \ldots, y_m).
\end{aligned}$$

This lemma says that h^* induces a function, also denoted h^*, on the projective space $\mathbb{P}_m(k)$.

(8.5) Lemma. *For* $(y_0, \ldots, y_m) \in \mathbb{Z}^{m+1} - \{0\}$ *without common factors, i.e., a* $\mathbb{Z}$-*reduced representative of a point in* $\mathbb{P}_m(\mathbb{Q})$, *the height*

$$h(y_0, \ldots, y_m) = h^*(y_0, \ldots, y_m).$$

Proof. Since each y_j is in $\mathbb{Z}$, it follows that a p-adic valuation satisfies $|y_j|_p \leq 1$ or $\log|y_j|_p \leq 0$. Since at least one y_j is not divisible by p, this means that $|y_j|_p < 1$ or $\log|y_j|_p < 0$, and, hence, we obtain $\max\{\log|y_o|_p, \ldots, \log|y_m|_p\} = 0$. We conclude that the p-adic valuations in the sum defining h^* do not contribute anything, and, therefore,

$$\begin{aligned} h^*(y_0, \ldots, y_m) &= \sum_{p \in V(\mathbb{Q})} \max\{\log|y_o|_p, \ldots, \log|y_m|_p\} \\ &= \max\{\log|y_0|_\infty, \ldots, \log|y_m|_\infty\}. \end{aligned}$$

This proves the lemma.

In view of the above two lemmas the next definition is a natural extension of (6.3).

(8.6) Definition. For a global field k the canonical height h on $\mathbb{P}_m(k)$ is defined by

$$h(P) = \sum_{v \in V(k)} \max\{\log|y_o|_v, \ldots, \log|y_m|_v\},$$

where $P = (y_0, \ldots, y_m)$.

Since $\mathbb{P}_1(k) = k \cup \{(0, 1)\}$ where $\infty = (0, 1)$ and a in k is identified with (l, a), the canonical height on k is the function $h(a) = \sum_{v \in V(k)} \max\{0, \log|a|_v\}$. The proof that h is a height in the technical sense of (6.2) follows the lines of (6.8).

7

Galois Cohomology and Isomorphism Classification of Elliptic Curves over Arbitrary Fields

In Chapter 3 we saw that $j(E)$ is an isomorphism invariant for elliptic curves defined over algebraically closed fields. In this chapter we describe all elliptic curves over a given field k which becomes isomorphic over k_s the separable algebraic closure of k, up to k isomorphism. This is done using the Galois group of k_s over k and its action on the automorphism group of the elliptic curve over k_s. The answer is given in terms of a certain first Galois cohomology set closely related to quadratic extensions of the field k.

We introduce basic Galois cohomology which is used to analyse how a group G acts on a group E. The zeroth cohomology group $H^0(G, E)$ is equal to E^G, the subgroup of E consisting of all elements of E left fixed by the action of G. The first cohomology set $H^1(G, E)$ is a set with base point, i.e., $H^1(G, E)$ is a pointed set, and it has an abelian group structure when E is an abelian group. For an exact sequence of G-groups there is a six-term exact sequence involving the three zeroth cohomology groups and the three first cohomology sets. This six-term sequence is used for most of the elementary calculations of Galois cohomology.

§1. Galois Theory: Theorems of Dedekind and Artin

We give a short resume of the basic properties of Galois groups and extensions which are used in the study of elliptic curves. For details the reader should consult the book of E. Artin, *Galois Theory,* Notre Dame Mathematical Lectures.

If K is an extension field of a field F, then K is a vector space over F from the multiplication on K, and we denote the dimension $\dim_F K$ of K over F by $[K : F]$. For a group G of automorphisms of a field K, we denote by $\mathrm{Fix}(G)$ the fixed elements of K under G, i.e., the set of $a \in K$ with $s(a) = a$ for all $s \in G$. Then $\mathrm{Fix}(G)$ is a subfield of K.

(1.1) Definition. A field extension K over F is Galois provided there is a group of automorphisms G of K such that $F = \mathrm{Fix}(G)$. The group G is denoted by $\mathrm{Gal}(K/F)$ and is called the Galois group of the extension K over F.

The following result is the basic property of Galois extensions, and it leads immediately to the Galois correspondence between subfields of K containing F and subgroups of $\mathrm{Gal}(K/F)$.

(1.2) Theorem. (1) (Dedekind) *Let $u_1, \ldots, u_n$ be field morphisms $K \to L$ over F, and suppose that there is an L-linear relation $a_1u_1 + \cdots + a_nu_n = 0$ where the u_1 are distinct. Then all coefficients $a_i = 0$ for $i = 1, \ldots, n$.*

(2) (Artin) *If $\Gamma \subset \mathrm{Aut}(K)$ is a finite subsemigroup, and if F is the set of all $x \in K$ with $u(x) = x$ for all $u \in \Gamma$, then $n = \#\Gamma = [K : F]$, K is separable over F, and Γ is a subgroup of* $\mathrm{Aut}(K)$.

Proof. (1) Choose an element $y \in K$ with $u_{n-1}(y) \neq u_n(y)$. Evaluate the L-linear relation at x times $u_n(y)$ and at xy, and we obtain

$$0 = a_1u_1(x)u_n(y) + \cdots + a_nu_n(x)u_n(y)$$

and also

$$\begin{aligned} 0 &= a_1u_1(xy) + \cdots a_nu_n(xy) \\ &= a_1u_1(x)u_1(y) + \cdots + a_nu_n(x)u_n(y). \end{aligned}$$

Subtracting the two resulting formulas, we eliminate one term

$$0 = a_1u_1(x)\left[u_1(y) - u_n(y)\right] + \cdots + a_{n-1}u_{n-1}(x)\left[u_{n-1}(y) - u_n(y)\right]$$

where not all coefficients are zero. With this algebraic operation we reduce the number of nonzero terms in an L-linear relation by at least one, and inductively this means that $u_1, \ldots, u_n$ is L-linearly independent.

As an application of (1), the sum $u_1 + \cdots + u_n : K \to L$ is nonzero. If K is separable over F, that is, for some $L \supset K$ the set E of field morphisms $K \to L$ over F has the property that $x \in K$ satisfies $u'(x) = u''(x)$ for all $u', u'' \in E$ implies $x \in F$. Then the trace

$$tr_{K/F}(x) = \sum_{u \in E} u(x)$$

is defined $tr_{K/F} : K \to F$, and it is a nonzero F-linear form.

(2) Show that any set $\alpha_0, \ldots, \alpha_n \in K$ is F-linearly dependent. Consider n equations in $x_0, \ldots, x_n$ of the form $\sum_{j=0}^{n} x_j u^{-1}(\alpha_j) = 0$ for $u \in \Gamma$. Thus there exists elements $c_0, \ldots, c_n \in K$ with $Tr_{K/F}(c_0) \neq 0$ and

$$\sum_{j=0}^{n} c_j u^{-1}(\alpha_j) = 0 \qquad \text{for } u \in \Gamma.$$

Applying u to the relation, we have $\sum_{j=0}^{n} u(c_j)\alpha_j = 0$ for $u \in \Gamma$, and summing over $u \in \Gamma$, we have $\sum_{j=0}^{n} Tr(c_j)\alpha_j = 0$ over F. Hence $[K : F] \leq n$, and therefore we have $\#\Gamma = n = [K : F]$ using Dedekind's theorem.

(1.3) Remark. Let K be a finite Galois extension of F with Galois group $\mathrm{Gal}(K/F)$. The function which assigns to a subgroup H of $\mathrm{Gal}(K/F)$ the subfield $\mathrm{Fix}(H)$ of K containing F is a bijection from the set of subgroups of $\mathrm{Gal}(K/F)$ onto the set of subfields L of K which contain F. The inverse function is the function which assigns to such a subfield L of K the subgroup G_L of all s in $\mathrm{Gal}(K/F)$ which restrict to the identity on L. These functions are inclusion reversing, and K over L is a Galois extension with $G_L = \mathrm{Gal}(K/L)$. Further, the extension L over F is a Galois extension if and only if $\mathrm{Gal}(K/L)$ is a normal subgroup of $\mathrm{Gal}(K/F)$, and in this case the quotient group $\mathrm{Gal}(K/F)/\mathrm{Gal}(K/L)$ is isomorphic to the Galois group $\mathrm{Gal}(L/F)$ by restriction of automorphism from K to L.

Now we are led to the basic problem of determinig all Galois extensions K of a given field F.

An element x of K is algebraic over a subfield F provided $P(x) = 0$ for some nonzero polynomial $P(X)$ in $F[X]$. Associated to x over F is a (unique) minimal polynomial which has minimal degree and leading coefficient one among the polynomials $P \neq 0$ with $P(x) = 0$. An extension K over F is algebraic provided every x in K is algebraic. Every finite extension K over F is algebraic.

(1.4) Definition. An extension K over F is normal provided it is algebraic and for every x in K the minimal polynomial of x over F has all its roots in K.

A finite extension K over F is normal if and only if K is generated by the roots of a polynomial with coefficients in F. Every Galois extension K of F is seen to be normal, because for x in K the conjugates $s(x)$, where $s \in \mathrm{Gal}(K/F)$, are finite in number and the minimal polynomial for x divided $Q(X) = \prod(X - s(x))$. Note $Q(X)$ has coefficients in F since they are invariant under $\mathrm{Gal}(K/F)$.

(1.5) Definitions. A polynomial $P(X)$ over a field F is called separable provided its irreducible factors do not have repeated roots, An element x in an extension field K over F is separable provided it is the root of a separable polynomial over F. An extension K over F is separable provided every element of K is separable over F.

With our definition a separable extension is an algebraic extension. In characteristic zero every algebraic extension is separable.

(1.6) Remark. An algebraic field extension is Galois if and only if it is normal and separable. Over a field F, which is of characteristic zero or is finite, an algebraic extension is Galois if and only if it is normal.

(1.7) Definition. A field F is algebraically closed provided any algebraic element x in an extension K of F is in F. An algebraic closure $\overline{F}$ of F is an algebraic extension F which is algebraically closed.

The algebraic closure $\overline{F}$ over F is a normal extension which is a Galois extension if F is either a field of characteristic zero or a finite field. The extension is usually infinite but it is the union of all finite subextensions, and indeed all finite normal subextensions, and is also the direct (inductive) limit

$$\overline{F} = \varinjlim_{K/F \text{ finite normal}} K.$$

The union of all finite separable subextensions is a subfield F_s of $\overline{F}$ which is a Galois extension of F contained in $\overline{F}$. This is also the direct limit

$$F_s = \varinjlim_{K/F \text{ finite Galois}} K.$$

The Galois group $\mathrm{Gal}(F_s/F)$ maps to each $\mathrm{Gal}(K/F)$, where K is a finite Galois subextension of F_s, and we have an isomorphism onto the projective limit of finite groups

$$\mathrm{Gal}(F_s/F) \to \varprojlim_{K/F \text{ finite Galois}} \mathrm{Gal}(K/F).$$

This projective limit has the limit topology in which it is compact and this compact topology is transferred to $\mathrm{Gal}(F_s/F)$ making the Galois group a compact topological group.

(1.8) Remark. The Galois correspondence of (1.3) becomes the assertion that the function which assigns to a closed subgroup H of $\mathrm{Gal}(F_s/F)$ the subfield $\mathrm{Fix}(H)$ of F_s containing F is a bijection from the set of closed subgroups of $\mathrm{Gal}(F_s/F)$ onto the set of subfields of F_s containing F.

(1.9) Example. For $F = \mathbb{F}_q$ the finite field of q elements there exists exactly one extension (up to isomorphism) of degree n over $\mathbb{F}_q$, namely $\mathbb{F}_{q^n}$. It is a Galois extension with cyclic Galois group $\mathbb{Z}/n$ with 1 in $\mathbb{Z}/n$ corresponding to the Frobenius automorphism $a \mapsto q^q$. The Galois group $\mathrm{Gal}(\overline{F}_q/\mathbb{F}_q)$ is $\hat{\mathbb{Z}}$ topologically generated by the Frobenius automorphism, and $\hat{\mathbb{Z}}$ is $\varprojlim_n \mathbb{Z}/n$, the completion of $\mathbb{Z}$ for the topology given by subgroups of finite index.

§2. Group Actions on Sets and Groups

We consider three types of objects: (1) sets, (2) pointed sets, i.e., sets with base point $*$, or (3) groups. For such an object E we have the group $\mathrm{Aut}(E)$ of automorphism which:

(1) for a set E consists of all bijections $E \to E$, i.e., permutations of E;
(2) for a pointed set E consists of all bijections preserving the base point; and
(3) for a group E consists of all group automorphisms.

(2.1) Definition. Let G be a group, and let E be an object, i.e., a set, a pointed set, or a group. A left group action of G on E is a homomorphism $G \to \mathrm{Aut}(E)$.

For $s \in G$ we denote the action of s on $x \in E$ by sx. Then the group homomorphism condition becomes

$$1x = x \quad \text{and} \quad s(tx) = (st)x \qquad \text{for all } s, t \in G,\ x \in E.$$

For a pointed set we also require $sx_0 = x_0$, where x_0 is the base point. When E is an additive group, we have $s(x + y) = sx + sy$, and when E is a multiplicative group, we write frequently ${}^s x$ instead of sx and the automorphism condition has the form ${}^s(xy) = {}^s x {}^s y$.

A right G action on E is an antihomomorphism $G \rightarrow \operatorname{Aut}(E)$, and the action of $s \in G$ on $x \in E$ is written xs or x^s. The antihomomorphism condition is $x1 = x$ and $(xs)t = x(st)$ so that $sx = xs^{-1}$ is a left action associated with the right action.

Let G be a group. We can speak of left and right G-sets, pointed G-sets, and G-groups, i.e., the corresponding object a set, a pointed set, or group together with the corresponding G action.

(2.2) Definition. A morphism $f : E \rightarrow E'$ of G-objects is a morphism of objects $E \rightarrow E'$ together with the G-equivariance property $f(sx) = sf(x)$ (or $f(xs) = f(x)s$) for all $s \in G, x \in E$.

(2.3) Example. Let D be a G-subgroup of a G-group E, so $sD = D$ for all $s \in G$. Then D has by restriction a G-group structure and the inclusion $D \rightarrow E$ is a morphism of G-groups. Now form the quotient pointed sets $D\backslash E$ of all right D-cosets Dx and E/D of all left D-cosets xD, where $x \in E$. The base point is the identity coset D. The G action on E induces a G action on the quotient pointed sets $D\backslash E$ and E/D such that the projections $E \rightarrow D\backslash E$ and $E \rightarrow E/D$ are morphisms of pointed G-sets. When D is a normal subgroup of E, then $E/D = D\backslash E$ is a G-group, and $E \rightarrow E/D$ is a morphism of G-groups. We write

$$1 \rightarrow D \rightarrow E \rightarrow D\backslash E \rightarrow 1 \quad \text{and} \quad 1 \rightarrow D \rightarrow E \rightarrow E/D \rightarrow 1$$

as short exact sequences of pointed sets, where $D \rightarrow E$ is injective and $\operatorname{im}(D \rightarrow E)$ is the kernel of the surjection $E \rightarrow D\backslash E$ or $E \rightarrow E/D$.

(2.4) Definition. Let E be a G-object. The subobject of fixed elements E^G consists of all $x \in E$ with $sx = x$ (or $xs = x$) for all $s \in G$.

If E is a G-set (resp. pointed G-set, G-group), then E^G is a subset (resp. pointed subset, subgroup) of E. We denote E^G also by $H^0(G, E)$ meaning the zeroth cohomology object of G with values in E. With this notation we anticipate the first cohomology set $H^1(G, E)$, where E is a G-group.

(2.5) Examples. Let k'/k be a Galois extension with Galois group $G = \operatorname{Gal}(k'/k)$. Then the additive group k' is a G-group and $(k')^G = H^0(G, k') = k$. The matrix groups $\mathrm{GL}_n(k')$ and $\mathrm{SL}_n(k')$ are also G-groups, and the subgroups fixed by the action are $\mathrm{GL}_n(k')^G = \mathrm{GL}_n(k)$ and $\mathrm{SL}_n(k')^G = \mathrm{SL}_n(k)$. Also det: $\mathrm{GL}_n(k') \rightarrow \mathrm{GL}_1(k')$ is a morphism of G-groups. Finally, if E is an elliptic curve defined over k, then $E(k')$ is a G-group with $E(k')^G = E(k)$. This can be seen directly with affine coordinates or from the fact that $\mathbb{P}_n(k')^G = \mathbb{P}_n(k)$ which is also checked by looking on each affine piece defined by $y_j \neq 0$.

(2.6) Proposition. *Let G be a group, and let E be a G-group with a G-subgroup A. Applying the fixed element functor to the exact sequence $1 \to A \to E \to E/A \to 1$, we obtain an exact sequence of pointed sets*

$$1 \to A^G \to E^G \to (E/A)^G.$$

Proof. Since $A^G \to E^G$ is the restriction of a monomorphism $A \to E$, it is a monomorphism. Next, we calculate

$$\begin{aligned}\ker\left(E^G \to (E/A)^G\right) &= \ker(E \to E/A) \cap E^G \\ &= \operatorname{im}(A \to E) \cap E^G = \operatorname{im}\left(A^G \to E^G\right).\end{aligned}$$

This proves the proposition.

The above simple proposition brings up the question of the surjectivity of $E^G \to (E/A)^G$. If we try to establish surjectivity, we would consider a coset xA fixed by G, that is, ${}^s(xA) = xA$ for all $s \in G$. Then we must determine whether or not xA can be represented by $x \in E$ with ${}^sx = x$. All we know is that ${}^sx \in xA$ for all s which means that ${}^sx = xa$ for some $a \in A$. We can view any coset X in E/A as a right A-set with the multiplication $E \times E \to E$ on E inducing the action $X \times A \to X$ which is G-equivariant, that is, on which ${}^s(xa) = {}^sx{}^sa$. Observe that X has the property that for two points $x, x' \in X$ there exists a unique $a \in A$ with $xa = x'$. This property says that X is a right A-principal homogeneous G-set, and these objects are considered in the next section. Returning to the question of the exact sequence (2.6), we will study the question of when the right A-set X has a G-invariant point in this context.

§3. Principal Homogeneous G-Sets and the First Cohomology Set $H^1(G, A)$

In this section G denotes a group. Following our analysis at the end of the previous section on cosets relative to a G-subgroup A, we make the following definition.

(3.1) Definition. Let A be a G-group. A principal homogeneous G-set X over A is a right A-set X with a left G-set structure such that:

1. The right A action on X defined $X \times A \to X$ is G-equivariant, i.e., ${}^s(xa) = {}^sx{}^sa$ for all $s \in G$, $x \in X$, and $a \in A$.
2. For any two points $x, x' \in X$ there exists $a \in A$ with $x_a = x'$, and further $a \in A$ is unique with respect to this property.

If A is a G-subgroup of a G-group E, then any coset $X \in E/A$ which is G-invariant, that is, $s(X) = X$ for all $s \in G$, is an example of a principal homogeneous G-set over A as we observed at the end of the previous section. In particular, A acting on itself by group multiplication is a principal homogeneous G-set, and in this case, there is at least one fixed element, the identity.

(3.2) Definition. Let A be a G-group, and let X and X' be two principal homogeneous G-sets over A. A morphism $f : X \to X'$ of principal homogeneous sets is a function which is both G- and A-equivariant.

Since for $f(x) = x'$ with $x \in X, x' \in X'$ in (3.2), we can write any $y \in X$ as $y = xa$ for a unique $a \in A$, the morphism f is given by the formula $f(y) = f(xa) = f(x)a = x'a$, and hence f is completely determined by one value $f(x) = x'$. Moreover, f is a bijection with inverse $f^{-1} : X' \to X$, and f^{-1} is a morphism of principal homogeneous spaces. Since every morphism is an isomorphism which is determined by its value at one point, it seems sensible to try to classify principal homogeneous sets by looking at the G action on one point.

(3.3) Remarks. Let X be a principal homogeneous G-set over the G-group A. Choose $x \in X$, and for each $s \in G$ consider ${}^s x \in X$. There is a unique $a_s \in A$ with ${}^s x = xa_s$. Then $s \mapsto a_s$ defines a function $G \to A$, but it is not in general a group morphism. Instead it satisfies a "twisted" homomorphism condition. To see what it is, we make explicit the associativity of the G action. The relation ${}^s({}^t x) = {}^{st}x$ shows that

$$ {}^s({}^t x) = {}^s(xa_t) = {}^s x({}^s a_t) = xa_s\,{}^s(a_t) = xa_{st}, $$

and thus this function a_s must satisfy

$$ \text{(CC)} \qquad a_{st} = a_s\,{}^s(a_t) \qquad \text{for all } s, t \in G, $$

which is the cocycle formula or cocycle condition. Observe the (CC) implies that $a_1 = 1$.

Next, choose a second point $x' \in X$, and write ${}^s x' = x'a'_s$. For a unique $c \in A$ we have $x' = xc$, and we have the calculation

$$ {}^s x' = {}^s x\,{}^s c = xa_s\left({}^s c\right) = x'\left(c^{-1}a_s\,{}^s c\right). $$

Thus a_s for x and a'_s for x' are related by the coboundary formula

$$ \text{(CB)} \qquad a'_s = c^{-1}a_s({}^s c) \qquad \text{for all } s \in G. $$

Now we formally consider functions satisfying (CC) and relations (CB) between these functions, and this leads us to the pointed sets $H^1(G, A)$.

(3.4) Definition. Let A be a G-group. An A-valued G-cocycle is a function a_s from G and A satisfying

$$ \text{(CC)} \qquad a_{st} = a_s\,{}^s a_t \qquad \text{for all } s, t \in G. $$

Let $Z^1(G, A)$ denote the pointed set of all A-valued G-cocycles with base point the cocycle $a_s = 1$ for all $s \in G$.

Note that by putting $s = t = 1$ in (CC), we obtain $a_1 = a_1\,{}^1 a_1 = a_1^2$ or $a_1 = 1$ as with a homomorphism.

(3.5) Definition. Two A-valued G-cocycles a_s and a'_s are cobounding (or cohomologous) provided there exists $c \in A$ with

(CB) $$a'_s = c^{-1} a_s ({}^s c) \qquad \text{for all } s \in G.$$

Let $H^1(G, A)$ denote the quotient of $Z^1(G, A)$ under this equivalence relation. This is the first cohomology set of G with values in A.

If $a'_s = c^{-1} a_s({}^s c)$ and $a''_s = d^{-1} a'_s({}^s d)$ for all $s \in G$, then $a''_s = (cd)^{-1} a_s\, {}^s(cd)$ and $a_s = c a'_c\, {}^s(c^{-1})$ show that cobounding is an equivalence relation. Moreover, a cocycle $a_s \in Z^1(G, A)$ represents the base point of $H^1(G, A)$ if and only if there exists $c \in A$ with $a_s = c^{-1}({}^s c)$ for all $s \in G$. Let $[a_s]$ denote the cohomology class of a_s in $H^1(G, A)$.

Observe that $Z_1(G, A)$ is a right A-set under $(a_s, c) \mapsto c^{-1} a_s({}^s c)$, and $H^1(G, A)$ is just the quotient of $Z^1(G, A)$ under this right A action.

(3.6) Remark. If A is an abelian G-group written additively, then cocycles, a_s are functions satisfying

$$a_{st} = a_s + s a_t,$$

and they form an abelian group $Z^1(G, A)$ by adding their values. The group $Z^1(G, A)$ has a subgroup $B^1(G, A)$ of all coboundaries $b_s = s(c) - c$ for $c \in A$. The quotient group is the first cohomology group

$$H^1(G, A) = \frac{Z^1(G, A)}{B^1(G, A)}.$$

Further, if G acts trivially on A, then $B^1(G, A) = 0$, and a cocycle is just a homomorphism so that $H^1(G, A) = \mathrm{Hom}(G, A)$.

Now we relate the pointed set $H^1(G, A)$ to the pointed set $\mathrm{Prin}(G, A)$ of isomorphism classes of principal homogeneous G-sets over A.

(3.7) Proposition. *Let X and X' be two principal homogeneous G-sets with points $x \in X$ and $x' \in X'$ where ${}^s x = x a_s$ and ${}^s x' = x' a_s$. Then there exists an isomorphism $f : X \to X'$ with $f(x) = x'$ if and only if $a_s = a'_s$ for all $s \in G$. Further, X and X' are isomorphic if and only if a_s and a'_s are cohomologous.*

Proof. If f is an isomorphism, then it follows that $f({}^s x) = {}^s f(x) = {}^s x' = x' a'_s$ and $f({}^s x) = f(x a_s) = f(x) a_s = x' a_s$. Thus $a_s = a'_s$ follows whenever f exists. Conversely, we can define f by the formula $f(xc) = f(x)c$ from X to X' as an A-mosphism. Since $a_s = a'_s$, the A-morphism f is also a G-morphism.

For the second assertion we compare $f(x)$ and x' where ${}^s f(x) = f(x) a_s$ and ${}^s x' = x' a'_s$ as in (3.3). This proves the proposition.

(3.8) Corollary. *A principal homogeneous space X over A is isomorphic to A over A if and only if X^G is nonempty, that is, if X has a G-fixed point. This is also equivalent to the cocycle a_s for X being a coboundary, namely $a_s = c^{-1}({}^s c)$ for some $c \in A$.*

Proof. We apply (3.7) noting that ${}^s1 = 1$ for all $s \in G$. For the second assertion observe that $a_s = 1$ when ${}^sx = xa_s$ for a fixed point x of X.

(3.9) Theorem. *The function which assigns to a principal homogeneous space X the cohomology class of a_s where ${}^sx = xa_s$ for some $x \in X$ defines a bijection* $\text{Prin}(G, A) \to H^1(G, A)$.

Proof. By (3.7) this is a well-defined injection. For surjectivity we have to prove that a class of a cocycle a_s in $H^1(G, A)$ defines a principal homogeneous set X. For given a_s we require $X = A$ as a right A-set, but we "twist" the action of G on X. Namely 1 in $A = X$ has image ${}^s1 = a_s$ by definition in X. Hence, in X the action ${}^su$ must equal ${}^s(1u) = {}^s1{}^su = a_s{}^su$, where the second ${}^su$ is calculated in A. Clearly, in X we have ${}^s(ua) = a_s{}^s(ua)\ (a_s{}^su){}^sa = {}^su{}^sa$ and $X \times A \to X$ is G-equivariant. Since a_s is a cocycle, the action satisfies ${}^{st}u = {}^s({}^tu)$ in X. This completes the construction of X giving the cocycle a_s and proves the theorem.

(3.10) Definition. The principal homogeneous G-set X constructed from A and the cocycle a_s is called the twisted form of A by a_s.

§4. Long Exact Sequence in G-Cohomology

Now we return to the exact sequence (2.6) at the end of §2 and show how to extend it two or three terms. To do this, we need some definitions. In this section G denotes a group.

(4.1) Definition. Let $f : A \to A'$ be a morphism of G-groups. Then $f_* = H^0(f) : H^0(G, A) = A^G \to A'^G = H^0(G, A')$ is the restriction of f and $f_* = H^1(f) : H^1(G, A) \to H^1(G, A')$ is given by $f_*[a_s] = [f(a_s)]$. These are called the induced coefficient morphism in cohomology.

Now we will relate H^0 and H^1 in a short exact sequence.

(4.2) Definition. Let E be a G-group with G-subgroup $A \subset E$. The boundary function $\delta : H^0(G, E/A) = (E/A)^G \to H^1(G, A)$ is defined by

$$\delta(X) = \left\{ \begin{array}{l} \text{cohomology class associated to the coset } X \text{ viewed as a} \\ \text{principal homogeneous } G\text{-set over } A. \end{array} \right.$$

Hence for $X = xA$ and ${}^sx = xa_s$, it follows that $\delta(X) = [a_s]$.

(4.3) Theorem. *Let E be a G-group with G-subgroup $A \subset E$ and corresponding short exact sequence*

$$1 \to A \xrightarrow{i} E \to E/A \to 1.$$

Then the five-term sequence of pointed sets is exact

$$1 \to H^0(G, A) \xrightarrow{i_*} H^0(G, E) \to H^0(G, E/A) \xrightarrow{\delta} H^1(G, A) \xrightarrow{i_*} H^1(G, E).$$

Further, if A is a normal subgroup of E, then

$$H^1(G, A) \to H^1(G, E) \to H^1(G, E/A)$$

is an exact sequence of pointed sets so that the entire six-term sequence is exact.

Proof. We have already shown that the image equals the kernel in A^G and E^G, see (2.6). For exactness at $(E/A)^G = H^0(G, E/A)$ observe that $\delta(X) = 1$ in $H^1(G, A)$ is equivalent to the coset X having the form xA with $s(x) = x$ for all $s \in G$ by (3.8). In other words, it is equivalent to the principal homogeneous G-set X being in the image of $H^0(G, E) \to H^0(G, E/A)$.

To show exactness at $H^1(G, A)$, we consider $[a_s] \in H^1(G, A)$ whose image in $H^1(G, E)$ is trivial. This means for $A \subset E$ that there exists $e \in E$ with $a_s = e^{-1}({}^se)$ for all $s \in G$. Then ${}^se = ea_s$ and we have $[a_s] = \delta(eA)$, where $eA \in H^0(G, E/A) = (E/A)^G$. Clearly $\delta(eA)$ in $H^1(G, A)$ has trivial image in $H^1(G, E)$.

When A is a normal subgroup of E, we prove exactness at $H^1(G, E)$. Let $[a_s] \in H^1(G, E)$ have trivial image in $H^1(G, E/A)$, that is, suppose there exists $e \in E$ with $a_s = e^{-1}({}^se)b_s$ with $b_s \in A$ for all $s \in G$. Since $({}^se)A = A({}^se)$, we have $c_s \in A$ with

$$a_s = e^{-1}({}^se)b_s = e^{-1}c_s\,{}^se \qquad \text{for all } s \in G.$$

Then $c_s \in Z^{-1}(G, A)$ defining $[c_s] \in H^1(G, A)$ which maps to $[a_s]$ under i_*. This proves the theorem.

(4.4) Remark. When A is commutative, it is possible to define $H^i(G, A)$ for all $i \geq 0$ algebraically. The six-term sequence of (4.3) becomes seven term for A abelian and normal in E. It has the form

$$1 \to H^0(G, A) \xrightarrow{i_*} H^0(G, E) \to H^0(G, E/A) \xrightarrow{\delta} H^1(G, A) \xrightarrow{i_*}$$
$$H^1(G, E) \to H^1(G, E/A) \xrightarrow{\delta} H^2(G, A).$$

If both A and E are commutative we have an exact triangle or long exact sequence for all H^i of A, E, and E/A of the form

$$\begin{array}{ccc} H^*(G, A) & \xrightarrow{i_*} & H^*(G, E) \\ & {\scriptstyle\delta}\nwarrow \quad \swarrow & \\ & H^*(G, E/A) & \end{array}$$

where δ has degree $+1$.

(4.5) Remark. Let $f : G' \to G$ be a group homomorphism. If E is a G-group with action ${}^sx$, then E is also a G'-group with action ${}^{s'}x = {}^{f(s')}x$ and f induces a natural morphism $H^0(G, E) = E^G \to E^{G'} = H^0(G', E)$ denoted f^* and a natural morphism $H^1(G, E) \to H^1(G', E)$ also denoted $f^*([a_s]) = [a_{f(s')}]$.

§5. Some Calculations with Galois Cohomology

Our interest in $H^1(G, A)$ will be for $G = \mathrm{Gal}(k'/k)$ for some Galois extension k'/k, and in this case we will speak of Galois cohomology groups especially in the case of H^1.

(5.1) Proposition. *For a finite Galois extension k'/k we have*

$$H^1\left(\mathrm{Gal}(k'/k), \mathrm{GL}_n\left(k'\right)\right) = 0.$$

Proof. Let a_s be a 1-cocycle, and for any $u \in M_n(k')$ form the "Poincaré series"

$$b = \sum_{t \in G} a_t t(u) \in M_n\left(k'\right) \text{ for } G = \mathrm{Gal}\left(k'/k\right).$$

We calculate for $s \in \mathrm{Gal}(k'/k)$ using the cocycle relation

$$s(b) = \sum_{t \in G} s\left(a_t\right) st(u) = a_s^{-1} \sum_{t \in G} a_{st} st(u) = a_s^{-1} b.$$

Hence a_s is a coboundary if we can choose u such that b is invertible. For $n = 1$ the linear independence of automorphisms gives an element u with $b \neq 0$.

For $n > 1$ we consider the linear transformation $B : k'^n \to k'^n$ defined by the matrix equation $B(x) = \sum_{t \in G} a_t t(x)$.

Assertion: The $B(x)$ generate the k'-vector space k'^n. For otherwise, there exists a linear form f with $f(B(x)) = 0$ for all x, then for all scalars c, we have

$$0 = f(B(x)) = \sum_{t \in G} a_t f(t(c)t(x)) = \sum_{t \in G} t(c) f\left(a_t t(x)\right).$$

This is a linear relation between all $t(c), t \in G$ as c varies over k'. The linear independence of the elements t of G implies that each $f(a_t t(x)) = 0$, and since the a_t are invertible, we deduce that $f = 0$ which is a contradiction.

Choose $x_1, \ldots, x_n \in k'^n$ such that the $B(x_1), \ldots, B(x_n)$ are linearly independent, and choose $u \in M_n(k')$ with $ue_i = x_i$ for $i = 1, \ldots, n$. For the matrix $b = \sum_{t \in G} a_t t(u)$, it follows that $b(e_i) = B(x_i)$ which implies that $b \in \mathrm{GL}_n(k')$. This proves the proposition.

For a cyclic group G of order n with generator s and $x \in A$, the function $G \to A$ defined inductively by the relations

$$a_1 = 1, \quad a_s = x, \ldots, a_{s^i} = x\left(a_{s^{i-1}}\right), \ldots$$

is a 1-cocycle on G with values in the abelian group A if and only if

$$1 = N(x) = x \cdot s(x) \cdot s^2(x) \ldots s^{n-1}(x).$$

This observation and previous proposition leads to the next corollary.

(5.2) Corollary (Hilbert's "Theorem 90"). *Let k'/k be a finite cyclic extension with generator s of $\mathrm{Gal}(k'/k)$. For $x \in k'^*$ the norm $N_{k'/k}(x) = 1$ if and only if there exists $y \in k'^*$ with $x = y/s(y)$. In other words, the following sequence is exact*

$$1 \to k^* \to k'^* \xrightarrow{f} k'^* \xrightarrow{N_{k'/k}} k^*,$$

where $f(y) = y/s(y)$.

(5.3) Corollary. *For a finite Galois extension k'/k we have*

$$H^1\left(\mathrm{Gal}(k'/k), \mathrm{SL}_n(k')\right) = 0.$$

induces the exact sequence

$$\mathrm{GL}_n(k) = H^0\left(\mathrm{Gal}, \mathrm{GL}_n(k')\right) \xrightarrow{\det} k^* = H^0\left(\mathrm{Gal}, \mathrm{GL}_1\right) \xrightarrow{\delta} H^1\left(\mathrm{Gal}, \mathrm{SL}_n\right) \to 1.$$

Since det : $\mathrm{GL}_n(k) \to k^*$ *is surjective and* $H^1(\mathrm{Gal}, \mathrm{GL}_n) = 0$, *we deduce that* $H^1(\mathrm{Gal}(k'/k), \mathrm{SL}_n(k')) = 1$.

Let $\mu_n(k)$ denote the nth roots of unity contained in k^*, and observe that $\mu_n(k')$ is a $\mathrm{Gal}(k'/k)$ submodule of k'^* for any Galois extension k'/k. For n prime to the characteristic of k and k' separably algebraically closed, the following sequence is exact

$$1 \to \mu_n(k') \to k'^* \xrightarrow{n} k'^* \to 1.$$

Applying the exact cohomology sequence, using (4.5), and using

$$H^0(\mathrm{Gal}(k'/k), k'^*) = k^*,$$

we have the next proposition.

(5.4) Proposition (Kummer Sequence). *For k' a separably algebraically closed Galois extension of k and n prime to the characteristic of k the sequence*

$$1 \to \mu_n(k) \to k^* \xrightarrow{n} k^* \to H^1(\mathrm{Gal}(k'/k), \mu_n(k'/k)) \to 1$$

is exact, and we have an isomorphism

$$k^*/(k^*)^n \to H^1(\mathrm{Gal}(k'/k), \mu_n(k')).$$

Further, if k contains the nth roots of unity, then

$$k^*/(k^*)^n \to H^1(\mathrm{Gal}(k'/k), \mu_n(k')) = \mathrm{Hom}(\mathrm{Gal}(k'/k), \mu_n(k')).$$

is an isomorphism.

For example, in characteristic different from 2, the nonzero elements of $k^*/(k^*)^2$ correspond to the subgroups of index 2 in $\mathrm{Gal}(k'/k)$, that is to quadratic extensions of k.

Finally, we make a remark about infinite Galois extensions K/k. The Galois group $\mathrm{Gal}(K/k)$ is the projective limit of all finite quotients $\mathrm{Gal}(k'/k)$, where k'/k denotes a finite Galois extension contained in K. In this way $\mathrm{Gal}K(K/k)$ is a compact totally disconnected topological group with $\mathrm{Gal}(K/k')$ forming an open neighborhood basis of the identity. The only $G(K/k)$-modules that we consider are those A such that $A = \bigcup_{k'/k} A^{\mathrm{Gal}(K/k')}$ as k' goes through finite extensions of k and where $\mathrm{Gal}(K/k')$ is the kernel $\mathrm{Gal}(K/k) \to \mathrm{Gal}(k'/k)$. Then one can prove that $H^*(\mathrm{Gal}(K/k), A)$ is the inductive limit of $H^*(\mathrm{Gal}(k'/k), A^{\mathrm{Gal}(K/k')})$. We use the notation $H^*(k, A)$ for $H^*(\mathrm{Gal}(k_s/k), A)$, where k_s is a separable algebraic closure of k.

§6. Galois Cohomology Classification of Curves with Given j-Invariant

In 3, §2, the invariant $j(E)$ of an elliptic curve E was introduced. In 3(3.2), 3(4.2), 3(5.2) and 3(8.3), we saw that two curves E and E' are isomorphic over an algebraically closed field k if and only if $j(E) = j(E')$. Now we take up the question of the classification of elliptic curves over a perfect field k in the following form. For E an elliptic curve over a perfect field k, we wish to describe all elliptic curves E'/k, up to isomorphism over k, which become isomorphic to E over the algebraic closure $\bar{k}$, or in other words, all E' over k, up to isomorphism over k, with $j(E) = j(E')$. The answer given in the next theorem is in terms of the first Galois cohomology group of $\mathrm{Gal}(\bar{k}/k)$ with values in the automorphism group of the elliptic curve.

(6.1) Theorem. *Let E be an elliptic curve over a perfect field k. We have a base point preserving bijection*

$$\left\{ \begin{matrix} \textit{Isomorphism classes over } k \\ \textit{of elliptic curves, } E'/k \textit{ with} \\ j(E) = j(E'). \end{matrix} \right\} \to H^1\left(Gal\left(\bar{k}/k\right),\ \mathrm{Aut}_{\bar{k}}(E)\right)$$

defined by choosing any isomorphism $u : E \to E'$ over $\bar{k}$ and assigning to the class of E' over k the cohomology class determined by the cocycle $a_s = u^{-1}({}^s u)$.

Proof. The 1-cocycle $a_s = u^{-1}({}^s u)$ is a cocycle by the calculation

$$\begin{aligned} a_{st} &= u^{-1}\left({}^{st}u\right) = u^{-1}\left({}^{s}u\right)\left[\left({}^{s}u\right)^{-1}\left({}^{st}u\right)\right] = u^{-1}\left({}^{s}u\right)\left[\left(u^{-1}\right)\left({}^{t}u\right)\right] \\ &= a_s \cdot {}^s a_t \qquad \text{for all } s, t \text{ in } \mathrm{Gal}(\bar{k}/k). \end{aligned}$$

If $v : E \to E'$ is a second isomorphism over $\bar{k}$, then $h = v^{-1}u$ in $\mathrm{Aut}_{\bar{k}}(E)$ and from the relation $vh = u$ we have a coboundary relation $a_s = u^{-1}({}^s u) = h^{-1}v^{-1}({}^s v)({}^s h) = h^{-1}b_s({}^s h)$, where b_s is the cocycle $b_s = v^{-1}({}^s v)$. Thus the cohomology class $[a_s] \in H^1(\mathrm{Gal}(\bar{k}/k), \mathrm{Aut}_{\bar{k}}(E))$ is well defined. Conversely, if

$a_s = h^{-1}b_s({}^s h)$, where $a_s = u^{-1}({}^s u)$ and $b_s = v^{-1}({}^s v)$, then the coboundary relation leads to $vhu^{-1} = {}^s v^s h^s(u^{-1}) = {}^s(vhu^{-1})$ and $vhu^{-1} : E' \to E''$ is an isomorphism over k between the two elliptic curves compared with E by $u : E \to E'$ and $v : E \to E''$ both isomorphisms over $\overline{k}$.

For $t = [a_s] \in H^1(\mathrm{Gal}(\overline{k}/k), \mathrm{Aut}_{\overline{k}}(E))$ we define a twisted curve ${}^t E$ by the requirement that ${}^t E(k)$ is the set of all (x, y) in $E(\overline{k})$ satisfying $({}^s x, {}^s y) = {}^s(x, y) = a_s(x, y)$. The difficulty is in showing that ${}^t E(k)$ really is the set of k-points on an elliptic curve over k. We work out in detail that case where the cocycle has values in the subgroup $\{+1, -1\}$ contained in $\mathrm{Aut}_{\overline{k}}(E)$. Note it is equal to $\mathrm{Aut}_{\overline{k}}(E)$ for $j(E) \neq 0, 12^3$. In this case t is a homomorphism $t : \mathrm{Gal}(\overline{k}/k) \to \mathbb{Z}/2\mathbb{Z}$. Such homomorphisms are in one-to-one correspondence with quadratic extensions $k_t = k(\sqrt{a})$ of k with nontrivial automorphism s satisfying $s(x + \sqrt{a}y) = x - \sqrt{a}y$. If E is given by $y^2 = f(x)$, a cubic polynomial $f(x)$ over k, then ${}^t E$ is given by the equation $ay^2 = f(x)$, and the isomorphism $u : E \to {}^t E$ is given by

$$u(x, y) = \left(x, \frac{y}{\sqrt{a}}\right).$$

In terms of the general construction the relation

$$-\left(x' + \sqrt{a}x'', y' + \sqrt{a}y''\right) = a_s(x, y) = {}^s(x, y) = \left(x' - \sqrt{a}x'', y' - \sqrt{a}y''\right)$$

becomes in this case $x'' = 0$ and $y' = 0$ since $-(x, y) = (x, -y)$. Finally, $(x', \sqrt{a}y'')$ is on ${}^t E(k)$ if and only if (x', ay'') is on $E(k)$. The other cases, including characteristic 2, are left to be checked by the reader.

(6.2) Summary. If E and E' are two elliptic curves over k with $j(E) = j(E') \neq 0, 12^3$, then there is a quadratic extension k' of k with E and E' isomorphic over k'.

(6.3) Remark. In 3(6.3) we found that there are five elliptic curves over $\mathbb{F}_2$ up to isomorphism of which two have $j = 1$ and three have $j = 0$. The two curves with $j = 1$ are isomorphic over a quadratic extension and $\mathrm{Aut}(E)$ is the two element group. These are the curves E_1 and E_2, see 3(6, Ex. 1). We leave it to the reader to explain in terms of Galois cohomology the results in Exercises 1–5 of 3, §6 concerning over which field the pairs of curves over $\mathbb{F}_2$ are isomorphic.

8

Descent and Galois Cohomology

Central to the proof of the Mordell theorem is the idea of descent which was present in the criterion for a group to be finitely generated, see 6(1.4). This criterion was based on the existence of a norm, which came out of the theory of heights, and the finiteness of the index $(E(\mathbb{Q}) : 2E(\mathbb{Q}))$, or more generally $(E(k) : nE(k))$. In this chapter we will study the finiteness of these indices from the point of view of Galois cohomology with the hope of obtaining a better hold on the rank of $E(\mathbb{Q})$, see 6(3.3). These indices are orders of the cokernel of multiplication by n, and along the same lines, we consider the cokernel of the isogeny $E[a, b] \xrightarrow{\varphi} E[-2a, a^2 - 4b]$.

There is a new version of the descent procedure when the index is studied for larger and larger n, that is, n equals m^i, powers of a fixed number m which is usually a prime. We are missing an important result at this stage of the book, namely that multiplication by n is surjective on $E(k)$ for k separable algebraically closed and n prime to the characteristic of k. In the case of $m = 2$ we know by 1(4.1) that multiplication by 2 is surjective for certain fields k, in particular, separable algebraically closed fields in characteristic different from 2.

We begin by considering some examples of homogeneous curves over an elliptic curve E.

§1. Homogeneous Spaces over Elliptic Curves

Let E be an elliptic curve over k, and let k_s be a separable algebraic closure of k with Galois group $\mathrm{Gal}(k_s/k)$. By 7(3.9) we have a natural bijection

$$\mathrm{Prin}\,(\mathrm{Gal}(k_s/k), E(k_s)) \to H^1\,(\mathrm{Gal}(k_s/k), E(k_s))\,.$$

In the context of elliptic curves E over k, a homogeneous space X is a curve over k together with a map $X \times E \to X$ over k defining a principal action. Over the separable algebraic closure k_s there is an isomorphism $E \to X$, and this means that we are considering certain curves X over k which become isomophic to E over k_s.

(1.1) Galois Action on the Homogeneous Space. Following 7(3.9) we can describe the action ${}^s(x, y)$ of $\mathrm{Gal}(k_s/k)$ on $X(k_s)$. This can be identified with $E(k_s)$ the locus of a cubic equation in normal form. The action is given by the formula

$${}^s(x, y) = (sx, sy) + a_s,$$

where (sx, sy) is the action of $s \in \mathrm{Gal}(k_s/k)$ on (x, y) a point of the defining cubic equation and $[a_s]$ is the cohomology class corresponding to X. Finally, we remark that the homogeneous space X corresponds to the zero cohomology class if and only if $X(k) = H^0(\mathrm{Gal}(k_s/k), X(k_s))$ is nonempty.

(1.2) Notations. Since cohomology for the group $G = \mathrm{Gal}(k_s/k)$ arises so frequently, we introduce a special notation, namely $H^i(k,\ )$ for $H^i(\mathrm{Gal}(k_s/k),\ )$. Implicitly, a choice has been made for a separable algebraic closure, and in the case of a perfect field the separable algebraic closure k_s is the algebraic closure.

(1.3) Notations. Let T be any group of order 2 with trivial $\mathrm{Gal}(k_s/k)$ action, the only action possible. A morphism u is defined on $H^1(k, T) = \mathrm{Hom}(\mathrm{Gal}(k_s/k), T)$ with values in $k^*/(k^*)^2$ with the property the quadratic extension corresponding to $\ker(t)$ in $\mathrm{Gal}(k_s/k)$ is generated by the square root of the image $u(t)$. This construction was used in 7(6.1).

If $E(k)$ has an element of order 2, then we have a morphism $T \to E(k_s)$ of $\mathrm{Gal}(k_s/k)$-groups. This induces a morphism

$$H^1(k, T) = \mathrm{Hom}\,(\mathrm{Gal}\,(k_s/k)\,, T) \to H^1\,(k, E\,(k_s))\,.$$

Hence to each quadratic extension k_t of k corresponding to a nonzero element $t \in H^1(k, T) = \mathrm{Hom}(\mathrm{Gal}(k_s/k), T)$ there is a curve P_t with $P_t(k_t) = E(k_t)$.

(1.4) The Case of the Curve $E = E[a, b]$. When $E(k)$ has an element of order 2, we might as well transform the curve by translation so that (0,0) is the point of order 2 and the curve has the form $E = E[a, b]$ which we have studied at length explicitly. The two-element group in question is $T = \{0(0, 0)\}$, and for $t \in H^1(k, T)$ we denote the corresponding quadratic extension by k_t again. The quotient $\mathrm{Gal}(k_t/k)$ of $\mathrm{Gal}(k_s/k)$ has two elements 1 and s where we write frequently $\bar{x}$ for sx. The image of t in the cohomology group $H^1(k, E(k_s))$ corresponds to a curve P_t where $P_t(k_t) = E(k_t)$ with Galois action given by the following formula where we employ 4(5.4):

$$s_{P_t}(x, y) = (\bar{x}, \bar{y}) + (0, 0) = \left(\frac{b}{\bar{x}}, -\frac{b\bar{y}}{\bar{x}^2}\right).$$

(1.5) Remarks. The element corresponding to P_t in $H^1(k, E(k_s))$ is zero if and only if s_{P_t} has a fixed point by 7(3.8), or, equivalently, $\mathrm{Gal}(k_t/k)$ has a fixed point on $P_t(k_t)$ or $\mathrm{Gal}(k_s/k)$ has a fixed point on $P_t(k_s)$. This is also equivalent to the assertion that $P_t(k)$ is nonempty.

Explicitly in terms of coordinates, a fixed point (x, y) in $E(k_t)$ is a point satisfying the relation $(x, y) = (b/\overline{x}, -b\overline{y}/\overline{x}^2)$. This fixed point relation is equivalent to

(FP) $$x\overline{x} = b, \quad \overline{x}y = \frac{b\overline{y}}{\overline{x}} = -x\overline{y}.$$

Now we analyze these fixed-point relations further with the notations $k_t = k(c)$, where c is a square root of $u(t)$, and $x = x' + cx''$, $y = y' + cy''$, and $\overline{x}y = cz$ with $z \in k$ since $\overline{x}y = -x\overline{y}$. Setting $d^2 = c$, we observe that $x\overline{x} = b$ if and only if $b = x'^2 - dx''^2$. Next, we calculate using the equation of the curve $E = E[a, b]$

$$\begin{aligned} dz^2 = (cz)^2 = (\overline{x}y)^2 &= \overline{x}^2\left(x^3 + ax^2 + bx\right) \\ &= b^2x + b^2a + b^2\overline{x} = b^2\left(2x' + 2a\right). \end{aligned}$$

This yields the first relation between x and z

(1) $$x' = \frac{d}{2b^2}z^2 - \frac{a}{2}.$$

Further for x'' we have

$$dx''^2 = x'^2 - b = \frac{1}{4}\left(\frac{d}{b^2}z^2 - a\right)^2 - b.$$

From $\overline{x}y = (x' - cx'')(y' + cy'') = (x'y' - dx''y'') + (x'y'' - x''y')c$, we deduce the relations

(2) $$x'y' = dx''y'' \text{ and } z = x'y'' - x''y'.$$

(1.6) Assertion. Let $t \in H^1(k, T)$ be given by the quadratic extension k_t and let P_t be the corresponding curve as above. The curve P_t over k can be described as the locus of the quartic equation

$$dx''^2 = \frac{1}{4}\left(\frac{d}{b^2}z^2 - a\right)^2 - b.$$

In terms of (x, y) we recover x' from the relation (1) in terms of z, and we recover y' and y'' from the two linear relations (2) in terms of the other variables.

(1.7) Remark. In (1.6) substitute $M = z/b$, $N = 2x''$, $d' = d$, and $d'' = a^2 - 4b$. Then the quartic equation for P_t becomes

$$N^2 = d'M^4 - 2aM^2 + d''$$

with $d'd'' = a^2 - 4b$.

(1.8) Remark. Returning to (1.5), we point out that the statement: the element P_t in $H^1(k, E(k_s))$ is zero if and only if $P_t(k)$ is nonempty is related to the transformation in 4(3.1) of a quartic $y^2 = f_4(x)$ to a cubic $y^2 = f_3(x)$ given a simple root of $f_4(x)$.

§2. Primitive Descent Formalism

As indicated in the chapter introduction, the descent procedure will come from the long exact sequence in Galois cohomology. It is applied to an isogeny $\psi : E \to E'$ all defined over a field k such that over a Galois extension k_1 of k the group homomorphism $\psi : E(k_1) \hookrightarrow E'(k_1)$ is surjective leading to the following exact sequence of $\mathrm{Gal}(k_1/k)$-modules:

$$0 \to P \to E(k_1) \xrightarrow{\psi} E(k_1) \to 0.$$

where P is the kernel of ψ.

(2.1) Remark. An application of the six-term long exact sequence in Galois cohomology 7(4.3) yields the exact sequence of pointed sets

$$0 \to H^0(G, P) \to H^0(G, E(k_1)) \xrightarrow{\psi} H^0(G, E(k_1)) \xrightarrow{\delta} H^1(G, P) \to$$
$$H^1(G, E(k_1)) \xrightarrow{\psi} H^1(G, E(k_1)).$$

where $G = \mathrm{Gal}(k_1/k)$.

Here the symbol ψ is used for $H^0(\psi)$ and $H^1(\psi)$. The middle term is contained in a short exact sequence where ${}_\psi H^1(G, E(k_1))$ denotes the kernel of $\psi : H^1(G, E(k_1)) \to H^1(G, E(k_1))$, namely

$$0 \to \frac{E(k)}{\psi E(k)} \xrightarrow{\delta} H^1(G, P) \to_\psi H^1(G, E(k_1)) \to 0.$$

In most applications $k_1 = k_s$ the separable algebraic closure of k, and then $G = \mathrm{Gal}(k_1/k)$ is the full Galois group and $H^i(G, E(k_s))$ is denoted by $H^i(k, E(k_s))$ and $H^1(G, P)$ by $H^1(k, P)$.

(2.2) Examples. We make use of two examples of ψ. Firstly, we consider ψ equal to multiplication by n prime to the characteristic of k where P is isomorphic to $(\mathbb{Z}/n\mathbb{Z})^2$ with trivial Galois action exactly when ${}_nE(k_s) \subset E(k)$. Secondly, we consider ψ equal to $\varphi : E[a, b] \to E[-2a, a^2 - 4b]$ as in 4(5.2) with $P = T = \{0, (0, 0)\}$. The two related short exact sequences are

$$0 \to E(k)/nE(k) \xrightarrow{\delta} H^1(k, {}_nE(k_s)) \to_n H^1(k, E(k_s)) \to 0$$

and

$$0 \to E'(k)/\varphi E(k) \xrightarrow{\delta} H^1(k, {}_\varphi E(k_s)) \to_\varphi H^1(k, E(k_s)) \to 0.$$

Now we return to the analysis started in 6(3.1) with the curve $E[a, b]$ over the rational numbers $\mathbb{Q}$ and relate it to the Galois cohomology formalism above. The isogeny appears in exact two sequences for $E = E[a, b]$ and $E' = E[-2a, a^2 - 4b]$

$$E(\mathbb{Q}) \xrightarrow{\varphi} E'(\mathbb{Q}) \begin{array}{l} \xrightarrow{\delta} H^1\left(\mathbb{Q}, {}_{\varphi}E\left(\overline{\mathbb{Q}}\right)\right) = \mathrm{Hom}\left(\mathrm{Gal}\left(\overline{\mathbb{Q}}/\mathbb{Q}\right), \mathbb{Z}/2\mathbb{Z}\right) \\ \xrightarrow{\alpha} \mathbb{Q}^*/(\mathbb{Q}^*)^2. \end{array}$$

The homomorphism α is given in 4(5.6) and considered further in 6(3.1) in this context, and the homomorphism δ is defined in 7(4.2).

Again let $x = m/e^2$ and $y = n/e^3$ be the coordinates of (x, y) on $E[a, b]$ with rational coefficients in lowest terms. For $n = 0$ we have x values given by roots of $x(x^2 + ax + b) = 0$, and for $x = 0$ the value $\alpha(0, 0) = b(\mathrm{mod}\, \mathbb{Q}^{*2})$, but otherwise there is a second root x' with $xx' = b$.

For $n \neq 0$ we have $m \neq 0$ and are led to the study of the congruence properties of $m, n,$ and e. From the equation

$$n^2 = m^3 + am^2e^2 + bme^4 = m\left(m^2 + ame^2 + be^4\right)$$

we introduce the greatest common divisor $b_1 = (m, b)$ and factor $m = b_1m_1$ and $b_1b_2 = b$ where $(m_1, b_2) = 1$. The equation becomes

$$\begin{aligned} n^2 &= b_1m_1\left(b_1^2m_1^2 + ab_1m_1e^2 + b_1b_2e^4\right) \\ &= b_1^2m_1\left(b_1m_1^2 + am_1e^2 + b_2e^4\right). \end{aligned}$$

Thus b_1^2 divides n^2 and so b_1 divides n giving a factorization $n = b_1n_1$. Putting this relation into the equation and dividing out b_1^2, we obtain the relation

$$n_1^2 = m_1\left(b_1m_1^2 + am_1e^2 + b_2e^4\right),$$

where the two terms m_1 and $b_1m_1^2 + am_1e^2 + b_2e^4$ in the factorization of n_1^2 are relatively prime since $(b_2, m_1) = 1$ and $(e, m_1) = 1$.

We can always choose the sign of b_1 so that m_1 is positive, and the above form of the equation implies that m_1 is a square which we write $m_1 = M^2$. Then M divides n_1 from the above equation, and we can write $n_1 = MN$, with all factorizations taking place in the integers $\mathbb{Z}$.

(2.3) Assertion. We can summarize the above discussion for a rational point (x, y) on $E[a, b]$ with $y \neq 0$ in terms of a representation as a quotient of integers in the form

$$x = \frac{b_1M^2}{e^2} \quad \text{and} \quad y = \frac{b_1MN}{e^3}.$$

For x and y reduced to lowest terms we have $(M, e) = (N, e) = (b_1, e) = 1$. Also, since $(b_2, m_1) = 1$, we have $(b_2, M) = (M, N) = 1$. The equation for M and N is

$$N^2 = b_1M^4 + aM^2e^2 + b_2e^4.$$

Conversely, any solution of this quartic equation yields a rational point on $E[a, b]$.

This quartic equation came from a down to earth factorization of the coordinates of a rational point on $E[a, b]$, but in §1 we also derived quartic equations used to describe principal homogeneous spaces representing elements in a Galois cohomology group. The Galois cohomology theory is contained in the following exact sequence:

$$0 \to \frac{E'(\mathbb{Q})}{\varphi E(\mathbb{Q})} \to H^1(\mathbb{Q}, T) \to {}_{\varphi}H^1\left(\mathbb{Q}, E\left(\overline{\mathbb{Q}}\right)\right) \to 0$$

considered in (2.2), where $E = E[a, b]$ and $E' = E[-2a, a^2 - 4b]$. Elements t in $H^1(\mathbb{Q}, T)$ correspond to quadratic extensions of $\mathbb{Q}$ and their image in $H^1(\mathbb{Q}, E(\overline{Q}))$ is described by a homogeneous space P_t, a curve of genus 1, defined by the quartic given in (1.7)

$$N^2 = d'M^4 - 2aM^2 + d'',$$

where $d'd'' = a^2 - 4b$. This curve associated with $E[-2a, a^2 - 4b]$ is exactly the same curve as the quartic above

$$N^2 = b_1 M^4 + aM^2e^2 + b_2e^4$$

associated with $E[a, b]$. The element t in $H^1(\mathbb{Q}, T)$ comes from a point in $E'(\mathbb{Q})/\varphi E(\mathbb{Q})$ if and only if this quartic in (1.7) has a rational point, i.e., the principal homogeneous space is trivial.

The methods described above do allow one to calculate the rank of certain elliptic curves over $\mathbb{Q}$.

(2.4) Example. Consider the curve $E = E[0, p]$ with equation $y^2 = x^3 - px$ for p a prime number. Then E' is $E[0, 4p]$ with equation $y^2 = x^3 + 4px$. For the first curve the divisors of $b = -p$ are $b' = \pm 1, \pm p$, and the corresponding quartic equations are

$$N^2 = M^4 - pe^4,\ N^2 = -M^4 + pe^4,\ N^2 = pM^4 - e^4,\ N^2 = -pM^4 + e^4.$$

For $p \equiv 3 \pmod 4$, -1 is not a square mod p so that the second and third equations have no solution mod p, and, hence, no solution in the integers. This means that for $p \equiv 3 \pmod 4$

$$\operatorname{im}(\alpha \text{ on } E) = \{1, -p\} \mod \left(\mathbb{Q}^*\right)^2.$$

For the second curve the divisors of $b = 4p$ are $b' = \pm 1, \pm 2, \pm 4, \pm p, \pm 2p, \pm 4p$, and the corresponding quartic equations are

$$N^2 = M^4 + 4pe^4, \quad N^2 = 2M^4 + 2pe^4, \quad N^2 = 4M^4 + pe^4,$$

where we excluded the cases where both factors of $4p$ are negative since N^2 is positive. For $p \equiv \pm 3 \pmod 8$, 2 is not a square mod p so that the second equation has no solution mod p in this case, and, hence, no solution over the integers. This means that for $p \equiv \pm 3 \pmod 8$

$$\operatorname{im}(\alpha \text{ on } E') = \{1, 4p\} \mod \left(\mathbb{Q}^*\right)^2.$$

The above discussion together with the exact sequence in 4(5.7) shows that certain curves have only finitely many rational points, and with the theory of 5 we are able to determine completely the Mordell group of rational points.

(2.5) Theorem. *For a prime number* $p \equiv 3 \pmod 8$ *the groups of rational points on the elliptic curves with equations*

$$y^2 = x^3 - px \quad \text{and} \quad y^2 = x^3 + 4px$$

are all equal to the group with two elements $\{0, (0, 0)\}$. *In particular, the rank of these groups is zero.*

Proof. The fact that the groups are finite follows from the above analysis of the quartic curves and their rational points together with 4(5.7) showing that $E(\mathbb{Q})/2E(\mathbb{Q})$ is a group of two elements with nonzero element of the class of (0, 0).

It remains to show that there is no odd torsion in $E(\mathbb{Q})$. For this we use the Nagell–Lutz theorem, 5(5.1). Since the discriminant of $y^2 = x^3 + bx$ is -2^6b^3, the above curves have bad reduction only at the primes p and 2. Mod 3 the curves become either the curve given by $y^2 = x^3 - x$ or by $y^2 = x^3 + x$. In both cases there are exactly four elements on these curves over $\mathbb{F}_3$. Thus by 5(5.1) there is no odd torsion, for it would have to map injectively into a group with four elements. We are left only with the case $p = 3$. The curves with equations $y^2 = x^3 - 3x$ and $y^2 = x^3 + 12x$ both reduce to $y^2 = x^3 + 2x$ mod 5, and this curve has only two points over $\mathbb{F}_5$. Again there is no odd torsion. This completes the proof of the theorem.

§3. Basic Descent Formalism

In this section we return to the Mordell–Weil theorem asserting that $E(k)$ is finitely generated for an elliptic curve E over a number field k. We use the notations of 6(8.2) for the family $V(k)$ of absolute values v of k, and we also use the fact that multiplication by any m is surjective $E(\bar{k}) \to E((\bar{k})$, see 12(3.6). By (2.1) and (2.2) the short exact sequence of $\operatorname{Gal}(\bar{k}/k)$-modules

$$0 \to {}_mE(\bar{k}) \to E(\bar{k}) \to E(\bar{k}) \to 0$$

leads to the long exact sequence for $G = \operatorname{Gal}((\bar{k}/k)$

$$E(k) \xrightarrow{m} E(k) \to H^1\left(k, {}_mE(\bar{k})\right) \to H^1\left(k, E(\bar{k})\right) \xrightarrow{m} H^1\left(k, E(\bar{k})\right)$$

which reduces to the short exact sequence

$$0 \to E(k)/mE(k) \to H^1\left(k, {}_mE(\bar{k})\right) \to {}_mH^1\left(k, E(\bar{k})\right) \to 0.$$

(3.1) Notations. Let E be an elliptic curve over a number field k. For each place v of k choose an embedding $\bar{k} \to \bar{k}_v$, and, hence, the decomposition subgroup $G_v = \mathrm{Gal}(\bar{k}_v/k_v)$ inside $G = \mathrm{Gal}(\bar{k}/k)$. This induces a morphism of the short exact sequence associated with cohomology

$$\begin{array}{ccccccc} 0 \to & \dfrac{E(k)}{mE(k)} & \to & H^1(G, {}_mE(\bar{k})) & \xrightarrow{f'} & {}_mH^1(G, E(\bar{k})) & \to 0 \\ & \downarrow & & & \searrow f & \downarrow f'' & \\ 0 \to \prod_v & \dfrac{E(k_v)}{mE(k_v)} & \to & \prod_v H^1(G_v, {}_mE(\bar{k})) & \longrightarrow & \prod_v {}_mH^1(G_v, E(\bar{k})) & \to 0. \end{array}$$

(3.2) Definition. The Selmer group for m-descent $S^{(m)} = S^{(m)}(E)$ is the kernel of f defined $H^1(G, {}_mE(\bar{k})) \to \prod_v {}_mH^1(G_v, E(\bar{k}))$, and the Tate–Šarafevič group for m-descent $Ш_m = Ш_m(E)$ is the kernel of f'' defined ${}_mH^1(G, E(\bar{k})) \to \prod_v {}_mH^1(G_v, E(\bar{k}))$.

The group $E(k)/mE(k)$, which is our main interest, is the kernel of f', and the following proposition is the exact sequence for the three kernels of the three morphisms in a composite $f = f''f'$.

(3.3) Proposition. *The mapping between the cohomology short exact sequences in (3.1) yields the short exact sequence*

$$0 \to \frac{E(K)}{mE(K)} \to S^{(m)}(E) \to Ш_m(E) \to 0$$

The finiteness of $E(k)/mE(K)$ will follow from the finiteness of $S^{(m)}(E)$ and the group $S^{(m)}(E)$ can be computed in some important cases. Consider the situation where K is large enough so that ${}_mE(K) = {}_mE(\bar{K})$. Then we have

$$H^1\left(G, {}_mE\left(\bar{K}\right)\right) = \mathrm{Hom}\left(G, {}_mE\left(\bar{K}\right)\right) = \mathrm{Hom}\left(G, \left(\frac{\mathbb{Z}}{m\mathbb{Z}}\right)^2\right),$$

and each element corresponds to an abelian extension L_x/K.

(3.4) Assertion. Let x in $H^1(G, {}_mE(K))$ have as corresponding abelian extension L_x/K. If x is in the Selmer group $S^{(m)}(E)$, then L_x is unramified outside the set S of places v where either E has bad reduction or $v(m) > 0$.

Proof. Let x project as in (3.1) to (x_v) in $\prod_v H^1(G_v, {}_mE(K_v))$. Since $f(x) = 0$, there exists $P_v \bmod {}_mE(K)$ mapping to x_v. Let $P_v = mQ_v$ in $E(\bar{K})$ and form the cocycle $s \to s(Q_v) - Q_v$ to obtain a representative in the cohomology class of x_v. The field L_x is generated by the coordinates of Q_v and the inertia group I_v acts trivially on L_x for v outside S. Hence L_x is unramified outside of S, and this proves the assertion.

Now we appeal to a basic finiteness result in algebraic number theory.

(3.5) Theorem. *Let K be an algebraic number field, and let S be a finite set of places of K. Then the set of isomorphism classes of extensions L of K of degree less than some fixed n and unramified outside of S is finite.*

Proof. By Lang [1970, p. 121, Theorem 5] there are only finitely many number fields with given discriminants d_L. At all places v outside S we have $v(d_L) = v(d_K)$, and for places v inside S the order of ramification in the discriminant is bounded by n. This proves the finiteness assertion.

With (3.4) and (3.5) we have the weak Mordell–Weil theorem again from these homological considerations.

(3.6) Theorem. *The Selmer group $S^{(m)}(E)$ and, hence, also $E(K)/mE(K)$ is finite for an elliptic curve E over a number field.*

Now consider the descent sequence (3.3) for m and powers m^n of m and make the comparison between the two sequences.

(3.7) Assertion. We have the following commutative diagram which compares the m and m^n descent sequences:

$$\begin{array}{ccccccccc}
E(K) & \xrightarrow{m^n} & E(K) & \longrightarrow & S^{(m^n)}(E) & \longrightarrow & \text{Ш}_{m^n}(E) & \longrightarrow & 0 \\
\big\downarrow{\scriptstyle m^{n-1}} & & \big\downarrow{\scriptstyle \mathrm{id}} & & \big\downarrow{\scriptstyle \alpha_n} & & \big\downarrow{\scriptstyle m^{n-1}} & & \\
E(K) & \xrightarrow{m} & E(K) & \xrightarrow{\alpha} & S^{(m)}(E) & \xrightarrow{m^n} & \text{Ш}_m(E) & \xrightarrow{m^n} & 0
\end{array}$$

Moreover, we have the inclusions

$$\mathrm{im}(\alpha) \subset \cdots \subset \mathrm{im}\,(\alpha_n) \subset \cdots \subset \mathrm{im}\,(\alpha_2) \subset \mathrm{im}\,(\alpha_1) = S^{(m)}(E),$$

and there is equality $\mathrm{im}(\alpha_n) = \mathrm{im}(\alpha)$ if and only if $m^{n-1}\text{Ш}_{m^n}(E) = 0$.

The construction of $S^{(m)}(E)$ approximating $E(K)/mE(K) \subset S^{(m)}(E)$ is sometimes called the "first m-descent", the construction of $S^{(m^2)}(E)$ and $E(K)/mE(K) \subset \mathrm{im}(\alpha_2) \subset S^{(m)}(E)$ is called the "second m-descent", and, more generally, the construction of $S^{(m^n)}(E)$ and $E(K)/mE(K) \subset \mathrm{im}(\alpha_n) \subset S^{(m)}(E)$ is called the "nth m-descent". The question of whether or not the nth m-descent gives $E(K)/mE(K)$ is regulated by $\text{Ш}_{m_n}(E)$ from the last assertion of (3.7), and hence we turn to properties of the Tate–Šafarevič group. Since $\text{Ш}_m \subset {}_mH^1(G, E(\bar{K}))$, we have an inductive system $\text{Ш}_m \hookrightarrow \text{Ш}_{m'}$ for $m \mid m'$.

(3.8) Definition. The Tate–Šafarevič group $\text{Ш} = \text{Ш}(E)$ of an elliptic curve E over a number field K is $\varinjlim_m \text{Ш}_m(E)$.

From the total Tate–Šafarevič group $\text{Ш}(E)$ we can recover the corresponding $\text{Ш}_m(E)$ as the m torsion elements in $\text{Ш}(E)$. The condition of (3.7) for the nth m-descent to give $E(K)/mE(K)$ would be assured if we had the following fundamental conjecture.

(3.9) Conjecture. *The group* $\text{Ш}(E)$ *is finite for every elliptic curve over a number field.*

See 17(8.2) for recent developments.

(3.10) Remark. The Selmer group and the Tate–Šafarevič group can be defined for any nonzero homomorphism or isogeny $\varphi : E \rightarrow E'$ with kernel T and corresponding exact sequence

$$0 \rightarrow T \rightarrow E\left(\bar{K}\right) \rightarrow E'\left(\bar{K}\right) \rightarrow 0.$$

For by 7(4.3) for $G = \text{Gal}(\bar{K}/K)$ we obtain

$$0 \rightarrow \frac{E'(K)}{\varphi E(K)} \rightarrow H^1(G, T) \rightarrow {}_{\varphi}H^1\left(G, E\left(\bar{K}\right)\right) \rightarrow 0.$$

As in (3.1), we have a commutative diagram

$$\begin{array}{ccccccc}
0 \rightarrow & \dfrac{E'(K)}{\varphi E(K)} & \longrightarrow & H^1(G, {}_{\varphi}E(\bar{K})) & \xrightarrow{f'} & {}_{\varphi}H^1(G, E(K)) & \longrightarrow 0 \\
& \downarrow & & \downarrow & \searrow^{f} & \downarrow{f''} & \\
0 \rightarrow & \prod_v \dfrac{E'(K)}{\varphi E(K_v)} & \longrightarrow & \prod_v H^1(G_v, {}_{\varphi}E(\bar{K}_v)) & \longrightarrow & \prod_v {}_{\varphi}H^1(G_v, E(\bar{K})) & \rightarrow 0.
\end{array}$$

(3.11) Definition. The Selmer group for the φ-descent $S^{(\varphi)}(E)$ is the kernel of f defined $H^1(G, {}_{\varphi}E(\bar{K})) \rightarrow \prod_v {}_{\varphi}H^1(G_v, E(\bar{K}))$, and the Tate–Šafarevič group for φ-descent $\text{Ш}_{\varphi} = \text{Ш}_{\varphi}(E)$ is the kernel of f'' defined ${}_{\varphi}H^1(G, E(\bar{K})) \rightarrow \prod_v {}_{\varphi}H^1(G_v, E(\bar{K}))$.

(3.12) Proposition. *The mapping between the cohomology short exact sequences in* (3.10) *yields the short exact sequence*

$$0 \rightarrow \frac{E'(K)}{\varphi E(K)} \rightarrow S^{(\varphi)}(E) \rightarrow \text{Ш}_{\varphi}(E) \rightarrow 0.$$

(3.13) Remark. It is possible to describe the Selmer group $S^{(\varphi)}(E)$ and $\text{Ш}_{\varphi}(E)$ in terms of the quartic in (1.7) or (2.3) too. An element t in $H^1(k, T)$ determines an element in the Selmer group $S^{(\varphi)}(E)$ if and only if the corresponding quartic

$$N^2 = d'M^4 - 2aM^2 + d''$$

has a solution in each completion k_v of the global field k. It determines a nonzero element in $\text{Ш}_{\varphi}(E)$ if and only if it does not have a global solution in k.

9

Elliptic and Hypergeometric Functions

For an elliptic curve E over the complex numbers $\mathbb{C}$ we have already observed in (7.3) of the Introduction that the group $E(\mathbb{C})$ is a compact group isomorphic to the product of two circles. This assertion ignores the fact that the complex structure on $E(\mathbb{C})$ comes from a representation of $E(\mathbb{C})$ as a quotient group $\mathbb{C}/L$ where L is a lattice in the complex plane, that is, a discrete subgroup on two free generators $L = \mathbb{Z}\omega_1 + \mathbb{Z}\omega_2$.

In the first part of this chapter we start with a lattice L in the complex plane and, using the theory of elliptic functions, construct an elliptic curve E over $\mathbb{C}$ and an analytic isomorphism $\mathbb{C}/L \rightarrow E(\mathbb{C})$. In the remainder of the chapter we treat the inverse problem. Starting with the elliptic curve in Legendre form $E_\lambda : y^2 = x(x-1)(x-\lambda)$, and using hypergeometric functions, we construct a lattice L_λ and an analytic isomorphism $E_\lambda(\mathbb{C}) \rightarrow \mathbb{C}/L_\lambda$. In this way the group of complex points on an elliptic curve is isomorphic, in the sense of complex analysis, to a one-dimensional complex torus.

§1. Quotients of the Complex Plane by Discrete Subgroups

Recall that all discrete subgroups of $\mathbb{R}$ are of the form $\mathbb{Z}a$ for $a \geq 0$. When a is nonzero, it is the minimal strictly positive element in the discrete group. In the next proposition we have the analogous result for the complex numbers $\mathbb{C}$.

(1.1) Proposition. *A discrete subgroup Γ of $\mathbb{C}$ is of the form $\mathbb{Z}\omega$ for some complex number ω or $\mathbb{Z}\omega_1 + \mathbb{Z}\omega_2$, where ω_1 and ω_2 are nonzero complex numbers with ω_1/ω_2 a nonreal complex number.*

Note that $\mathbb{R}$-linear independence of ω_1 and ω_2 in $\mathbb{C}$ is equivalent to ω_1 and ω_2 nonzero with the quotient ω_1/ω_2 nonreal.

Proof. If $\Gamma = 0$, then $\Gamma = \mathbb{Z} \cdot 0$. Otherwise for r large enough the compact disc $|z| \leq r$ contains elements of $\Gamma - \{0\}$ and only a finite number since Γ is discrete. Let ω_1 be a nonzero element of Γ with minimal absolute value.

We are left with the case where there exists $\omega \in \Gamma - \mathbb{Z}\omega_1$. Among these ω, let ω_2 be one with a minimal absolute value. Observe that ω_2/ω_1 is not real, for if it were, there would exist an integer n such that $n < \omega_2/\omega_1 < n+1$, and hence $|\omega_2 - n\omega_1| < |\omega_1|$ contradicting the minimal character of $|\omega_1|$. This is the argument which shows that discrete subgroups of $\mathbb{R}$ are generated by one element as remarked above.

It remains to show that $\Gamma = \mathbb{Z}\omega_1 + \mathbb{Z}\omega_2$. Since ω_2/ω_1 is nonreal, it follows that $\mathbb{C} = \mathbb{R}\omega_1 + \mathbb{R}\omega_2$. For $\omega \in \Gamma$ we can write $\omega = a\omega_1 + b\omega_2$ with $a,\ b \in \mathbb{R}$ and find integers m, n with $|a - m| \leq 1/2$ and $|b - n| \leq 1/2$. Since $\omega, \omega_1, \omega_2 \in \Gamma$, we have

$$\omega' = \omega - m\omega_1 - n\omega \in \Gamma,$$

where $|\omega'| \leq (1/2)|\omega_1| + (1/2)|\omega_2| \leq |\omega_2|$, and the first inequality is strict since ω_2 is not a real multiple of ω_1. From the minimal character of $|\omega_2|$ we deduce that $\omega' \in \mathbb{Z}\omega_1$ and, hence, $\omega \in \mathbb{Z}\omega_1 + \mathbb{Z}\omega_2$. This proves the proposition.

(1.2) Definition. A lattice L in the complex numbers $\mathbb{C}$ is a discrete subgroup of the form $L = \mathbb{Z}\omega_1 + \mathbb{Z}\omega_2$, where ω_1 and ω_2 are linearly independent over $\mathbb{R}$. A complex torus T is a quotient group $\mathbb{C}/L$ of the complex plane $\mathbb{C}$ by a lattice with projection usually denoted $p : \mathbb{C} \to T = \mathbb{C}/L$.

A function f on a complex torus $T = \mathbb{C}/L$ is equivalent to a function g on $\mathbb{C}$ which is L-periodic, that is, $g(z + \omega) = g(z)$ for all $\omega \in L$. Such an L-periodic g corresponds to f on $T = \mathbb{C}/L$ where $g = fp$.

(1.3) Definition. Two lattices L and L' in $\mathbb{C}$ are equivalent provided there exists $\lambda \in \mathbb{C}^* = \mathbb{C} - \{0\}$ with $\lambda L = L'$, that is, L and L' are homothetic. Multiplication by λ defined $\mathbb{C} \to \mathbb{C}$ induces an isomorphism $T = \mathbb{C}/L \to \mathbb{C}/L' = T'$, also denoted by λ, defined by the commutative diagram

$$\begin{array}{ccc} \mathbb{C} & \xrightarrow{\lambda} & \mathbb{C} \\ \downarrow & & \downarrow \\ T = \mathbb{C}/L & \xrightarrow{\lambda} & \mathbb{C}/L' = T'. \end{array}$$

(1.4) Remark. Any complex number λ satisfying $\lambda L \subset L'$ induces a homomorphism, also denoted $\lambda : T = \mathbb{C}/L \to T' = \mathbb{C}/L'$ as in (1.3). Either $\lambda = 0$ or $\lambda : T \to T'$ is surjective with kernel isomorphic to $\lambda^{-1}L'/L$ or $L'/\lambda L$. Such a nonzero morphism is called an isogeny, see 12(3.3) also.

(1.5) Example. Let $L = \mathbb{Z}\omega_1 + \mathbb{Z}\omega_2$ be a lattice where $\mathbf{Im}(\omega_1/\omega_2) > 0$. Then L is equivalent to $(1/\omega_2)L = \mathbb{Z}\tau + \mathbb{Z} = L_\tau$, where $\tau = \omega_1/\omega_2$ is in the upper half plane $\mathfrak{H}$ of $z \in \mathbb{C}$ with $\mathbf{Im}(z) > 0$. Observe that for each integer k we have $L_\tau = L_{\tau+k}$.

Exercises

1. Let Γ be a subgroup of $\mathbb{R}^n$. Prove that Γ is discrete if and only if $\Gamma = \mathbb{Z}\omega_1 + \cdots + \mathbb{Z}\omega_r$, where $\omega_1, \ldots, \omega_r$ are $\mathbb{R}$-linearly independent (so $r \leq n$). Further prove that $\mathbb{R}^n/\Gamma$ is compact if and only if $r = n$.

2. Let Γ be a subgroup of $\mathbb{R}^n$. Prove that Γ is discrete if and only if the canonical $\mathbb{R}$-linear map $\mathbb{R} \otimes_Z \Gamma \to \mathbb{R}^n$ is an injection.
3. Show that the analytic homomorphisms of the complex numbers C to itself are given by multiplication by a complex number, that is, they have the form $f(z) = \lambda z$.

§2. Generalities on Elliptic Functions

In this section L is a lattice in $\mathbb{C}$.

(2.1) Definition. An elliptic function f (with respect to L) is a meromorphic function on $\mathbb{C}$ such that $f(z+\omega) = f(z)$ for all $z \in \mathbb{C}$, $\omega \in L$.

For $L = \mathbb{Z}\omega_1 + \mathbb{Z}\omega_2$ the L-periodicity condition $f(z) = f(z+\omega)$ for all $\omega \in L$, $z \in \mathbb{C}$ is equivalent to $f(z) = f(z+\omega_1)$ and $f(z) = f(z+\omega_2)$ for all $z \in \mathbb{C}$.

(2.2) Remark. A holomorphic elliptic function f is constant. For f factors by continuous functions $\mathbb{C} \to \mathbb{C}/L \to \mathbb{C}$, and since $\mathbb{C}/L$ is compact, f if bounded. By Liouville's theorem a bounded holomorphic function on $\mathbb{C}$ is constant.

We will study elliptic functions by integration around a fundamental parallelogram, i.e., a set consisting of all elements of the form $z_0 + a\omega_1 + b\omega_2$ where $0 \le a, b < 1$, $a, b \in \mathbb{R}$, and $L = \mathbb{Z}\omega_1 + \mathbb{Z}\omega_2$.

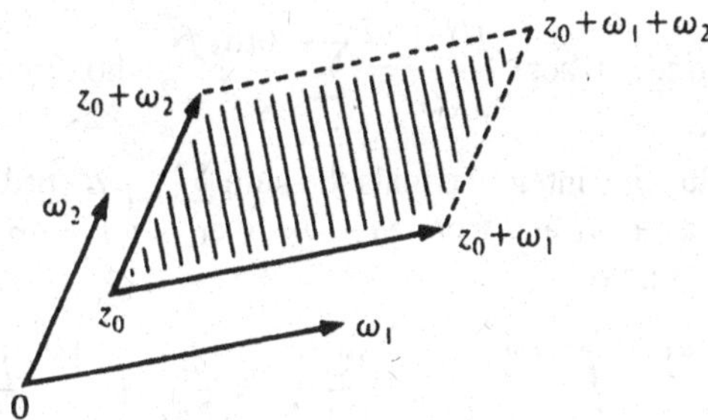

For a fundamental parallelogram P its boundary is denoted ∂P, and it is given a counterclockwise orientation.

(2.3) Theorem. *Let $f(z)$ be an elliptic function with no poles on the boundary of the fundamental parallelogram P. Then the sum of the residues in P is zero.*

Proof. We calculate by Cauchy's theorem

$$\frac{1}{2\pi i}\int_{\partial P} f(z)dz = \sum \mathrm{Res}(f).$$

The integrals on the opposite sides of ∂P cancel each other since

$$f(z)dz = f(z+\omega_1)dz, \quad f(z)dz = f(z+\omega_2)dz$$

and the directions of integration are reversed. Thus the integral, and hence, also the sum are zero.

(2.4) Corollary. *An elliptic function cannot have only one simple pole* mod L.

For a meromorphic function $f(z)$ we denote $m = \operatorname{ord}_a f(z)$ when $f(z) = (z-a)^m g(z)$, where $g(z)$ is holomorphic near a and $g(a) \neq 0$. When $\operatorname{ord}_a f(z) > 0$ the point a is a zero of f and when $\operatorname{ord}_a f(z) < 0$, it is a pole of f.

(2.5) Theorem. *Let $f(z)$ be an elliptic function with no zeros or poles on the boundary of the fundamental parallelogram P. Then*

$$\sum_{a \in P} \operatorname{ord}_a f(z) = 0,$$

and

$$\sum_{a \in P} a \cdot \operatorname{ord}_a f(z) \equiv 0 \pmod{L}.$$

Proof. The zeros and poles of f are the simple poles of f'/f, and the multiplicities are the residues of f'/f counted positive for zeros and negative for poles. The first relation follows from (2.3) applied to the elliptic function f'/f which is an elliptic function.

The second relation follows by considering the integral

$$\frac{1}{2\pi i}\int_{\partial P} z\frac{f'(z)}{f(z)}dz, \quad \text{where } \frac{f'(z)}{f(z)} = \sum \frac{\operatorname{ord}_a f}{z-a} + \text{holomorphic function.}$$

By the residue calculus this integral equals the sum $\sum_{a\in P} a \cdot \operatorname{ord}_a(f)$. Consider the part of P from z_0 to $z_0+\omega_1$ and from $z_0+\omega_2$ to $z_0+\omega_1+\omega_2$, and calculate this part of the integral around a

$$\frac{1}{2\pi i}\left[\int_{z_0}^{z_0+\omega_1} - \int_{z_0+\omega_2}^{z_0+\omega_1+\omega_2}\right] z\frac{f'(z)}{f(z)}dz = \frac{\omega_2}{2\pi i}\int_a^{a+\omega_1}\frac{f'(z)}{f(z)}dz.$$

Except for the factor ω_2 the right-hand expression is the winding number around 0 of the closed curve parametrized by $f(a+t\omega_1)$ for $0 \le t \le 1$. Hence this part of the integral is in $\mathbb{Z}\omega_2$ and by a similar argument the other part is in $\mathbb{Z}\omega_1$. Hence the above integral around ∂P representing the sum of $a \cdot \operatorname{ord}_a(f)$ over P is in L. This proves the theorem.

(2.6) Remarks. The set of all elliptic functions M_L associated with the lattice L forms a field under the usual operations of addition and multiplication of functions. The field of complex numbers is always the subfield of the constant elliptic functions M_L, and at this point these are the only examples of elliptic functions. In view of the previous two theorems the simplest nonconstant elliptic functions would have either a single double pole with residue zero in $1/2L$ or two simple poles whose sum is in the lattice with two residues adding to zero. There are constructions for elliptic functions having such simple singularities which are due to Weierstrass and Jacobi, respectively.

(2.7) Remark. An elliptic function $f(z)$ associated with L is a function on the complex plane $\mathbb{C}$, but the periodicity condition is equivalent to f defining a function f^* on the quotient torus $T = \mathbb{C}/L$ where for the projection $p : \mathbb{C} \to \mathbb{C}/L$ $f^* p = f$. In this way we view M_L as the field of meromorphic functions on the torus $T = \mathbb{C}/L$ with quotient complex structure under the bijection which assigns to f the quotient function f^*.

(2.8) Remark. If λ is a complex number, satisfying $\lambda L \subset L'$, then λ defines a morphism $\lambda : T = \mathbb{C}/L \to T' = \mathbb{C}/L'$, the function which assigns to $f \in M_{L'}$ the elliptic function $f(\lambda z) \in M_L$ is an embedding of $M_{L'}$ as a subfield of the field M_L of elliptic functions on $T = \mathbb{C}/L$. When λ is an isomorphism, then this map $M_{L'} \to M_L$ is an isomorphism of fields. We will see later that T and T' are isomorphic if and only if M_L and $M'_{L'}$ are isomorphic fields. For the above embedding $i_\lambda : M_{L'} \to M_L$ the degree of the field extension is given by the following formula

$$[M_L : M_{L'}] = \deg(\lambda) = \left[L' : \lambda L\right].$$

§3. The Weierstrass ℘-Function

In order to prove the convergence of certain infinite expressions defining elliptic functions, we will need the following lemma.

(3.1) Lemma. *For a lattice L and a real number $s > 2$, the following infinite sum converges absolutely*

$$\sum_{\omega \in L - 0} \frac{1}{\omega^s}.$$

Proof. Let $L = \mathbb{Z}\omega_1 + \mathbb{Z}\omega_2$ where ω_2/ω_1 is nonreal. Since L is discrete, there exists $c > 0$ such that for all integers n_1, n_2 we have $|n_1\omega_1 + n_2\omega_2| \geq c(|n_1| + |n_2|)$. Considering the $4n$ pairs (n_1, n_2) with $|n_1| + |n_2| = n$, we obtain

$$\sum_{\omega \neq 0} \frac{1}{|\omega|^s} \leq \frac{4}{c^s} \sum_{1 \leq n} \frac{1}{n^{s-1}} < +\infty$$

for $s - 1 > 1$.

We wish to construct an elliptic function $f(z)$ with the origin as the only pole and z^{-2} as singular part. Since $f(z) - f(-z)$ is an elliptic function with no singularities, we see that $f(z)$ will be an even function. Up to a constant, it will have a Laurent development of the form $z^{-2} + a_2 z^2 + a_4 z^4 + \cdots + a_{2k} z^{2k} + \cdots$. Also it is zero at $\omega_1/2$ for $L = \mathbb{Z}\omega_1 + \mathbb{Z}\omega_2$.

(3.2) Definition. The Weierstrass ℘-function and ζ-function associated with a lattice L are given by the infinite sums with and without notation for L

$$\wp(z; L) = \wp(z) = \frac{1}{z^2} + \sum_{\omega \in L-0} \left[\frac{1}{(z-\omega)^2} - \frac{1}{\omega^2} \right],$$

$$\zeta(z; L) = \zeta(z) = \frac{1}{z} + \sum_{\omega \in L-0} \left[\frac{1}{z-\omega} + \frac{1}{\omega} + \frac{z}{\omega^2} \right].$$

Since

$$\frac{1}{(z-\omega)^2} - \frac{1}{\omega^2} = \frac{2\omega z - z^2}{\omega^2(z-\omega)^2}$$

and

$$\frac{1}{z-\omega} + \frac{1}{\omega} + \frac{z}{\omega^2} = \frac{z^2}{\omega^2(z-\omega)}$$

both expressions behave like $1/\omega^3$ for $|\omega|$ large. Thus by Lemma (3.1) the above infinite sums converge uniformly on compact subsets of $\mathbb{C} - L$ and define meromorphic functions on $\mathbb{C}$ with, respectively, a simple double pole and a simple first-order pole at each $\omega \in L$. Observe that the derivative $\zeta'(z) = -\wp(z)$, and we have the following expression

$$\wp'(z; L) = \wp'(z) = -2 \sum_{\omega \in L-0} \frac{1}{(z-\omega)^3}$$

for the derivative of the $\wp$-function. It is clear from the formula that $\wp'(z)$ is an elliptic function with a third-order pole at each $\omega \in L$ having zero residue. From the formula for $\wp(z)$ it follows that $\wp(z)$ is an even function, i.e., $\wp(z) = \wp(-z)$, and $\wp'(z)$ is clearly an odd function. Since also $\wp'(z + \omega_i) = \wp'(z)$ we deduce that $\wp(z+\omega_i) = \wp(z)+c_i$ for $i = 1, 2$, where c_i are constant, and by setting $z = -\omega_i/2$, we calculate

$$\wp\left(\frac{\omega_i}{2}\right) = \wp\left(\frac{-\omega_i}{2} + \omega_i\right) = \wp\left(\frac{-\omega_i}{2}\right) + c_i = \wp\left(\frac{\omega_i}{2}\right) + c_i.$$

Thus each $c_i = 0$ and $\wp(z)$ is an elliptic function. The Weierstrass zeta function, which should not be confused with the Riemann zeta function, is not an elliptic function by (2.4). See the exercises for further properties.

With the elliptic functions $\wp(z)$ and $\wp'(z)$ we obtain all others in the sense made precise in the next theorem.

(3.3) Theorem. *For a lattice L the field M_L of elliptic functions is*

$$\mathbb{C}(\wp(z, L), \wp'(z, L)),$$

the field generated by $\wp$ and $\wp'$ over the constants, and the field of even elliptic function is $\mathbb{C}(\wp(z, L))$.

Proof. Every elliptic function $f(z)$ is the sum of an even and an odd elliptic function

$$f(z) = \frac{f(z) + f(-z)}{2} + \frac{f(z) - f(-z)}{2}.$$

Since $\wp'(z)$ times an odd elliptic function is an even elliptic function, it suffices to show that $\mathbb{C}(\wp(z))$ is the field of even elliptic functions.

If $f(z)$ is an even elliptic function, then $\mathrm{ord}_0 f(z) = 2m$ is even and $f(z) = \wp(z)^{-m} g(z)$, where $g(z)$ is an even elliptic function with no zeros or poles on the associated lattice L. If a is a zero of $\wp(z) - c$, then so is $\omega - a$ for $\omega \in L$, and if a is a zero or pole of $g(z)$, then so is $\omega - a$. If $2a \in L$, then the zero (or pole) is of order at least 2 since $g'(-z) = -g'(z)$ and so $g'(a) = g'(-a) = -g'(a)$. Thus

$$g(z) = c \cdot \frac{\prod_i (\wp(z) - \wp(a_i))}{\prod_i (\wp(z) - \wp(b_i))},$$

where $\{a_i, \omega - a_i\}$ are the zeros of $g(z)$ and $\{b_i, \omega - b_i\}$ are the poles of $g(z)$ in a fixed fundamental domain. This proves the theorem.

Let M_L denote the field of elliptic meromorphic functions for the lattice L. This can be viewed as the field of meromorphic functions on the complex torus $\mathbb{C}/L = T$ as in (2.7).

(3.4) Definition. A divisor D on $\mathbb{C}/L$ is a finite formal integral linear combination $D = \sum_u n_u(u)$ of points u in $\mathbb{C}/L$, and its degree $\deg(D) = \sum_u n_u$ is an integer. The group $\mathrm{Div}(\mathbb{C}/L)$ of all divisors on $\mathbb{C}/L$ is the group of all divisors. A principal divisor D is one of the form $(f) = \sum_u \mathrm{ord}_u(f)(u)$, where f is a nonzero meromorphic function in M_L. We denote by $\mathrm{Div}_0(\mathbb{C}/L)$ and $\mathrm{Div}_p(\mathbb{C}/L)$, respectively, the subgroups of divisors of degree zero and principal divisors.

We have an exact sequence

$$0 \to \mathrm{Div}_0(\mathbb{C}/L) \to \mathrm{Div}(\mathbb{C}/L) \overset{\deg}{\to} \mathbb{Z} \to 0,$$

and $\mathrm{Div}_p(\mathbb{C}/L)$ is a subgroup of $\mathrm{Div}_0(\mathbb{C}/L)$ by (2.5).

(3.5) Theorem (Abel–Jacobi). *The function* $f(\sum_u n_u(u)) = \sum_u n_u \cdot u$ *from* $\mathrm{Div}(\mathbb{C}/L)$ *to* $\mathbb{C}/L$ *induces a function also denoted* $f : \mathrm{Div}_0(\mathbb{C}/L)/\mathrm{Div}_p(\mathbb{C}/L) \to \mathbb{C}/L$ *which is an isomorphism of groups.*

Proof. Since $u = f((u) - (0))$ for any u in $\mathbb{C}/L$, the group homomorphism is surjective. The subgroup $\mathrm{Div}_p(\mathbb{C}/L)$ is carried to zero by f from (2.5). The construction following the proof of Theorem (3.3) above shows that $\sum_u n_u \cdot u = 0$ in $\mathbb{C}/L$ is sufficient for the existence of an elliptic function f with $(f) = \sum_u n_u(u)$. This proves the theorem.

Another proof that $\mathrm{Div}_p(\mathbb{C}/L)$ is the kernel of f results by using the σ-function introduced in Exercises 3, 4, and 5 at the end of this section.

Exercises

1. Show that $\zeta(z+\omega) = \zeta(z) + \eta(\omega)$, where $\eta(\omega)$ are constants.
2. For $\eta(\omega_i) = \eta_i$ derive Legendre's relation

$$\eta_1\omega_2 - \eta_2\omega_1 = 2\pi i$$

by considering $\int_{\partial P} \zeta(z)dz$, where P is a fundamental parallelogram whose boundary is disjoint from L.
3. Show that the following infinite product converges and represents an entire function:

$$\sigma(z) = z \prod_{\omega \in L-0} \left(1 - \frac{z}{\omega}\right) \cdot \exp\left[\frac{z}{\omega} + \frac{1}{2}\left(\frac{z}{\omega}\right)^2\right].$$

Show also that $\zeta(z) = \sigma'(z)/\sigma(z) = (d/dz)\log\sigma(z)$.
4. For $L = \mathbb{Z}\omega_1 + \mathbb{Z}\omega_2$ and $\zeta(z+\omega_i) = \zeta(z) + \eta_i$ show that

$$\sigma(z+\omega_i) = -\sigma(z)\exp\left[\eta_i\left(z + \frac{\omega_i}{2}\right)\right].$$

5. Show that any elliptic function can be written as

$$c\prod_{i=1}^{n} \frac{\sigma(z-a_i)}{\sigma(z-b_i)},$$

where c is a constant.

§4. The Differential Equation for $\wp(z)$

Since $\wp'(z)$ is an odd elliptic function, its square $\wp'(z)^2$ is an even elliptic function, and by (3.3) it is a rational function in $\wp(z)$. In this section we will prove that this rational function is a cubic polynomial. To do this, we consider the Laurent development of $\zeta(z, L)$, $\wp(z, L)$, and $\wp'(z, L)$ at the origin.

From the geometric series we have

$$\frac{1}{z-\omega} = -\frac{1}{\omega}\sum_{0\le n}\left(\frac{z}{\omega}\right)^n$$

and

$$\frac{1}{z-\omega} + \frac{1}{\omega} + \frac{z}{\omega^2} = -\frac{1}{\omega}\sum_{2\le n}\left(\frac{z}{\omega}\right)^n$$

which converge for $|z| \le |\omega|$. Thus

$$\begin{aligned}\zeta(z, L) &= \frac{1}{z} + \sum_{\omega\in L-0}\left[\frac{1}{z-\omega} + \frac{1}{\omega} + \frac{z}{\omega^2}\right]\\ &= \frac{1}{z} - \sum_{\omega\in L-0, 2\le n}\frac{z^n}{\omega^{n+1}}\end{aligned}$$

$$= \frac{1}{z} - \sum_{2\le k} G_k(L) z^{2k-1},$$

where $G_k(L) = \sum_{\omega\in L-0} \omega^{-2k}$ converges for $k \le 2$ by (3.1). Observe that the odd powers for $k \ge 2$ sum to zero

$$\sum_{\omega\in L-0} \omega^{-2k+1} = 0.$$

Since $\wp(z) = -\zeta'(z)$ we derive the Laurent series expansions for $\wp(z)$ and $\wp'(z)$ by differentiation.

(4.1) Laurent Series Expansions.

$$\zeta(z, L) = \frac{1}{z} - \sum_{2\le k} G_k(L) z^{2k-1},$$

$$\boxed{\wp(z, L) = \frac{1}{z^2} + \sum_{2\le k} G_k(L)(2k-1) z^{2k-2},}$$

$$\wp'(z, L) = \frac{-2}{z^3} + \sum_{2\le k} G_k(L)(2k-1)(2k-2) z^{2k-3},$$

where $G_k(L) = \sum_{\omega\in L-0} \omega^{-2k}$.

In order to derive the differential equation for $\wp(z)$, we write out the first few terms of the expansions at 0 for the elliptic functions $\wp(z)$, $\wp'(z)$, and various combinations of these functions:

$$\begin{aligned}
\wp(z) &= \frac{1}{z^2} + 3G_2 z^2 + 5G_3 z^4 + \cdots, \\
\wp'(z) &= -\frac{2}{z^3} + 6G_2 z + 20G_3 z^3 + \cdots, \\
\wp'(z)^2 &= \frac{4}{z^6} - \frac{24G_2}{z^2} - 80G_3 + \cdots, \\
4\wp(z)^3 &= 4\wp(z)\left(\frac{1}{z^4} + 6G_2 + 10G_3 z^2 + \cdots\right) \\
&= \frac{4}{z^6} + \frac{36G_2}{z^2} + 60G_3 + \cdots, \\
60G_2\wp(z) &= \frac{60G_2}{z^2} + 180\,(G_2)^2 z^2 + \cdots,
\end{aligned}$$

Hence the following equation

$$\boxed{\wp'(z)^2 = 4\wp(z)^3 - 60G_2\wp(z) - 140G_3}$$

is a relation between elliptic functions for the lattice L with all poles at points of L. The relation is an equality because the difference between the two sides is an elliptic function without poles which is zero at 0, and, hence, the difference is zero.

(4.2) Differential Equation for $\wp(z)$.

$$\boxed{\wp'(z, L)^2 = 4\wp(z, L)^3 - g_2(L)\wp(z, L) - g_3(L),}$$

where $g_2(L) = 60G_2(L)$ and $g_3(L) = 140G_3(L)$. Thus the points $(\wp(z), \wp'(z)) \in \mathbb{C}^2$ lie on the curve defined by the cubic equation with $g_2 = g_2(L)$ and $g_3 = g_3(L)$

$$y^2 = 4x^3 - g_2x - g_3.$$

For a basis $L = \mathbb{Z}\omega_1 + \mathbb{Z}\omega_2$ and $\omega_3 = \omega_1 + \omega_2$, let $e_i = \wp(\omega_i/2)$, where $i = 1, 2, 3$. Then the elliptic function $\wp(z) - e_i = f(z)$ has a zero at $\omega_i/2$ which must be of even order by (2.5), and thus

$$f'\left(\frac{\omega_i}{2}\right) = \wp'\left(\frac{\omega_i}{2}\right) = 0$$

for $i = 1, 2, 3$. By comparing zeros and poles, we obtain the factorization of (4.2):

$$\wp'(z)^2 = 4\left(\wp(z) - e_1\right)\left(\wp(z) - e_2\right)\left(\wp(z) - e_3\right),$$

where e_1, e_2, and e_3 are the roots of $4x^3 - g_2x - g_3$. Since $\wp(z)$ takes the value e_i with multiplicity 2 and has only one pole of order 2 modulo L, we see that $e_i \neq e_j$ for $i \neq j$. We are led to the following result since the cubic $4x^3 - g_2x - g_3$ has distinct roots.

(4.3) Theorem. *The function $h : \mathbb{C}/L \to E(\mathbb{C})$, where E is the elliptic curve over $\mathbb{C}$ with equation*

$$wy^2 = 4x^3 - g_2(L)w^2x - g_3(L)w^3$$

and $h(z \bmod L) = (1, \wp(z), \wp'(z))$ for $z \notin L$, $h(0 \bmod L) = (0, 0, 1)$, is an analytic group isomorphism.

Proof. Clearly $h(z \bmod L) = (0, 0, 1) \in E(\mathbb{C})$ is the zero element of $E(\mathbb{C})$ if and only if $z \bmod L = 0 \in \mathbb{C}/L$, and $h(z \bmod L) = (z^3 : z^3\wp(z) : z^3\wp'(z))$ is analytic at $0 \in \mathbb{C}/L$ with values in the projective plane.

To see that $h : \mathbb{C}/L - 0 \to E(\mathbb{C}) - 0$ is an analytic isomorphism, we consider $(x, y) \in E(\mathbb{C}) - 0$. There are two zeros z_1, z_2 of the function $\wp(z) - x$ with $\wp'(z_1) = -\wp'(z_2) = \pm y$ from the equation of E. Thus $h(-z \bmod L) = -h(z \bmod L)$ since $z_1 + z_2 \in L$ by (3.5). Note further that $z_1 \equiv z_2 \bmod L$ if and only if $\pm\wp'(z_1) = y = 0$. Hence h is an analytic isomorphism commuting with the operation of taking inverses.

Finally, to see that h or equivalently h^{-1} preserves the group structure, we study the intersection points of the line $y = \lambda x + \nu$ with the cubic curve E as in 1, §1. The elliptic function $f(z) = \wp'(z) - (\lambda\wp(z) + \nu)$ has a pole of order 3 at each point of

L, and by (2.5) there are three zeros z_1, z_2, and z_3 of $f(z)$ with $z_1 + z_2 + z_3 \in L$. Thus under h^{-1} the three intersection points of $y = \lambda x + \nu$, which add to zero on E, are transformed to three points $z_1 + L, z_2 + L$, and $z_3 + L \in \mathbb{C}/L$ which add to zero on $\mathbb{C}/L$. This proves the theorem.

Using the above considerations in proving h^{-1} is a group morphism and the techniques of 1, §2, we can derive the addition formulas for $\wp(z)$. For $\wp'(z_i) = \lambda\wp(z_i) + \nu$, where $i = 1, 2, 3$, the equation

$$4x^3 - (\lambda x + \nu)^2 - g_2 x - g_3 = 0$$

has three roots, and the sum of the roots is related to the coefficient of x^2 by the relation

$$\frac{\lambda^2}{4} = \wp(z_1) + \wp(z_2) + \wp(z_3).$$

Given z_1, z_2, choose $z_3 = -(z_1 + z_2) \bmod L$, and for $\wp(z_1) \neq \wp(z_2)$, we have $\wp'(z_1) - \wp'(z_2) = \lambda(\wp(z_1) - \wp(z_2))$. Since $\wp$ is an even function, this gives the following first division formula. The second formula is obtained by letting z_1 and z_2 approach z in the limit so z_3 becomes $-2z$.

(4.4) Addition Formulas for the Weierstrass $\wp$-Function.

(1) $$\wp(z_1 + z_2) = -\wp(z_1) - \wp(z_2) + \frac{1}{4}\left(\frac{\wp'(z_1) - \wp'(z_2)}{\wp(z_1) - \wp(z_2)}\right)^2.$$

(2) $$\wp(2z) = -2\wp(z) + \frac{1}{4}\left(\frac{\wp''(z)}{\wp'(z)}\right)^2.$$

In 3(2.3) we introduced the notion of an admissible change of variable carrying one normal form of the equation of an elliptic curve into another or equivalently defining an isomorphism between two elliptic curves.

(4.5) Remark. An admissible change of variable defining an isomorphism $f : E' \to E$, where E is given by $y^2 = 4x^3 - g_2 x - g_3$ and E' by $y'^2 = 4x'^3 - g_2' x' - g_3'$, has the form $xf = u^2 x'$ and $yf = u^3 y'$. Moreover, we have by 2(2.4) the relations $g_2 = u^4 g_2'$ and $g_3 = u^6 g_3'$.

In (1.3) we considered the equivalence between two lattices L and L' together with the corresponding isomorphism defined between their related tori $T = \mathbb{C}/L$ and $T' = \mathbb{C}/L'$. Under this isomorphism there are transformation relations for the Weierstrass $\wp$-function and its related coefficient functions.

(4.6) Proposition. *For two equivalent lattices L and $L' = \lambda L$ the following relations hold:*

$$\wp(z, L) = \lambda^2 \wp(\lambda z, \lambda L), \quad \wp'(z, L) = \lambda^3 \wp'(\lambda z, \lambda L),$$

$$G_{2k}(L) = \lambda^{2k} G_{2k}(\lambda L), \quad g_2(L) = \lambda^4 g_2(\lambda L), \quad \textit{and} \quad g_3(L) = \lambda^6 g_3(\lambda L).$$

Proof. These formulas follow easily from the definitions. For example, calculate

$$\begin{aligned}\wp'(\lambda z, \lambda L) &= -2 \sum_{\lambda\omega \in \lambda L - 0} \frac{1}{(\lambda z - \lambda\omega)^3} = \lambda^{-3}(-2) \sum_{\omega \in L - 0} \frac{1}{(z-\omega)^3} \\ &= \lambda^{-3}\wp'(z, L).\end{aligned}$$

Now we can make precise how the isomorphism h in (4.3) relates to isomorphisms between complex tori and the corresponding elliptic curves.

(4.7) Theorem. *For two equivalent lattices L and $L' = \lambda L$ we have the following commutative diagram where h is defined in* (4.3)*:*

$$\begin{array}{ccc} \mathbb{C}/\lambda L & \xrightarrow{\lambda^{-1}} & \mathbb{C}/L \\ h\downarrow & & h\downarrow \\ E'(\mathbb{C}) & \xrightarrow{f} & E(\mathbb{C}). \end{array}$$

The curve E is defined by $y^2 = 4x^3 - g_2x - g_3$, the curve E' by $y'^2 = 4x'^3 - g_2'x - g_3'$, and the isomorphism f by the admissible change of variable with $u = \lambda$.

Proof. The relation $\wp(\lambda^{-1}z, L) = \lambda^2\wp(z, \lambda L)$ corresponds to $xf = \lambda^2 x'$ under f, the relation $\wp'(\lambda^{-1}z, L) = \lambda^3\wp'(z, \lambda L)$ corresponds to $yf = \lambda^3 y'$ under f, the relation $g_2(L) = \lambda^4 g_2(\lambda L)$ corresponds to $g_2 = \lambda^4 g_2'$, and the relation $g_3(L) = \lambda^6 g_3(\lambda L)$ corresponds to $g_3 = \lambda^6 g_3'$. This proves the theorem.

(4.8) Remark. From 3(3.5) the discriminant of the cubic $x^3 + ax + b$ is $\mathrm{Disc}(x^3 + ax + b) = 27b^2 + 4a^3$ and it is also the discriminant of $(2x)^3 + a(2x) + b$ or $4x^3 + ax + b/2$. Hence $\mathrm{Disc}(4x^3 - g_2x - g_3) = 4(27g_3^2 - g_2^3)$. For a lattice L the corresponding elliptic curve defined by $y^2 = 4x^3 - g_2(L)x - g_3(L)$ we defne $\Delta(L) = g_2(L)^3 - 27g_3(L)^2$ and

$$j(L) = 12^3 \frac{g_2(L)^3}{\Delta(L)}.$$

Then $j(L) = j(E)$ where E is the elliptic curve defined by the Weierstrass equation $y^2 = 4x^3 - g_2(L)x - g_3(L)$. Since $g_2(L)^3 = (\lambda^4 g_2(\lambda L))^3 = \lambda^{12} g_2(\lambda L)^3$ and $\Delta(L) = \lambda^{12}\Delta(\lambda L)$, we have $j(L) = j(\lambda L)$.

(4.9) Summary Remark. The j-function has the very basic property that it classifies elliptic curves up to isomorphism, that is, two curves E and E' over $\mathbb{C}$ are isomorphic if and only if $j(E) = j(E')$. This function will also come up in an essential way in Chapter 11 on modular functions.

Exercises

1. Prove the following determinantal relation

$$\begin{vmatrix} \wp(z_1) & \wp'(z_1) & 1 \\ \wp(z_2) & \wp'(z_2) & 1 \\ \wp(z_1+z_2) & -\wp'(z_1+z_2) & 1 \end{vmatrix} = 0.$$

2. For $\sigma(z)$ as introduced in Exercise 3, §3, show that

$$\wp(z_1) - \wp(z_2) = -\frac{\sigma(z_1 - z_2)\sigma(z_1 + z_2)}{\sigma(z_1)^2\sigma(z_2)^2}$$

and

$$\wp'(x) = -\frac{\sigma(2z)}{\sigma(z)^4}.$$

3. Next derive the relation

$$\frac{\wp'(z_1)}{\wp(z_1) - \wp(z_2)} = \zeta(z_1 - z_2) + \zeta(z_1 + z_2) - 2\zeta(z_1).$$

4. Derive the addition formula

$$\zeta(z_1 + z_2) = \zeta(z_1) + \zeta(z_2) + \frac{1}{2}\frac{\wp'(z_1) - \wp'(z_2)}{\wp(z_1) - \wp(z_2)}.$$

From this formula derive the addition formulas in (4.4).

§5. Preliminaries on Hypergeometric Functions

In order to find the complex torus associated with an elliptic curve, we will use some special functions which we introduce in this section.

(5.1) Definition. The gamma function $\Gamma(s)$ is defined for $\mathbf{Re}(s) > 0$ by the following integral:

$$\Gamma(s) = \int_0^\infty x^s e^{-x} \frac{dx}{x}.$$

Due to the exponential factor e^{-x} the integral at ∞ converges for all s and for $\mathbf{Re}(s) > 0$ the factor x^{s-1} is integrable near 0.

(5.2) Remark. For $\mathbf{Re}(s) > 1$ an easy integration by parts, which we leave as an exercise, gives the relation

$$\boxed{\Gamma(s) = (s-1)\Gamma(s-1).}$$

Since $\Gamma(1) = \int_0^\infty e^{-x} dx = 1$, we prove inductively that for a natural number n we have

$$\boxed{\Gamma(n) = (n-1)!.}$$

Hence $\Gamma(s)$ is a generalization or analytic continuation of the factorial. The analytic function $\Gamma(s)$ defined for $\mathbf{Re}(s) > 0$ can be continued as a meromorphic function in the entire plane $\mathbb{C}$ with poles on the set $-\mathbb{N}$ of negatives of natural numbers. Using (5.2), we derive the relation

$$\Gamma(s) = \frac{\Gamma(s+n)}{s(s+1)(s+2)\cdots(s+n-1)}$$

for $\mathbf{Re}(s) > 0$, but the right-hand side is a meromorphic function for $\mathbf{Re}(s) > -n$, and it defines a meromorphic continuation of $\Gamma(s)$ to the half plane $\mathbf{Re}(s) > -n$. This works on all half planes $\mathrm{Re}(s) > -n$ and hence on all of $\mathbb{C}$.

(5.3) Remark. By change of variable $x = t^2$, $dx = 2t\,dt$, and $dx/x = 2(dt/t)$, we have the following integral formula for the gamma function:

$$\Gamma(s) = 2\int_0^\infty t^{2s}e^{-1}\frac{dt}{t}.$$

(5.4) Proposition. *For $a, b > 0$ we have*

$$2\int_0^{\pi/2} \cos^{2a-1}\theta \cdot \sin^{2b-1}\theta\, d\theta = \frac{\Gamma(a)\Gamma(b)}{\Gamma(a+b)}.$$

Proof. We calculate

$$\begin{aligned}\Gamma(a)\Gamma(b) &= 4\int_0^\infty\int_0^\infty \xi^{2a-1}\eta^{2b-1}\exp\left[-\left(\xi^2+\eta^2\right)\right]d\xi\, d\eta \\ &= 2\int_0^\infty r^{2a+2b}e^{-r^2}\frac{dr}{r}\cdot 2\int_0^{\pi/2}\cos^{2a-1}\theta\,\sin^{2b-1}\theta\, d\theta\end{aligned}$$

using the transformation $d\xi\, d\eta = r\, dr\, d\theta$ to polar coordinates. This proves the proposition.

As a special case $a = b = 1/2$, we have $2\int_0^{\pi/2} d\theta = \Gamma(1/2)\Gamma(1/2)/\Gamma(1)$ and hence $\Gamma(1/2) = \sqrt{\pi}$. The gamma function prolongs the factorial, and the binomial coefficient is prolonged by the formula

$$\binom{s}{n} = \frac{s(s-1)\cdots(s-n+1)}{n!}.$$

The corresponding binomial series is

$$(1+x)^s = \sum_{0\le n}\binom{s}{n}x^n$$

which converges for $|x| < 1$ and $\mathbf{Re}(s) > 0$.

(5.5) Proposition. *The relation*

$$2\int_0^{\pi/2} \sin^{2m}\theta\, d\theta = \pi(-1)^m\binom{-1/2}{m}$$

holds for every natural number m.

Proof. By (5.4) we have

$$\begin{aligned} 2\int_0^{\pi/2} \sin^{2m}\theta\, d\theta &= \frac{\Gamma\left(\frac{1}{2}\right)\Gamma\left(m+\frac{1}{2}\right)}{\Gamma(m+1)} \\ &= \frac{\Gamma\left(\frac{1}{2}\right)\left(m-\frac{1}{2}\right)\left(m-\frac{3}{2}\right)\cdots\frac{1}{2}\Gamma\left(\frac{1}{2}\right)}{m(m-1)\cdots 2\cdot 1} \\ &= \pi(-1)^m\binom{-\frac{1}{2}}{m}. \end{aligned}$$

This proves the proposition.

In order to define the hypergeometric series, we will need the following notation: $(a)_n$ where $(a)_0 = 1$ and $(a)_n = a(a+1)\cdots(a+n-1) = (a+n-1)(a)_{n-1}$.

(5.6) Definition. The hypergeometric series for $a, b \in \mathbb{C}$ and $c \in \mathbb{C} - \mathbb{N}$ is given by

$$F(a,b,c;z) = \sum_{0\le n} \frac{(a)_n(b)_n}{n!(c)_n} z^n.$$

An easy application of the ratio test shows that $F(a,b,c;z)$ is absolutely convergent for $|z| < 1$ and uniformly convergent for $|z| \le r < 1$, and, hence, it represents an analytic function on the unit disc called the hypergeometric function.

(5.7) Elementary Properties of the Hypergeometric Series.

(1) $$F(a,b,c;z) = F(b,a,c;z).$$

(2) $$F(a,b,b;z) = (1-z)^{-a}.$$

(3) $$F(a,a,1;z) = \sum_{0\le n}\binom{-a}{n}^2 z^n.$$

We use the relations $(1)_n = n!$ and $(a)_n/n! = (-1)^n\binom{-a}{n}$ to verify (2) and (3). We will make use of (3) for the case $a = 1/2$.

$$F\left(\frac{1}{2},\frac{1}{2},1;z\right) = \sum_{0\le n}\binom{-\frac{1}{2}}{n}^2 z^n.$$

(5.8) Hypergeometric Differential Equation. The hypergeometric function F satisfies the following differential equation called the hypergeometric differential equation:

$$z(1-z)\frac{d^2}{dz^2}F + [c-(a+b+1)z]\frac{d}{dz}F - abF = 0.$$

For the case of $F(1/2, 1/2, 1; z) = \sum_{0\le n} \binom{1/2}{n}^2 z^n$ this function satisfies

$$z(1-z)\frac{d^2}{dz^2}F + (1-2z)\frac{d}{dz}F - \frac{1}{4}F = 0.$$

This hypergeometric function can be used to evaluate the following elliptic integral.

(5.9) Theorem. *For a complex number λ with $|\lambda| < 1$*

$$2\int_0^{\pi/2}\left(1-\lambda\sin^2\theta\right)^{-1/2} d\theta = \pi F\left(\frac{1}{2}, \frac{1}{2}, 1; \lambda\right).$$

Proof. Using the binomial series, we have

$$\left(1-\lambda\sin^2\theta\right)^{-1/2} = \sum_{0\le n}\binom{-\frac{1}{2}}{n}(-\lambda)^n \sin^{2n}\theta.$$

By (5.5) and integrating term by term, we obtain

$$\begin{aligned}2\int_0^{\pi/2}\left(1-\lambda\sin^2\theta\right)^{-1/2} d\theta &= \sum_{0\le n}\binom{-\frac{1}{2}}{n}(-1)^n\lambda^n\left[\pi(-1)^n\binom{-\frac{1}{2}}{n}\right]\\ &= \pi\sum_{0\le n}\binom{-\frac{1}{2}}{n}^2\lambda^n = \pi F\left(\frac{1}{2}, \frac{1}{2}, 1; \lambda\right).\end{aligned}$$

This proves the theorem.

Exercises

1. Show that $\Gamma(s)\Gamma(1-s) = \pi/\sin \pi s$ for $0 < \mathbf{Re}(s) < 1$, and prove that $\Gamma(s)$ prolongs to a mermorphic function on $\mathbb{C}$ using this functional equation.
2. Show that the following formulas hold for $|z| < 1$.
   (a) $zF(1, 1, 2; -z) = \log(1+z)$.
   (b) $zF(1/2, 1, 3/2; -z^2) = \tan^{-}1(z)$.
   (c) $zF(1/2, 1/2, 3/2; z^2) = \sin^{-1}(z)$.
3. Show that $(d/dz)F(a, b, c; z) = (ab/c)F(a+1, b+1, c+1; z)$.
4. Show that $T_n(z) = F(n, -n, 1/2; (1-z)/2)$ is a polynomial of degree n.

§6. Periods Associated with Elliptic Curves: Elliptic Integrals

Let E be an elliptic curve defined over the complex numbers by an equation $y^2 = f(x)$, where $f(x)$ is a cubic equation in x. We wish to determine a complex torus $\mathbb{C}/L$ with $L = \mathbb{Z}\omega_1 + \mathbb{Z}\omega_2$ such that $(\wp, \wp') : \mathbb{C}/L \to E(\mathbb{C})$ is an analytic isomorphism. The lattice L is determined by generators ω_1 and ω_2. Our goal is to describe ω_1 and ω_2 in terms of E using the fact that these generators are examples of periods of integrals

$$\omega_1 = \int_{C_1} \theta \quad \text{and} \quad \omega_2 = \int_{C_2} \theta,$$

where C_1 and C_2 are suitably chosen closed curves on $E(\mathbb{C})$ and

$$\theta = \frac{dx}{f'(x)} = \frac{dx}{2y} = \frac{1}{2}\,dz \quad (x = \wp(z))$$

is the invariant differential on the cubic curve. In terms of the period parallelogram these integrals are just the integral of dz over a side of the parallelogram, hence a period of the lattice. The suitably chosen closed paths on $E(\mathbb{C})$ are determined by considering $x(z) = \wp(z)$ mapping a period parallelogram onto the Riemann sphere and looking at the four ramification points corresponding to the half periods $\omega_1/2, \omega_2/2, (\omega_1+\omega_2)/2$, and 0.

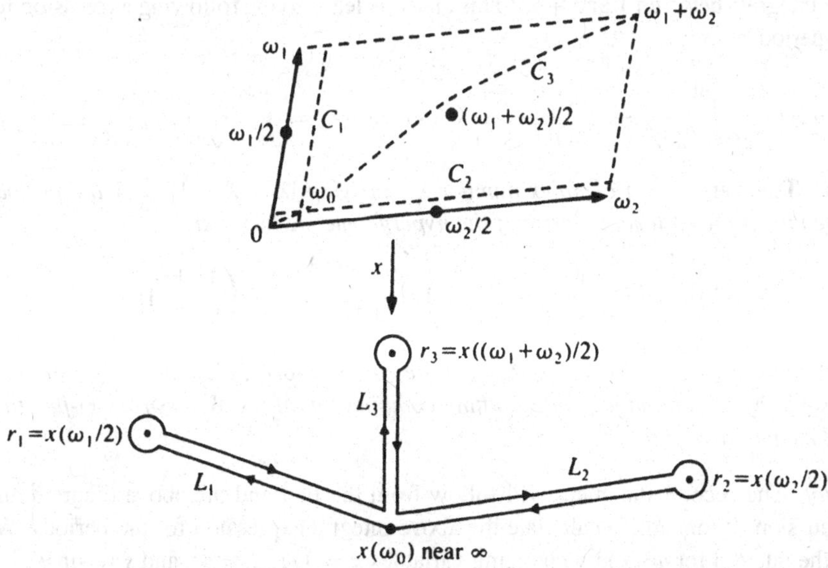

Implicit in the above construction is a basis of the 2-division points $(r_1, 0)$, $(r_2, 0)$ corresponding to ω_1, ω_2. For explicit calculations we normalize the roots of $f(x)$ by transforming r_1 to 0, r_2 to 1, and r_3 to some λ. The equation for the elliptic curve is in Legendre form $E_\lambda : y^2 = x(x-1)(x-\lambda)$ for $\lambda \in \mathbb{C} - \{0, 1\}$. We know from 1(6.3) that there are six possible choices of λ, namely

$$\lambda, \quad 1-\lambda, \quad \frac{1}{\lambda}, \quad \frac{1}{1-\lambda}, \quad \frac{\lambda-1}{\lambda}, \quad \frac{\lambda}{\lambda-1},$$

and we will choose the one of these six λ with the property $|\lambda| < 1$ and $|\lambda - 1| < 1$. We will calculate $\omega_1(\lambda)$ and $\omega_2(\lambda)$ for λ in this open set. For this consider the invariant differential on E

$$\theta = \frac{dx}{2y} = \frac{dx}{2\sqrt{x(x-1)(x-\lambda)}}.$$

The period integrals take the following form:

$$\omega_1(\lambda) = \int_{C_1} \theta = \int_{L_1} \frac{dx}{2\sqrt{x(x-1)(x-\lambda)}}.$$
$$\omega_2(\lambda) = \int_{C_2} \theta = \int_{L_2} \frac{dx}{2\sqrt{x(x-1)(x-\lambda)}}.$$

Now move ω_0 in the figure to 0 so that $x(\omega_0)$ goes to $\pm\infty$ on the real axis. The paths L_1 and L_2 are along the real axis $-\infty$ to near 0 for L_1 and from near 1 to $+\infty$. The cubic in the complex integrand of θ changes argument of 2π while turning around 0 for $\omega_1(\lambda)$ or 1 for $\omega_2(\lambda)$ which means that as the path L_1 deforms down to 0 becoming two integrals between $-\infty$ and 0 both integrals give half the contribution to the period. The same assertion applies to L_2 as it deforms down to 1 becoming two integrals between 1 and $+\infty$. This analysis leads to the following expression for the periods:

$$\omega_1(\lambda) = \int_{-\infty}^{0} \frac{dx}{\sqrt{x(x-1)(x-\lambda)}} \quad \text{and} \quad \omega_2(\lambda) = \int_{1}^{\infty} \frac{dx}{\sqrt{x(x-1)(x-\lambda)}}.$$

(6.1) Theorem. *For a complex number λ satisfying $|\lambda|$, $|\lambda - 1| < 1$ the periods have the following form in terms of the hypergeometric function*

$$\omega_1(\lambda) = i\pi F\left(\frac{1}{2}, \frac{1}{2}, 1; 1-\lambda\right), \quad \omega_2(\lambda) = \pi F\left(\frac{1}{2}, \frac{1}{2}, 1; \lambda\right).$$

For the lattice $L_\lambda = \mathbb{Z}\omega_1(\lambda) + \mathbb{Z}\omega_2(\lambda)$ the complex tori $\mathbb{C}/L_\lambda$ and $E_\lambda(\mathbb{C})$ are isomorphic by a map made from an affine combination of the Weierstrass $\wp$-function and its derivative.

Proof. The second statement will follow from the first and the above diagram and discussion. It remains to calculate the above integral expression for the periods. We do the integral for $\omega_2(\lambda)$ by changing variables $x = 1/t$, $t = s^2$, and $s = \sin\theta$

$$\omega_2(\lambda) = \int_1^{\infty} \frac{dx}{\sqrt{x(x-1)(x-\lambda)}} = \int_1^0 \frac{-dt/t^2}{\sqrt{(1/t)(1/t-1)(1/t-\lambda)}}$$
$$= \int_0^1 \frac{dt}{\sqrt{t(1-t)(1-\lambda t)}} = 2\int_0^1 \frac{ds}{\sqrt{(1-s^2)(1-\lambda s^2)}}$$

$$= \int_0^{\pi/2} \frac{d\theta}{\sqrt{1-\lambda\sin^2\theta}} = \pi F\left(\frac{1}{2}, \frac{1}{2}, 1; \lambda\right) \qquad \text{by (5.9).}$$

Next we calculate the integral for $\omega_1(\lambda)$ using the change of variable $y = -x + 1$ or $x = 1 - y$

$$\begin{aligned}
\omega_1(\lambda) &= \int_{-\infty}^{1} \frac{dx}{\sqrt{x(x-1)(x-\lambda)}} \\
&= \int_1^{\infty} \frac{dy}{\sqrt{-(y-1)(-y)((y-(1-\lambda))}} \\
&= i\int_1^{\infty} \frac{dy}{\sqrt{y(y-1)(y-(1-\lambda))}} \\
&= i\pi F\left(\frac{1}{2}, \frac{1}{2}, 1; 1-\lambda\right)
\end{aligned}$$

by the first change of variable used to calculate $\omega_2(\lambda)$. This proves the theorem.

Thus we have shown how to pass from tori to elliptic curves with the elliptic function $\wp$ and from elliptic curves over $\mathbb{C}$ to tori with the hypergeometric function $F(1/2, 1/2, 1; \lambda)$.

There is one case where the calculations of the periods are particularly agreeable, namely $\lambda = 1/2, -1$, or 2. Since for $\lambda = 1/2,\ 1 - \lambda = 1/2$, we have $\omega_1(1/2) = i\omega_2(1/2)$ and the lattice is of the form $\mathbb{Z}[i] \cdot \Omega = (\mathbb{Z} \cdot i + \mathbb{Z}) \cdot \Omega$. To determine Ω for one of these curves all isomorphic over $\mathbb{C}$, we calculate

$$\begin{aligned}
\omega_2(-1) = \Omega &= \int_0^1 \frac{dt}{\sqrt{t(1-t^2)}} = \int_0^{\pi/2} \frac{d\theta}{\sqrt{\sin\theta}} \\
&= \int_0^{\pi/2} \sin^{2(1/4)-1}\theta \cos^{2(1/2)-1}\theta\, d\theta \\
&= \frac{\Gamma\left(\frac{1}{4}\right)\Gamma\left(\frac{1}{2}\right)}{\Gamma\left(\frac{3}{4}\right)} \qquad \text{by (5.4).}
\end{aligned}$$

(6.2) Proposition. *The period lattice for the elliptic curve given by $y^2 = x^3 - x$ is $\mathbb{Z}[i] \cdot \Omega$, where*

$$\Omega = \sqrt{\pi}\frac{\Gamma\left(\frac{1}{4}\right)}{\Gamma\left(\frac{3}{4}\right)}.$$

As we have remarked before, with elliptic functions we map from a torus to a cubic curve, and with the hypergeometric function we can assign to a cubic $y^2 = x(x-1)(x-\lambda)$ the corresponding torus. The projection $w : x : y$ to $w : x$ from the cubic curve given by

$$wy^2 = (x - e_1 w)\left(x - e^2 w\right)(x - e_3 w)$$

to the projective line is a 2 to 1 map, in general, which can be pictured roughly with the following

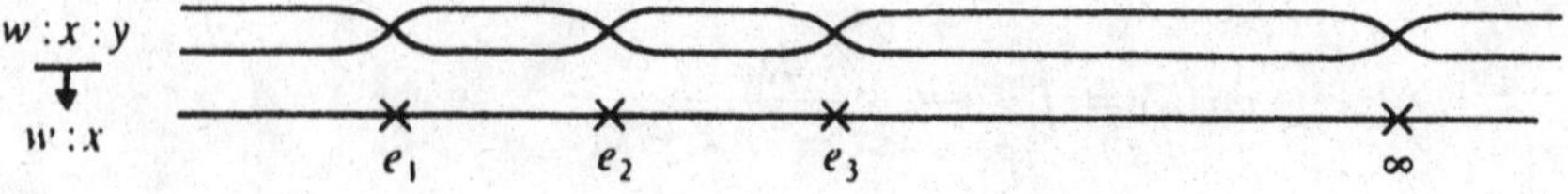

The projection mapping has four values over which there is double ramification e_1, e_2, e_3, and ∞.

If the elliptic curve is viewed as the parallelogram identified on the opposite edges with certain paths on the curve represented in the parallelogram, then we can draw two images in the complex plane C contained in the Reimann sphere $\mathbb{P}_1(\mathbb{C})$.

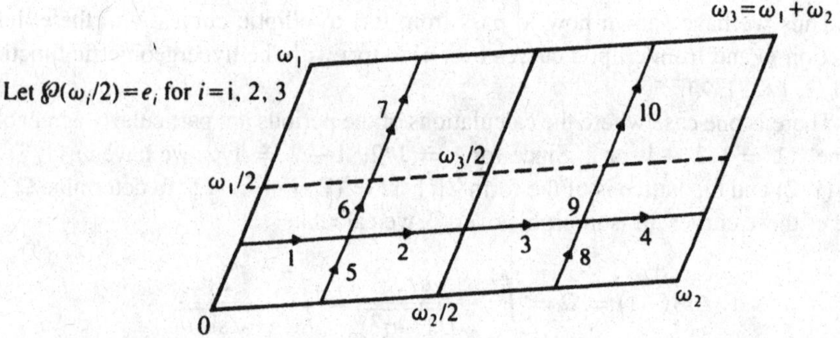

The first image is of the left-hand parallelogram under $\wp(z)$ and the second of the right-hand parallelogram under $\wp(z)$.

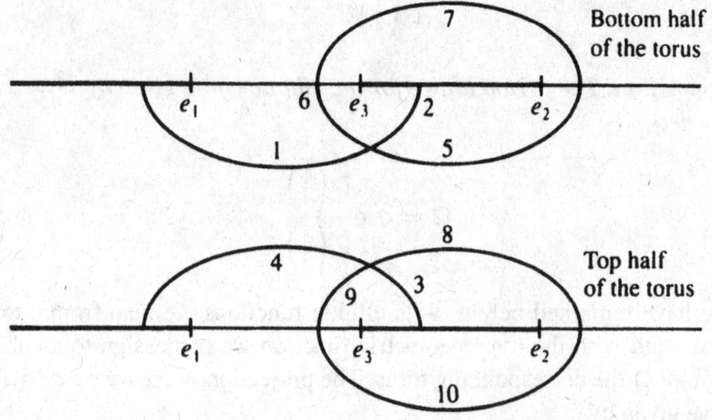

The torus can be obtained by opening up the slits $\overline{\infty, e_1}$ and $\overline{e_3, e_2}$ in the two planes and identifying opposite edges of the slits.
The left-hand parallelogram maps to the bottom half of the torus, the right-hand parallelogram to the top half of the torus, the bottom half parallelogram to the front of the torus, and the top half paralellogram to the back of the torus.

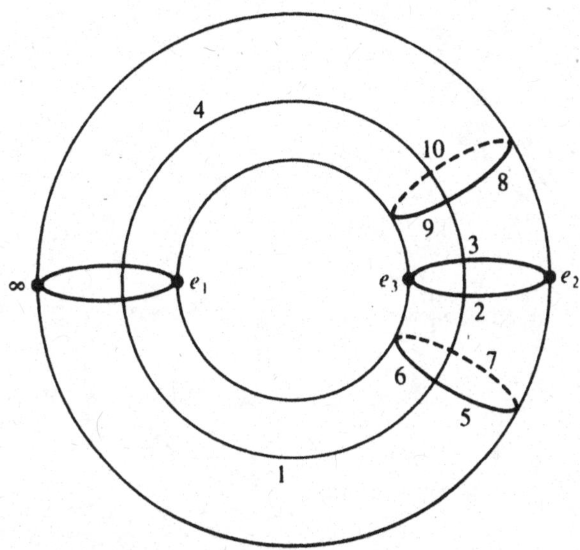

The reader is invited to express the arclength of the ellipse in terms of the integrals in this section.

10

Theta Functions

Quotients of theta functions provide another source of elliptic functions. They are defined for a lattice L of the form $L_\tau = \mathbb{Z}\tau + \mathbb{Z}$ with $\mathbf{Im}(\tau) > 0$. This is no restriction, because every lattice L is equivalent to some such L_τ. Since these functions $f(z)$ are always periodic in the sense $f(z) = f(z+1)$, we will consider their expansions in terms of $q_z = e^{2\pi i z}$ where $f(z) = f^*(q_z)$ and f^* is defined on $\mathbb{C}^* = \mathbb{C} - \{0\}$. In §1 we consider various expansions in the variable $q = q_z$ of functions introduced in the previous chapter.

Under the change of variable z to $z + \tau$ theta functions are not periodic but periodic up to a specific factor. On the other hand they are holomorphic on the entire plane. There are four specific theta functions which will give an embedding of a torus into $\mathbb{P}_3(\mathbb{C})$ such that the image of the torus is the locus of intersection of two quadric surfaces in this three-dimensional space. This is another representation of an elliptic curve as a curve in three space, see §8 of introduction.

An important feature of the theta-function picture of elliptic curves is that this approach extends to the p-adic case while the Weierstrass definitions do not. This was discovered by John Tate, and we give an introduction to Tate's theory of p-adic theta functions.

The basic reference for the first four sections is the first chapter of the book by D. Mumford, *Theta Functions I* (Birkhäuser Boston).

§1. Jacobi q-Parametrization: Application to Real Curves

(1.1) Remark. For a lattice L with basis ω_1 and ω_2 with $\tau = \omega_1/\omega_2$ where $\mathbf{Im}(\tau) > 0$ we have $L = \omega_2 L_\tau$ where $L_\tau = \mathbb{Z}\tau + \mathbb{Z}$. Now multiplication by ω_2 carries $\mathbb{C}/L$ onto $\mathbb{C}/L_\tau$ isomorphically and substitution $f(z) \mapsto f(\omega_2 z)$ carries elliptic functions for L isomorphically onto the field of elliptic functions for L_τ. These considerations were taken up in greater detail in 9(1.5) and 9(2.8).

In 11(1.4) we will determine how unique the invariant τ is among the lattices L_τ equal to a given L up to a nonzero complex scalar, but for now it suffices to observe that $L_\tau = L_{\tau+k}$ for any integer k. Hence the exponential $e^{2\pi i\tau} = q$ or q_τ, has importance for elliptic functions associated with the lattice L_τ.

(1.2) Remark. Under the exponential map $z \mapsto e^{2\pi i z} = w$ defined $\mathbb{C} \to \mathbb{C}^*$ with kernel $\mathbb{Z}$, the lattice $L = \mathbb{Z}\tau + \mathbb{Z}$ in $\mathbb{C}$ is mapped to $q^{\mathbb{Z}}$ in $\mathbb{C}^*$, the infinite cyclic subgroup generated by $q = e^{2\pi i\tau}$ with $|q| < 1$. Hence, by passing to the quotient torus, we have an analytic isomorphism $E_\tau = \mathbb{C}/L_\tau \to \mathbb{C}^*/q^{\mathbb{Z}}$.

Now we study the form of the Weierstrass equation and the elliptic functions in terms of $w = e^{2\pi i z}$ and $q = e^{2\pi i\tau}$. These expressions in the new parameters are derived by using thc following expansion from complex analysis

$$\sum_{n\in\mathbb{Z}} \frac{1}{(\zeta+n)^2} = \frac{\pi^2}{\sin^2 \pi\zeta}.$$

For $T = e^{2\pi i\zeta}$ this becomes

$$\sum_{n\in\mathbb{Z}} \frac{1}{(\zeta+n)^2} = (2\pi i)^2 \frac{1}{(e^{in\zeta} - e^{-in\zeta})^2} = (2\pi i)^2 \frac{e^{2\pi i\zeta}}{(1-e^{2\pi i\zeta})^2}$$
$$= (2\pi i)^2 \frac{T}{(1-T)^2} = (2\pi i)^2 \sum_{1\le n} nT^n.$$

From the chain rule we have $d/d\zeta = 2\pi i T(d/dT)$ and hence $(d/d\zeta)^{k-2} = (2\pi i)^{k-2}(T(d/dT)^{k-2}$. Applying this to the previous relation, we obtain the following power series relation.

(1.3) Remark. For $T = e^{2\pi i\zeta}$ we have the follwing expansion related to the series $g_k(T) = \sum_{1\le n} n^{k-1}T^n$:

$$(k-1)!(-1)^k \sum_n \frac{1}{(\zeta+n)^k} = (2\pi i)^k \sum_{1\le n} n^{k-1}T^n = (2\pi i)^k g_k(T).$$

Note that $g_{k+1}(T) = (T[d/dT])g_k(T)$ and $g_k(1/T) = (-1)^k g_k(T)$. Two important special cases used later are

$$g_2(T) = \frac{T}{(1-T)^2} \quad \text{and} \quad g_3(T) = T\frac{1+T}{(1-T)^3}.$$

(1.4) *q*-Expansion of Eisenstein Series. The following modular functions, called Eisenstein series, arose previously as coefficients of the differential equation for the Weierstrass $\wp$-function

$$G_{2k}(\tau) = \sum_{(m,n)\ne(0,0)} \frac{1}{(m\tau+n)^{2k}}$$
$$= 2\sum_{1\le n} \frac{1}{n^{2k}} + 2\sum_{1\le m}\sum_n \frac{1}{(m\tau+n)^{2k}}$$

which for $q = e^{2\pi i\tau}$ becomes, by (1.3),

$$2\zeta(2k) + 2\frac{(2\pi i)^{2k}}{(2k-1)!}\sum_{m,n<1} n^{2k-1}q^{mn}$$

where $\zeta(s) = \sum_{1\le n}(1/n^s)$ is the Riemann zeta function. For $\sigma_s(n) = \sum_{d|n} d^s$ this formula takes the more precise form, called the q-expansion of the Eisenstein series,

$$G_{2k}(\tau) = 2\zeta(2k) + 2\frac{(2\pi i)^{2k}}{(2k-1)!}\sum_{1\le n}\sigma_{2k-1}(n)q^n$$
$$= 2\zeta(2k) + 2\frac{(2\pi i)^{2k}}{(2k-1)!}\sum_{1\le n}\frac{n^{2k-1}q^n}{1-q^n}.$$

We introduce $E_{2k}(\tau)$ by $G_{2k}(\tau) = 2\zeta(2k)E_{2k}(\tau)$. The cases $2k = 4$ and 6 are considered in greater detail using

$$\zeta(4) = \frac{\pi^4}{90} = \frac{\pi^4}{2\cdot 3^2\cdot 5} \quad \text{and} \quad \zeta(6) = \frac{\pi^6}{945} = \frac{\pi^6}{3^3\cdot 5\cdot 7},$$

reference J.-P. Serre, *Course in Arithmetic*, p. 91. We have

$$G_4(\tau) = \frac{\pi^4}{45} + \frac{2(2\pi i)^4}{3!}\sum_{1\le n}\sigma_3(n)q^n$$
$$= \frac{(2\pi i)^4}{2^4\cdot 45}\left\{1 + 240\sum_{1\le n}\sigma_3(n)q^n\right\} = \frac{(2\pi i)^4}{720}E_4(\tau)$$

hence, $g_2(\tau) = 60G_4(\tau) = [(2\pi i)^4/12]E_4(\tau)$, and also

$$G_6(\tau) = \frac{2\pi^6}{945} + \frac{2(2\pi i)^6}{5!}\sum_{1\le n}\sigma_s(n)q^n$$
$$= -\frac{(2\pi i)^6\cdot 2}{2^6\cdot 945}\left\{1 - 504\sum_{1\le n}\sigma_5(n)q^n\right\} = \frac{-(2\pi i)^6\cdot 2}{2^6 3^3 5\cdot 7}E_6(\tau)$$

hence, $g_3(\tau) = 140G_6(\tau) = -[(2\pi i)^6/2^3 3^3]E_6(\tau)$.

(1.5) q-Expansions of Elliptic Functions. For $q = e^{2\pi i\tau}$ and $w = e^{2\pi i\tau}$ we have the following expansions using (1.4):

$$\wp(z,\tau) = \frac{1}{z^2} + \sum_{(m,n)\ne(0,0)}\left\{\frac{1}{(z-m\tau-n)^2} - \frac{1}{m\tau+n)^2}\right\}$$
$$= \sum_n \frac{1}{(z-n)^2} - 2\sum_{1\le n}\frac{1}{n^2} + \sum_{m\ne 0}\sum_n\left\{\frac{1}{(z-m\tau-n)^2} - \frac{1}{(m\tau+n)^2}\right\}$$

$$= (2\pi i)^2 \left[\frac{w}{(1-w)^2} - \frac{2\zeta(2)}{(2\pi i)^2} + \sum_{m\neq 0} \left\{ \frac{wq^m}{(1-wq^m)^2} - \frac{q^m}{(1-q^m)^2} \right\} \right].$$

$$\wp(z,\tau) = (2\pi i)^2 \left\{ \sum_m \frac{wq^m}{(1-wq^m)^2} + \frac{1}{12} - \sum_{m\neq 0} \frac{q^m}{(1-q^m)^2} \right\},$$

and

$$\wp'(z,\tau) = -2 \sum_{(n,m)} \frac{1}{(z+m\tau+n)^3}$$

$$= -2 \sum_n \frac{1}{(z+n)^3} - 2 \sum_{m\neq 0} \frac{1}{z+m\tau+n)^3}$$

$$= (2\pi i)^3 \left\{ \frac{w+w^2}{(1-w^3)} + \sum_{m\neq 0} \frac{wq^m(1+wq^m)}{(1-wq^m)^3} \right\},$$

$$\wp'(z,\tau) = (2\pi i)^3 \sum_m \frac{wq^m(1+wq^m)}{(1-wq^m)^3}.$$

Now we take the equation for the Weierstrass function and divide it by $4(2\pi i)^6$ to obtain

$$\left(\frac{\wp'(z)}{2(2\pi i)^3} \right)^2 = \left(\frac{\wp(z)}{(2\pi i)^2)} \right)^3 - \frac{g_2(\tau)}{4(2\pi i)^4} \left(\frac{\wp(z)}{(2\pi i)^2} \right) - \frac{g_3(\tau)}{4(2\pi i)^6}.$$

In summary we have the following result.

(1.6) Theorem. *The Weierstrass equation divided by* $4(2\pi i)^6$ *in terms of the variables* $w = e^{2\pi i z}$ *and* $q = e^{2\pi i \tau}$ *is*

$$E_q: \quad Y^2 = X^3 - e_2(q)X - e_3(q),$$

where

$$X = X(w,q) = \sum_m \frac{wq^m}{(1-wq^m)} + \frac{1}{12} - 2 \sum_{1\le m} \frac{q^m}{(1-q^m)^2},$$

$$Y = Y(w,q) = \frac{1}{2} \sum_m \frac{wq^m(1+wq^m)}{(1-wq^m)^3},$$

$$e_2(q) = \frac{1}{48} E_4(\tau) = \frac{1}{48} \left\{ 1 + 240 \sum_{1\le n} \sigma_3(n) q^n \right\},$$

$$e_3(q) = -\frac{2}{12^3} E_6(\tau) = -\frac{2}{12^3} \left\{ 1 - 504 \sum_{1\le n} \sigma_5(n) q^n \right\}.$$

(1.7) Remark. If q is real, then E_4 is defined over the real numbers $\mathbb{R}$ in the sense that the cubic equation has real coefficients. Observe that the following are equivalent for $q = e^{2\pi i\tau}$:

(1) q is real.
(2) $\mathbf{Re}(\tau)$ is a half integer.
(3) $\tau + \bar{\tau}$ is an integer.
(4) $\mathbb{Z}\tau + \mathbb{Z}$ is stable under complex conjugation.

Finally we have the discriminant, see sec 9(4.8),

$$\Delta(\tau) = g_2(\tau)^3 - 27g_3(\tau)^2$$

with $\Delta(\tau) \neq 0$ for τ in the upper half plane, and the j-function $j(E)$ of a complex torus $j(\mathbb{C}/\mathbb{Z}\tau + \mathbb{Z}) = j(\tau)$, where

$$j(\tau) = 12^3 \frac{g_2(\tau)^3}{\Delta(\tau)}.$$

(1.8) *q*-Expansions of $\Delta(\tau)$ and $j(\tau)$. The following expansions are given in J. P. Serre, *Course in Arithmetic*, pp. 90 and 95:

$$\Delta(\tau) = (2\pi)^{12} q \prod_{n=1}^{\infty} (1 - q^n)^{24},$$

$$j(\tau) = \frac{1}{q} + 744 + \sum_{n=1}^{\infty} c(n) q^n,$$

where $q = e^{2\pi i\tau}$ and $c(1) = 2^2 \cdot 3^3 \cdot 1823 = 196884$, $c(2) = 2^{11} \cdot 5 \cdot 2099 = 21493760$. The coefficients $c(n)$ are integers which can be seen from the above expression of j as a quotient of g_2 and Δ.

§2. Introduction to Theta Functions

Implicit in the formalism of the previous section is the Jacobi theory of theta functions. The Weierstrass theory shows that a complex torus is the locus of solutions of a cubic equation in $\mathbb{P}_2(\mathbb{C})$. The Jacobi theory shows that a complex torus is the intersection of two quadrics (degree 2 surfaces) in $\mathbb{P}_3(\mathbb{C})$. We sketch this theory which is carried out in detail in Mumford, *Theta Functions* (Birkhäuser Boston), Chapter 1.

Just as the Weierstrass theory revolves around the function $\wp(z)$, the Jacobi theory centers on the basic theta function.

(2.1) Definition. The basic theta function $\theta(z, \tau)$ defined on $\mathbb{C} \times \mathfrak{h}$ is given by the series

$$\theta(z, \tau) = \sum_{n \in \mathbb{Z}} \exp(2\pi i n z + \pi i n^2 \tau).$$

For $|\mathbf{Im}(z)| < c$ and $\mathbf{Im}(\tau) > \varepsilon$, we have the following inequality

$$|\exp(2\pi i n z + \pi i n^2 \tau)| < (e^{-\pi\varepsilon})^{n^2}(e^{2\pi c})^n.$$

If n' is chosen such that

$$(e^{-\pi\varepsilon})^{n'} e^{2\pi c} < 1,$$

then the inequality

$$|\exp(2\pi i n z + \pi i n^2 \tau)| < (e^{-\pi\varepsilon})^{n(n-n')}$$

shows that the series is uniformly majorized on this set by a convergent geometric series. Thus θ converges on $\mathbb{C} \times \mathfrak{H}$.

Unlike the meromorphic function $\wp(z, \tau)$ of z, which is periodic with respect to the lattice $\mathbb{Z}\tau + \mathbb{Z}$, the holomorphic function $\theta(z, \tau)$ of z is only quasiperiodic. This is made precise in the next proposition which is proved by a straightforward change of variable.

(2.2) Proposition. *For two integers a and b we have*

$$\theta(z + a\tau + b, \tau) = \exp(-\pi i a^2 \tau - 2\pi i a z)\theta(z, \tau) \quad \textit{and} \quad \theta(-z, \tau) = \theta(z, \tau).$$

In particular $\theta(z + 1, \tau) = \theta(z, \tau)$ *and* $\theta(z + \tau, \tau) = \exp(-\pi i \tau - 2\pi i z)\theta(z, \tau)$.

Thus the zeros of the function $\theta(z, \tau)$ in the variable z form a periodic set with respect to the lattice $\mathbb{Z}\tau + \mathbb{Z}$ which is symmetric around 0. To count these zeros, we apply the previous proposition to deduce $(\theta'/\theta)(z + 1, \tau) = (\theta'/\theta)(z, \tau)$ and $(\theta'/\theta)(z + \tau, \tau) = (\theta'/\theta)(z, \tau) - 2\pi i$, and to apply the argument principle to the period parallelogram, we use the following result.

(2.3) Proposition. *The integral* $(1/2\pi i)\int(\theta'/\theta)(z, \tau)\,dz = 1$.

Thus by the argument principle there is one zero and, due to symmetry about 0, it is at one of the following points: 0, $1/2$, $\tau/2$, or $(\tau + 1)/2$. We return to this in (2.7).

(2.4) Remark. The two periodicity conditions

$$\theta(z + 1, \tau) = \theta(z, \tau) \quad \text{and} \quad \theta(z + \tau, \tau) = \exp(-\pi i \tau - 2\pi i z)\theta(z, \tau)$$

lead to two q-expansions for $q = e^{2\pi i z}$ and $q_\tau = e^{\pi i \tau}$, namely

$$\theta(z, \tau) = \sum_n (e^{\pi i n^2 \tau}) q^n \quad \textit{and} \quad \theta(z, \tau) = \sum_n (e^{2\pi i z}) q_\tau^{n^2}.$$

Now using suitable translates of $\theta(z, \tau)$, we will be able to embed the complex torus $\mathbb{C}/\mathbb{L}$ into projective space such that the image is an intersection of algebraic hypersurfaces.

(2.5) Definition. For two rational numbers a, b we define

$$\begin{aligned}\theta_{a,b}(z,\tau) &= \exp[\pi i a^2\tau + 2\pi i a(z+b)]\theta(z+a\tau+b,\tau)\\ &= \sum_{n\in\mathbb{Z}}[2\pi i(n+a)(z+b)+\pi i(a+n)^2\tau].\end{aligned}$$

For a strictly positive integer N we define a space $V_N(\tau)$ as the vector space of all entire functions $f(z)$ such that

$$f(z+N) = f(z) \quad and \quad f(z+\tau) = \exp(-2\pi i N z - \pi i N^2\tau)f(z).$$

(2.6) Proposition. *The vector space $V_N(\tau)$ has dimension N^2 over the complex numbers and a set of $\theta_{a,b}(z,\tau)$ where (a,b) runs over representatives of $\{[(1/N)\mathbb{Z}]/\mathbb{Z}\}^2$ is a basis of $V_N(\tau)$. Each $f(z)$ in $V_N(\tau)$ has N^2 zeros in the period parallelogram $N(\mathbb{Z}\tau+\mathbb{Z})$.*

Proof. For $f(z) \in V_n(\tau)$ the condition $f(z) = f(z+N)$ is equivalent to the existence of an expansion of the form

$$\begin{aligned}f(z) &= \sum_{n\in\mathbb{Z}} a_n \exp\left(\frac{2\pi i}{N}nz\right)\\ &= \sum_{n\in\mathbb{Z}} b_n \exp[(2\pi i N n z + i\pi n^2\tau)/N^2], \quad \text{where} \quad b_n = b_n(\tau).\end{aligned}$$

Since $f(z) = f(z+\tau)\exp(2\pi i N z + i\pi N^2\tau)$ by definition a straight-forward change of variable shows that $b_n = b_{n+N^2}$ for all n and is independent of τ. Hence the dimension of $V_N(\tau)$ is at most N^2.

For nonzero $f(z) \in V_N(\tau)$ the assertion about the number of zeros follows as in (2.3) using $f(z+N) = f(z)$ and $f(z+\tau) = c\cdot\exp(-2\pi i N z)f(z)$, where c is a constant.

Finally the remainder of the assertions follow from the linear independence of the functions $\theta_{a,b}(z,\tau)$ as given in the proposition. In Mumford's book, see Chapter 1, §3, this is provided by showing that the first N^2 Fourier coefficients of $f(z) \in V_N(\tau)$ can be realized as the Fourier coefficients of linear combinations of $0_{a,b}(z,\tau)$. This completes the sketch of the proof of the proposition.

(2.7) Example. For $N = 1$, the space $V_1(\tau)$ is one dimensional and has $\theta(z,\tau)$ as a basis element. The zero of $\theta(z,\tau)$ is at the center point of the fundamental parallelogram $(1/2)\tau + 1/2$ since $\theta(z,\tau) = \theta(-z,\tau)$, see Mumford's book.

§3. Embeddings of a Torus by Theta Functions

(3.1) Notations. For any basis $f_0(z), \ldots, f_m(z)$ of $V_N(\tau)$ we define a map $f_{(N)}$: $\mathbb{C}/N(\mathbb{Z}\tau+\mathbb{Z}) \to \mathbb{P}_m(\mathbb{C})$ by the relation $f_{(N)}(z) = f_0(Nz) : \cdots : f_m(Nz)$, where $m = N^2 - 1$.

The map $f_{(N)}$ is well defined by N up to a linear change of variable corresponding to a linear relation between the two bases of the vector space $V_N(\tau)$.

(3.2) Proposition. *For $N > 1$ the map $f_{(N)}$ is an embedding.*

Proof. If there is a pair of points where $f_{(N)}$ has the same value or if there is a point where $df_{(N)}$ has a zero, then there is a second by the translation with some element of $(1/N^2)L_\tau$. Take $q = N^2 - 3$ points all distinct and distinct from these points z_j mod NL_τ. By solving linear equations in c_i for $f = \sum_i c_i f_i$, we can find $f \in V_N(\tau)$, $f \neq 0$, with $f(z_j) = 0$ and $f = 0$ at the double points. Then either $f = 0$ at $N^2 + 1$ points or $f = 0$ at $N^2 - 1$ points with two double zeros. Thus $f = 0$ which is a contradiction from which we deduce that $f_{(N)}$ is injective and $df_{(N)}$ is nonzero at all points. This proves the proposition.

(3.3) Remark. The image of $f_{(N)}$ is a closed complex curve in complex projective space. By a theorem of Chow this image is an algebraic curve. Note that the intersection of $\mathrm{im}(f_{(N)})$ with a plane $z_j = 0$ for the basis of translates of the basic theta function is on the N^2 images of the N-division points L_τ/NL_τ under $f_{(N)}$.

This we explain in the case of $N = 2$. We have the following table for the functions $\theta_{a,b}(z, \tau)$ together with the set of zeros described in terms of the lattice L_τ.

(3.4) Example. The 2-division point theta functions consist of the following four functions:

$$\theta_{0,0}(z, \tau) = \theta(z, \tau), \qquad \text{zeros} = \frac{\tau + 1}{2} + L_\tau,$$

$$\theta_{0,1/2}(z, \tau) = \theta(z + \frac{1}{2}, \tau), \qquad \text{zeros} = \frac{\tau}{2} + L_\tau,$$

$$\theta_{1/2,0}(z, \tau) = \exp\left(\frac{1}{4}\pi i\tau + \pi i z\right)\left(z + \frac{\tau}{2}, \tau\right), \qquad \text{zeros} = \frac{1}{2} + L_\tau,$$

$$\theta_{1/2,1/2}(z, \tau) = \qquad \text{zeros} = L_\tau.$$

$$\exp\left[\frac{1}{4}\pi i\tau + \pi i\left(z + \frac{1}{4}\right)\right]\theta\left(z + \frac{\tau + 1}{2}, \tau\right),$$

With the 2-division theta functions we obtain an embedding of the complex torus into the three-dimensional projective space where the image is the intersection of two quadratic surfaces whose equations are given precisely in the next theorem.

(3.5) Theorem. *The curve $\mathrm{im}(f_{(N)})$ in $\mathbb{P}_3(\mathbb{C})$ is the intersection of the two quadratic surfaces with equations*

$$A^2x_0^2 = B^2x_1^2 + C^2x_2^2,$$
$$A^2x_3^2 = C^2x_1^2 - B^2x_2^2,$$

where the mapping function is given by

$$x_0 = \theta_{0,0}(2z, \tau), \quad x_1 = \theta_{0,1/2}(2z, \tau), \quad x_2 = \theta_{1/2,0}(2z, \tau), \quad x_3 = \theta_{1/2,1/2}(2z, \tau),$$

and the coefficients are given by

$$A = \theta_{0,0}(0, \tau) = \theta_3(\tau) = \sum_n q^{n^2},$$

$$B = \theta_{0,1/2}(0, \tau) = \theta_4(\tau) = \sum_n (-1)^n q^{n^2},$$

$$C = \theta_{1/2,0}(0, \tau) = \theta_2(\tau) = \sum_n q^{(n+1/2)^2},$$

with $q = e^{i\pi\tau} = q_\tau$. *We have the additional relation* $A^4 = B^4 + C^4$.

This last relation is called Jacobi's identity between the "theta constants" A, B, and C.

The proof of this theorem is contained in §5 of the first chapter of Mumford's book. The derivation of the relations is rather involved. Knowing that $\mathrm{im}(f_{(2)})$ is contained in the curve E defined by the two quadratic equations, we can show that $\mathrm{im}(f_{(2)}) = E$. First, by 2(3.1) a plane $a_0x_0 + \cdots + a_3x_3 = 0$ intersects E in at most four points, but this plane intersects $\mathrm{im}(f_{(2)})$ at points where

$$a_0\theta_{0,0}(2z, \tau) + a_1\theta_{0,1/2}(2z, \tau) + a_2\theta_{1/2,0}(2z, \tau) + a_3\theta_{1/2,1/2}(2z, \tau) = 0.$$

There are four such points mod $2L_\tau$ by (2.6). This completes our remarks on the proof of the theorem.

§4. Relation Between Theta Functions and Elliptic Functions

(4.1) Remark. If $f(z)$ is a holomorphic function satisfying for L_τ the two conditions $f(z) = f(z+1)$ and $f(z) = e^{az+b} f(z+\tau)$, then the second logarithmic derivative $(d^2/dz^2) \log f(z)$ is a doubly periodic function for the lattice L_τ and is meromorphic on the plane. In particular, it is an elliptic function. For example, there is a constant c such that the Weierstrass $\wp$-function is given by

$$\wp(z, L_\tau) = -\frac{d^2}{dz^2} \log \theta_{1/2,1/2}(z, \tau) + c.$$

Referring to 9(3.2) and the exercises at the end of the section, we see that $\theta_{1/2,1/2}(z, \tau)$ is essentially the sigma function $\sigma(z, L_\tau)$ and the logarithmic derivative is the zeta function $\zeta(z, L_\tau)$. This leads to two constructions for elliptic functions related to the exercises in Chapter 9, §3 using holomorphic $f(z)$ satisfying the theta periodicity conditions of (4.1).

(4.2) Remark. For $a_1 + \cdots + a_k = b_1 + \cdots + b_k$ in $\mathbb{C}$, the function $f(z)$ satisfying the theta periodicity conditions of (4.1) gives an elliptic function

$$\prod_{j=1}^{k} \frac{f(z-a_j)}{f(z-b_j)}.$$

If $f = \theta$, then this elliptic function has zeros at $a_j + (1/2)(1+\tau)$ and poles at $b_j + (1/2)(1+\tau)$, and if $f = \theta_{1/2,1/2}$, then this elliptic function has zeros at a_j and poles at b_j. For $f = \theta$ or $\theta_{1/2,1/2}$ we can represent every elliptic function in the above form.

This representation should be compared with the factorization

$$f(z_0 : z_1) = \prod_{i=0}^{k} \frac{a_j z_1 - a'_j z_0}{b_j z_1 - b'_j z_0}$$

of a meromorphic (rational) function on $\mathbb{P}_1(\mathbb{C})$ with zeros a'_j/a_j and poles b'_j/b_j.

(4.3) Remark. For $a_1, \ldots, a_k$ in $\mathbb{C}$ and $c_1 + \cdots + c_k = 0$ the function $f(z)$ satisfying the theta periodicity conditions of (4.1) gives an elliptic function

$$\sum_{j=1}^{k} c_j \frac{d}{dz} \log f(z - a_j).$$

If $f = \theta$, then this elliptic function has simple poles at $a_j + (1/2)(1+\tau)$ with residues c_j, and if $f = \theta_{1/2,1/2}$, then this elliptic function has simple poles at a_j with residues c_j for $j = 1, \ldots, k$.

§5. The Tate Curve

We begin by referring to (1.6) where the curve

$$E_q : \quad Y^2 = X^3 - e_2(q)X - e_3(q)$$

was derived from the Weierstrass differential equation by change of variable $w = e^{2\pi i z}$ and $q = e^{2\pi i \tau}$, and for $|q| < 1$ the mapping

$$w \mapsto (X(w,q), Y(w,q)) = \varphi(w)$$

defines an isomorphism $\varphi : \mathbb{C}^*/q^{\mathbb{Z}} \to E_q(\mathbb{C})$, where $\varphi(q^{\mathbb{Z}}) = 0$. We wish to see that the formulas which described the curve E_q over the complex numbers works for a general local field K. To do this, we introduce a small change of variables from those used in (1.6)

$$x = x(w,q) = X(w,q) - \frac{1}{12} \quad \text{and} \quad y = Y - \frac{1}{2}x.$$

(5.1) Proposition. *In terms of $x(w,q)$ and $y(w,q)$ the equation of the cubic becomes*

$$E_q: \quad y^2 + xy = x^3 - h_2(q)x - h_3(q),$$

where

$$x = x(w,q) = \sum_m \frac{wq^m}{(1-wq^m)^2} - 2\sum_{m\ge 1}\frac{q^m}{(1-q^m)^2}$$

$$y = y(w,q) = \sum_m \frac{(wq^m)^2}{(1-wq^m)^3} + \sum_{m\ge 1}\frac{q^m}{(1-q^m)^2}$$

$$h_2(q) = \frac{1}{48}(E_4(\tau)-1) = 5\sum_{1\le n}\sigma_3(n)q^n = 5\sum_{1\le n}\frac{n^3q^n}{1-q^n}$$

$$h_3(q) = \frac{1}{12^3}(1-3E_4(\tau)+2E_6(\tau)) = -\frac{1}{12}\sum_{1\le n}(5\sigma_3(n)+7\sigma_5(n))q^n$$

$$= -\frac{1}{12}\sum_{1\le n}\frac{(5n^3+7n^5)q^n}{1-q^n}$$

Proof. The previous equation from (1.6) becomes

$$\left(y+\frac{x}{2}\right)^2 = \left(x+\frac{1}{12}\right)^3 - \frac{1}{48}E_4(\tau)\left(x+\frac{1}{12}\right) + \frac{2}{12^3}E_6(\tau)$$

or

$$y^2 + xy = x^3 + \frac{1}{48}x + \frac{1}{12^3} - \frac{1}{48}E_4(\tau)x - \frac{3}{12^3}E_4(\tau) + \frac{2}{12^3}E_6(\tau).$$

The remainder of the calculation is left to the reader.

Observe that

$$\frac{T}{(1-T)^2} = \frac{1}{T-2+T^{-1}} = \frac{T^{-1}}{(1-T^{-1})^2}$$

and that

$$T\frac{d}{dT}\left(\frac{T}{(1-T)^2}\right) = \frac{T+T^2}{(1-T)^3}$$

so that

$$\frac{T+T^2}{(1-T)^3} = -\frac{T^{-1}+T^{-2}}{(1-T^{-1})^3}.$$

(5.2) Remark. Let K be a non-Archimedean field and $|q| < 1$. The series expansions for $x(w,q)$ and $y(w,q)$ can be rewritten as

$$x(w,q) = \frac{w}{(1-w)^2} + \sum_{1\le n}\left(\frac{wq^n}{(1-wq^n)^2} + \frac{q^nw^{-1}}{(1-q^nw^{-1})^2} - 2\frac{q^n}{(1-q^n)^2}\right),$$

$$y(w,q) = \frac{w^2}{(1-w)^3} + \sum_{1\le n}\left(\frac{(wq^n)^2}{(1-wq^n)^3} + \frac{(q^n w^{-1})^2}{(1-q^n w^{-1})^3} - \frac{q^n}{(1-q^n)^2}\right).$$

Comparing with the geometric series, we see that these two series converge absolutely for any w in the multiplicative group K^* and uniformly on any annulus $0 < r \le |w| \le r'$. An examination of the series expansions yields the functional equations

$$x(qw,q) = x(w,q) = x(w^{-1},q),$$

and

$$y(qw,q) = y(w,q) \quad \text{and} \quad y(w^{-1},q) + y(w,q) = -x(w,q).$$

The series giving $h_2(q)$ and $h_3(q)$ converge since the coefficents are integers.

(5.3) Remark. The modular invariant is given by

$$\begin{aligned}\Delta = 12^3(g_2^3 - 27g_3^2) &= 12^3\left[(4h_2 + \frac{1}{12})^3 - 4\left(4h_3 + \frac{1}{3}h_2 + \frac{1}{6^3}\right)^2\right]\\ &= h_2^2 - h_3 - 72h_2h_3 - 432h_3^2 + 64h_2^3\\ &= q - 24q^2 + 252q^3 + \cdots.\end{aligned}$$

Since $\Delta \equiv q (\operatorname{mod} q^2)$, it follows that $\Delta(q) \ne 0$ for any $q \ne 0$. The absolute j-invariant is given by

$$\begin{aligned}j = \frac{(12g_2)^3}{\Delta} = \frac{(1+48h_2)^3}{\Delta} &= \frac{1 + 240q + 2160q^2 + \cdots}{q - 24q^2 + 252q^3 + \cdots}\\ &= \frac{1}{q}(1 + 744q + 196884q^2 + \cdots).\end{aligned}$$

(5.4) Lemma. *Let K be a complete discrete value field, and let $f(t) = a_0 + a_1 t + \cdots$ be a power series with $|a_j| \le 1$ for all j. Then the map $q \mapsto j(q) = 1/q + f(q)$ is a bijection of the set of q with $0 < |q| < 1$ onto the set of all j in K with $|j| > 1$.*

Proof. Since $x^n - y^n = (x-y)(x^{n-1} + x^{n-2}y + \cdots + y^{n-1})$, we have $|f(q) - f(q')| \le |q - q'|$. Clearly $|j(q)| = |1/q + f(q)| = |1/q| > 1$, and $j(q)$ is injective because

$$|j(q) - j(q')| = \left|\frac{q'-q}{q'q} + f(q) - f(q')\right| = \frac{|q'-q|}{|q'q|} > |q'-q|.$$

To show that $j(q)$ is surjective, we consider $|j| > 1$ and solve for q in the relation $j = 1/q + f(q) = [1 + qf(q)]/q$ or $q = [1 + qf(q)]/j = T(q)$. The function T is defined $X \to X$, where X is the complete metric space of all q with $0 < |q| < 1$. Since

$$|T(q) - T(q')| = \frac{|qf(q) - q'f(q')|}{|j|} \le \frac{1}{|j|}|q - q'|$$

with $1/|j| < 1$, the elementary contraction fixed point theorem applies, and there is a solution to the equation $q = T(q)$ which in turn maps to the given j. This proves the lemma.

Observe that $j = 1/q + f(q)$ with $f(q)$ in $R[[q]]$ as above can be inverted with a formal series $g(x)$ in $R[[x]]$ in the sense that $q = 1/j + g(1/j)$. This observation can also be used to prove the lemma.

(5.5) Definition. The Tate curve E_q over a local field K is defined for all q in K with $0 < |q| < 1$ by the equation

$$y^2 + xy = x^3 - h_2(q)x - h_3(q),$$

where the coefficients $h_2(q)$ and $h_3(q)$ are given by the convergent series

$$h_2(q) = 5\sum_{1 \le n} \frac{n^3 q^n}{1 - q^n} \quad \text{and} \quad h_3(q) = -\frac{1}{12}\sum_{1 \le n} \frac{(5n^3 + 7n^5)q^n}{1 - q^n}.$$

Of course the formulas were derived from other formulas over the complex numbers, see (5.1), and we noted that they converge as series in a non-Archimedean field K.

(5.6) Remark. By (5.3) the curve E_q is an elliptic curve with bad reduction since $\Delta \equiv q (\bmod q^2)$ and $|q| < 1$. The reduced curve $\bar{E}_q$ is given by $y^2 + xy = x^3$ since the series expansions have the form

$$h_2(q) = 5q + \cdots \quad \text{and} \quad h_3(q) = -q + \cdots,$$

and $(\bar{E}_q)_{\mathrm{ns}}$ is the multiplicative group, see 3(7.2). Since $|j(E_q)| > 1$, the singularity is not removed by ground field extension, and this is related to the terminology of semistable reduction, see 5(7.1). By (5.4) for each nonintegral j in K, so $|j| > 1$, there is a unique Tate curve E_q with $j = j(E_q)$. If E is an elliptic curve over K with $|j(E)| > 1$, then E is isomorphic to the Tate curve E_q over a quadratic extension of K where q is chosen such that $j(E) = j(E_q)$.

The remarkable and useful feature of these curves E_q of Tate is that, as in the case of the complex numbers, the group of points $E_q(K)$ can be viewed as a "torus," namely $E_q(K)$ is isomorphic to the p-adic torus $K^*/q^{\mathbb{Z}}$. One proof of this result is sketched in the next theorem, and another version is a consequence of the study of the p-adic theta functions contained in the next section.

(5.7) Theorem (Tate). *The function $\varphi_q : K^*/q^{\mathbb{Z}} \to E_q(K)$ defined by*

$$\varphi_q(w) = \begin{cases} 0 = \text{the origin of } E_q(K) & \text{for } w \text{ in } q^{\mathbb{Z}}, \\ (x(w,q), y(w,q)) & \text{for } w \text{ in } K^* - q^{\mathbb{Z}}, \end{cases}$$

where $x(w,q)$ and $y(w,q)$ are given in (5.1) or (5.2) is an isomorphism of groups.

Remarks. The first point of the proof is that $\varphi_q(w)$ is on the curve $E_q(K)$ in the plane. From the classical theory $\varphi_q(w)$ is on the curve with equation $y^2 + xy = x^3 - h_2(q)x - h_3(q)$ so that we have a formal identity over the ring of power series in q with rational coefficients by varying q and keeping $|q| < |w| < |q|^{-1}$. In particular, the formal identity specializes to the conclusion that $\varphi_q(w)$ is on $E_q(K)$ for all $w \in K^*/q^{\mathbb{Z}}$.

If $ww' = 1$ in $K^*/q^{\mathbb{Z}}$, then $x(w, q) = x(w', q)$ and

$$y(w, q) + y(w', q) = -x(w, q)$$

by (5.2). Thus $\varphi_q(w') = -\varphi_q(w)$ since both points are on the vertical line $x = x(w, q)$ where it intersects $E_q(K)$.

Even though the argument can be worked out directly, we postpone now the remainder of the proof of the theorem to the next section where we use some of the theory of q-periodic meromorphic functions on K^*, see (6.11).

The reader can check that the isomorphism φ_q carries the quotient of the p-adic filtration on $K^*/q^{\mathbb{Z}}$ coming from K^* onto the canonical p-adic filtration on E_q defined in (4.1). Also, for an algebraic field extension L over K we have a commutative diagram of morphisms with the horizontal morphism being isomorphisms

$$\begin{array}{ccc} L^*/q^{\mathbb{Z}} & \xrightarrow{\varphi_q} & E_q(L) \\ \cup & & \cup \\ K^*/q^{\mathbb{Z}} & \xrightarrow{\varphi_q} & E_q(K). \end{array}$$

(5.8) Remark. Restricting φ_q to the N-torsion elements or N-division points, we obtain an isomorphism

$$\varphi_q : {}_N(K^*/q^{\mathbb{Z}}) \to {}_N E_q(K).$$

Observe that these isomorphic groups are isomorphic to $(\mathbb{Z}/N\mathbb{Z})^2$ if and only if K contains a primitive Nth root of unity and an Nth root of q. In this case one primitive Nth root of unity and one Nth root of q forms a basis of ${}_N(K^*/q^{\mathbb{Z}})$ over $\mathbb{Z}/N\mathbb{Z}$ and their image under φ_q is a basis of ${}_N E_q(K)$ over $\mathbb{Z}/N\mathbb{Z}$.

(5.9) Remarks. If u is an automorphism of the field K with $u(q) = q$, then u acts on the quotient $K^*/q^{\mathbb{Z}}$ and thus on $E_q(K)$, and the isomorphism φ_q is u-equivariant, i.e., $\varphi_q(u(x)) = u(\varphi_q(x))$. In terms of the Nth root of unity, Nth root of q basis, the automorphism u will have a matrix representation

$$\begin{pmatrix} a & b \\ 0 & 1 \end{pmatrix},$$

where a is in the automorphism group of Nth roots of unity contained in $(\mathbb{Z}/N\mathbb{Z})^*$ and b is in $\mathbb{Z}/N\mathbb{Z}$ viewed as the group of Nth roots of unity with a chosen Nth root of unity. For the separable algebraic closure K_s of K the action

$$\varphi_N : \mathrm{Gal}(K_s/K) \to \mathrm{Aut}({}_N E_q) \cong \mathrm{GL}_2(\mathbb{Z}/N\mathbb{Z})$$

is through the subgroup of matrices of the form

$$\begin{pmatrix} * & * \\ 0 & 1 \end{pmatrix}.$$

In particular it is never surjective.

This property of the Galois action plays an important role in the global case.

(5.10) Remark. Let N be prime to the characteristic of K. The field $K({}_N E_q)$ generated over K by the coordinates of the N-division points is exactly $K(\zeta_N, q^{1/N})$ the field generated by a primitive Nth root ζ_N of 1 and an Nth root $q^{1/N}$ of q.

Returning to 5(7.4), we summarize the situation with the following statement.

(5.11) Remark. For E over K with $v(j(E)) < 0$, we form E_q for $q \in K^*$ such that $v(q) = -v(j(E))$ from the q-expansion of $j(q)$, and $j(q) = j(E)$. Then E and E_q are isomorphic over a quadratic extension of K. Ogg [1967, p. 5] shows that E and E_q are isomorphic over K if and only if the reduced curve $\bar{E}$ is the multiplicative group. Otherwise it is the additive group in the case where k is algebraically closed corresponding to a cusp.

§6. Introduction to Tate's Theory of p-Adic Theta Functions

Over the complex numbers analytic functions are defined by convergent power series locally. The same is true for a complete non-Archimedian field K as considered in 5(1.1), but there is no analytic continuation since every triangle is isosceles. For K^* we use the following special definition of holomorphic function.

(6.1) Definition. A K-valued function f on K^* is holomorphic provided it is of the form $f(z) = \sum_{n\in\mathbb{Z}} a_n z^n$, where $a_n \in K$ and the series converges absolutely for all z in K^*.

The convergence condition is equivalent to saying that $|a_n|r^n$ is bounded in n for each real $r > 0$. These functions form a ring which we denote by H_K. In order to study f near a point c in K^* we consider $f_c(w) = f(c(1+w))$ and expand this function of w as power series which converges for small $|w|$

$$f_c(w) = f(c) + b_1 w + b_2 w^2 + \cdots .$$

We can define $\mathrm{ord}_c(f)$ the order of the zero of f at c by the relations $\mathrm{ord}_c(f) = 0$ provided $f(c) \neq 0$ and $n = \mathrm{ord}_c(f)$ provided $f(c) = 0$, $b_n \neq 0$, and $b_i = 0$ for $i < n$. For each c the function that assigns to $f(z)$ in H_K the power series f_c in $K[[w]]$ is an embedding of rings.

(6.2) Remarks and Notations. The ring H_K is an integral domain, and we can form the field of fractions M_K of meromorphic functions on K^*. A nonzero $f(z)$ in H_K has only a finite number of zeros in any annulus $r_1 \le |z| \le r_2$. The only functions $f(z)$ with no zeros on $\bar{K}^*$ are monomials of the form $a_n z^n$. This is a result which one proves by looking at the Newton polygon of the $f(z)$ and studying the corners in this polygon. The method is explained in the book by E. Artin, *Algebraic Numbers and Algebraic Functions* (1968).

Now fix an element q in K with $0 < |q| < 1$. The object is to form the quotient $A_q = K^*/q^{\mathbb{Z}}$ of K^* by the discrete subgroup $q^{\mathbb{Z}}$ and find an elliptic curve E_q with $A_q = E_q(K)$. This we did with specific equations in the previous section from the Weierstrass theory of elliptic functions. Now we outline a theta function approach to the problem.

(6.3) Notations. Let $M_{K,q}$ denote the subfield of $f(z)$ in M_K satisfying the condition that $f(z) = f(qz)$.

Then $M_{K,q}$ is the field of meromorphic functions on A_q. Analogously, if $F(u)$ is a complex elliptic function with periods 1 and τ, then the function $f(z) = F(e^{2\pi i z})$ is a complex valued meromorphic function on $\mathbb{C}^*$ satisfying $f(z) = f(qz)$, where $q = e^{2\pi i \tau}$. As with elliptic functions we see that $H_K \cap M_{K,q}$ is the field of constants.

Again by analogy with the classical case over the complex numbers we consider functions which are holomorphic on $\mathbb{C}^*$, almost periodic under multiplication by q, and with periodic set of zeros.

(6.4) Definition. A theta function $f(z)$ of type cz^r is a holomorphic $f(z)$ in H_K satisfying $f(z) = cz^r f(qz)$ for z in K^*.

A theta function is always defined relative to a given quasiperiod q and type cz^r. We denote by $H_{K,q}(cz^r)$ the vector space over K of all theta functions relative to q of type cz^r. Observe that elements of $M_{K,q}$ can be constructed as quotients of two elements of $H_{K,q}(cz^r)$. Hence the dimension of this space is important.

(6.5) Proposition. *We have*

$$\dim_K H_{K,q}(cz^r) = \begin{cases} r & \text{if } r > 0, \\ 0 & \text{if } r < 0, \\ 0 & \text{if } r = 0 \text{ and } c \notin q^{\mathbb{Z}}, \\ 1 & \text{if } r = 0 \text{ and } c \in q^{\mathbb{Z}}. \end{cases}$$

Proof. Let $f(z) = \sum_n a_n z^n$ be holomorphic on K^*. Form the function

$$cz^r f(qz) = \sum_n c a_n q^n z^{n+r}$$

and use

$$f(z) = \sum_n a_{n+r} z^{n+r}.$$

Then we see that $f(z) = cz^r f(qz)$ if and only if $a_{n+r} = ca_n q^q$ for all integers n. Check the case $r = 0$ directly. For $r \neq 0$ we see that we can pick the coefficients $a_0, \ldots, a_{r-1}$ arbitrarily. As for convergence, the quasiperiodicity condition becomes

$$a_{n+mr} = a_n c^m q^{mn+r(m^2-m)/2} \to 0 \quad for \quad r > 0$$

since the m^2 term dominates, and we have convergence in this case and no convergence for $r < 0$. This proves the proposition.

(6.6) Notation. A basic theta function of type $-z^{-1}$ with zeros exactly q^Z is

$$\varphi(z) = (1-z)\prod_{1\le m}(1-q^m z)\prod_{1\le m}\left(1-\frac{q^m}{z}\right).$$

The fact that $-z\varphi(qz) = \varphi(z)$ is an easy shuffling of the terms. As an expansion in powers z^n of z, the function $\varphi(x)$ has the form

$$\varphi(z) = \sum_n (-1)^n q^{(n^2-n)/2} z^n.$$

Observe that $\varphi(z, c) = \varphi_c(z) = \varphi(z/c)$ is a theta function of type $-c^{-1}z$ with zeros the set $cq^{\mathbb{Z}}$.

(6.7) Remark. If $\prod_{1\le i\le k} a_i = \prod_{1\le i\le k} b_i$ in the field K, then the quotient

$$\frac{\prod_{1\le i\le k}\varphi(z, a_i)}{\prod_{1\le i\le k}\varphi(z, b_i)}$$

is a theta function of type $(-1)^k(\prod_{1\le i\le k} a_i)z^k/(-1)^k(\prod_{1\le i\le k} b_i)z^k = 1$, thus it is an elliptic function in $M_{K,q}$. The zeros are the a_i's and the poles are the b_i's. Conversely, from the fact that $H_K \cap M_{K,q}$ is the field of constants, we deduce that for any f in $M_{K,q}$ with zeros $a_i \cdot q^{\mathbb{Z}}$ with $i = 1, \ldots, k$ and poles $b_j \cdot q^{\mathbb{Z}}$ with $j = 1, \ldots, m$, then $k = m$ and for suitable representatives a_i and b_j the quotient $(\prod_{1\le j\le k} a_i)/(\prod_{1\le i\le k} b_j)$ equals 1. This is the Abel–Jacobi theorem in this context.

The previous remark, the Abel–Jacobi theorem, is a multiplicative criterion for the existence of a q-periodic meromorphic function with given zeros and poles. The additive description of q-periodic meromorphic functions with prescribed polar expansions is the Riemann approach to the existence of q-periodic meromorphic functions and the Riemann–Roch theorem is the answer in the form of a dimension formula. An important special case for the K-vector space $L(n)$ of $f \in M_{K,q}$ with only poles on $q^{\mathbb{Z}}$ and having order at most n is studied by observing that $f(z) \mapsto \varphi(z)^n f(z)$ is an isomorphism $L(n) \to H_{K,q}((-1)^n z^n)$ with inverse $g(z) \mapsto \varphi(z)^{-n} g(z)$. Hence from (6.5) we deduce the following dimension formula.

(6.8) Proposition (Riemann–Roch). *The vector space $L(n)$ of q-periodic functions with pole of order at most n on $q^{\mathbb{Z}}$ has dimension given by*

$$\dim_K L(n) = \begin{cases} 1 & for\ n = 0, \\ n & for\ n \ge 1. \end{cases}$$

The one-dimensional spaces $L(0)$ and $L(1)$ reduce to constant functions K, and hence there is no q-periodic meromorphic function with only a simple pole on $q^{\mathbb{Z}}$, i.e., a simple pole at each point of $q^{\mathbb{Z}}$. Then $L(2)$ has dimension 2 over K and is, for example, a direct sum $K \oplus Kx(w, q)$ where $x(w, q)$ was introduced in (5.1) and (5.2). Any function in $L(2) - K$ must have a double pole on $q^{\mathbb{Z}}$ which can be seen directly from the expansion for $x(w, q)$. The function $x(w, q)$ also satisfies the evenness condition of $x(1/w, q) = x(w, q)$. By (6.7) any nonconstant $f(z)$ and $L(2)$ has two zeros in the region $|q| < |z| \leq 1$, and they are of the form

(1) $z_0, q/z_0$ if $|q| < |z_0| < 1$, or
(2) $z_0, 1/z_0$ if $|z_0| = 1$.

Using a multiplicative version of the argument in 5(3.3), we establish as in the classical case the following for the subfield H of even $f(z)$ in $M_{K,q}$, i.e., $f(z)$ satisfying $f(z) = f(1/z)$ the following theorem.

(6.9) Theorem. *The subfield H of even functions in $M_{K,q}$ is a rational function field in one variable $K(x(w, q))$ over K generated by $x(w, q)$ and $M_{K,q}$ is a quadratic extension of H.*

The inclusions of $K = L(0) = L(1) \subset L(2) \subset L(3) \subset \cdots \subset L(n) \subset \cdots$ together with the dimension formula used in 6(1.1) to derive the normal form of the cubic equation for an abstract curve E of genus 1 embedded into $\mathbb{P}_2$. Implicit in the normal form is the fact that the function field $k(E)$ of E is a quadratic extension of a rational function field in one variable over k. The same considerations apply to the p-adic torus, as they did for the complex torus, to yield an embedding into $\mathbb{P}_2$. In fact, for any x in $L(2) - L(1)$ and y in $L(3) - L(2)$ we have an embedding $(1 : x : y) : K^*/q^{\mathbb{Z}} \to \mathbb{P}_2(K)$. Looking at it in terms of any basis f_0, f_1, f_2 of $H_{K,q}(-z^3)$, we also have an embedding $(f_0 : f_1 : f_2) : K^*/q^{\mathbb{Z}} \to \mathbb{P}_2(K)$. The two approaches can be related by choosing, for example, $f_0 = \varphi^3$, $f_1 = \varphi^3 x$, and $f_2 = \varphi^3 y$.

(6.10) Remark. Returning to the functions $x(w, q)$ and $y(w, q)$ in (5.1) and (5.2) where x is in $L(2) - L(1)$ and y in $L(3) - L(2)$, we can apply the above considerations. The function y will generate the quadratic extension $M_{K,q}$ over $H = K(x)$, but, further, by (5.2) we have the relation $y(w, q) + y(1/w, q) = -x(w, q)$, and an easy check shows that $y(w, q)^2 + x(w, q)y(w, q)$ is invariant under w changing to $1/w$. Since this expression has a pole of at most order 6 and since it is in $K[x]$, we deduce that $y^2 + xy$ is a cubic polynomial in x with leading coefficient 1. This almost recovers the equation which was derived by classical expansions.

(6.11) Remark. Finally we complete the proof of Theorem (5.7) concerning $\varphi_q : K^*/q^{\mathbb{Z}} \to E_q(K)$, where $\varphi_q(w) = (1 : x(w, q) : y(w, q))$ and the assertion that $(f_0 : f_1 : f_2) : K^*/q^{\mathbb{Z}} \to \mathbb{P}_2(K)$ is an embedding is contained in the argument. To show that φ_q is bijective, consider $(\bar{x}, \bar{y})$ in $E_q(K)$ and recall by (6.7) that $x(w, q) - \bar{x}$ has two zeros w and w' in $K^*/q^{\mathbb{Z}}$ with ww' in $q^{\mathbb{Z}}$ and $\varphi_q(w') = -\varphi_q(w)$. Now $y(w, q) - \bar{y}$ has three zeros v, v', and v'' none in $q^{\mathbb{Z}}$ with $vv'v''$ in $q^{\mathbb{Z}}$. Hence only one w or w' is a zero of $y(w, q) - \bar{y}$, and this is the unique $\bar{w}$ in $K^*/q^{\mathbb{Z}}$ with $\varphi_q(\bar{w}) = (\bar{x}, \bar{y})$. Therefore φ_q is bijective.

The last step is to show that $\varphi_q(w)$ is a group homomorphism. We saw in (5.7) that it preserved inverses. Now consider $w_1 w_2 w_3 = 1$ in $K^*/q^{\mathbb{Z}}$. By (6.7) there exists $f(w)$ in $L(3)$ with a triple pole on $q^{\mathbb{Z}}$ and three zeros w_1, w_2, and w_3. Moreover, $f(w) = ay(w, q) + bx(w, q) + c$ with a, b, and c in K and $a \neq 0$. If $(x_i, y_i) = \varphi_q(w_i)$, then we have relations

$$a(y_1 - y_2) + b(x_1 - x_2) = 0 \quad \text{and} \quad a(y_2 - y_3) + b(x_2 - x_3) = 0.$$

These relations of proportionality show that (x_1, y_1), (x_2, y_2), and (x_3, y_3) are on a line, and, hence, they add to zero on $E_q(K)$. This completes the proof of Theorem (5.7).

(6.12) Remark. There is an isogeny cyclic of degree m defined $E_{q^m} \to E_q$ with induced field morphism $K(E_q) \to K(E_{q^m})$ given by the inclusion of the q-periodic meromorphic functions into the field of q^m-periodic meromorphic functions. This is clear in the context of p-adic tori just as questions of isogenies of complex elliptic curves reduce to inclusion properties of lattices.

A general reference for p-adic theta functions is the book by Peter Roquette, *Analytic Theory of Elliptic Functions over Local Fields*.

11

Modular Functions

Every elliptic curve E over the complex numbers $\mathbb{C}$ corresponds to a complex torus $\mathbb{C}/L_\tau$ with $L_\tau = \mathbb{Z}\tau + \mathbb{Z}$ and $\tau \in \mathfrak{H}$, the upper half plane, where $\mathbb{C}/L_\tau$ is isomorphic to the complex torus of complex points $E(\mathbb{C})$ on E. In the first section we show easily that $\mathbb{C}/L_\tau$ and $\mathbb{C}/L_{\tau'}$ are isomorphic if and only if there exists

$$\begin{pmatrix} a & b \\ c & d \end{pmatrix} \in \mathbb{SL}_2(\mathbb{Z})$$

with

$$\tau' = \frac{a\tau + b}{c\tau + d}.$$

Then for this action of $\mathbb{SL}_2(\mathbb{Z})$ on $\mathfrak{H}$ we see that $\mathbb{SL}_2(\mathbb{Z}) \setminus \mathfrak{H}$ can be identified with isomorphism classes of elliptic curves over $\mathbb{C}$. Such a space is called a moduli space for elliptic curves, and this and related moduli spaces, called modular curves, are considered in §2. These modular curves are closely related to the existence of torsion points on elliptic curves over $\mathbb{Q}$.

Modular functions are functions on the upper half plane $\mathfrak{H}$ satisfying certain transformations laws under the action of $\mathbb{SL}_2(\mathbb{Z})$. They include $j(\tau)$, $g_2(\tau)$, $g_3(\tau)$, and $\Delta(\tau)$ considered in 9(4.8), 10(1.4), and 10(1.8).

§1. Isomorphism and Isogeny Classification of Complex Tori

(1.1) Remark. Let $\begin{pmatrix} a & b \\ c & d \end{pmatrix}$ be a 2 by 2 real matrix and z a complex number. The imaginary part of the transform is given by

$$\mathbf{Im}\left(\frac{az + b}{cz + d}\right) = \frac{(ad - bc)\mathbf{Im}(z)}{|cz + d|^2}.$$

In particular, the group $\mathbb{GL}_2^+(\mathbb{R})$ of all real nonsingular $\begin{pmatrix} a & b \\ c & d \end{pmatrix}$ with $ad - bc > 0$ acts on the upper half plane $\mathfrak{H}$. By restriction any subgroup of $\mathbb{GL}_2^+(\mathbb{R})$ also acts on $\mathfrak{H}$. We will be especially interested in the cases of $\mathbb{SL}_2(\mathbb{Z})$ and $\mathbb{GL}_2^+(\mathbb{Q})$.

Recall from 9(1.3) and 9(1.4) the definition of isomorphism and isogeny. This leads to an equivalence relation.

(1.2) Definition. Two complex tori $T = \mathbb{C}/L$ and $T' = \mathbb{C}/L'$ are isomorphic (resp. isogenous) provided there exists an isomorphism (resp. isogeny) $\lambda : T \to T'$ induced by multiplication $\lambda : \mathbb{C} \to \mathbb{C}$ for $\lambda \in \mathbb{C}$ with $\lambda L = L'$ (resp. $\lambda L \subset L'$).

(1.3) Remark. For an isogeny $\lambda : T \to T'$ with $\lambda L \subset L'$ the index $N = [L' : \lambda L]$ is finite and $(N/\lambda)L' \subset L$ defines an isogeny $N/\lambda : T' \to T$ showing that the isogeny relation is reflexive.

(1.4) Theorem. *For τ, τ' in the upper half plane $\mathfrak{H}$ we have:*

(1) *The tori T_τ and $T_{\tau'}$ are isomorphic if and only if there exists $s \in \mathbb{SL}_2(\mathbf{Z})$ with $s(\tau) = \tau'$.*

(2) *The tori T_τ and $T_{\tau'}$ are isogenous if and only if there exists $s \in \mathbb{GL}_2^+(\mathbb{Q})$ with $s(\tau) = \tau'$.*

Proof. For an isomorphism we need a complex number λ with $\lambda L_{\tau'} = L_\tau$ and for an isogeny we need a λ with $\lambda L_{\tau'} \subset L_\tau$. Then there exists an integral matrix $\begin{pmatrix} a & b \\ c & d \end{pmatrix}$ with $\lambda\tau' = a\tau + b$ and $\lambda = c\tau + d$, and by division we obtain $\tau' = (a\tau + b)/(c\tau + d)$. Since τ, τ' are in the upper half plane,

$$ad - bc = \det \begin{pmatrix} a & b \\ c & d \end{pmatrix} > 0,$$

and further

$$\det \begin{pmatrix} a & b \\ c & d \end{pmatrix} = 1$$

if and only if $\lambda L_{\tau'} = L_\tau$.

Conversely, the relation can be written, after clearing denominators in the entries of s if necessary, as

$$\tau' = \frac{a\tau + b}{c\tau + d},$$

where a, b, c, and d are integers. Let $\lambda = c\tau + d$ so that $\lambda\tau' = a\tau + b$. Then $\lambda L_{\tau'} \subset L_\tau$ with $\lambda L_{\tau'} = L_\tau$ if and only if

$$\det \begin{pmatrix} a & b \\ c & d \end{pmatrix} = 1.$$

This completes the proof of the theorem.

(1.5) Remark. From 9(4.8) we know that $j(L) = j(\lambda L)$, and denoting again $j(\tau) = j(L_\tau)$, we have

$$j(\tau) = j\left(\frac{a\tau + b}{c\tau + d}\right) \qquad \text{for all} \begin{pmatrix} a & b \\ c & d \end{pmatrix} \in \mathbb{SL}_2(\mathbb{Z})$$

in view of 1.4. In 10(1.8) we have an expansion of the function $j(\tau)$ in terms of $q = e^{2\pi i\tau}$ since $j(\tau) = j(\tau + 1) = j\left(\begin{pmatrix} 1 & 1 \\ 0 & 1 \end{pmatrix}\tau\right)$.

Now we return to 3(4.2) where the isomorphism of an elliptic curve were determined and make explicit these isomorphisms in terms of the complex torus $T_\tau = \mathbb{C}/L_\tau$.

(1.6) Remark. There are two special values of $j(E)$ where $\mathrm{Aut}(E)$ is larger than (± 1):

(1) $j(E) = 12^3$. In this case $\mathrm{Aut}(E) = \{\pm 1, \pm i\}$. This is the case where $\tau = i$, $g_3(i) = 0$, $E(\mathbb{C})$ is isomorphic to $\mathbb{C}/\mathbb{Z}i + \mathbb{Z}$, and $y^2 = 4x^3 - g_2(i)x$ is the cubic equation for E.

(2) $j(E) = 0$. In this case $\mathrm{Aut}(E) = \{\pm 1, \pm\rho, \pm\rho^2\}$, the group of sixth roots of unity where $\rho^2 + \rho + 1 = 0$. This is the case where $\tau = \rho$, $g_2(\rho) = 0$, $E(\mathbb{C})$ is isomorphic to $\mathbb{C}/\mathbb{Z}_\rho + \mathbb{Z}$, and $y^2 = 4x^3 - g_3(\rho)$ is the cubic equation for E.

These cases play a special role in the next two sections.

(1.7) Summary Remarks. For two values τ and τ' in the upper half plane the following are equivalent:

(1) There exists s in $\mathbb{SL}_2(\mathbb{Z})$ with $s(\tau) = \tau'$.
(2) The tori T_τ and $T_{\tau'}$ are isomorphic.
(3) $j(\tau) = j(\tau')$.

§2. Families of Elliptic Curves with Additional Structures

In (1.4) we saw that every elliptic curve E over $\mathbb{C}$ viewed as a complex torus is isomorphic to $T_\tau = \mathbb{C}/L_\tau$ where L_τ is the lattice $\mathbb{Z}\tau \oplus \mathbb{Z}$, and, further, T_τ and $T_{\tau'}$ are isomorphic if and only if $s(\tau) = \tau'$ for some $s \in \mathbb{SL}_2(\mathbb{Z})$. Therefore, the points of the quotient $\mathbb{SL}_2(\mathbb{Z}) \setminus \mathfrak{H}$ parametrize the isomorphism classes of elliptic curves. In a situation like this, it is natural to try and define a family $p : X \to B$ of elliptic curves over $B = \mathbb{SL}_2(\mathbb{Z}) \setminus \mathfrak{H}$ where for each $b = \tau \pmod{\mathbb{SL}_2(\mathbb{Z})} \in B$, the fibre $p^{-1}(b)$ is an elliptic curve isomorphic to $E(\tau)$. The natural place to start is with the family $q : Y \to \mathfrak{H}$ where $q^{-1}(\tau)$ is $E(\tau)$. This family is the quotient of $\mathfrak{H} \times \mathbb{C} \to \mathfrak{H}$ by the relation: (τ, z) and (τ', z') are equivalent if and only if $\tau = \tau'$ and $z - z' \in L_\tau$. Clearly $q^{-1}(\tau) = \{\tau\} \times (\mathbb{C}/L_\tau)$ is isomorphic to $E(\tau)(\mathbb{C})$. But if we try to identify further (τ, z) and (τ', z') where $s(\tau) = \tau'$ for $s \in \mathbb{SL}_2(\mathbb{Z})$ so as to obtain the desired family over B, we run into difficulty relating z and z' because there is not a unique isomorphism $E(\tau) \to E(\tau')$. In fact, the number of isomorphisms $E(\tau) \to E(\tau')$ is usually two but can be four or six.

This leads us to consider elliptic curves together with an additional structure. When the identity is the only automorphism, we will be able to construct a family over a quotient of $\mathfrak{H}$ such that each isomorphism class appears exactly once in the family. We say then that the additional structure rigidifies the elliptic curve (or complex torus).

(2.1) Additional Structures. For an integer $N \geq 1$ and an elliptic curve $E = \mathbb{C}/L$, the subgroup ${}_NE$ of points $x \in E$ with $Nx = 0$ is given by

$$_NE(\mathbb{C}) = \ker\left(E \xrightarrow{N} E\right) = \left(\frac{1}{N}L\right)/L \subset \mathbb{C}/L.$$

This group ${}_NE$ is isomorphic to $(\mathbb{Z}/N\mathbb{Z})^2$. In fact, we can remark that this holds over any field of characteristic prime to N. This leads to three kinds of additional structures on E:

(1) A cyclic subgroup C of order N in E. An isomorphism $u : (E, C) \to (E', C')$ is an isomorphism of elliptice curves $u : E \to E'$ such that for the cyclic subgroups $u(C) = C'$.
(2) A point P of order N on E. An isomorphism $u : (E, P) \to (E', P')$ is an isomorphism of elliptic curves $u : E \to E'$ such that $u(P) = P'$ for the points of order N.
(3) A basis (P, Q) of ${}_NE$ as a $\mathbb{Z}/N\mathbb{Z}$ module. An isomorphism $u : (E, P, Q) \to (E', P', Q')$ is an isomorphism of elliptic curves $u : E \to E'$ such that $u(P) = P'$ and $u(Q) = Q'$ for the basis elements.

(2.2) Examples. For the elliptic curve $E(\tau) = \mathbb{C}/L_\tau$, where $L_\tau = \mathbb{Z}\tau \oplus \mathbb{Z}$, consider the special examples:

(1) The set $C(\tau) = (\mathbb{Z}\tau + (1/N)\mathbb{Z})(\mathbb{Z}\tau + \mathbb{Z}) \subset \mathbb{C}/L_\tau = E(\tau)$ is a cyclic subgroup of order N equal to the image of $\{0, 1/N, 2/N, \ldots, (N-1)/N\}$ along the real axis into $\mathbb{C}/L_\tau$.
(2) The point $P(\tau) = (1/N) \mod L_\tau \in \mathbb{C}/L_\tau = E(\tau)$ is of order N. Moreover, the point $P(\tau)$ generates the cyclic group $C(\tau)$.
(3) The points $P(\tau) = (1/N) \mod L_\tau$, $Q(\tau) = (\tau/N) \mod L_\tau$ is a basis for the subgroup ${}_NE(\tau)$.

In order to compare $(E(\tau), C(\tau))$, $(E(\tau), P(\tau))$, and $(E(\tau), P(\tau), Q(\tau))$ for different values of τ in the upper half plane, we will use the following congruence subgroups of $\mathbb{SL}_2(\mathbb{Z})$ introduced by Hecke.

(2.3) Definition. For an integer $N \geq 1$ the subgroups $\Gamma(N) \subset \Gamma_1(N) \subset \Gamma_0(N) \subset \mathbb{SL}_2(\mathbb{Z})$ are defined by the following congruences on entries of an element

$$\begin{pmatrix} a & b \\ c & d \end{pmatrix} \in \mathbb{SL}_2(\mathbb{Z}) :$$

(1) $\begin{pmatrix} a & b \\ c & d \end{pmatrix} \in \Gamma_0(N)$ if and only if $c \equiv 0 \pmod N$.

(2) $\begin{pmatrix} a & b \\ c & d \end{pmatrix} \in \Gamma_1(N)$ if and only if $c \equiv 0 \pmod N$ and $a \equiv d \equiv 1 \pmod N$.

(3) $\begin{pmatrix} a & b \\ c & d \end{pmatrix} \in \Gamma(N)$ if and only if $b \equiv c \equiv 0 \pmod N$ and $a \equiv d \equiv 1 \pmod N$.

The subgroup $\Gamma(N)$ is called the full congruence subgroup of level N and $\Gamma_0(N)$ the "Nebentypus" congruence subgroup of level N.

The canonical group morphism $\mathbb{SL}_2(\mathbb{Z}) \to \mathbb{SL}_2(\mathbb{Z}/N\mathbb{Z})$ induced by $\mathbb{Z} \to \mathbb{Z}/N\mathbb{Z}$, reduction mod N, has kernel equal to $\Gamma(N)$. The subgroup $\Gamma_0(N)$ is the inverse image of the subgroup of matrices of the form $\begin{pmatrix} * & * \\ 0 & * \end{pmatrix}$ in $\mathbb{SL}_2(\mathbb{Z}/N\mathbb{Z})$ under the canonical reduction morphism, and $\Gamma_1(N)$ is the inverse image of the subgroup of matrices of the form $\begin{pmatrix} 1 & * \\ 0 & 1 \end{pmatrix}$ in $\mathbb{SL}_2(\mathbb{Z}/N\mathbb{Z})$. In the next section we take up the question of the calculation of the indices of these subgroups in $\mathbb{SL}_2(\mathbb{Z})$ and will use these remarks at that time.

(2.4) Theorem. *We use the notations of* (2.2) *for the subgroup $C(\tau)$ and the points $P(\tau)$ and $Q(\tau)$ of the elliptic curve $E(\tau)$.*

(1) *Let (E, C) be an elliptic curve E over $\mathbb{C}$ and C a cyclic subgroup of order N. There exists an isomorphism $u : (E, C) \to (E(\tau), C(\tau))$ for some τ in the upper half plane. There is an isomorphism $u : (E(\tau), C(\tau)) \to (E(\tau'), C(\tau'))$ for $\tau, \tau' \in \mathfrak{H}$ if and only if there exists $s \in \Gamma_0(N)$ such that $s(\tau) = \tau'$.*

(2) *Let (E, P) be an elliptic curve E over $\mathbb{C}$ and P a point of order N on E. There exists an isomorphism $u : (E, P) \to (E(\tau), P(\tau))$ for some τ in the upper half plane. There is an isomorphism $u : (E(\tau), P(\tau)) \to (E(\tau'), P(\tau'))$ for $\tau, \tau' \in \mathfrak{H}$ if and only if there exists $s \in \Gamma_1(N)$ such that $s(\tau) = \tau'$.*

(3) *Let (E, P, Q) be an elliptic curve E over $\mathbb{C}$ and (P, Q) a basis of ${}_NE$. There exists an isomorphism $u : (E, P, Q) \to (E(\tau), P(\tau), Q(\tau))$ for some τ in the upper half plane. There is an isomorphism $u : (E(\tau), P(\tau), Q(\tau)) \to (E(\tau'), P(\tau'), Q(\tau'))$ for $\tau, \tau' \in \mathfrak{H}$ if and only if there exists $s \in \Gamma(N)$ such that $s(\tau) = \tau'$.*

Proof. For an elliptic curve $E = \mathbb{C}/L$ we can choose a basis of the lattice $L = \mathbb{Z}\omega_1 + \mathbb{Z}\omega_2$ such that $C = (\mathbb{Z}\omega_1 + \mathbb{Z}\omega_2/N)/L$ for case (1). $P = \omega_2/N \pmod L$ for case (2), and $P = \omega_2/N \pmod L$ and $Q = \omega_1/N \pmod L$ for case (3). If u is multiplication by ω_2^{-1} and $\tau = \omega_1/\omega_2$, then $u : E \to E(\tau)$ defines an isomorphism $u : (E, C) \to (E(\tau), C(\tau))$ for case (1), $u : (E, P) \to (E(\tau), P(\tau))$ for case (2), and $u : (E, P, Q) \to (E(\tau), P(\tau), Q(\tau))$ for case (3).

For the second part of each of the assertions (1), (2), and (3) we know by 9(1.3) that there is an isomorphism $u : E(\tau) \to E(\tau')$ if and only if $u^{-1}\tau' = a\tau + b \cdot 1$ and $u^{-1}1 = c\tau + d \cdot 1$ where

$$\begin{pmatrix} a & b \\ c & d \end{pmatrix} \in \mathbb{SL}_2(\mathbb{Z}).$$

This is just the condition $u^{-1} = L_{\tau'} = L_\tau$. For assertion (1) observe that $uC = C'$ if and only if for $u^{-1}(1/N) = (c/N)\tau + d/N$ the difference

$$u^{-1}(1/N) - d/N = (c/N)\tau \in L_\tau, \quad \text{i.e.,} \quad c \equiv 0 \pmod{N}.$$

For assertion (2) observe that $uP(\tau) = P(\tau')$ if and only if $c \equiv 0 \pmod{N}$ and $d \equiv 1 \pmod{N}$. Further, these two conditions imply $a \equiv 1 \pmod{N}$ for a matrix in $\mathbb{SL}_2(\mathbb{Z})$. For assertion (3) consider also $u^{-1}(\tau/N) = (a/N)\tau + b/N$, and we deduce that $uP(\tau) = P(\tau')$ and $uQ(\tau) = Q(\tau')$ if and only if $a \equiv d \equiv 1 \pmod{N}$ and $b \equiv c \equiv 0 \pmod{N}$. This proves the theorem.

Since an automorphism $u : E \to E$ with $u \neq \pm 1$ is either $u = \pm i$ for E isomorphic to $E(i)$ or $u = \pm\rho, \pm\rho^2$ for E isomorphic to $E(\rho)$, we deduce immediately the following proposition and see how elliptic curves are rigidified with these extra structures.

(2.5) Proposition. *For $N \geq 2$ we have:*

(1) $\mathrm{Aut}(E, C) = \{\pm 1\}$ *for all* $N \geq 2$ *and* $\mathrm{Aut}(E, P) = \mathrm{Aut}(E, P, Q) = \{\pm 1\}$ *for* $N = 2$.
(2) $\mathrm{Aut}(E, P) = \mathrm{Aut}(E, P, Q) = \{1\}$ *for* $N \geq 3$.

In the cases where the automorphism group reduces to the identity, the construction mentioned at the end of the first paragraph works. In fact, when all automorphism groups reduce to $\{\pm 1\}$ the construction goes through on a twofold covering of the base space. We carry this out after introducing the open modular curves.

(2.6) Definition. The affine modular curves are orbit spaces defined as the following quotients of the upper half plane

$$Y_0(N) = \Gamma(N) \setminus \mathfrak{H}, \quad Y_1(N) = \Gamma_1(N) \setminus \mathfrak{H}, \quad Y(N) = \Gamma(N) \setminus \mathfrak{H}.$$

Since we have the subgroup relations $\Gamma(N) \subset \Gamma_1(N) \subset \Gamma_0(N) \subset \mathbb{SL}_2(\mathbb{Z})$, we have natural mappings of quotient Riemann surfaces

$$Y(N) \to Y_1(N) \to Y_0(N) \to \mathbb{SL}_2(\mathbb{Z}) \setminus \mathfrak{H}.$$

(2.7) Universal Family of Elliptic Curves with a Point of Order $N > 2$. Let $\bar{Y}_1(N)$ denote the quotient of the subset of $(\tau, z, p) \in \mathfrak{H} \times \mathbb{C} \times \mathbb{C}$, where $P(\tau) = p \pmod{L_\tau}$ is of order N, by the equivalence relation:

(τ, z, p) and (τ', z', p') are equivalent if and only if $s(\tau) = \tau'$ for $s \in \Gamma_1(N)$

and

$$uz \pmod{L_\tau} = z' \pmod{L_\tau}$$

where

$$u : (E(\tau), P(\tau)) \to (E(\tau'), P(\tau')) \text{ is the unique isomorphism.}$$

The projection $(\tau, z, p) \mapsto \tau$ induces a projection $\pi : \bar{Y}_1(N) \to Y_1(N)$ where the fibre $\pi^{-1}(\tau \bmod \Gamma_1(N)) = (E(\tau), P(\tau))$. For each elliptic curve (E, P) with a point P of order N, there is a unique fibre $\pi^{-1}(\tau \bmod \Gamma_1(N))$ isomorphic to (E, P). The family π is in this sense the universal family of elliptic curves together with a point of order N.

(2.8) Universal Family of Elliptic Curves with a Basis of the N-Division Points for $N > 2$. Let $\bar{Y}(N)$ denote the quotient of the subset of $(\tau, z, p, q) \in \mathfrak{H} \times \mathbb{C} \times \mathbb{C} \times \mathbb{C}$, where $p \bmod L = P(\tau)$ and $q \bmod L = Q(\tau)$ form a basis of the N-division points, by the equivalence relation:

(τ, z, p, q) and (τ', z', p', q') are equivalent if and only if $s(\tau) = \tau'$ for $s \in \Gamma(N)$

and

$$uz \ (\text{mod } L_\tau) = z' \ (\text{mod } L_{\tau'})$$

where

$u : (E(\tau), P(\tau), Q(\tau)) \to (E(\tau'), P(\tau'), Q(\tau'))$ is the unique isomorphism.

The projection $(\tau, z, p, q) \to \tau$ induces a projection $\pi : \bar{Y}(N) \to Y(N)$ where the fibre $\pi^{-1}(\tau \bmod \Gamma(N)) = (E(\tau), P(\tau), Q(\tau))$. For each elliptic curve (E, P, Q) with a basis (P, Q) of the N-division points, there is a unique fibre $\pi^{-1}(\tau \bmod \Gamma(N))$ isomorphic to (E, P, Q). The family $\pi : \bar{Y}(N) \to Y(N)$ is, in this sense, the universal family of elliptic curves together with a basis of the N-division points.

§3. The Modular Curves $X(N)$, $X_1(N)$, and $X_0(N)$

In this section we will see that we can systematically add a finite set of points to the open modular curves $Y(N)$, $Y_1(N)$, and $Y_0(N)$ to obtain complete curves. In the language of Riemann surfaces these completed Riemann surfaces are compact. The new points added are called cusps and are special points on the completed curves.

This completion process for the complex plane $\mathbb{C}$ is achieved by adding one point at ∞ to obtain $\mathbb{C} \cup \{\infty\} = \mathbb{P}_1(C)$, the Riemann sphere. The group $\mathbb{GL}_2(\mathbb{C})$ acts naturally on the Riemann sphere by fractional linear transformations

$$\begin{pmatrix} a & b \\ c & d \end{pmatrix}(z) = \frac{az+b}{cz+d}.$$

and, in fact, the quotient $\mathbb{GL}_2(\mathbb{C})/\mathbb{C}^*$ is the full automorphism group of $\mathbb{P}_1(\mathbb{C})$ as a Riemann surface.

(3.1) Remarks. The j-function $j : \mathfrak{H} \to \mathbb{C}$ where $j(\tau)$ is the j-value of the elliptic curve with complex points $\mathbb{C}/L_\tau$ induces functions on certain quotients of $\mathfrak{H}$. Let $\Gamma \subset \mathbb{SL}_2(\mathbb{Z})$ denote the subgroup $\pm\begin{pmatrix} 1 & b \\ 0 & 1 \end{pmatrix}$, that is, the subgroup of elements with $c = 0$. The horizontal maps in the following diagram are quotient maps:

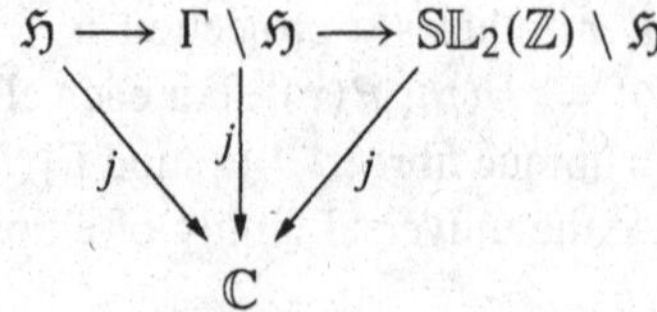

In this case $\Gamma \setminus \mathfrak{H} \xrightarrow{q} \mathfrak{U}^0$, the set of all q in $\mathbb{C}$ with $0 < |q| < 1$ where $q(\tau) = e^{2\pi i\tau}$. Then $j(q)$ has a first-order pole at 0, and hence it extends to the unit disc $\mathfrak{U} \to \mathbb{P}_1(\mathbb{C})$. This completion of $\Gamma \setminus \mathfrak{H}$ from $\mathfrak{A}^0$ to $\mathfrak{A}$ induces a completion of the modular curve $\mathbb{SL}_2(\mathbb{Z}) \setminus \mathfrak{H}$.

(3.2) Assertion/Definition. There is a unique topology on $\mathfrak{H}^* = \mathfrak{H} \cup \mathbb{P}_1(\mathbb{Q})$ such that the action of $\mathbb{SL}_2(\mathbb{Z})$ is continuous, $\mathfrak{H}$ is an open subset of $\mathfrak{H}^*$, and the neighborhoods of ∞ are supersets of sets consisting of ∞ together with all $\tau \in \mathfrak{H}$ with $\mathbf{Im}(\tau) > t$ for some t. This topology is called the horocycle topology and the quotient space $\mathbb{SL}_2(\mathbb{Z}) \setminus \mathfrak{H}^*$ is the Riemann sphere as a space. The $\mathbb{SL}_2(\mathbb{Z})$ action on $\mathbb{P}_1(\mathbb{Q})$ is the restriction from $\mathbb{P}_1(\mathbb{C})$.

Just as we completed $\mathbb{C}$ to the Riemann sphere, we can now complete open modular curves $Y_0(N)$, $Y_1(N)$, and $Y(N)$.

(3.3) Definition. The modular curves as spaces are defined to be quotients of $\mathfrak{H}^*$ with the horocycle topology

$$X_0(N) = \Gamma_0(N) \setminus \mathfrak{H}^*, \quad X_1(N) = \Gamma_1(N) \setminus \mathfrak{H}^*, \quad X(N) = \Gamma(N) \setminus \mathfrak{H}^*.$$

The natural mappings

$$X(N) \to X_1(N) \to X_0(N) \to \mathbb{SL}_2(\mathbb{Z}) \setminus \mathfrak{H}^* = \mathbb{P}_1(\mathbb{C})$$

exhibit the modular curves as ramified coverings and define a unique structure as a compact Riemann surface on each curve. The points $\Gamma_0(N) \setminus \mathbb{P}_1(\mathbb{Q}) = X_0(N) - Y_0(N)$, $\Gamma_1(N) \setminus \mathbb{P}_1(\mathbb{Q}) = X_1(N) - Y_1(N)$, and $\Gamma(N) \setminus \mathbb{P}_1(\mathbb{Q}) = X(N) - Y(N)$ are called the cusps of the respective modular curve. For more details, see Shimura's book [1971], Chapter 1, §(1.1), (1.3).

Since the ramification of $\mathfrak{H}^* \to \mathbb{P}_1(\mathbb{C}) = X(1)$ is all over the points with j-values ∞, 12^3, and 0 corresponding to the orbits $\mathbb{P}_1(\mathbb{Q})$, $\mathbb{SL}_2(\mathbb{Z})i$, and $\mathbb{SL}_2(\mathbb{Z})\rho$, the ramification of any of the maps or their composites $X(N) \to X_1(N) \to X_0(N) \to X(1)$ is over ∞, 12^3, and 0. If we can calculate the number of sheets of these coverings and the ramification numbers, then the Riemann–Hurwitz relation will give the Euler number and genus of the modular curves. The number of sheets of the coverings is given by the index of the various modular groups in $\mathbb{PSL}_2(\mathbb{Z}) = \mathbb{SL}_2(\mathbb{Z})/\{\pm I\}$.

(3.4) Proposition. *The homomorphism* $\mathbb{SL}_2(\mathbb{Z}) \to \mathbb{SL}_2(\mathbb{Z}/N\mathbb{Z})$ *induced by reduction* mod N *of the coefficients of matrices is surjective.*

Proof. We begin with an assertion: if c, d, and N are integers without a common prime factor, then there exists an integer k with c and $d + Nk$ relatively prime, i.e., $(c, d + Nk) = 1$.

To prove the assertion, we write $c = c'c''$ where $(c', N) = 1, (c', c'') = 1$, and if p divides c'', then p divides N. There are integers u and v with $1 = uc' + vN$, so that $1 - d = (1 - d)uc' + (1 - d)vN$, or $1 = [(1 - d)u]c' + (d + kN)$ where $k = (1 - d)v$. This means that $(c', d + Nk) = 1$ and finally $(c, d + Nk) = 1$ since if p divides c'', then p divides n and p does not divide $d + Nk$ where $(c, d, N) = 1$.

Now given

$$\begin{pmatrix} \bar{a} & \bar{b} \\ \bar{c} & \bar{d} \end{pmatrix} \in \mathbb{SL}_2(\mathbb{Z}/N\mathbb{Z})$$

we can lift $\bar{c}$ and $\bar{d}$ to integers c and d such that $(c, d, N) = 1$ from the determinant condition mod N. By the above assertion we can replace d by $d + Nk$ such that $(c, d) = 1$ where (c, d) reduces to $(\bar{c}, \bar{d})$. Now lift $\bar{a}$ and $\bar{b}$ to a and b and look for $a + uN$ and $b + vN$ such that

$$\det \begin{pmatrix} a + uN & b + vN \\ c & d \end{pmatrix} = 1,$$

i.e., $ud - vc = (1 - ad + bc)/N$ which is an integer. Since $(c, d) = 1$, there is a solution u, v and we change a to $a + Nu$ and b to $b + Nv$ to obtain a matrix

$$\begin{pmatrix} a & b \\ c & d \end{pmatrix} \in \mathbb{SL}_2(\mathbb{Z})$$

projecting to the given matrix $\mathbb{SL}_2(\mathbb{Z}/N\mathbb{Z})$. This proves the proposition.

From the previous proposition we see that the indices

$$(\Gamma(1) : \Gamma(N)) = \# \mathbb{SL}_2(\mathbb{Z}/N\mathbb{Z}), (\Gamma_0(N) : \Gamma(N)) = \# B_2(\mathbb{Z}/N\mathbb{Z}),$$

and $(\Gamma_1(N) : \Gamma(N)) = \# B_2'(\mathbb{Z}/N\mathbb{Z})$, where B_2 and B_2' are, respectively, the subgroups of elements of the form $\begin{pmatrix} * & * \\ 0 & * \end{pmatrix}$ and $\begin{pmatrix} 1 & * \\ 0 & 1 \end{pmatrix}$ in $\mathbb{SL}_2$. For $(N', N'') = 1$ the isomorphism $\mathbb{Z}/N'N''\mathbb{Z} \to \mathbb{Z}/N'\mathbb{Z} \oplus \mathbb{Z}/N''\mathbb{Z}$ given by the Chinese remainder theorem induces an isomorphism

$$\mathbb{SL}_2(\mathbb{Z}/N'N''\mathbb{Z}) \to \mathbb{SL}_2(\mathbb{Z}/N'\mathbb{Z}) \oplus \mathbb{SL}_2(\mathbb{Z}/N''\mathbb{Z})$$

which restricts to a corresponding isomorphism for the subgroups B_2 and B_2'. Hence the three indices are multiplicative for relatively prime values of N. This reduces calculations to prime powers $N = p^a$. Since the number of bases in $\mathbb{F}_q^2$ is $(q^2 - 1)(q^2 - q)$ and is also $\# \mathbb{GL}_2(\mathbb{F}_q)$, it follows that $\# \mathbb{SL}_2(\mathbb{F}_q) = q(q^2 - 1)$, $\# B_2(\mathbb{F}_q) = q(q - 1)$, and $\# B_2'(\mathbb{F}_q) = q$. Since $I + p\mathbb{M}_2(\mathbb{Z}/p^a\mathbb{Z})$ is the kernel of the morphism $\mathbb{GL}_2(\mathbb{Z}/p^a\mathbb{Z}) \to \mathbb{GL}_2(\mathbb{Z}/p\mathbb{Z})$ under reduction $\mathbb{Z}/p^a\mathbb{Z} \to \mathbb{Z}/p\mathbb{Z}$, we deduce that $\# \ker(\mathbb{SL}_2(\mathbb{Z}/p^a\mathbb{Z}) \to \mathbb{SL}_2(Z/p\mathbb{Z})) = (p^{a-1})^3$, $\# \ker(B_2(\mathbb{Z}/p^a\mathbb{Z}) \to B_2(\mathbb{Z}/p\mathbb{Z})) = (p^{a-1})^2$, and also $\# \ker(B_2'(\mathbb{Z}/p^a\mathbb{Z}) \to (B_2'(\mathbb{Z}/p\mathbb{Z})) = p^{a-1}$. Now we can summarize the above calculations with the following formulas for indices between related modular groups.

(3.5) Proposition. *For the modular groups $\Gamma(N) \subset \Gamma_1(N) \subset \Gamma_0(N)$ all contained in $\Gamma(1) = \mathbb{SL}_2(\mathbb{Z})$, we have:*

(1) $(\Gamma(1) : \Gamma(N)) = N^3 \prod_{p|N}(1 - 1/p^2)$,
(2) $(\Gamma(1) : \Gamma_1(N)) = N^2 \prod_{p|N}(1 - 1/p^2)$,
(3) $(\Gamma(1) : \Gamma_0(N)) = N \prod_{p|N}(1 + 1/p)$,

Note that $\pm I$ is the only part of $\Gamma(1)$ acting as the identity on the upper half plane. With this small modification we can calculate the order of the coverings arising with modular curves from this previous proposition.

(3.6) Proposition. *For the ramified coverings of modular curves $X(N) \to X_1(N) \to X_0(N) \to \mathbb{P}_1(\mathbb{C}) = X(1)$ the indices of the coverings are given by the following formulas:*

$$(1) \qquad [X(N) : X(1)] = \begin{cases} 6 & \text{for } N = 2, \\ \dfrac{1}{2}N^3 \displaystyle\prod_{p|N}(1 - 1/p^2) & \text{for } N > 2, \end{cases}$$

$$(2) \qquad [X_1(N) : X(1)] = \begin{cases} 3 & \text{for } N = 2, \\ \dfrac{1}{2}N^2 \displaystyle\prod_{p|N}(1 - 1/p^2) & \text{for } N > 2, \end{cases}$$

$$(3) \qquad [X_0(N) : X(1)] = \begin{cases} 3 & \text{for } N = 2, \\ N \displaystyle\prod_{p|N}(1 + 1/p) & \text{for } N > 2. \end{cases}$$

Finally we use the calculation of the order of coverings to show how to calculate the genus $g(X)$ of the Euler number $e(X)$ of a modular curve X, where $e(X) = 2 - 2g(X)$. For a ramified covering of compact Riemann surfaces with m sheets $p : X \to Y$ with m sheets and ramification numbers e_x for $x \in X$. This means that locally p maps x to $p(x)$ in Y and in local coordinates $p(z) = z^{e_x}$. Except for a finite number of x we know that $e_x = 1$. Using a triangulation of Y with vertices at ramification points of Y, one can lift the triangulation to X and deduce rather quickly the following relation.

(3.7) Hurwitz's Relation. For a ramified covering $p : X \to Y$ of compact Riemann surfaces with m sheets the Euler numbers of X and Y are related by $e(X) = me(Y) - \sum_{x \in X}(e_x - 1)$. In the case where e_x is independent of the x with $p(x) = y$ it is denoted e_y, and the relation takes the following form:

$$e(X) = m\left(e(Y) - \sum_{y \in Y}\left(1 - \frac{1}{e_y}\right)\right).$$

The Euler number $e(X) = 2 - 2g(X)$ where $g(X)$ is the genus of X.

For the three types of modular curves $X \to S^2 = \mathbb{P}_1(\mathbb{C}) = Y$ we have $e(S^2) = 2$ and all the ramification is over the three points $\infty = j(\infty), 0 = j(\rho)$, and $12^3 = j(i)$. There is a major difference between the ramified coverings $X(N) \to X(1)$ and either $X_1(N) \to X(1)$ or $X_0(N) \to X(1)$. For in the first case all x projecting to a given y in $X(1)$ has the same e_x value e_y while this is not the case for the other two coverings. For $X(N) \to X(1)$ the ramification indices are $e_0 = 3, e_{1728} = 2$, and $e_\infty = N$, and therefore we obtain, using the Hurwitz relation,

$$\begin{aligned} e(X(N)) &= [X(N) : X(1)]\left(2 - \left(1 - \frac{1}{e_0}\right) - \left(1 - \frac{1}{e_{1728}}\right)\left(1 - \frac{1}{e_\infty}\right)\right) \\ &= [X(N) : X(1)]\left(\frac{1}{3} + \frac{1}{2} + \frac{1}{N} - 1\right) \\ &= [X(N) : X(1)]\left(\frac{1}{N} - \frac{1}{6}\right). \end{aligned}$$

This leads to the following proposition.

(3.8) Proposition. *The genus of $X(N)$ is given by the following formula*

$$g(X(N)) = 1 + [X(N) : X(1)]\left(\frac{N-6}{12N}\right).$$

The corresponding results for $X_0(N)$ and $X_1(N)$ are more difficult, and we just quote the results, referring the reader to the related literature, see Ogg [1969] and Shimura [1971].

(3.9) Assertion. The genus of the curves $X_1(N)$ and $X_0(N)$ is given in the following formulas:

$$g(X_1(N)) = 1 + [X_1(N) : X(1)] - \frac{1}{2}\sigma^*(N),$$

where

$$\sigma^*(N) = \begin{cases} \dfrac{1}{2}\sum_{d|n} \varphi(d)\varphi(N/d) & \text{for } N > 4, \\ 3 & \text{for } N = 4. \end{cases}$$

$$g(X_0(N)) = 1 + [X_0(N) : X(1)] - \frac{1}{2}\sigma(N) - \frac{1}{4}\mu_{1728}(N) - \frac{1}{3}\mu_0(N),$$

$$\sigma(N) = \sum_{d|n} \varphi(d)\varphi(N/d),$$

$$\mu_{1728}(N) = \begin{cases} 0 & \text{for } 4 \mid N, \\ \prod_{p|N}(1 + (-4/p)) & \text{otherwise} \end{cases}$$

$$\mu_0(N) = \begin{cases} 0 & \text{for } 2 \mid N \text{ or } 9 \mid N, \\ \prod_{p|N}(1 + (-3/p)) & \text{otherwise.} \end{cases}$$

§4. Modular Functions

The subject of modular functions is treated in the book by Koblitz, Chapter III [1984] and there is an excellent introduction in the book by Serre, Chapter 7 [1977]. We will just give an introduction to show the relation with functions that have already arisen in the theory of elliptic curves.

Modular functions are holomorphic functions on the upper half plane $\mathfrak{H}$ which have certain periodic properties under the substitution

$$z \mapsto \frac{az+b}{cz+d} \quad \text{for} \begin{pmatrix} a & b \\ c & d \end{pmatrix} \text{ contained in a certain subgroup of } \mathbb{SL}_2(\mathbb{Z}).$$

As with elliptic functions and theta functions this invariance condition will include a periodicity under translation z to $z+N$. This leads to a replacing of the variable in the upper half plane with an exponential q as in the first section of the previous chapter.

(4.1) Remark. If $f(z)$ is a holomorphic function on the upper half plane $\mathfrak{H}$ (resp. the complex plane $\mathbb{C}$) satisfying the periodicity relation $f(z) = f(z+N)$ for N positive real, then $f(z) = f^*(q_N)$, where $q_N = \exp(2\pi iz/N)$ and $f^*(q)$ is holomorphic for $0 < |q| < 1$ (resp. $q \neq 0$). For a periodic $f(z) = f(z+N)$ holomorphic on $\mathfrak{H}$, we will speak of its behavior at ∞ in terms of the behavior of f^* at 0.

The function $f^*(q)$ has a Laurent or Fourier expansion

$$f^*(q) = \sum_{n\in\mathbb{Z}} c_n q^n.$$

(4.2) Definition. A holomorphic function $f(z)$, which is periodic $f(z) = f(z+N)$ is meromorphic at ∞ provided it has a pole at ∞, that is, $c_N = 0$ for $n < n_0$ and is holomorphic at ∞ provided $c_n = 0$ for $n < 0$. In this case the value $f(\infty) = f^*(0) = c_0$.

(4.3) Definition. A modular function $f(z)$ for the group $\mathbb{SL}_2(\mathbb{Z})$ of weight m is a meromorphic function on the upper half plane $\mathfrak{H}$ satisfying

$$f\left(\frac{az+b}{cz+d}\right) = (cz+d)^m f(z) \quad \text{for all} \begin{pmatrix} a & b \\ c & d \end{pmatrix} \in \mathbb{SL}_2(\mathbb{Z}),$$

and $f(z)$ is meromorphic at ∞. A modular form is a modular function which is holomorphic on $\mathfrak{H}$ and at ∞, and it is a cusp form provided $f(\infty) = 0$ also.

(4.4) Examples. The Eisenstein series $(2k \geq 4)$

$$G_{2k}(z) = \sum_{(m,n)\neq(0,0)} \frac{1}{(mz+n)^{2k}}$$

satisfying the relation

$$G_{2k}\left(\frac{az+b}{cz+d}\right) = (cz+d)^{2k} \sum_{(m,n)\neq(0,0)} \frac{1}{[(ma+nc)z+(mb+nd)]^{2k}}$$
$$= (cz+d)^{2k} G_{2k}(z),$$

and having q-expansions given in 10(1.4), are modular forms of weight $2k$. Since $g_2(z) = 60G_4(z)$ has weight 4 and $g_3(z) = 140G_6(z)$ has weight 6, the modular form $\Delta(z) = g_2(z)^3 - 27g_3(z)^2$ has weight 12 and is a cusp form from 10(1.8). The function

$$j(z) = 12^3 \frac{g_2(z)^3}{\Delta(z)}$$

is a modular function of weight zero with a simple pole at ∞.

(4.5) Remarks. Since $\begin{pmatrix} -1 & 0 \\ 0 & -1 \end{pmatrix}$ is in $\mathbb{SL}_2(\mathbb{Z})$ and

$$f(z) = f\left(\frac{-z+0}{0-1}\right) = (-1)^k f(z),$$

we have modular forms only for even weight. In the first section, Chapter 7 of Serre [1977] it is proven that the two elements $\begin{pmatrix} 1 & 1 \\ 0 & 1 \end{pmatrix}$ and $\begin{pmatrix} 0 & 1 \\ -1 & 0 \end{pmatrix}$ generate $\mathbb{SL}_2(\mathbb{Z})$. Thus a holomorphic function $f(z)$ on the upper half plane is a modular form of weight $2k$ if and only if $f(z) = f(z+1)$, $f(-1/z) = z^{2k} f(z)$, and $f(z)$ is holomorphic at ∞.

(4.6) Remark. Let M_k be the complex vector space of modular forms of weight $2k$. This space has a basis consisting of $G_4^a G_6^b$ where $4a + 6b = 2k$. This is proved by both Serre [1977] and Koblitz [1984] using a relation between $2k$ and the number of zeros of $f(z)$ in a fundamental domain.

The following growth property of Fourier coefficients of modular forms is due to Hecke, see Serre [1977, p. 94].

(4.7) Assertion. Let $f(z)$ be a modular function of weight $2k$ having q-expansion $f(z) = \sum_{n\geq 0} a_n q^n$. If $f(z)$ is a cusp form, then there is a constant C such that $|a_n| \leq Cn^k$, and if $f(z)$ is not a cusp form, then there are constants A and B with

$$An^{2k-1} \leq |a_n| \leq B_n^{2k-1}.$$

The result for Eisenstein series follows from the q-expansions given 10(1.4). For a cusp form we have an inequality of the form $|f(z)| \leq My^{-k}$ on the upper half plane which gives the estimate. The case of a general modular form follows from the decomposition $f(z) = cG_{2k}(z) + f^*(z)$, where $f^*(z)$ is a cusp form.

The concept of a modular form for $\mathbb{SL}_2(\mathbb{Z})$ generalizes to any subgroup of $\mathbb{SL}_2(\mathbb{Z})$, see §8. We will only be interested in congruence subgroups Γ, i.e., subgroups Γ such that for some N the inclusions $\Gamma(N) \subset \Gamma \subset \mathbb{SL}_2(\mathbb{Z})$ hold. Examples arise naturally in the theory of theta functions, see Mumford, Chapter 1, Theorem 7.1, for a formula for $\theta[z/(c\tau+d), (a\tau+b)/(c\tau+d)]$ in terms of $\theta(z,\tau)$ up to factors of the square root of $c\tau+d$. The theta functions $\theta_{0,0}, \theta_{1/2,0}, \theta_{1,1/2}$, and $\theta_{1/2,1/2}$ are modular forms for $\Gamma(4)$, see Mumford again.

(4.8) Remark. A modular function $f(z)$ of weight zero for a congruence subgroup Γ defines a meromorphic function on $X(\Gamma) = \Gamma\backslash(\mathfrak{H}\cup\mathbb{P}_1(\mathbb{Q}))$ which is holomorphic on $\Gamma\setminus\mathfrak{H}$ away from the cusps. A modular function $f(z)$ of weight 2 for Γ has an interpretation on $X(\Gamma)$ as a differential since

$$d\left(\frac{az+b}{cz+d}\right) = \frac{dz}{(cz+d)^2},$$

and hence $f(z)\,dz$ is invariant by the action of the group Γ on $\mathfrak{H}$. It is the cusp forms which extend to holomorphic differentials on the compactified space $X(\Gamma)$. This is seen by looking in local coordinates near the cusp, see Shimura's book.

§5. The L-Function of a Modular Form

Recall from 9(5.1) that the gamma function is given by the integral

$$\Gamma(s) = \int_0^\infty x^s e^{-x}\frac{dx}{x},$$

and hence we have by change of variable

$$\int_0^\infty y^s e^{-2\pi ny}\frac{dy}{y} = \int_0^\infty \left(\frac{x}{2\pi n}\right)^s e^{-x}\frac{dx}{x} = (2\pi)^{-s}\frac{\Gamma(s)}{n^s}.$$

(5.1) Notations. Holomorphic functions $f(z)$ on the upper half plane satisfying $f(z) = f(z+1)$ and holomorphic at ∞ have a q-expansion $f(z) = \sum_{n\geq 0} a_n q^n$, where $q = e^{2\pi iz}$. We form the Dirichlet series

$$L_f(s) = \sum_{1\leq n}\frac{a_n}{n^s}.$$

(5.2) Remarks. If $|a_n| \leq An^{c-1}$ for constants A and c, then the Dirichlet series $L_f(s)$ converges for $\mathbf{Re}(s) > c$, and the related function has an integral representation

$$L_f^*(s) = (2\pi)^{-2}\Gamma(s)L_f(s) = \int_0^\infty y^s(f(iy) - a_0)\frac{dy}{y}$$

for $\mathbf{Re}(s) > c$. This arises by multiplying the r-integral by a_n and summing up.

The bijection between functions $f(z)$ as in (5.1) and certain Dirichlet series L_f or L_f^* is elementary, but Hecke went further and showed how the modular form symmetry condition $f(-1/z) = z^{2k} f(z)$ corresponds to a functional equation for L^*_f. This arises by dividing the line of integration in two parts and applying the relation $f(i/y) = (-1)^k y^{2k} f(iy)$ where $\mathbf{Re}(s) > 2k$.

$$\begin{aligned}
L^*_f(s) &= \int_1^\infty t^s(f(it) - a_0)\frac{dt}{t} + \int_0^1 t^s(f(it) - a_0)\frac{dt}{t} \\
&= \int_1^\infty t^s(f(it) - a_0)\frac{dt}{t} + \int_1^\infty t^{-s}\left(f\left(\left(\frac{i}{t}\right) - a_0\right)\frac{dt}{t}\right. \\
&= \int_1^\infty (t^s + (-1)^k t^{2k-s})(f(it) - a_0)\frac{dt}{t} \\
&\quad - a_0 \int_1^\infty t^{-s}\frac{dt}{t} + a_0(-1)^k \int_1^\infty t^{2k-s}\frac{dt}{t}.
\end{aligned}$$

We have used the convergence of $\int_1^\infty t^{-s}(dt/t) = 1/s$ for $\mathbf{Re}(s) > 0$. We deduce the formula

$$L^*_f(s) = \int_1^\infty (t^s + (-1)^k t^{2k-s})(f(it) - a_0)\frac{dt}{t} - \frac{a_0}{s} - \frac{(-1)^k a_0}{2k - s}$$

which holds for $\mathbf{Re}(s) > 2k$.

(5.3) Theorem (Hecke). *For a modular form $f(z)$ of weight $2k$ the L-function $L^*_f(s)$ extends to meromorphic function on the entire plane $\mathbb{C}$ satisfying the functional equation*

$$L^*_f(2k - s) = (-1)^k L^*_f(s)$$

and having poles only at most at $s = 0$ and $s = 2k$ with residues $-a_0$ and $-(-1)^k a_0$, respectively.

Proof. The verification of all assertions follow directly from the above formula preceding the theorem. Observe that the integrals converge for all values of s from the exponential decay of the terms in $f(it) - a_0 = \sum_{n\geq 1} a_n e^{-2\pi nt}$.

(5.4) Remark. There is a converse theorem which says that if $f(z)$ as in (5.1) has an L-function satisfying the conclusions of 5.3 for related twisted L-functions together with a certain growth condition, then $f(z)$ is a modular form.

Unfortunately, the most interesting L-functions for the theory of elliptic curves do not come from modular functions over $\mathbb{SL}_2(\mathbb{Z})$ but from modular functions of weight 2 over $\Gamma_0(N)$. The above analysis does not apply directly because $\begin{pmatrix} 0 & 1 \\ -1 & 0 \end{pmatrix}$ is not in $\Gamma_0(N)$. For $X_0(N)$ and the group $\Gamma_0(N)$ we do have an important substitute.

(5.5) Remark. Conjugation by $\begin{pmatrix} 0 & 1 \\ -N & 0 \end{pmatrix}$ is seen to carry $\Gamma_0(N)$ to itself and induces a map $w_N : X_0(N) \to X_0(N)$ preserving the cusps and $Y_0(N)$. It is an involution, i.e., $w_N^2 = I$. In terms of the description in (2.4) of points of $Y_0(N)$ the involution w_N carries the pair of an elliptic curve E and cyclic subgroup C on order N to the pair $(E/C,\ {}_N E/C)$.

(5.6) Remark. On the space of modular forms $M_{2k}(N)$ of weight $2k$ for the group $\Gamma_0(N)$ we have an operator W_N given by

$$W_N(f)(z) = N^{-k}z^{-2k}f\left(-\frac{1}{Nz}\right)$$

with $W_N^2 = I$. Hence there is a corresponding $\pm$ eigenspace decomposition

$$M_{2k}(N) = M_{2k}(N)^+ \oplus M_{2k}(N)^-.$$

For eigenfunctions of W_N in $M_{2k}(N)$ we define an L-function with the Mellin transform and a power of N as follows:

$$\begin{aligned} L_f^*(s) &= (2\pi)^{-s}\Gamma(s)N^{s/2}L_f(s) \\ &= \int_0^\infty N^{s/2}t^s(f(it) - a_0)\frac{dt}{t}. \end{aligned}$$

By dividing the path of integration at $1/\sqrt{N}$, we prove, using an argument similar to the one in (5.3), the following extension of (5.3) which reduces to (5.3) for $N = 1$.

(5.7) Theorem. *For a modular form $f(z)$ of weight $2k$ for $\Gamma_0(N)$ with $W_N(f) = \varepsilon f$ ($\varepsilon = \pm 1$) the L-function $L_f^*(s)$ extends to a meromorphic function on the entire plane $\mathbb{C}$ satisfying the functional equation*

$$L_f^*(2k - s) = \varepsilon(-1)^k L_f^*(S)$$

and having poles only at most at $s = 0$ and $s = 2k$ with residues multiples of a_0.

(5.8) Remark. For applications to elliptic curves it is weight 2 modular forms which play a basic role, and by (4.8) these forms $f(z)$ which are cuspidal correspond to holomorphic differentials $f(z)dz$ on $X_0(N)$. For $W_N(f) = \varepsilon f$ the L-function satisfies

$$L_f^*(2 - s) = -\varepsilon L_f^*(s).$$

§6. Elementary Properties of Euler Products

This section is designed to give background material on Dirichlet series which have product decompositions generalizing the classical Euler product of $\zeta(s)$, the Riemann zeta function

$$\prod_p \frac{1}{1 - p^{-s}} = \zeta(s) = \sum_{1 \le n} \frac{1}{n^s}.$$

(6.1) Definition. An Euler product of degree n is a formal Dirichlet series of the form

$$\sum_{1\le n} \frac{a_n}{n^s} = \prod_p \frac{1}{P_p(p^s)},$$

where $P_p(T) \in 1 + T\mathbb{C}[T]$ is a polynomial of degree $\le n$, defined for each rational prime number p. The coefficients a_n are complex numbers.

When we speak of an Euler product of degree n, it is usually the case that all but a finite number of $P_p(T)$ have degree exactly n. In the case of the Riemann zeta function $P_p(T) = 1 + T$ for all primes p. We will be primarily interested in linear and quadratic Euler products, that is, those of degree 1 and 2.

(6.2) Remark. Euler products are defined to be indexed by rational primes p. These include products indexed by the places v of a number field F in the following sense. For polynomials $Q_v(T) \in 1 + T\mathbb{C}[T]$ of degree $\le m$ we can form

$$\prod_v \frac{1}{Q_v(q_v^{-s})},$$

where q_v is the number of elements in the residue class field at v. By defining the polynomial $P_p(T) = \prod_{p|v} Q_v(T^{n(v)})$ where $q_v = p^{n(v)}$, we see that this product decomposition can be rearranged into an Euler product

$$\prod_v \frac{1}{Q_v(q_v^{-s})} = \prod_p \frac{1}{P_p(p^{-s})}$$

as formal Dirichlet series. The degree of the polynomials $P_p(T)$ is at most $m[F : \mathbb{Q}]$. These products defined by polynomials $Q_v(T)$ could be called Euler products over the number field F, and the above discussion together with elementary decomposition properties of primes in number fields shows that an Euler product of degree m over F can be assembled into an Euler product of degree $m[F : \mathbb{Q}]$ (over $\mathbb{Q}$).

Now we consider convergence properties of Euler products. By analogy with the classical case, we expect a convergence assertion in some right half plane.

(6.3) Remark. If the coefficients a_n satisfy $|a_n| \le cn^b$, then the Dirichlet series $\sum_{1\le n}(a_n/n^s)$ converges absolutely in the right half plane $\mathbf{Re}(s) > 1 + b$. This is essentially Proposition 8 in Chapter 6 of J.-P. Serre, *Course in Arithmetic*. In fact, §§2 and 3 of this chapter in Serre's book give a general background for this section on Euler products.

In the case of a linear Euler product

$$\prod_p \frac{1}{1 - a_p p^{-s}} = \sum_{1\le n} \frac{a_n}{n^s}$$

there is a formula for the nth coefficient a_n in terms of the prime factorization $n = p(1)^{m(1)} \cdots p(r)^{m(r)}$ of n, namely

$$a_n = a_{p(1)}^{m(1)} \cdots a_{p(r)}^{m(r)}.$$

Hence, if each $|a_p| \le p^b$ for all primes where b is a real number, then $|a_n| \le n^b$ and the previous remark applies, that is, we have absolute convergence of the Dirichlet series and the Euler product expansion is the half plane $\mathbf{Re}(s) > 1 + b$.

A convergence assertion for Euler products of degree n can be formulated in terms of the factorization into linear factors

$$P_p(T) = (1 - \alpha_p(1)T) \cdots (1 - \alpha_p(n)T)$$

with each $\alpha_p(j)$ equal to either the reciprocal of a root of $P_p(T)$ or to 0 which happens in the case when deg $P_p(T) < n$.

(6.4) Proposition. *An Euler product of degree n*

$$\prod_p \frac{1}{P_p(p^{-s})}$$

converges absolutely for $\mathbf{Re}(s) > 1+b$ *if the absolute values of each of the reciprocal roots in bounded* $|\alpha_p(j)| \le p^b$ *where*

$$P_p(T) = (1 - \alpha_p(1)T) \ldots (1 - \alpha_p(n)T).$$

The proof results immediately by applying the above discussion for the absolute convegrence of the linear Euler products.

(6.5) Corollary. *A quadratic Euler product*

$$\prod_p \frac{1}{1 - a_p p^{-s} + p^{c-2s}}$$

converges for $\mathbf{Re}(s) > 1 + c/2$ *when* $|a_p| \le 2p^{c/2}$ *for each* p.

Proof. The inequality on the absolute values of the coefficients is equivalent to the assertion that the roots of the quadratic polynomials are purely imaginary, and, equivalently, have the same absolute value $p^{c/2}$. Now the proposition applies.

(6.6) Proposition. *An Euler product of degree n over a number field K of degree* $[K : Q]$

$$\prod_v \frac{1}{Q_v(q_v^{-s})}$$

converges absolutely for $\mathbf{Re}(s) > 1+b$ *if the absolute values of each of the reciprocal roots is bounded* $|\alpha_v(j)| \le q_v^b$ *where* $Q_v(T) = (1 - \alpha_v(1)T) \ldots (1 - \alpha_v(n)T)$.

Proof. This assertion reduces to 6.4 by using the observation that all of the terms with q_v^{-s} factors and $q_v = p^m$, where $m = n(v)$, decompose

$$\frac{1}{1 - aq^{-s}} = \prod_{\alpha^m = a} \frac{1}{1 - \alpha p^{-s}}$$

and the fact that the inequality $|a| \le q^b$ is equivalent to $|\alpha| \le p^b$.

We can ask which Dirichlet series $\sum_{1 \le n}(c_n/n^s)$ have an Euler product decomposition. One necessary condition is that it is multiplicative in the sense that $c_{mn} = c_m c_n$ whenever $(m, n) = 1$, i.e., the natural numbers are relatively prime. Necessary and sufficient conditions are in general more difficult to formulate, but we will do the cases of linear and quadratic Euler products.

(6.7) Remark. A Dirichlet series $\sum_{1 \le n}(c_n/n^s)$ has a linear product decomposition if and only if it is strictly multiplicative, i.e., the relation $c_m c_n = c_{mn}$ holds for all natural numbers m, n. This assertion was taken up when we considered the convergence properties of linear Euler products.

(6.8) Remark. A Dirichlet series $\sum_{1 \le n}(c_n/n^s)$ has a quadratic Euler product decomposition if and only if:

(1) It is multiplicative.
(2) There constants b_p for each rational prime p such that

$$c_{p^{n+1}} = c_p c_{p^n} - b_p c_{p^{n-1}} \quad \text{for } n \ge 1.$$

In particular, $c_1 = c_{p^0} = 1$. Note that we are reduced to the previous example if and only if all $b_p = 0$. The degree 2 Euler factors are defined by the quadratic polynomials

$$P_p(T) = 1 - c_p T + b_p T^2.$$

§7. Modular Forms for $\Gamma_0(N)$, $\Gamma_1(N)$, and $\Gamma(N)$

The aim of this section is to organize the various relations between modular forms for these three classes of groups, and then introduce the concept of new forms and state some basic properties. In order to describe modular forms for subgroups of $\mathbb{SL}_2(\mathbb{Z})$ or more generally of $\mathbb{SL}_2(\mathbb{R})$ we use the following notation.

(7.1) Notation. Let f be a function on the upper half plane $\mathfrak{H}$, and let $\gamma = \begin{pmatrix} a & b \\ c & d \end{pmatrix} \in \mathbb{SL}_2(\mathbb{R})$. We define the function $f|_k\gamma$ by the formula $(f|_k\gamma)(\tau) = (c\tau + d)^{-k} f(\gamma(\tau))$ where $\gamma(\tau) = (a\tau + b)/(c\tau + d)$.

(7.2) Definition. Let Γ be a subgroup of $\mathbb{SL}_2(\mathbb{Z})$ containing $\Gamma(N)$ for some N. A modular form (resp. cusp form) of weight k for Γ is a complex valued homomorphic function on $\mathfrak{H}$ satisfying:

(1) For all $\gamma \in \Gamma$ we have $f|_k\gamma = f$, and
(2) The function is homomorphic at certain points of the boundary of the upper half plane in the following sense: for all $\sigma \in \mathbb{SL}_2(\mathbb{Z})$ the function $f|_k\sigma$ is periodic $(f|_k\sigma)(\tau) = (f|_k\sigma)(\tau + N)$, and Fourier series $(f|_k\sigma)(z) = \sum_{n\in\mathbb{Z}} a_n(q_N)^n$ where $q_N = \exp(2\pi i\tau/N)$ has coefficients $a_n = 0$ for $n < 0$ (resp. $n \leq 0$).

For the groups $\Gamma(N) \subset \Gamma_1(N) \subset \Gamma_0(N) \subset \mathbb{SL}_2(\mathbb{Z})$ introduced in (2.3) we denote the corresponding complex vector spaces of modular forms of weight k and their inclusions by

$$M_k(\Gamma_0(N)) \subset M_k(\Gamma_1(N)) \subset M_k(\Gamma(N)).$$

and the corresponding subspaces of cusp forms of weight k by

$$M_k^0(\Gamma_0(N)) \subset M_k^0(\Gamma_1(N)) \subset M_k^0(\Gamma(N))$$

where $M_k^0(\Gamma) \subset M_k(\Gamma)$.

The $\Gamma(N)$ modular forms which are $\Gamma_1(N)$ modular forms are characterized in the following assertion.

(7.3) Assertion. Let f be a modular form of weight k for $\Gamma(N)$. Then f is a modular form for $\Gamma_1(N)$ if and only if $f(\tau) = f(\tau + 1)$ and the Fourier series reduces to $f(\tau) = \sum_{0\leq n} a_n q^n$ where $q = q_1 = e^{2\pi i\tau}$.

This follows from the fact that $\Gamma_1(N)$ is generated by $\Gamma(N)$ and the single matrix $\begin{pmatrix} 1 & 1 \\ 0 & 1 \end{pmatrix}$.

For the characterization of the $\Gamma_1(N)$ modular forms which are $\Gamma_0(N)$ modular forms we use the following description of cosets.

(7.4) Proposition. *The function* $\psi\begin{pmatrix} a & b \\ c & d \end{pmatrix} = d \ (mod\ N)$ *induces a bijection* $\psi : \Gamma_0(N)/\Gamma_1(N) \to (\mathbb{Z}/N\mathbb{Z})^*$.

Proof. This follows from the formula $\begin{pmatrix} a & b \\ 0 & d \end{pmatrix}\begin{pmatrix} 1 & x \\ 0 & 1 \end{pmatrix} = \begin{pmatrix} a & ax+b \\ 0 & d \end{pmatrix}$ where it is clear that d (mod N) is an invariant.

(7.5) Definition. Let $G(N)$ denote the multiplicative group of units $(\mathbb{Z}/N\mathbb{Z})^*$, and let $\hat{G}(N)$ denote the group of Dirichlet characters mod N, that is, morphism $\varepsilon : G(N) \to \mathbb{C}^*$. A character ε is called even (resp. odd) provided $\varepsilon(-1) = +1$ (resp. -1).

(7.6) Remark. The group $\Gamma_0(N)$ acts on $M_k(\Gamma_1(N))$ by $f|_k\gamma$ for $\gamma \in \Gamma_0(N)$, and using (7.4), we see that this depends only on $\psi(\gamma\Gamma_1(N)) = d$ (mod N). We write $f|_k\gamma = f|R_d$ for d (mod N) $\in G(N)$. In particular the group $G(N)$ acts on the complex vector space $M_k(\Gamma_1(N))$. For example $f|R_{-1} = (-1)^k f$.

(7.7) Assertion. For the action of $G(N)$ on $M_k(\Gamma_1(N))$ there is a Fourier decomposition parametrized by the characters

$$M_k(\Gamma_1(N)) = \bigoplus_{\varepsilon \in \hat{G}(N)} M_k(\Gamma_1(N))_\varepsilon$$

where $M_k(\Gamma_1(N))_\varepsilon$ is the space of $f \in M_k(\Gamma_1(N))$ such that $f|R_d = \varepsilon(d)f$ for all $d \in G(N)$. For the unit Dirichlet character $\varepsilon = 1$ we have $M_k(\Gamma_1(N))_1 = M_k(\Gamma_0(N))$.

This condition for $f \in M_k(\Gamma(N))$ to be in $M_k(\Gamma_1(N))_\varepsilon$ can be formulated as follows that f satisfies the formula

$$f\left(\frac{a\tau + b}{c\tau + d}\right) = \varepsilon(d)(c\tau + d)^k f(z) \qquad \text{for all } \gamma = \begin{pmatrix} a & b \\ c & d \end{pmatrix} \in \Gamma_0(N).$$

(7.8) Definition. The elements of $M_k(\Gamma_1(N))_\varepsilon$ for $\varepsilon \in \hat{G}(N)$ are called modular forms of type (k, ε) for $\Gamma_0(N)$.

Every modular form f of weight k for $\Gamma_1(N)$ has a unique sum decomposition of modular forms f_ε of type (k, ε) for $\Gamma_0(N)$.

§8. Hecke Operators: New Forms

For the role of Hecke operators in this context of modular forms for $\Gamma_0(N)$ we recommend the treatment in Deligne and Serre [1974]. Also Serre [1977, Chapter 7] for $N = 1$ and Koblitz [1982, Chapter III] for the general case.

(8.1) Definition. Let p be a prime number. For a modular form $f(q) = \sum_{n \geq 0} a_n q^n$ of type (k, ε) for $\Gamma_0(N)$, the Hecke operator T_p has the following effect on the q-expansion at ∞:

$$f|T_p = \sum a_{pn} q^n + \varepsilon(p) p^{k-1} \sum a_n q^{pn} \qquad \text{if } p \nmid N,$$
$$f|U_p = \sum a_{pn} q^n \qquad \text{if } p \mid N.$$

These operators preserve cusp forms. There is also the notion of "new form" which means roughly that it does not come from a modular form for $\Gamma_0(N')$ with $N'|N$ but $N' < N$.

Since the operators can be shown to commute with each other there is the possibility of a simultaneous eigenform for all T_p and U_p. In this case we state the following theorem.

(8.2) Theorem. *Let f be a new cusp form of type (k, ε) for $\Gamma_0(N)$ normalized so $a_1 = 1$, and assume that f is an eigenform for T_p and U_p with corresponding eigenvalues a_p. Then the related Dirichlet series $L_f(s) = \sum_{n \geq 1} a_n n^{-s}$ has an Euler product expansion of the form*

$$L_f(s) = \prod_{p|N} \frac{1}{(1 - a_p p^{-s})} \cdot \prod_{p \nmid N} \frac{1}{(1 - a_p p^{-s} + \varepsilon(p) p^{k-1-2s})}.$$

This leaves the broader question of whether or not there exist simultaneous eigenfunctions for infinitely many operators $T(p)$. We sketch the argument that these functions exist.

(8.3) Petersson Inner Product. This inner product is defined on $M^0_{2k}(N)$ by the integral

$$(f|g)_{2k} = \int\int_{Y_0(N)} f(z)\overline{g(z)}y^{2k}\frac{dxdy}{y^2}.$$

Each $T(p)$ for p not dividing N is Hermitian with respect to this inner product, and hence $M^0_{2k}(N)$ has a basis of simultaneous eigenfunctions for all but a finite number of $T(p)$. In the case $N = 1$ we deduce that the entire space $M^0_{2k}(1)$ has an orthonormal basis of functions which are eigenfunctions for all Hecke operators $T(p)$.

(8.4) Remarks. The operators $T(p)$ or U_p for which $p|N$ are not even necessarily normal. There is a subspace $M^0_{2k}(N)^{\text{old}}$ generated by $f(N_2z)$, where $N = N_1N_2$ and f is a simultaneous eigenfunction for all $T(p)$ on $M^0_{2k}(N_1)$, where p does not divide N_1. On the orthogonal complement $M^0_{2k}(N)^{\text{pr}}$ of $M^0_{2k}(N)^{\text{old}}$ in $M^0_{2k}(N)$ there is a simultaneous basis of eigenfunctions for all $T(p)$. For a proof of this see either Koblitz [1984] or Li [1975]. The elements of $M^0_{2k}(N)^{\text{pr}}$ are called "new forms."

(8.5) Definition. The orthogonal complement $M^0_{2k}(N)^{\text{pr}}$ of $M^0_{2k}(N)^{\text{old}}$ is called the space of new forms.

§9. Modular Polynomials and the Modular Equation

In this section we consider briefly a polynomial related to families of elliptic curves. It is the modular polynomial $\Phi_N(x, y)$ with integral coefficients which serves in some sense as an equation for $X_0(N)$ as a scheme. It is symmetric in x and y, that is, $\Phi_N(x, y) = \Phi_N(y, x)$, of degree $N\prod_{p|N}(1 + 1/p) = \psi(N) = (\Gamma : \Gamma_0(N))$ in each variable, for $\psi(N) = (\Gamma : \Gamma_0(N))$ see 11(3.5).

(9.1) Notation. We define the following sets of matrices in $M_2(\mathbb{Z})$ the set of two by two integral matrices. Denote by

$D(N)$ the set of $\begin{pmatrix} a & b \\ c & d \end{pmatrix} \in M_2(\mathbb{Z})$ with $ad - bc = N$ and g.c.d.$(a, b, c, d) = 1$,

and

$S(N)$ the set of $\begin{pmatrix} a & b \\ 0 & d \end{pmatrix} \in D(N)$ with $d > 0$ and $0 \le b < d$.

(9.2) Remark. The inclusion $S(N) \to D(N)$ induces a function

$$S(N) \to \mathbb{SL}_2(\mathbb{Z}) \setminus D(N)$$

which is a bijection. As another exercise for the reader, check that $\#S(N) = N\prod_{p|N}(1 + 1/p) = \psi(N)$.

With this set $S(N)$ we define the modular polynomial as a norm as follows in terms of the modular j-function on the upper half plane.

(9.3) Definition. The modular polynomial $\Phi_N(x) \in \mathbb{Z}[j][x]$ is given by the following product

$$\Phi_N(x) = \prod_{w \in S(N)} (x - jw)$$

where the composite jw is a function on the upper half plane.

(9.4) Remark. It is an exercise to show that $\Phi_N(x)$ is a polynomial function in j and x. We also use the notation $\Phi_N(x, j)$ or just $\Phi_N(x, y)$ with y instead of j. It has the property that $\Phi_N(x, y) = \Phi_N(y, x)$, and its degree is the cardinality $\#S(N)$.

Now we restrict ourselves to $N = \ell$ a prime number.

(9.5) Assertion. The polynomial $\Phi_\ell(x, y) = 0$ is an equation for the curve $X_0(\ell)$, and this polynomial has the property that $\Phi_\ell(j(\tau), j(\ell\tau)) = 0$ for the transcendental function $j(\tau)$. We can interpret the zeros of $\Phi_\ell(j(E), y)$ as the j-invariants $j(E/C)$ indexed by pairs (E, C) associated the $\ell + 1$ different cyclic subgroups $C \subset E$ of order ℓ. Multiply roots of $\Phi_\ell(j(E), y)$ correspond to isomorphic E/C' and E/C'' for two pairs (E, C') and (E, C'').

In other terms the E/C are the $\ell + 1$ isogeny classes $E \to E'$ with kernel cyclic of order ℓ.

12

Endomorphisms of Elliptic Curves

The isomorphism classification of elliptic curves was previously carried out completely in terms of the Weierstrass equation with a minimum of algebraic geometry. This was possible because an isomorphism between two curves with equation in normal form is given by simple formulas. The situation with homorphisms is not so easy and more algebraic geometry is needed. The exception to this is multiplication by 2 and the 2-isogeny, see 4(5.2).

In this chapter we will just sketch the basic results and leave the reader to study the details in the general context of abelian varieties as explained in D. Mumford's book on the subject. Homomorphisms are very elementary in the case of complex tori and there all details will be supplied.

The ring of endomorphisms contains the integers as a subring, and the two are usually equal. In exceptional situations it is larger than the ring of integers; this is the case of complex multiplication. The term complex multiplication comes from the classical setting of complex tori where homomorphisms are induced by multiplication with a complex number, see 9(1.4). In characteristic zero, e.g., over the complex numbers, the endomorphism ring of a curve with complex multiplication is a subring of finite index in the ring of integers of an imaginary quadratic number field. Complex tori of dimension one having complex multiplication are isomorphic to a torus with a period lattice contained in an imaginary quadratic number field.

In the next chapter we consider curves with complex multiplication in characteristic p. These are exactly the curves defined over a finite field.

The reader should refer to Silverman [1981], Chapters III and IV.

§1. Isogenies and Division Points for Complex Tori

For two complex tori $T = \mathbb{C}/L$ and $T' = \mathbb{C}/L'$ the set if isogenies (resp. isomorphisms) $\lambda : T \to T'$ can be identified with the set of complex numbers λ with $\lambda L \subset L'$ (resp. $\lambda L = L'$), see 9(1.4).

(1.1) Remarks. A nonzero analytic homomorphism $\lambda : \mathbb{C}/L \to \mathbb{C}/L'$ is surjective, and $\ker(\lambda) = \lambda^{-1}L'/L \cong L'/\lambda L$ is a finite subgroup of order $n = [L' : \lambda L]$, the index of λL in L'. Note that $nL' \subset \lambda L \subset L'$ and $(n/\lambda) : T' = \mathbb{C}/L' \to \mathbb{C}/L = T$ is an isogeny which composed with λ gives $n : T \to T$, i.e., multiplication by the integer n.

(1.2) Definition. An isogeny between two complex tori, or elliptic curves, is a nonzero analytic homomorphism. For complex tori the degree, denoted $\deg(\lambda)$, is the index $[L' : \lambda L]$ for $\lambda : T = \mathbb{C}/L \to \mathbb{C}/L' = T'$. The dual isogeny $\hat{\lambda}$ of λ is defined to be $\hat{\lambda} = n/\lambda : T' \to T$. The dual of 0 is defined to be 0. See also 9(1, Ex. 3).

(1.3) Proposition. *The function* $\lambda \mapsto \hat{\lambda}$ *is a group morphism* $\mathrm{Hom}(T, T') \to \mathrm{Hom}(T', T)$ *satisfying* $\hat{\hat{\lambda}} = \lambda$, $\deg(\lambda) = \deg(\hat{\lambda}) = n$, $\hat{\lambda}\lambda = n$ *in* $\mathrm{End}(T)$, *and* $\lambda\hat{\lambda} = n$ *in* $\mathrm{End}(T')$. *For* $\lambda \in \mathrm{Hom}(T', T)$ *and* $\mu \in \mathrm{Hom}(T, T')$ *we have* $\widehat{(\mu\lambda)} = \hat{\lambda}\hat{\mu}$. *The involution* $\lambda \mapsto \hat{\lambda}$ *of the ring* $\mathrm{End}(T)$, *called the Rosati involution, satisfies* $\hat{n} = n$. *Finally,* $\deg(n) = n^2$ *and the function* $\deg : \mathrm{End}(T) \to \mathbb{Z}$ *is a positive quadratic function where* $\deg(\lambda) = \lambda\hat{\lambda} = |\lambda|^2$.

Proof. Let $a(L)$ denote the area of a period parallelogram associated with the lattice L. Then we have $a(\lambda L) = |\lambda|^2 a(L)$ and for $L'' \subset L'$ the index $[L' : L''] = a(L'')/a(L')$. Hence for an isogeny $\lambda : \mathbb{C}/L \to \mathbb{C}/L'$ the degree $n = \deg(\lambda) = [L' : \lambda L]$ equals $|\lambda|^2[a(L)/a(L')]$ which shows that the dual isogeny is given by the formula

$$\hat{\lambda} = \frac{a(L)}{a(L')}\bar{\lambda}$$

from which the proposition follows directly.

There is a formula for $a(L)$ when $L = \mathbb{Z}\omega_1 + \mathbb{Z}\omega_2$, namely

$$a(L) = \frac{1}{2}\,|\omega_1\bar{\omega}_2 - \omega_2\bar{\omega}_1|\,.$$

If ω_1, ω_2 is changed to $\omega_1' = a\omega_1 + b\omega_2$, $\omega_2' = c\omega_1 + d\omega_2$, then

$$|\omega_1'\bar{\omega}_2' - \omega_2'\bar{\omega}_1'| = \left|\det\begin{pmatrix} a & b \\ c & d \end{pmatrix}\right| \cdot |\omega_1\bar{\omega}_2 - \omega_2\bar{\omega}_1|\,,$$

and, therefore, this expression for $a(L)$ is independent of a basis of L. From this formula $a(\lambda L) = |\lambda|^2 a(L)$. Finally, for L_τ we calculate directly that $a(L_\tau) = (1/2)|\tau - \bar{\tau}| = |\mathbf{Im}(\tau)|$, which is height times base, namely the area.

(1.4) Definition. For a natural number N and N-division point on an elliptic curve $E(k)$ or a complex torus T is a solution x to the equation $Nx = 0$. The N-division points of E (resp. T) form a subgroup denoted by ${}_NE$ (resp. ${}_NT$).

In the case of a complex torus $T = \mathbb{C}/L$ this subgroup ${}_NT = [(1/N)L]/L$ is isomorphic to L/NL. If ω_1, ω_2 is a basis of L over $\mathbb{Z}$, then ${}_NT$ is isomorphic to $(\mathbb{Z}\omega_1/N + \mathbb{Z}\omega_2/N)/L$. Thus the group of N-division points on a complex torus T has order N^2 and is the direct sum of two cyclic groups of order N. The group ${}_NT$ is also the kernel of the isogeny $N : T \to T$, and L_λ is the lattice with

$$ {}_\lambda L/L = \ker(\lambda : \mathbb{C}/L \to \mathbb{C}/L'). $$

(1.5) Remark. If $\lambda : T = \mathbb{C}/L \to T' = \mathbb{C}/L'$ is an isogeny of degree N, then $\ker(\lambda) \subset {}_NT$ and $\ker(\lambda = {}_\lambda L/L$ where $L \subset {}_\lambda L \subset (1/N)L$, $[{}_\lambda L : L] = N$, and $[(1/N)L : {}_\lambda L] = N$. Further, $\lambda : T \to T'$ factors by $T = \mathbb{C}/L \xrightarrow{l} \mathbb{C}/{}_\lambda L \xrightarrow{\lambda} \mathbb{C}/L' = T$, where $\mathbb{C}/{}_\lambda L \xrightarrow{\lambda} \mathbb{C}/L'$ is an isomorphism. Thus up to isomorphism, each isogeny of $T = \mathbb{C}/L$ of degree N is given by a lattice $L^* \supset L$ with $[L^* : L] = N$ and has the form $\mathbb{C}/L \xrightarrow{l} \mathbb{C}/L^*$. A cyclic isogeny is one with the kernel a cyclic group.

§2. Symplectic Pairings on Lattices and Division Points

Let L be a lattice in $\mathbb{C}$, and choose a basis such that $L = \mathbb{Z}\omega_1 + \mathbb{Z}\omega_2$ satisfies $\mathbf{Im}(\omega_1/\omega_2) > 0$. The determinant gives rise to a symplectic pairing e or e_L defined

$$ e : L \times L \to \mathbb{Z} $$

which is given by the formula

$$ e(a\omega_1 + b\omega_2, c\omega_1 + d\omega_2) = ad - bc = \det \begin{pmatrix} a & b \\ c & d \end{pmatrix}. $$

A change of basis from ω_1, ω_2 to ω_1', ω_2' is given by a 2 by 2 matrix of determinant 1, and the pairing associated with the second basis is the same as the pairing defined by the first basis.

(2.1) Remark. The function $e : L \times L \to \mathbb{Z}$ is uniquely determined by the requirements that:

(1) e is biadditive, or $\mathbb{Z}$-bilinear,
(2) $e(x, x) = 0$ for all x in L,
(2)$'e(x, y) = -e(y, x)$ for all x, y in L, and
(3) for some basis ω_1, ω_2 of L with $\mathbf{Im}(\omega_1/\omega_2) > 0$ we have $e(\omega_1, \omega_2) = 1$. This holds then for any oriented basis of L.

Note that under (1) the two assretions (2) and (2)$'$ are equivalent. The assertion (3) will hold for any such basis when it holds for one, and by (1), (2), (2)$'$ the above formula for e in terms of the determinant is valid and shows the uniqueness.

Observe that $e : L \times L \times \mathbb{Z}$ defines by reduction modulo m a symplectic pairing $e_m : L/mL \times L/mL \rightarrow \mathbb{Z}/m$, and for m dividing n, we have a commutative diagram

$$\begin{array}{ccc} L/nL \times L/nL & \xrightarrow{e_n} & \mathbb{Z}/n \\ \downarrow & & \downarrow \\ L/mL \times L/mL & \xrightarrow{e_m} & \mathbb{Z}/m. \end{array}$$

(2.2) Remark. The symplectic pairings e and e_m are nondegenerate. This means for example that for any linear map $u : L \rightarrow \mathbb{Z}$ there exists a unique y in L with $u(x) = e(x, y)$ for all x in L.

Using the definitions going into the formula for the dual isogeny in (1.2), we can obtain a useful formula for $e(x, y)$. First, let $\mathrm{sgn}(x, y)$ equal $+1, 0, -1$ when $\mathbf{Im}(x/y)$ is $> 0, = 0$, and < 0, respectively. Then it is easy to check that

$$e(x, y) = \mathrm{sgn}(x, y) \cdot \frac{a(\mathbb{Z}x + \mathbb{Z}y)}{a(L)}.$$

(2.3) Remark. With this formula for e, we show that for any isogeny $\lambda : E = \mathbb{C}/L \rightarrow E' = \mathbb{C}/L'$ with $\lambda L \subset L'$ and dual isogeny $\hat{\lambda} : E' \rightarrow E$ that the relation $e_{L'}(\lambda x, x') = e_L(x, \hat{\lambda} x')$ holds for x in L and x' in L'. For we have the following inclusions between lattices

$$\begin{array}{ccc} \lambda L & \subset & L' \\ \cup & & \cup \\ \mathbb{Z}\lambda x + \mathbb{Z}nx' & \subset & \mathbb{Z}\lambda x + \mathbb{Z}x'. \end{array}$$

Now calculate

$$\begin{aligned} e_{L'}(x, \hat{\lambda} x') &= \mathrm{sgn}(x, \hat{\lambda} x') \cdot \frac{a(\mathbb{Z}x + \mathbb{Z}(n/\lambda)x')}{a(L)} \\ &= \mathrm{sgn}(\lambda x, x') \cdot \frac{a(\mathbb{Z}\lambda x + \mathbb{Z}nx')}{a(\lambda L)} \\ &= \mathrm{sgn}(\lambda x, x')[\lambda L : \mathbb{Z}\lambda x + \mathbb{Z}nx'] \\ &= \mathrm{sgn}(\lambda x, x')[L' : \mathbb{Z}\lambda x + \mathbb{Z}x'] \\ &= e_{L'}(\lambda x, x'). \end{aligned}$$

This verifies the formula.

The above discussion takes place on latices. Now we reinterpret the pairing on the division points ${}_N E(\mathbb{C}) = (I/N)L/L \subset E(\mathbb{C})$ contained in the complex points of the curve. This will lead later to the algebraic definition of the symplectic pairing which is due to A. Weil. The results over the complex numbers are summarized in the following proposition.

(2.4) Proposition. *Let E and E' be elliptic curves over $\mathbb{C}$. The isomorphism ${}_N E(\mathbb{C}) = [(1/N)L]/L \to L/NL$, given by multiplication by N, transfers the symplectic pairing on L to a nondegenerate pairing*

$$e_N : {}_N E(\mathbb{C}) \times {}_N E(\mathbb{C}) \to \mathbb{Z}/N\mathbb{Z}.$$

For an isogeny $\lambda : E \to E'$ we have

$$e_N(\lambda x, x') = e_N(x, \hat{\lambda} x')$$

for x in ${}_N E(\mathbb{C})$ and x' in ${}_N E'(\mathbb{C})$.

A final statement about the complex case for the reader with a background in homology theory. On $H_1(E(\mathbb{C}), \mathbb{Z}) = L$, where $E(\mathbb{C}) = \mathbb{C}/L$, the symplectic homology intersection pairing on the real surface $E(\mathbb{C})$ is $e : L \times L \to \mathbb{Z}$ in (2.1).

§3. Isogenies in the General Case

(3.1) Remark. For a lattice L we denote by $\mathrm{El}(L)$ the field of elliptic functions on $\mathbb{C}/L$. If $\lambda : \mathbb{C}/L \to \mathbb{C}/L'$ is an isogeny, then $f(z) \mapsto f(\lambda z)$ defines an embedding of $\mathrm{El}(L')$ into $\mathrm{El}(L)$ as a subfield. The group $\lambda^{-1}L'/L$ acts on the field $\mathrm{El}(L)$ by translation of variables in the function. The fixed field is equal to $\mathrm{El}(L')$, or, more precisely, the image of $\mathrm{El}(L')$ in $\mathrm{El}(L)$. Hence $\mathrm{El}(L)/\mathrm{El}(L')$ is a Galois extension of fields with Galois group $\lambda^{-1}L'/L = \ker(\lambda)$, and by the theorem of Artin

$$[\mathrm{El}(L) : \mathrm{El}(L')] = \#(\lambda^{-1}L'/L) = [L' : \lambda L].$$

In fact, the embedding $\mathrm{El}(L')$ into $\mathrm{El}(L)$ determines $\lambda : \mathbb{C}/L \to \mathbb{C}/L'$, and this gives us the clue as to how to formulate the notion of isogeny in the general case. Further, note that although an isogeny $\mathbb{C}/L \to \mathbb{C}/L'$ is additive, it suffices for it to be just analytic preserving 0 by the remarks preceding (1.1).

(3.2) Definition. For an elliptic curve E defined over a field k by the Weierstrass equation $f(X, Y) = 0$ the function field $k(E)$ of E over k is the field of fractions of the ring $k[X, Y]/(f)$.

The field $k(E)$ can also be described as the quadratic extension $k(x, y)$ of the field of rational function in one variables $k(x)$ where y satisfies the quadratic equation $f(x, y) = 0$ in y over $k(x)$. In the case $E(\mathbb{C}) = \mathbb{C}/L$ over the complex numers we know there is an isomorphism between $\mathbb{C}(E)$ and $\mathrm{El}(L)$ by 9(3.3), for $\mathrm{El}(L)$ is generated by $\wp$ and $\wp'$ and $\wp'$ satisfies a quadratic equation of Weierstrass type over the rational function field $\mathbb{C}(\wp)$.

Now we survey the theory of isogenies over an arbitrary field and suggest reader go to Mumford, *Abelian Varieties*.

(3.3) Definition. Let E and E' be two elliptic curves over k. An isogeny $\lambda : E \to E'$ is a nonzero rational map over k with $\lambda(0) = 0$. This means λ is given by an embedding $k(E) \leftarrow k(E')$ which we usually view as an inclusion. The degree of λ, denoted $\deg(\lambda)$, is $[k(E) : k(E')]$.

In arbitrary characteristic, the degree of a field extension is the product of the separable degree and the purely inseparable degree so that

$$\begin{aligned}\deg(\lambda) = [k(E) : k(E')] &= [k(E) : k(E')]_{\mathrm{s}}[k(E) : k(E')]_{\mathrm{i}} \\ &= \deg(\lambda)_{\mathrm{s}} \deg(\lambda)_{\mathrm{i}}.\end{aligned}$$

The separable degree $\deg(\lambda)_{\mathrm{s}}$ is the order of the group of geometric points in $\ker(\lambda) = \lambda^{-1}(0)$ as a variety while $\deg(\lambda)_{\mathrm{i}}$ is related to the scheme theoretical structure of $\ker(\lambda)$. Multiplication by p in characteristic p is always inseparable.

The condition $\lambda(0) = 0$ can be understood in terms of the function field $k(E)$ $k(x', y')$, where x', y' satisfy a Weierstrass equation. On E the functions x' and y', i.e., viewed in $k(E)$, must have a pole at 0 for $\lambda(0) = 0$ to hold.

In order to see that in isogeny λ is automatically additive and to define the dual isogeny $\hat{\lambda}$, we have to look at divisors on E. These are finite formal sums $\sum n_P P$ of points on E over an algebraically closed field k, and the degree of a divisor is given by

$$\deg\left(\sum n_P P\right) = \sum n_P, \qquad \text{an integer.}$$

If f is in the multiplicative group $k(E)^*$, then the divisor (f) of zeros and poles is defined, and, as on the projective line, we have

$$\deg(f) = 0.$$

We have three groups $\mathrm{Div}(E) \supset \mathrm{Div}_0(E) \supset \mathrm{Div}_1(E)$ of all divisors, all divisors of degree 0, and all divisors of fnctions. The sum function $s : \mathrm{Div}_0(E) \to E(k)$ where $s(\sum n_P P) = \sum n_P P$ in $E(k)$ has kernel $\mathrm{Div}_1(E)$ and is surjective, see 9(3.5) for the proof over the complex numbers. The function $P \mapsto P - 0$ is a cross-section of s. Thus $P_1 + P_2 + P_3 = 0$ in $E(k)$ if and only if $P_1 + P_2 + P_3 = -3 \cdot 0$ is the divisor of a function.

(3.4) Remarks/Definition. An isogeny $\lambda : E \to E'$ defines a group morphism $\lambda : \mathrm{Div}(E) \to \mathrm{Div}(E')$ by $\lambda(\sum n_P P) = \sum n_P \lambda(P)$, and the following diagram is commutative:

$$\begin{array}{ccc} \mathrm{Div}_0(E) & \xrightarrow{\lambda} & \mathrm{Div}_0(E') \\ \downarrow s & & \downarrow s \\ E(k) & \xrightarrow{\lambda} & E'(k) \end{array}$$

showing that λ is additive. Further, λ defines $\lambda^{-1} : \mathrm{Div}_0(E') \to \mathrm{Div}_0(E)$ by

$$\lambda^{-1}\left(\sum n_P P\right) = \sum n_P \lambda^{-1}(P),$$

where $\lambda^{-1}(P) = m(Q_1 + \cdots + Q_r)$ for $\{Q_1, \ldots, Q_r\} = \lambda^{-1}(P)$ set theoretically and $mr = \deg(\lambda)$. This induces $\hat{\lambda} : E' \to E$ the dual isogeny such that the following diagram is commutative:

$$\begin{array}{ccc} \mathrm{Div}_0(E') & \xrightarrow{\lambda^{-1}} & \mathrm{Div}_0(E) \\ \big\downarrow s & & \big\downarrow s \\ E'(k) & \xrightarrow{\hat{\lambda}} & E(k). \end{array}$$

Now the following extension of (1.5).

(3.5) Theorem. *The function $\lambda \mapsto \hat{\lambda}$ is a group morphism*

$$\mathrm{Hom}(E, E') \to \mathrm{Hom}(E', E)$$

satisfying $\hat{\hat{\lambda}} = \lambda$, $\deg(\lambda) = \deg(\hat{\lambda})$, $\hat{\lambda}\lambda = n$ in $\mathrm{End}(E)$, and $\lambda\hat{\lambda} = n$ in $\mathrm{End}(E')$. For λ in $\mathrm{Hom}(E, E')$ and μ in $\mathrm{Hom}(E', E'')$ we have $\widehat{\mu\lambda} = \hat{\lambda}\hat{\mu}$ and $\deg(\mu\lambda) = \deg(\mu) \cdot \deg(\lambda)$. The involution $\lambda \mapsto \hat{\lambda}$ of the ring $\mathrm{End}(E)$ satisfies $\hat{n} = n$ where $\deg(n) = n^2$. The degree functions $\deg : \mathrm{Hom}(E, E') \to \mathbb{Z}$ is a positive quadratic function.

The involution taking an isogeny to its dual is called the rosati involution on $\mathrm{End}(E)$. The assertion that $\deg(\lambda)$ is a positive quadratic function on $\mathrm{Hom}(E, E')$ means that

(a) $\deg \lambda \geq 0$ and by definition $\deg(\lambda) = 0$ for $\lambda = 0$.
(b) $\deg(m\lambda) = m^2 \deg \lambda$, and
(c) $\deg(\lambda + \mu) = \deg(\lambda) + (\hat{\lambda}\mu + \hat{\mu}\lambda) + \deg(\mu)$ where $(\lambda, \mu) \mapsto \hat{\lambda}\mu + \hat{\mu}\lambda$ is a biadditive function defined

$$\mathrm{Hom}(E, E') \times \mathrm{Hom}(E, E') \to \mathbb{Z}.$$

The remarks following (1.1) have the following extension.

(3.6) Theorem. *Let E be an elliptic curve over a field k.*

(1) *If k is separably closed and n is prime to the characteristic, then $E(k)$ is divisible by n, i.e., the map $n : E(k) \to E(k)$ is surjective.*
(2) *If k is separably closed and n is prime to the characteristic, then the subgroup ${}_nE(k)$ of n-division points is isomorphic to $\mathbb{Z}/n \times \mathbb{Z}/n$.*
(3) *If k is algebraically closed of characteristic $p > 0$, then $E(k)$ is divisible by p. Moreover, the p-division points form a group ${}_pE(k)$ is isomorphic to either $\mathbb{Z}/p$ or 0.*

The two possibilities for the group of p-division points will be considered further in the next chapter. A curve is called supersingular provided its group of p-division points over $\bar{k}$ reduces to 0.

The proofs of Theorem (3.5) and (3.6) can be found in the book by Mumford, *Abelian Varieties*. They are true for higher-dimensional complete group varieties, but messy proofs for elliptic curves can be carried out using the Weierstrass equation. The prevailing wisdom in the mathematics community is that the reader who has gotten this far in the theory of elliptic curves should start with the theory of abelian varieties. An other possibility is the see the books of Silverman. In this line our discussion in this chapter from §3 on is only a sketch of results designed to give the reader an overview of the results.

The pairing $e_N :_N E \times {}_N E \rightarrow \mathbb{Z}/N\mathbb{Z}$ of (2.4) has an algebraic meaning and an algebraic definition can be given. In fact, for any isogeny $\lambda : E \rightarrow E'$ there is a pairing $e_\lambda : \ker(\lambda) \times \ker(\hat{\lambda}) \rightarrow \mu_N(k)$, where $N = \#\ker(\lambda)$. This uses divisors as in (3.4). Recall that associated to λ is an inclusion $k(E') \rightarrow k(E)$, and for $x' \in \ker(\hat{\lambda})$ we have $\lambda^{-1}(x') = 0$, or as divisors $\lambda^{-1}((x') - (0)) = (f')$ for some $f' \in k(E)$, and thus $(f^N) = (f'\lambda)$. Now f^N is invariant by translates by $x \in E$, and for a general point y of E, we can define

$$e_\lambda(x, x') = \frac{f(y+x)}{f(y)} \in \mu_N(k).$$

(3.7) Remark. In the special case $\lambda = N$, there is another formula for the pairing. We need the notation $h(\sum_P n_P P) = \prod_P h(P)^{n_P}$ for a divisor $\sum_P n_P P$ which has no common prime factor with the divisor (h) of the function h. For $x, x' \in {}_N E$ choose divisors D and D' differing from $(x) - (0)$ and $(x') - (0)$ up to the divisor of a function and having no prime factors in common. Since $ND = (f)$ and $ND' = (f')$, it can be shown that $e_N(x, x') = f'(D)/f(D')$.

This definition works because of the:

Reciprocity Law. When (f) and (g) have no common prime factors, then

$$f((g)) = g((f)).$$

This reciprocity law holds for functions on $\mathbb{P}^1$ by a direct calculation for $f(x) = (x - a)/(x - b)$ and $g(x) = (x - c)/(x - d)$ and the multiplicative character of the formula. Then it holds on any curve by mapping onto the projective line by a finite ramified covering.

Now we summarize the basic properties of this pairing.

(3.8) Theorem. *Let E and E' be elliptic curves over a perfect field k and $\lambda : E \rightarrow E'$ an isogeny. The pairing $e_\lambda : \ker(\lambda) \times \ker(\hat{\lambda}) \rightarrow \mu_N(k)$ is biadditive and nondegenerate, and for an automorphism σ of k we have*

$$e_{\lambda^\sigma}(x^\sigma, x'^\sigma) = e_\lambda(x, x')^\sigma.$$

The special case of $e_N : {}_NE \times {}_NE \rightarrow \mu_N(k)$ is antisymmetric, and if M divides N, then the following diagram is commutative:

$$\begin{array}{ccc} {}_NE \times {}_NE & \xrightarrow{e_N} & \mu_N(k) \\ {\scriptstyle N:M}\big\downarrow & & \big\downarrow{\scriptstyle N:M} \\ {}_ME \times {}_ME & \xrightarrow{e_M} & \mu_M(k). \end{array}$$

Here $\ker(\lambda)$ is a scheme which is étale for $p \nmid \deg(\lambda)$.

(3.9) Remark. Observe that the relation $e_N(\lambda x, x') = e_N(x, \hat{\lambda}x')$ shows that $e_N(\lambda x, \lambda x') = e_N(x, \hat{\lambda}\lambda x') = \deg(\lambda)e_N(x, x')$.

§4. Endomorphisms and Complex Multiplication

Now we study the ring $\operatorname{End}(E) = \operatorname{Hom}(E, E)$ for an elliptic curve E over k. This ring is equipped with the Rosati involution and the degree map which is possitive and quadratic.

(4.1) Proposition. *The ring* $\operatorname{End}(E)$ *has no zero divisors.*

Proof. If $\mu\lambda = 0$, then $0 = \deg(\mu\lambda) = \deg(\mu)\deg(\lambda)$, and, therefore, either $\deg(\lambda) = 0$, so $\lambda = 0$, or $\deg(\mu) = 0$, so $\mu = 0$.

For an alternative proof over $\mathbb{C}$ the ring $\operatorname{End}(E)$ is the set of complex numbers λ with $\lambda L \subset L$ and, in particular, a subring of $\mathbb{C}$, hence without zero divisors.

(4.2) Definition. For an element λ in $\operatorname{End}(E)$ the trace is $T(\lambda = \lambda + \hat{\lambda}$ and the characteristic polynomial is

$$c_\lambda(t) = t^2 - T(\lambda)t + \deg(\lambda).$$

Note that in terms of the trace the quadratic condition (c) of (3.5) becomes $\deg(\lambda + \mu) = \deg(\lambda) + T(\hat{\lambda}\mu) + \deg(\mu)$.

(4.3) Proposition. *The trace of* λ *in* $\operatorname{End}(E)$ *is in the subring* $\mathbb{Z}$ *of* $\operatorname{End}(E)$ *and* $c_\lambda(t)$ *is in* $\mathbb{Z}[t]$. *Further,* $c_\lambda(\lambda) = 0$.

Proof. We calculate that $\deg(1 + \lambda) = (1 + \lambda)(1 + \hat{\lambda}) = 1 + (\lambda + \hat{\lambda}) + \lambda\hat{\lambda}$ is in $\mathbb{Z}$, and using the fact that $\deg(\lambda)$ is an integer, we deduce that $T(\lambda)$ is an integer. Finally we have $c_\lambda(\lambda) = \lambda^2 - (\lambda + \hat{\lambda})\lambda + \lambda\hat{\lambda} = 0$.

By the previous proposition every element of $\operatorname{End}(E)$ satisfies a quadratic equation over the subring $\mathbb{Z}$. This is a very strong restriction on $\operatorname{End}(E)$. Moreover, these quadratic equations have the following additional positivity property.

(4.4) Theorem (Hasse). *For* λ *in* End(E) *the characteristic polynomial has values* $c_\lambda(r) \geq 0$ *for any rational number and*

$$\left|\lambda + \hat{\lambda}\right| \leq 2\sqrt{\deg(\lambda)}.$$

Proof. For $r = n/m$ we see that

$$m^2 c_\lambda(n/m) = n^2 + nmT(\lambda) + m^2 \deg(\lambda) = \deg(n + m\lambda) \geq 0.$$

Hence, the discriminant $4\deg(\lambda) - T(\lambda)^2$ is positive. Now take the square root of $T(\lambda)^2 = (\lambda + \hat{\lambda})^2 \leq 4\deg(\lambda)$ to obtain the result.

(4.5) Definition. An elliptic curve E over k has complex multiplication provided $\mathbb{Z} \neq \mathrm{End}(E)$.

(4.6) Theorem. *Assume that* End(E) *is commutative, then either* End$(E) = \mathbb{Z}$ *or* End$(E) \otimes_{\mathbb{Z}} \mathbb{Q}$ *is an imaginary quadratic extension of* $\mathbb{Q}$*. In the second case* End(E) *is an order in the imaginary quadratic field* $\mathrm{End}^0(E) = \mathrm{End}(E) \otimes_{\mathbb{Z}} \mathbb{Q}$.

Proof. Every element in $\mathrm{End}^0(E)$ satisfies a quadratic equation over the subfield $\mathbb{Q}$ by (4.3), and this means that either $\mathrm{End}^0(E)$ has degree 1 over $\mathbb{Q}$, that is $\mathrm{End}^0(E) = \mathbb{Q}$, or $\mathrm{End}^0(E)$ has degree 2 over $\mathbb{Q}$, that is, $\mathrm{End}^0(E) = \mathbb{Q}(\alpha)$, where α is a quadratic irrationality. From the positivity of the quadratic equation for α, see (4.4), the field $\mathbb{Q}(\alpha)$ is an imaginary quadratic number field. By (4.1) and (4.3) the endomorphism ring embeds as a subring of the ring of integers in End(E). Moreover, it is of finite index in the ring of integers and it contains $\mathbb{Z}$, i.e., End(E) is an order in $\mathrm{End}^0(E)$. This proves the theorem.

(4.7) Proposition. *Let E be an elliptic curve over the complex numbers with complex points* $\mathbb{C}/L_\tau$*. Then* End(E) *is commutative. The curve has complex multiplication if and only if τ is an imaginary quadratic number. In the case of complex multiplication* $\mathrm{End}(\mathbb{C}/L_\tau) \otimes \mathbb{Q}$ *is* $\mathbb{Q}(\tau)$ *the imaginary quadratic number field obtained by adjoining τ to $\mathbb{Q}$, and as a subring of complex numbers* End(E) *is contained in* L_τ.

Proof. Since End(E) is naturally isomorphic to the subring of complex numbers λ satisfying $\lambda L_\tau \subset L_\tau$ it is commutative. For such a λ with $\lambda L_\tau \subset L_\tau = \mathbb{Z}\tau + \mathbb{Z}$ we see that $\lambda = a + b\tau$ is contained in L_τ and the relation $c + d\tau = \lambda\tau = a\tau + b\tau^2$ holds for integers a, b, c, and d. This shows that τ satisfies a quadratic equation over $\mathbb{Q}$. Conversely $a\tau^2 = b\tau + c$ with a, b, and c integers implies that $(a\tau)L_\tau \subset L_\tau$, and thus $\mathbb{C}/L_\tau$ has complex multiplication. Finally, $\mathrm{End}(\mathbb{C}/L_\tau) \otimes \mathbb{Q} = \mathbb{Q}(\tau)$ holds from the quadratic relations for τ over the rational numbers. This proves the proposition.

(4.8) Remark. For an elliptic curve over the complex numbers we can see from the above proposition that Aut(E) is $\{+1, -1\}$ if and only if $j \neq 0$ or 12^3. The case $j = 0$ is E isomorphic to $E(\rho)$ and $\mathrm{Aut}(E) = \{\pm 1, \pm\rho, \pm\bar{\rho}\}$, and the case $j = 12^3$ is E isomorphic to $E(i)$ and $\mathrm{Aut}(E) = \{\pm 1, \pm i\}$.

Finally we consider the determination of all elliptic curves isogenous to a given elliptic curve E with complex multiplication.

(4.9) Remarks. For an imaginary quadratic field K, let R denote the ring of integers in K. For a subring A of R note that $\mathrm{rk}(A) = \mathrm{rk}(R) = 2$ if and only if as abelian groups A is of finite index in R. Every such subring A has the form $R_f = \mathbb{Z} + fR$, where f is an integer $f \geq 1$ called the conductor of the order $R_f = A$.

We denote by $\mathrm{Pic}(A)$ the projective class group. It consists of isomorphism classes of projective modules L over A of rank 1 with abelian group structure given by tensor product of projective modules of rank 1. There is an embedding $L \to \mathbb{Q} \otimes_{\mathbb{Z}} L = K \otimes_A L = V$, and V is a one-dimensional vector space over K. A choice of basis element gives an embedding $L \to K$ as a fractional ideal, and two such embeddings differ up to multiplication by an element of K. Hence a second interpretation of $\mathrm{Pic}(A)$ is as fractional ideal classes of A in its field of fractions K when $A = R_f$.

For a number field K and an order A contained in the ring of integes R of K, the group $\mathrm{Pic}(A)$ is finite and its cardinality is the class number of A or of K in the case $A = R$. For the $A = R_f = \mathbb{Z} + fR$ we denote by $h_{K,f} = \#\mathrm{Pic}(R_f)$, the cardinality of Pic.

(4.10) Theorem. *Let $K \subset \mathbb{C}$ be a quadratic imaginary field. For each class $[L]$ in $\mathrm{Pic}(R_f)$, choose a representative L of $[L]$ and an embedding $L \to K \subset \mathbb{C}$. The function that assigns to $[L]$ the elliptic curve E_L with $E_L(C) = \mathbb{C}/L$ is a bijection of $\mathrm{Pic}(R_f)$ onto the set of isomorphism classes of elliptic curves E over $\mathbb{C}$ with $\mathrm{End}(E) = R_f$.*

Proof. Let E be an elliptic curve over $\mathbb{C}$ with $E(\mathbb{C}) = \mathbb{C}/L$ where L can be chosen to have a nonzero element in common with K. Since $R_f L = L$, where $R_f = \mathrm{End}(E)$, we have $L \subset K$ as a functional idea, and this ideal is defined up to scalar multiple. This proves the theorem since $\mathrm{End}_{R_f}(L) = \mathrm{End}(\mathbb{C}/L) = R_f$.

The ring $R_f = \mathbb{Z} + fR$ is called the order with conductor f in an imaginary quadratic field K.

(4.11) Proposition. *The j-invariant of $\mathbb{C}/L$ for any $[L]$ in $\mathrm{Pic}(R_f)$, denoted $j(L) = f(\mathbb{C}/L)$, is an algebraic number of degree $\leq h_{K,f}$.*

Proof. The group $\mathrm{Aut}(\mathbb{C})$ acts on the finite set $J_{K,f}$ of all $j(L)$ for $[L] \in \mathrm{Pic}(R_f)$, and this means that these numbers are algebraic numbers of degree less than the cardinality of the finite set.

(4.12) Remark. The question of j values of complex multiplication curves will be considered further in Chapter 13. For example $j(L)$ is an algebraic integer of degree exactly $h_{f,K}$ and $\mathrm{Aut}(\mathbb{C})$ acts transitively on the set $J_{K,f}$. For further information see Serre, "Complex Multiplication," in the Cassels and Frohlich book [1967]. We include a table of the j values of $L = R_f$ for the cases where $h_{K,f} = 1$ from Serre [1967]. Classically these values appear in Weber, *Algebra III*, see §§125–128 and Tabelle VI.

(4.13) Imaginary Quadratic Fields with Class Number 1.

$$
\begin{array}{lll}
K = \mathbb{Q}(\sqrt{-1}) & R = \mathbb{Z}[\sqrt{-1}] & j(\tau) = j(\sqrt{-1}) = (2^2 \cdot 3)^3 \\
 & R_2 = \mathbb{Z}[2\sqrt{-1}] & = j(2\sqrt{-1}) = (2 \cdot 3 \cdot 11)^3 \\
= \mathbb{Q}(\sqrt{-2}) & R = \mathbb{Z}[\sqrt{-2}] & = j(\sqrt{-2}) = (2^2 \cdot 5)^3 \\
= \mathbb{Q}(\sqrt{-3}) & R = \mathbb{Z}\left[\dfrac{-1+\sqrt{-3}}{2}\right] & = j\left(\dfrac{-1+\sqrt{-3}}{2}\right) = 0 \\
 & R_2 = \mathbb{Z}[\sqrt{-3}] & = j(\sqrt{-3}) = 2^4 \cdot 3^3 \cdot 5^3 \\
 & R_3 = \mathbb{Z}\left[\dfrac{-1+3\sqrt{-3}}{2}\right] & = j\left(\dfrac{-3+3\sqrt{-3}}{2}\right) \\
 & & = -2^{15} \cdot 3 \cdot 5^3 \\
= \mathbb{Q}(\sqrt{-7}) & R = \mathbb{Z}\left[\dfrac{-1+\sqrt{-7}}{2}\right] & = j\left(\dfrac{-1+\sqrt{-7}}{2}\right) \\
 & & = -(3 \cdot 5)^3 \\
 & R_2 = \mathbb{Z}[\sqrt{-7}] & = j(\sqrt{-7}) = (3 \cdot 5 \cdot 17)^3 \\
= \mathbb{Q}(\sqrt{-11}) & R = \mathbb{Z}\left[\dfrac{-1+\sqrt{-11}}{2}\right] & = j\left(\dfrac{-1+\sqrt{-11}}{2}\right) \\
 & & = -(2^5)^3 \\
= \mathbb{Q}(\sqrt{-19}) & R = \mathbb{Z}\left[\dfrac{-1+\sqrt{-19}}{2}\right] & = j\left(\dfrac{-1+\sqrt{-19}}{2}\right) \\
 & & = -(2^5 \cdot 3)^3 \\
= \mathbb{Q}(\sqrt{-43}) & R = \mathbb{Z}\left[\dfrac{-1+\sqrt{-43}}{2}\right] & = j\left(\dfrac{-1+\sqrt{-43}}{2}\right) \\
 & & = -(2^6 \cdot 3 \cdot 5)^3 \\
= \mathbb{Q}(\sqrt{-67}) & R = \mathbb{Z}\left[\dfrac{-1+\sqrt{-67}}{2}\right] & = j\left(\dfrac{-1+\sqrt{-67}}{2}\right) \\
 & & = -(2^5 \cdot 3 \cdot 5 \cdot 11)^3 \\
= \mathbb{Q}(\sqrt{-163}) & R = \mathbb{Z}\left[\dfrac{-1+\sqrt{-163}}{2}\right] & = j\left(\dfrac{-1+\sqrt{-163}}{2}\right) \\
 & & = -(2^6 \cdot 3 \cdot 5 \cdot 23 \cdot 29)^3
\end{array}
$$

(4.14) Remark. The class number of $\mathbb{Q}(\sqrt{-23})$ is three and

$$j\left(\frac{-1+\sqrt{-23}}{2}\right) = -\alpha^{12} 5^3 (2\alpha - 1)^3 (3\alpha + 2)^3,$$

where $\alpha^3 - \alpha - 1 = 0$

(4.15) Remark. In Gross and Zagier [1985], *On Singular Moduli*, other values of $j(\tau)$ are given, for example,

$$j\left(\frac{1+\sqrt{-163}}{2}\right) = -2^6 \cdot 3^6 \cdot 7^2 \cdot 11^2 \cdot 19^2 \cdot 127^2 \cdot 163.$$

§5. The Tate Module of an Elliptic Curve

Let ℓ denote a prime number throughtout this section. We have a sequence of reductions of $\mathbb{Z}$ by powers of ℓ, namely

$$\cdots \to \mathbb{Z}/\ell^{n+1} \xrightarrow{r_n} \mathbb{Z}/\ell^n \to \cdots \xrightarrow{r_2} \mathbb{Z}/\ell^2 \xrightarrow{r_1} \mathbb{Z}/\ell.$$

The projective limit of this sequence, denoted $\varprojlim_n \mathbb{Z}/\ell^n$ consists of all $a = (a_n)$ in the product $\prod_{1 \le n} \mathbb{Z}/\ell^n$ such that $r_n(a_{n+1}) = a_n$ for all $n \ge 1$. This projective limit is $\mathbb{Z}_\ell$ the ring of ℓ-adic integers. The field of fractions $\mathbb{Z}_\ell \otimes_{\mathbb{Z}} \mathbb{Q}$ is $\mathbb{Q}_\ell$ the field of ℓ-adic numbers. This approach is carried out in detail in Serre, *Course in Arithmetic*, Chapter II.

The sequence of reductions of $\mathbb{Z}$ can be described using subgroups of $\mathbb{Q}/\mathbb{Z}$ rather than quotients of $\mathbb{Z}$. Let ${}_N(\mathbb{Q}/\mathbb{Z})$ be the subgroup of all x in $\mathbb{Q}/\mathbb{Z}$ with $Nx = 0$. Then the equivalent sequence is the following where all the morphisms are multiplication by ℓ:

$$\cdots \xrightarrow{\ell} {}_{\ell^{n+1}}(\mathbb{Q}/\mathbb{Z}) \xrightarrow{\ell} {}_{\ell^n}(\mathbb{Q}/\mathbb{Z}) \xrightarrow{\ell} \cdots \xrightarrow{\ell} {}_{\ell^2}(\mathbb{Q}/\mathbb{Z}) \xrightarrow{\ell} {}_{\ell}(\mathbb{Q}/\mathbb{Z}).$$

A group having essentially the same structure as $(\mathbb{Q}/\mathbb{Z})$ is $\mu(k)$, the group of roots of unity in an algebraically closed field k. The subgroup $\mu_N(k)$ of x in $\mu(k)$ with $x^N = 1$ consists of the Nth roots of unity. It is cyclic and of order N when k is algebraically closed and N is prime to the characteristic of k. Further, if k_s denotes the seperable algebraic closure of k, then the Galois group $\mathrm{Gal}(k_s/k)$ operates on $\mu(k_s)$ and each $\mu_N(k_s)$. As above we have a sequence of cyclic groups of order ℓ^n, where ℓ is prime to the characteristic of k, namely

$$\cdots \xrightarrow{\ell} \mu_{\ell^{n+1}}(k) \xrightarrow{\ell} \mu_{\ell^n}(k) \xrightarrow{\ell} \cdots \xrightarrow{\ell} \mu_{\ell^2}(k) \xrightarrow{\ell} \mu_\ell(k).$$

The inverse limit is denote $\mathbb{Z}_\ell(1)(k)$ or simple $\mathbb{Z}_\ell(1)$ and it is a $\mathrm{Gal}(k_s/k)$-module.

(5.1) Definition. Let k_s be the separable algebraic closure of a field k of characteristic prime to ℓ. The Tate module of k_s^* is the inverse limit denoted $\mathbb{Z}_\ell(1)(k_s)$ or $\mathbb{Z}_\ell(1)$ together with its action of $\mathrm{Gal}(k_s/k)$.

The Tate module, or as it is sometimes called the Tate twist, is just $\mathbb{Z}_\ell$ as a limit group, but it is more, namely a Galois module.

There is a corresponding construction for elliptic curves E over a field k of characteristic different from ℓ using the division points ${}_N E(k_s)$ of E over the separable algebraic closure k_s of k. Again the Galois group $\mathrm{Gal}(k_s/k)$ acts on $E(k_s)$ and each ${}_N E(k_s)$ from its action on the x-and y-coordinates of a point.

(5.2) Definition. The Tate module $T_\ell(E)$ of an elliptic curve over k is the projective limit where ℓ is a prime unequal to the the characteristic ℓ

$$\cdots \xrightarrow{\ell} {}_{\ell^{n+1}}E(k_s) \xrightarrow{\ell} {}_{\ell^n}E(k_s) \xrightarrow{\ell} \cdots \xrightarrow{\ell} {}_{\ell^2}E(k_s) \xrightarrow{\ell} {}_{\ell}E(k_s)$$

together with the action of $\mathrm{Gal}(k_s/s)$ on the limit group. We define $V_\ell(E) = T_\ell(E) \otimes_{\mathbb{Z}} \mathbb{Q}$ together with the extended $\mathrm{Gal}(k_s/k)$ action.

Since each ${}_{\ell^n}E(k_s)$ is a free $\mathbb{Z}/\ell^n$-module of rank 2, the limit $T_\ell(E)$ is a free $\mathbb{Z}_\ell$-module of rank 2, and $V_\ell(E)$ is a $\mathbb{Q}_\ell$ vector space of dimension 2. The Galois action can be described as a representation

$$\mathrm{Gal}(k_s/k) \to \mathrm{GL}(T_\ell(E)) \subset \mathrm{GL}(V_\ell(E))$$

which is referred to as the two-dimensional ℓ-adic representation of $\mathrm{Gal}(k_s/k)$ associated to E over k.

In the case where $E = \mathbb{C}/L$ over $\mathbb{C}$, we have ${}_NE(\mathbb{C}) = (\mathbb{Z}/\mathbb{Z}N) \otimes L$ and $T_\ell = \varprojlim_n (\mathbb{Z}/\mathbb{Z}\ell^n) \otimes L = \mathbb{Z}_\ell \otimes L$.

(5.3) Remark. There is a symplectic structure on the Tate module of E. For $N = \ell^n$ and passing to the inverse limit of the pairings $e_{\ell^n} : {}_{\ell^n}E \times {}_{\ell^n}E \to \mu_{\ell^n}(k_s)$ as n approaches ∞, we obtain a nondegenerate symplectic pairing, also denoted e_ℓ,

$$e_\ell : T_\ell(E) \times T_\ell(E) \to \mathbb{Z}_\ell(1)$$

using (3.8). Further tensoring with $\mathbb{Q}$, we have a nondegenerate symplectic pairing

$$e_\ell : V_\ell(E) \times V_\ell(E) \to \mathbb{Q}_\ell(1).$$

For σ in $\mathrm{Gal}(k_s/k)$ the Galois properties of e_ℓ are contained in the relation

$$e_\ell(x^\sigma, x'^\sigma) = e_\ell(x, x')^\sigma$$

which highlights the necessity of using the Tate twist $\mathbb{Z}_\ell(1)$ instead $\mathbb{Z}_\ell$ for the image module of e_ℓ.

One of the great steps forward in the theory of elliptic curves came when it was realized that this Galois representation on the Tate module $T_\ell(E)$ contains many of the basic invariants of the isomorphism type of E. The isogeny invariants of E are contained in the study of the Galois representation $V_\ell(E)$. This is based on an understanding of how faithful an action $\mathrm{End}(E)$ has on $T_\ell(E)$ and $\mathrm{End}^0(E)$ has on $V_\ell(E)$. This is the subject of the next section.

§6. Endomorphisms and the Tate Module

Since every homomorphism $\lambda : E \to E'$ restricts to a group homomorphism ${}_NE \to {}_NE'$ for every N commuting with multiplication in the inverse system defining the Tate modules, we have a canonical homomorphism of groups

$$T_\ell : \mathrm{Hom}(E, E') \to \mathrm{Hom}_{\mathbb{Z}_\ell}(T_\ell(E), T_\ell(E'))$$

for each prime ℓ. Tensoring with $\mathbb{Z}_\ell$ we obtain homomorphism of $\mathbb{Z}_\ell$-modules also denoted with the same letter

$$T_\ell : \mathrm{Hom}(E, E') \otimes \mathbb{Z}_\ell \to \mathrm{Hom}_{\mathbb{Z}_\ell}(T_\ell(E), T_\ell(E')).$$

The basic result is the following which holds also for abelian varieties and where the proof follows the one in Mumford [1974, pp. 176–178].

(6.1) Theorem. *For a prime ℓ unequal to the ground field characteristic the natural map*

$$T_\ell : \mathrm{Hom}(E, E') \otimes \mathbb{Z}_\ell \to \mathrm{Hom}_{\mathbb{Z}_\ell}(T_\ell(E), T_\ell(E'))$$

is injective. Moreover, the group $\mathrm{Hom}(E, E')$ *is finitely generated and free abelian.*

Proof. Since every nonzero $E \to E'$ is surjective, the group $\mathrm{Hom}(E, E')$ is torsion free, and we can think of $\mathrm{Hom}(E, E')$ as a subgroup of in $\mathrm{Hom}^0(E, E')$.

Assertion. For any finitely generated subgroup M of $\mathrm{Hom}(E, E')$ the subgroup

$$\mathbb{Q}M \cap \mathrm{Hom}(E, E') = \{\lambda : E \to E' \text{ with } n\lambda \text{ in } M \text{ for some } n \neq 0\}$$

is again finitely generated.

To prove the assertion, we note that $\mathrm{Hom}(E, E') = 0$ when E and E' are not isogenous, and using the injection $\mathrm{Hom}(E, E') \to \mathrm{End}(E)$ induced by an isogeny $E' \to E$, we are reduced to the case $E = E'$. The norm $\deg : \mathrm{End}(E) \to \mathbb{N}$ where $\deg(\lambda) > 0$ if and only if $\lambda \neq 0$ extends to $\deg : \mathrm{End}^0(E) \to \mathbb{Q}$. Now $\mathbb{Q}M$ is a finite dimensional space, and the set if λ in $\mathrm{End}^0(E)$ with $N(\lambda) < 1$ is a neighborhood V of zero in $\mathbb{Q}M$. Since $U \cap \mathrm{End}(E) = (0)$, it follows that $\mathbb{Q}M \cap \mathrm{End}(E)$ is discrete in $\mathbb{QM}$, and thus it is finitely generated.

By the above assertion it suffices to proove that for ant finitely generated M in $\mathrm{Hom}(E, E')$ satisfying $M = \mathbb{Q}M \cap \mathrm{Hom}(E, E')$, the restricted homomorphism is a monomorphism

$$T_\ell : M \otimes \mathbb{Z}_\ell \to \mathrm{Hom}_{\mathbb{Z}_\ell}(T_\ell(E), T_\ell(E')).$$

Let $\lambda_1, \ldots, \lambda_m$ be a basis of the abelian group M. Since the right-hand side is $\mathbb{Z}_\ell$-free, we would have $T_\ell(c_1\lambda_1 + \cdots + c_m\lambda_m) = 0$ where we may assume that one c_j a unit in $\mathbb{Z}_\ell$ if T_ℓ were not injective. This would mean that there are integers $a_1, \ldots, a_m$ not all divisible by ℓ such that for $\lambda = a_1\lambda_1 + \cdots + a_m\lambda_m$ we have $T_\ell(\lambda)(T_\ell(E)) \subset T_\ell(E')$ and hence, $f({}_\ell E) = 0$ in E'. Then $\lambda = \ell\mu$, where μ is in $\mathrm{Hom}(E, E')$, and, since

$$\mu \text{ is in } \mathbb{Q}M \cap \mathrm{Hom}(E, E') = M,$$

we can decompose $\mu = b_1\lambda_1 + \cdots + b_m\lambda_m$. Now $\sum_{j=1}^m a_j\lambda_j = \ell\sum_{j=1}^m b_j\lambda_j$, which implies that the prime ℓ divides all a_j which is a contradiction. This proves that T_ℓ is injective.

Now $\operatorname{Hom}_{\mathbb{Z}_\ell}(T_\ell(E), T_\ell(E'))$ is of finite rank, in fact, of rank at most 4. By the injectivity of T_ℓ on $M \otimes \mathbb{Z}_\ell$, we see that $\operatorname{Hom}^0(E, E')$ has dimension at most four and finally that $\operatorname{Hom}(E, E')$ is finitely generated. This proves the theorem.

We can define $V_\ell(\lambda) = T_\ell(\lambda) \otimes_{\mathbb{Z}_\ell} \mathbb{Q}_\ell$, and then the theorem has the following immediate corollary.

(6.2) Corollary. *For a prime ℓ unequal to the ground field characteristic the natural map*

$$V_\ell : \operatorname{Hom}(E, E') \otimes \mathbb{Q}_\ell \to \operatorname{Hom}_{\mathbb{Q}_\ell}(V_\ell(E), V_\ell(E'))$$

is injective.

For elliptic curves E and E' defined over k the Galois group $\operatorname{Gal}(k_s/k)$ acts on $\operatorname{Hom}(E, E')$ on the right by the formula $\lambda^\sigma = \sigma^{-1}\lambda\sigma$. The fixed subgroup $\operatorname{Hom}(E, E')^{\operatorname{Gal}(k_s/k)} = \operatorname{Hom}_k(E, E')$ is the subgroup of homomorphisms defined over k. For λ in $\operatorname{Hom}(E, E')$ and x in $V_\ell(E)$ we have the relation

$$(\lambda x)^\sigma = \lambda^\sigma x^\sigma$$

for the right action x^σ equal to $\sigma^{-1}x$ from the left action.

(6.3) Remark. The monomorphism of (6.1) and (6.2) restrict to monomorphisms

$$\operatorname{Hom}_k(E, E') \otimes \mathbb{Z}_\ell \to \operatorname{Hom}_{\operatorname{Gal}(k_s/k)}(T_\ell(E), T_\ell(E')),$$
$$\operatorname{Hom}_k(E, E') \otimes \mathbb{Q}_\ell \to \operatorname{Hom}_{\operatorname{Gal}(k_s/k)}(V_\ell(E), V_\ell(E')).$$

Hence homomorphisms and isogenies over k are distinguished by their action on Tate modules viewed as Galois modules. In the next chapter we will see that there are cases where these monomorphisms are isomorphisms.

(6.4) Remark. The cokernel of the monomorphism T_ℓ is torsion free. This follows since every k-homomorphism $f : E \to E'$ equal to zero on the ℓ-division points is of the form ℓg for some k-homomorphism $g : E \to E'$.

§7. Expansions Near the Origin and the Formal Group

The formal group of an elliptic curve is used in the in the next two chapters, and appendix III. We give now a brief introduction. In the equation for an elliptic curve E in normal form we introduce new variables $t = -x/y$ and $s = -1/y$. The equation in the affine (t, s)-plane becomes

$$s = t^3 + a_1ts + a_2t^2s + a_3s^2 + a_4ts^2 + a_6s^3,$$

where zero on E has coordonates $(t, s) = (0, 0)$, and t is a local parameter at 0 on E. This (t, s)-plane version of the cubic equation generates a formal series relations for s as a function of t by iterated substitution of the six-term expression for s into each term of s on the right-hand side of the above equation. The formal expansion has the following form:

$$\begin{aligned} s &= t^3 + a_1t^4 + (a_1^2 + a_2)t^5 + (a_1^3 + 2a_1a_2 + a_3)t^6 \\ &\quad + (a_1^4 + 3a_1^2a_2 + 3a_1a_3 + a_2^2 + a_4)t^7 + \cdots \\ &= t^3(1 + A_1t + A_2t^2 + \cdots), \end{aligned}$$

where A_n is a polynomial of weight n in the a_i with positive integer coefficients.

(7.1) Proposition. *In terms of the local uniformizing parameter $t = -x/y$ at 0 on E the following formal expansions hold in $\mathbb{Z}[a_1, a_2, a_3, a_4, a_6][[t]]$:*

$$\begin{aligned} x &= t^{-2} - a_1t^{-1} - a_2 - a_3t - (a_4 + a_1a_3)t^2 - \cdots, \\ y &= -\frac{x}{t} = -t^{-3} + a_1t^{-2} + a_2t^{-1} + a_3 + \cdots, \\ \omega &= \left[1 + a_1t + (a_1^2 + a_2)t^2 + (a_1^3 + 2a_1a_2 + a_3)t^3 + \cdots\right]dt. \end{aligned}$$

The coefficients of t^i are isobaric of weight $i + 2$, $i + 3$, and i, respectively, for x, y, ω. Further, if the coefficients a_j are in a ring R, then the expansions lie in $R[[t]]$.

Proof. The formal expansions for x and y come from the expansion $s = t^3(1+A_1t+\cdots)$, $y = -1/s$, and $x-ty$, and it is clear that x and y are in $\mathbb{Z}[a_1, a_2, a_3, a_4, a_6][[t]]$. The formal expansion for ω is derived from this in two ways:

$$\omega = \frac{dx}{2y + a_1x + a_3} = \frac{-2t^{-3} + \cdots}{-2t^{-3} + \cdots}dt,$$

which has coefficients in $\mathbb{Z}[1/2, a_1, a_2, a_3, a_4, a_6][[t]]$, and

$$\omega = \frac{dy}{3x^2 + 2a_2x + a_4 - a_1y} = \frac{3t^{-4} + \cdots}{3t^{-4} + \cdots}dt,$$

which has coefficients in $\mathbb{Z}[1/3, a_1, a_2, a_3, a_4, a_6][[t]]$. This proves the assertion about the coefficients of the above expression.

The expansion $x(t)$ and $y(t)$ in the previous proposition are an algebraic analogue of the complex analytic expansions of $\wp(u)$ and $\wp'(u)$. In Chapter 13 we will see that these formal expressions have p-adic convergence properties.

We can go further to analyze the group law near 0 formally also. The line joining two points (t_1, s_1) and (t_2, s_2) in the (t, s)-plane has slope given by

$$\lambda = \frac{s_2 - s_1}{t_2 - t_1} = \frac{t_2^3 - t_1^3}{t_2 - t_1} + A_1\frac{t_2^4 - t_1^4}{t_2 - t_1} + \cdots$$

$$= t_2^2 + t_1 t_2 + t_1^2 + A_1 \left(t_2^3 + t_2^2 t_1 + t_2 t_1^2 + t_1^3 \right) + \cdots ,$$

where the coefficients A_j are determined in the above expansion of s in terms of t. Now let $v = s_1 - \lambda t_1 = s_2 - \lambda t_2$, and substitute $s = \lambda t + v$ into the Weierstrass cubic to obtain a cubic equation in t with roots t_1 and t_2. The negative of the coefficient of t^2 in the cubic is the sum of the three roots t_1, t_2, and t_3 can be expressed as a power series Φ_E in t_1 and t_2 with coefficients in $\mathbb{Z}[a_1, a_2, a_3, a_4, a_6]$. A calculation, which is left to the reader, yields the following result.

(7.2) Proposition. *If $(t_1, s_1) + (t_2, s_2) = (t_3, s_3)$ on the elliptic curve E in the (t, s)-plane, then formally $t_3 = \Phi_E(t_1, s_2)$ has the form*

$$\begin{aligned} t_3 &= -t_1 - t_2 + \frac{a_1\lambda + a_3\lambda^2 - a_2 v - 2a_4\lambda v - 3a_6\lambda^2 v}{1 + a_2\lambda + a_4\lambda^2 + a_6\lambda^3} \\ &= t_1 + t_2 - a_1 t_1 t_2 - a_2\left(t_1^2 t_2 + t_1 t_2^2\right) - 2a_3\left(t_1^3 t_2 + t_1 t_2^3\right) + \\ &\quad + (a_1 a_2 - 3a_3) t_1^2 t_2^2 + \cdots , \end{aligned}$$

where $\Phi_E(t_1, s_2)$ is in $\mathbb{Z}[a_1, a_2, a_3, a_4, a_6][[t_1, t_2]]$.

Observe that if the coefficients a_j of the model for E lie in a ring R, then $t_3 = \Phi_E(t_1, t_2)$ is in $R[[t_1, t_2]]$.

(7.3) Definition. A formal group law $F(X, Y)$ is a formal series $F(X, Y)$ in $R[[X, Y]]$ satisfying:

(1) $F(X, 0) = X, F(0, Y) = Y$,
(2) (associativity) $F(X, F(Y, Z)) = F(F(X, Y), Z)$ in $R[X, Y, Z]$.

Further, $F(X, Y)$ is a commutative formal group law provided $F(X, Y) = F(Y, X)$.

There exists $\theta(X) \in XR[[x]]$ with $F(X, \theta(X)) = 0$.

The formal series $\Phi_E(t_1, t_2)$ arising from the group law on an elliptic curve is a formal group law. The formal additive group law is $F(X, Y) = X + Y$ and the formal multiplicative group law is $F(X, Y) = X + Y + XY$.

Associated with the formal group $F(X, Y)$ over R is a sequence of formal series $[m](X)$ over R defined inductively by

$$[1](X) = X \quad \text{and} \quad [m](X) = F(X, [m-1](X)) \qquad \text{for } m > 1.$$

In general $[m](X) = mX + \cdots$ (higher-order terms). This is the formal multiplication by m in the formal group $F(X, Y)$. For example, if $F = \Phi_E$ for an elliptic curve E, then we have

$$[2](X) = 2X - a_1 X^2 - 2a_2 X^3 + (a_1 a_2 - 7a_3) X^4 + \cdots$$

and

$$[3](X) = 3X - 3a_1 X^2 + (a_1^2 - 8a_2) X^3 + 3(4a_1 a_2 - 13a_3) X^4 + \cdots .$$

If $pR = 0$, i.e., R has characteristic a prime p, then it can be shown that

$$[p](X) = c_1 X^{p^h} + c_2 X^{2p^h} + c_3 X^{3p^h} + \cdots,$$

when $h \geq 1$ is an integer called the height of the formal group. When $[p](X) = 0$, we set $h = \infty$. If $R = k$ an algebraically closed field, then the height classifies the formal group up to isomorphism.

In Chapter 13 we will see that the height of the formal group of an elliptic curve in characteristic p is either 1, which is the usual case, or 2, which is called the supersingular case. This is related to the Hasse invariant which can be defined as the coefficient of X^p in $[p](X)$ above.

(7.4) Remark. A formal group $F(X, Y)$ over R, like an elliptic curve, has an invariant differential $\omega(Y) = (D_1 F(0, Y))^{-1} dY$. Observe that the derivative $D_1 F(X, Y) =$ $1+$ higher-order terms, and hence the inverse power series is defined. The invariance property for a differential $A(X)dX$ means that for a variable T

$$A(X)dX = A(F(X, T)) D_1 F(X, T) dX.$$

For $A(Y) = (D_1 F(0, Y))^{-1}$ we calculate from the associative law using the chain rule the following relations:

$$(D_1 F)(F(X, Y), T) \cdot D_1 F(X, Y) = D_1 F(X, F(Y, T)).$$

Setting $X = 0$

$$(D_1 F)(Y, T) \cdot D_1 F(0, T) = D_1 F(0, F(Y, T))$$

which gives

$$D_1 F(0, Y)^{-1} dY = D_1 F(0, F(Y, T))^{-1} (D_1 F(Y, T)) dY.$$

(7.5) Example. The formal group $\Phi_E(X, Y)$ of an elliptic curve E is defined over $\mathbb{Z}[a_1, a_2, a_3, a_4, a_6]$ and the coefficients specialize to the field of definition k of E. The invariant differential of the formal group Φ_E is just the differential ω of E with the expansion given in (6.1).

(7.6) Definition. A formal logarithm for a formal group $F(X, Y)$ over R is a power series $L(X)$ in $R[[X]]$ with $L(X) = X+$ higher-order terms and $F(X, Y) = L^{-1}(L(X) + L(Y))$.

In other words $L(F(X, Y)) = L(X) + L(Y)$ which says that $L : F \to \mathbb{G}_a$ is an isomorphism onto the formal additive group. Then $[m]_F(X) = L^{-1}(mL(X))$ and from

$$L'(Y) \cdot (D_1 F)(0, Y) = L'(0) = 1,$$

we see that the invariant differential is given by

$$\omega(Y) = L'(Y)dY = dL(Y).$$

Thus from an expansion for $\omega(T) = (\sum_{n=1}^{\infty} a_n T^n)(dT/T)$ we have an expansion for the formal logarithm

$$L(T) = \sum_{n=1}^{\infty} \frac{a_n}{n} T^n, \qquad \text{where } a_1 = 1.$$

Hence the formal logarithm exists if and only if given the coefficients a_n of the invariant differential ω in R we can form a_n/n in R. In a $\mathbb{Q}$-algebra R this is always possible, and in general in characteristic zero it is possible in an extension ring $R \otimes \mathbb{Q}$ of R.

In characteristic $p > 0$ we can always form a_n/n for $n < p$, and thus the coefficient a_p plays a basic role especially for our applications in Chapter 13 to elliptic curves. we will use:

(7.7) Proposition. *The coefficient a_p in the expansion of the invariant differential of a formal group $F(X, Y)$ in characteristic p is the coefficent of X^p in the expansion of $[p]_F(X)$.*

13

Elliptic Curves over Finite Fields

In this chapter we carry further the algebraic theory of elliptic curves over fields of characteristic $p > 0$. We already pointed out that the p-division points in characteristic p form a group isomorphic to $\mathbb{Z}/p\mathbb{Z}$ or zero while the ℓ-division points form a group isomorphic to $(\mathbb{Z}/\ell\mathbb{Z})^2$ for $p \neq \ell$. Moreover, the endomorphism algebra has rank 1 or 2 in characteristic 0 but possibly also rank 4 in characteristic $p > 0$.

A key issue for elliptic curves in characteristic p is whether or not a curve can be defined over a finite field. A basic result of Deuring is the E has complex multiplication if and only if E can be defined over a finite field. Thus the elliptic curves E with rk(End(E)) equal to 2 or 4 are the curves defined over a finite field, and among the curves E in characteristic p the case rk(End(E)) is equal to 4 occurs if and only if the group ${}_pE(\bar{k})$ is zero. Curves E with these equivalent properties are called supersingular, and we go further to give ten characteriszations for a curve to be supersingular. Supersingular curves are all defined over $\mathbb{F}_p$ or $\mathbb{F}_{p^2}$ in characteristic p, and they form a single isogeny class of finitely many curves up to isomorphism. We derive a formula for the number of these curves for given p.

The number of rational points $\#E(\mathbb{F}_q)$ on an elliptic curve E over a finite field $\mathbb{F}_q$ is estimated by the Riemann hypothesis

$$|\#E(\mathbb{F}_q) - q - 1| \leq 2\sqrt{q}.$$

This inequality is equivalent to the assertion that the zeros of the zeta function of an elliptic curve E over $\mathbb{F}_q$ are all on the line $\mathbf{Re}(s) = 1/2$. This is the first topic taken up in this chapter.

The reader should refer to Silverman [1986], Chapter V in this chapter.

§1. The Riemann Hypothesis for Elliptic Curves over a Finite Field

In part 1 the group $E(\mathbb{Q})$ was studied for elliptic curves over the rational numbers $\mathbb{Q}$. A similar problem is the determination of $E(\mathbb{F}_q)$ for E an elliptic curve defined over a finite field $\mathbb{F}_q$. We may view $E(\mathbb{F}_q)$ as a subgroup of $E(k)$ where k is an algebraically closed field containing $\mathbb{F}_q$. Since E is taken to be defined by a cubic equation with coefficients in $\mathbb{F}_q$, the Frobenius map

$$\pi(x, y) = (x^q, y^q)$$

restricts to an endomorphism $\pi = \pi_E : E \to E$.

(1.1) Definition. Let E be an elliptic curve defined over a finite field $\mathbb{F}_q$. The Frobenius endomorphism $\pi_E : E \to E$ is given by $\pi_A(x, y) = (z^q, y^q)$

Then π_E is in $\mathrm{End}(E)$, has degree q, and is purely inseparable. Moreover, (x, y) is in $E(\mathbb{F}_q)$ if and only if $\pi_E(x, y) = (x, y)$. Since the differential of $1_E - \pi_E$ is the identity id_E, the difference endomorphism $1 - \pi_E$ in $\mathrm{End}(E)$ is separable. Also, we have that

$$(x, y) \text{ is in } E(\mathbb{F}_q) \quad \text{if and only if} \quad (x, y) \text{ is in} \ker(1_E - \pi_E).$$

From the general theory of separable endomorphisms see 12(4) we know that

$$N_1 = \#E(\mathbb{F}_q) = \deg(1 - \pi_E) = \deg(\pi_E) - \mathrm{Tr}(\pi_E) + 1 = 1 + q - \mathrm{Tr}(\pi_E).$$

Further, since $m^2 - mn\,\mathrm{Tr}(\pi) + n^2 q = \deg(m - n\pi) \geq 0$ for all m, n we see that $\mathrm{Tr}(\pi)^2 - 4\deg(\pi) \leq 0$, or $|\mathrm{Tr}(\pi)| \leq 2\sqrt{q}$. Hence, this simple argument, using properties of degrees of endomorpisms applied to the Frobenius endomorphism, leads to the following result.

(1.2) Theorem (Riemann Hypothesis for Elliptic Curves). *Let E be an elliptic curve defined over a finite field $\mathbb{F}_q$, and let $N_m = \#E(\mathbb{F}_{q^m})$. Then for all $m \geq 1$ we have*

$$|1 + q^m - N_m| \leq 2 \cdot q^{m/2}.$$

This theorem was conjectured by Artin in his thesis and proved by Hasse [1934]. Since $\mathbb{F}_{q^m} \subset \mathbb{F}_{q^n}$ if and only if m/n, we have in this case $E(\mathbb{F}_{q^m}) \subset E(\mathbb{F}_{q^n})$.

(1.3) Example. Consider the elliptic curve E defined by the equation $y^2 + y = x^3$. Then

$$E(\mathbb{F}_2) = \{\infty = 0, (0, 0), (0, 1)\}$$

and

$$E(\mathbb{F}_4) = \{\infty = 0, (0, 0), (0.1), (1, \omega), (1, \omega^2), (\omega, \omega), (\omega, \omega^2), (\omega^2, \omega), (\omega^2, \omega^2)\}.$$

where $\mathbb{F}_4 = \{0, 1, \omega, \omega^2\}$ with $\omega^2 + \omega + 1 = 0$. The group $E(\mathbb{F}_2)$ is isomorphic to $\mathbb{Z}/3$ and $E(\mathbb{F}_4)$ is isomorphic to $(\mathbb{Z}/3)^2$. For $q = 2$ the difference $N_1 - 1 - q = 3 - 1 - 2 = 0$ and $N_1 = 1 + q$, while for $q^2 = 2^2$, we have $N_2 - 1 - q^2 = 9 - 1 - 4 = 4 = 2q^{2/2} = 2.2$. Hence the inequality in the Riemann hypothesis is the best possible for one power of q and $N_m = 1 + q^m$ for another power.

(1.4) Example. The elliptic curve E defined by $y^2 + y = x^3 + x$ has five points over $\mathbb{F}_2$, namely

$$\{\infty = 0, (0,0), (0,1), (1,0), (1,1)\}.$$

Hence $E(\mathbb{F}_2)$ is isomorphic to $\mathbb{Z}/5$ and $N_1 = 5$. Note that $N_1 - 1 - q = 5 - 1 - 2 = 2 \le 2\sqrt{2}$.

(1.5) Definition. Let E be an elliptic curve defined over a finite field $\mathbb{F}_q$. The characteristic polynomial of Frobenius is

$$f_E(T) = \det(1 - \pi_E T) = 1 - (\mathrm{Tr}(\pi))T + qT^2 \quad \text{in } \mathbb{Z}[T],$$

and the zeta function $\zeta_E(s)$ is the rational function in q^s

$$\zeta_E(s) = \frac{f_E(q^{-s})}{(1-q^{-s})(1-q^{1-s})} = \frac{1 - (\mathrm{Tr}(\pi))q^{-s} + q^{1-2s}}{(1-q^{-s})(1-q^{1-s})}.$$

(1.6) Remark. The zeta function $\zeta_E(s)$ has poles at $s = 0$ and $s = 1$. The inequality in (1.2), called the Riemann hypothesis, is equivalent to the assertion that the roots of $f_E(T)$ are complex conjugates of each other. These roots have absolute value equal to $1/\sqrt{q}$. In turn this condition is equivalent to the assertion that the zeta function $\zeta_E(s)$ has zeros only on the line $\mathbf{Re}(s) = 1/2$.

For $f_E(q^{-s}) = 1 - (\mathrm{Tr}\,\pi)q^{-s} + q^{1-2s}$, we replace s by $1 - s$, and we obtain the relation

$$\begin{aligned} f_E(q^{-(1-s)}) &= 1 - (\mathrm{Tr}\,\pi)q^{s-1} + q^{2s-1} = q^{2s-1}(1 - (\mathrm{Tr}\,\pi)q^{-s} + q^{1-2s}) \\ &= q^{2s-1} f_E(q^{-s}). \end{aligned}$$

This functional equation for f_E and the invariance of the denominator of the zeta function under s changed to $1 - s$ yield the following immediately.

(1.7) Proposition. *The zeta function $\zeta_E(s)$ of E over $\mathbb{F}_q$ satisfies the functional equation*

$$\zeta_E(1-s) = q^{2s-1}\zeta_E(s).$$

There is a far reaching generalization of this zeta function to zeta functions for any projective variety over a finite field due to A. Weil. In fact, it is the focus of a vast conjectural program announced by Weil [1949], carried out up to the Riemann hypothesis primarily by Grothendieck and Artin [1963–64], and completed with Deligne's [1973] proof of the Riemann hypothesis for smooth projective varieties. Note, for example, the generalization of (1.7) is a consequence of Poincaré duality in ℓ-adic cohomology theory. For a survey of the whole story, see Katz [1976].

(1.8) Remark. The definition of $f_E(T)$ extends to singular cubic curves C with singularity c_0.

$$f_C(T) = \begin{cases} 1 - T & \text{if } c_0 \text{ is a node with rational tangents,} \\ 1 + T & \text{if } c_0 \text{ is a node with tangents quadratic over } k, \\ 1 & \text{if } c_0 \text{ is a cusp.} \end{cases}$$

Then for C either singular or nonsingular cubic

$$\#C_{\mathrm{ns}}(\mathbb{F}_q) = qf_C(q^{-1}).$$

We end this section by counting the points on a pair of elliptic curves E and E^t over $\mathbb{F}_q$. Let $g(x) \in k[x]$ be a cubic polynomial, and let t be a nonzero scaler in k. Then denote by $g_t(x) = t^3 g(x/t) \in k[x]$. We thank O. Foster for pointing this out to us.

(1.9) Definition. let E be an elliptic curve defined by $y^2 = g(x)$ where $g(x)$ is a cubic polynomial over a field k. The twist by a nonzero $t \in k$ is the elliptic curve E^t defined by the equation $y^2 = g_t(x)$ or $t^{-1}(y/t)^2 = g(x/t)$. It is isomorphic to curve with equation $t^{-1}y^2 = g(x)$.

In characteristic > 3 by (3.8) the elliptic curves E and E^t are isomorphic if and only if t is a nonzero square, and further $E^{t'}$ and $E^{t''}$ are isomorphic if and only if t'/t'' is a square.

(1.10) Proposition. *Let E be an elliptic curve over a finite field $\mathbb{F}_q$ of characteristic $p > 3$. Then up to isomorphism there is exactly one twist E^t where t is any nonsquare in $\mathbb{F}_q$. Moreover*

$$\#E(\mathbb{F}_q) + \#E^t(\mathbb{F}_q) = 2q + 2.$$

Proof. The first assertion follows from 3(8.3) and the fact that for a finite field k the quotient $k^*/(k^*)^2$ has two elements. Next we count the points (x, y) on the two curves. Firstly, note that $g(x) = 0$ if and only if $g_t(x) = 0$ giving one point on each curve for such a value of x. Secondly, note that the element $g(x)$ is a square (resp. a nonsquare) in k if and only if $g_t(x)$ is a nonsquare (resp. square) in $k = \mathbb{F}_q$, that is giving two points in either E or E^t for each such a value of x. When we add the two points at infinity for zero of the curves, we have

$$\#E(\mathbb{F}_q) + \#E^t(\mathbb{F}_q) = 2(\#\text{ of } x \in \mathbb{P}(\mathbb{F}_q)) = 2(q + 1).$$

This proves the proposition.

§2. Generalities on Zeta Functions of Curves over a Finite Field

Let C be an algebraic curve over a field $k_1 = \mathbb{F}_q$. We wish to study the number of points on C over $k_n = \mathbb{F}_{q^n}$ for every n. For every point P on $C(\bar{k}_1)$ we have P in

$C(k_n)$ for some n. For the minimal such n, we write $k_1(P) = k_n$ and $n = \deg(P)$, the local degree of P. If σ generates the cyclic Galois group $\mathrm{Gal}(k_n/k_1)$ of order n, then $\mathfrak{p} = P + P^\sigma + \cdots + P^{\sigma^{n-1}}$ is an example of a prime divisor over k_1 on C which is rational over k_1. In fact, all prime divisors on C over k_1 are of this form and the norm of the divisor is defined $N\mathfrak{p} = q^n$. Using multiplicative notation, in constrast to Chapter 12, §3, we denote a positive rational divisor over k_1 by

$$\mathfrak{a} = \mathfrak{p}_1^{n(1)} \cdots \mathfrak{p}_r^{n(r)},$$

where $n(1), \ldots, n(r)$ are natural numbers and $\mathfrak{p}_1, \ldots, \mathfrak{p}_r$ are prime divisors rational over k_1. Then norm of this $\mathfrak{a}$ is given by

$$N\mathfrak{a} = (N\mathfrak{p}_1)^{n(1)} \cdots (N\mathfrak{p}_r)^{n(r)}.$$

Through the above examples, we circumvent the general theory of rational divisors.

(2.1) Definition. Let C be an algebraic curve defined over $k_1 = \mathbb{F}_q$. The zeta function $\zeta_C(s)$ of C/k_1 is defined as a sum or as a product involving divisors rational over k_1

$$\zeta_C(s) = \sum_{\text{positive } \mathfrak{a}} (N\mathfrak{a})^{-s} = \prod_{\text{prime } \mathfrak{p}} (1 - (N\mathfrak{p})^{-s})^{-1}.$$

At this point these two formal expressions for the zeta function are equivalent from the unique decomposition of divisors as products of prime divisors. Now we begin a discussion which will show that this zeta function is the same zeta function as in Definition (1.5) in the case of an elliptic curve. We will also outline its basic properties for a general complete nonsingular curve.

Let A_m denote the number of positive divisors $\mathfrak{a}$ rational over k_1 with norm q^m on C and let P_m denote the number of prime divisors $\mathfrak{y}$ rational over k_1 with norm q^m on C. The equality of the two expressions in the definition of the zeta function leads to two expressions for the related function

$$Z_C(u) = \sum_{m=0}^{\infty} A_m u^m = \prod_{m=0}^{\infty} (1 - u^m)^{-P_m}$$

and $Z_C(q^{-2}) = \zeta_C(s)$. If N_m is the number of points on $C(k_m)$ where $k_m = \mathbb{F}_{q^m}$ as above, then

$$N_m = \sum_{d|m} d P_d$$

from the above description of prime divisors rational over k_m. Now calculate

$$\frac{d}{du} \log Z_C(u) = \frac{1}{u} \sum_{d=1}^{\infty} \frac{d P_d u^d}{1 - u^d} = \frac{1}{u} \sum_{d,d'=1}^{\infty} d P_d u^{dd'} = \frac{1}{u} \sum_{m=1}^{\infty} N_m u^m.$$

Hence we have with the above notations the following result.

(2.2) Proposition. *For $N_m = \#C(k_m)$, where C is an algebraic curve over k_1, the zeta function satisfies the following relation*

$$\zeta_C(s) = \exp\left(\sum_{m=1}^{\infty} \frac{N_m}{m} q^{-ms}\right).$$

(2.3) Remark. Since $N_m \le 1+q^m+q^{2m} = \#\mathbb{P}_2(k_m)$ for all m, the previous formula for the zeta function shows that it is a convergent series or convergent product for $\mathbf{Re}(s) > 2$. In fact, we can do much better and show that it is a rational function with poles at $s = 0$ and $s = 1$, and its zeros are on the line $\mathbf{Re}(s) = 1/2$.

Returning to the case of an elliptic curve E, we know that

$$N_m = 1 + q^m - \mathrm{Tr}(\pi^m) = 1 + q^m - \alpha^m - \alpha^{-m},$$

where α and $\bar{\alpha}$ are the two imaginary conjugate roots of the characteristic polynomial $\det(1 - \pi_E T)$ as an element of $\mathrm{End}(E)$. There is also another interpretation where α and $\bar{\alpha}$ are the eigenvalues of the inverse of Frobenius endomorphism acting on any Tate module $V_\ell(E)$ where ℓ is any prime number different from the characteristic p. This follows from the fact that the inverse of Frobenius has the same characteristic polynomial on $V_\ell(E)$. Using the above expression for N_m, we calculate the log of the zeta function

$$\begin{aligned}
\log \zeta_C(s) &= \sum_{m=1}^{\infty} (1 + q^m - \alpha^m - \bar{\alpha}^m) q^{-ms} \\
&= \sum_{m=1}^{\infty} q^{-ms} + \sum_{m=1}^{\infty} q^{-m(s-1)} - \sum_{m=1}^{\infty} (\alpha q^{-s})^m - \sum_{m=1}^{\infty} (\bar{\alpha} q^{-s})^m \\
&= -\log(1 - q^{-s}) - \log(1 - q^{1-s}) + \log[(1 - \alpha q^{-s})(1 - \bar{\alpha} q^{-s})].
\end{aligned}$$

Hence the exponential is the zeta function as a rational function of q^{-s}

$$\zeta_C(s) = \frac{1 - (\alpha + \bar{\alpha}) q^{-s} + q^{1-2s}}{(1 - q^{-s})(1 - q^{1-s})} = \frac{f_E(q^{-s})}{(1 - q^{-s})(1 - q \cdot q^{-s})},$$

where $f_E(T) = \det(1 - \pi_E T) = 1 - (\alpha + \bar{\alpha})T + qT^2 = (1 - \alpha T)(1 - \bar{\alpha} T)$.

(2.4) Remark. From the above calculation we see that the two definitions (1.5) and (2.1) for the zeta function of an elliptic curve yield the same function.

Further, the above discussion gives the framework for describing the result in the general case of any curve.

(2.5) Theorem. *Let C be a smooth projective curve of genus g over $\mathbb{F}_{q^m} = k_m$ for $m = 1$, and let $N_m = \#C(k_m)$. Then there are algebraic integers $\alpha_1, \alpha_2, \ldots, \alpha_{2g}$ with $|\alpha_j| = \sqrt{q}$ and*

$$N_m = q^m + 1 - \sum_{i=1}^{2g} \alpha_i^m .$$

Then zeta function is given by

$$\zeta_C(s) = \frac{P(q^{-s})}{(1 - q^{-s})(1 - q^{1-s})},$$

where $P(T) = (1 - \alpha_1 T) \ldots (1 - \alpha_{2g} T)$. *Further* ζ_C *satisfies a functional equation*

$$q^{g(2s-1)} \cdot \zeta_C(s) = \zeta_C(1 - s)$$

which is equivalent to the assertion that the α_j *pair off such that after reordering* $\bar{\alpha}_{2j-1} = \alpha_{2j}$ *and so* $\alpha_{2j-1}\alpha_{2j} = q$.

The above theorem is due to Weil [1948]. For other discussions see Monsky [1970] and Bombieri [1973]. The article by Katz [1976] puts this result in the perspective of general smooth projective varieties and outlines how the above form of the zeta function for curves has a natural general extension first conjectured A. Weil and finally completely proven by P. Deligne.

Observe that the zeta function for a smooth projective curve is a rational function of q^{-s}. Its poles are at $s = 0$ and $s = 1$, and its zeros are on the line $\mathbf{Re}(s) = 1/2$. This last assertion is equivalent to $|\alpha_j| = \sqrt{q}$ and is called the Riemann hypothesis for algebraic curves over a finite field. It is the most difficult statement in the above theorem to demonstrate.

For elliptic curves there were three ways to obtain the quadratic polynomial f_E which is the nontrivial factor in the zeta function. First, it is a quadratic equation for π_E^{-1} with constant term 1 satisfied in $\mathrm{End}^0(E)$ over $\mathbb{Q}$. Second, it is the characteristic polynomial of the inverse Frobenius of E acting on any Tate module $V_\ell(E)$, where ℓ is unequal to the characteristic p. Third, we can calculate N_1 directly to obtain $\mathrm{Tr}(\pi_E)$. In the case of general smooth curves C the nontrivial factor $P(T)$ of (2.5) in the zeta function can be obtained from the theory of correspondences on C which are certain divisors on the surface $C \times C$. In this way Grothendieck reproved the basic results for $\zeta_C(s)$ obtained earlier by A. Weil. Originally Weil looked at the Jacobian $J(C)$, an abelian variety of dimension g, and the corresponding Tate module $V_\ell(J(C))$. The charcteristic polynomial of π_C^{-1} on $V_\ell(J(C))$ with constant term 1 is the factor $P(T)$ of (2.5). In a more general context this Tate module is the first ℓ-adic cohomology group and for higher-dimensional smooth varieties all ℓ-adic cohomology groups together with the action of Frobenius must be bought into the analysis of the zeta function of the variety.

§3. Definition of Supersingular Elliptic Curves

In line with our general program of developing as much of the theory of elliptic curves as possible via the theory of cubic equations, we use the following definition of supersingular for elliptic curves.

(3.1) Definition. An elliptic curve E defined by a cubic equation $f(w, x, y)$ over a field k of characteristic p is supersingular provided the coefficient of $(wxy)^{p-1}$ in $f(w, x, y)^{p-1}$ is zero.

A supersingular elliptic curve is also said to have Hasse invariant 0 or height 2, otherwise the curve has nonzero Hasse invariant or height 1. The origin of these other terminologies will be clearer later. The concept was first studied by Hasse [1934] and it was referred to as the invariant A.

(3.2) Example. For $p = 2$ the equation of any elliptic curve E can be put in the following form relative to a point of order 3:

$$E_\alpha : wy^2 + w^2y + \alpha wxy = x^3.$$

Since $p - 1 = 2 - 1 = 1$, it follows that E_α is supersingular for exactly one curve when $\alpha = 0$, that is, the curve

$$y^2 + y = x^3.$$

This is the curve considered in (1.3) where the number of points is $p + 1 = 2 + 1$. This property is characteristic of supersingular curves defined over the prime field as we will see in §4.

(3.3) Example. For $p = 3$ the equation of E can be put in the form $0 = f(w, x, y) = x^3 + awx^2 + bw^2x + cw^3 - wy^2$. Since $p - 1 = 3 - 1 = 2$, we calculate $f(w, x, y)^2$ and note that the coefficient of $(wxy)^2$ is equal to $-2a$. Hence E is supersingular if and only if $a = 0$, i.e., if it has the form $y^2 = x^3 + bx + c$. Changing x to $\alpha x + \beta$, we change x^3 to $\alpha^3 x^3 + \beta^3$ and with a suitable choice of α and β the equation becomes $y^2 = x^3 - x$. This is the only supersingular elliptic curve in characteristic 3. Moreover, $E(\mathbb{F}_3) = \{\infty = 0(0, 0), (1, 0), (-1, 0)\}$ and this is isomorphic to $(\mathbb{Z}/2)^2$. The number of points is $p + 1 = 3 + 1 = 4$ as in (3.2).

(3.4) Lemma. *The coefficient of x^k in $(x - 1)^k(x - \lambda)^k$ is*

$$(-1)^k \sum_{j=0}^{k} \binom{k}{j}^2 \lambda^j.$$

Proof. Expand both $(x-1)^k = \sum_a \binom{k}{a}(-1)^{k-a}x^a$ and $(x-\lambda)^k = \sum_a \binom{k}{b}(-\lambda)^{k-b}x^b$. Hence the product is given by

$$(x - 1)^k(x - \lambda)^k = \sum_i \sum_{a+b=i} \lambda^{k-b} \binom{k}{a}\binom{k}{b} (-1)^{2k-i} x^i.$$

Hence the coeffiecient of x^k is $(-1)^k \sum_{j=0}^{k} \binom{k}{j}^2 \lambda^j$.

(3.5) Proposition. *In characteristic $p > 2$ the curve in Legendre form $E_\lambda : y^2 = x(x-1)(x-\lambda)$ is supersingular if and only if λ is a root of the Hasse invariant*

$$H_p(\lambda) = (-1)^m \sum_{i=0}^{m} \binom{m}{i}^2 \lambda^i, \quad \textit{where } m = \frac{p-1}{2}.$$

Proof. We must calculate the coefficient of $(wxy)^{p-1}$ in f^{p-1} where

$$f(w,x,y) = wy^2 - x(x-w)(x-\lambda w).$$

In $f(w,x,y)^{p-1}$ the term $(wxy)^{p-1}$ will appear only in the middle of the binomial expansion, i.e., the term $\binom{p-1}{m}(wy^2)^m[x(x-w)(x-\lambda w)]^m$. Observe that $\binom{p-1}{m} = (-1)^m \pmod p$ from the relation $(x+y)^{p-1}(x+y) = x^p + y^p \pmod p$. Since $2m = p-1$, the coefficient of $(wxy)^{p-1}$ is just the coefficient of $(xw)^m$ in $(x-w)^m(x-\lambda w)^m$ up to $(-1)^m$ and by Lemma (3.4) this is $H_p(\lambda)$ which proves the proposition.

In Chapter 9 we saw how the hypergeometric function $F(1/2, 1/2, 1; \lambda)$ played a basic role in describing the period lattice of an elliptic curve over **C**. This same function is related to the Hasse invariant $H_p(\lambda)$. To see this, observe that the coefficients

$$\binom{-\frac{1}{2}}{k} = \frac{\left(-\frac{1}{2}\right)\left(-\frac{3}{2}\right)\dots(1-2k/2)}{k!} = \left(\frac{-1}{2}\right)^k \frac{1\cdot 3\cdot 5\dots(2k-1)}{k!}$$
$$= (-1)^k \left(\frac{1}{2}\right)^k \binom{2k}{k} \quad \text{are in } \mathbb{Z}\left[\frac{1}{2}\right].$$

Hence reducing modulo p, we see that the formal series $F(1/2, 1/2, 1; \lambda)$ in $\mathbb{F}_p[[\lambda]]$ is defined, and moreover we denote by

$$G_p(\lambda) = \sum_{i=0}^{m} \binom{m}{i}^2 \lambda^i \equiv F\left(\frac{1}{2}, \frac{1}{2}, 1; \lambda\right) \mod (p, \lambda^p)$$

The polynomial $G_p(\lambda)$ is called the Deuring polynomial. Since for $m = (p-1)/2$, we have the congruence

$$\binom{m}{k} = \left(\frac{1}{2}\right)^k \frac{(p-1)(p-3)\dots(p-2k+1)}{k!} \equiv \left(\frac{-1}{2}\right)^k \frac{1\cdot 3\cdot 5\cdots(2k-1)}{k!}$$
$$= \binom{-\frac{1}{2}}{k} \pmod p \quad \text{for } k < p.$$

(3.6) Remark. In the ring of formal series $\mathbb{F}_p[[\lambda]]$ the following relation holds:

$$F\left(\frac{1}{2}, \frac{1}{2}, 1; \lambda\right) = G_p(\lambda)\cdot G_p(\lambda^p)\cdot G_p(\lambda^{p^2})\dots.$$

The function $G_p(\lambda)$ is a solution to the hypergeometric differential equation

$$\lambda(1-\lambda)\frac{d^2w}{d\lambda^2}+(1-2\lambda)\frac{dw}{d\lambda}+\frac{1}{4}w=0.$$

This can be seen by direct calculation or by using the assertion that all solutions of the differential equation in $\mathbb{F}_p[[\lambda]]$ are of the form $F(1/2,1/2,1;\lambda)u(\lambda^p)$, where u is a formal series. Since the polynomial $G_p(\lambda)$ is the solution of a second-order differential equation it follows from formal considerations that its roots are simple. This follows from the corresponding recurrence formula for the coefficients of a Taylor expansion around a root. Hence we have the following result.

(3.7) Proposition. *The Deuring polynomial $G_p(\lambda)=(-1)^m\sum\binom{m}{i}^2\lambda^i$ has m simple roots in the algebraic closure of $\mathbb{F}_p$.*

The difference between the regular and nonregular solutions of the hypergeometric differential equation is only a question of a sign since $G_p(\lambda)=(-1)^m G_p(1-\lambda)$. There is only one period up to a constant modulo p and it is a scalar multipole of $G_p(\lambda)$. In the supersingular case there are no nonzero periods. We will come back to this when we consider the various formulations of the notion of supersingular curve.

In later sections, using the theory of isogenies and formal groups, we will have as many as ten criterions for a curve to be supersingular. One of these can be derived now using the relation to the Deuring polynomial.

(3.8) Proposition. *An elliptic curve E is supersingular if and only if the invariant differential ω is exact.*

Proof. For $p=2$ we see that $\omega=dx/(\alpha x+1)$ when E has the form $y^2+y+\alpha xy=x^3$. Then ω is exact if and only if $\alpha=0$, i.e., E is supersingular.

For $p>2$ we have $\omega=dx/2y$, and we put E into the form $y^2=x(x-1)(x-\lambda)$. Then ω is exact if and only if $y^{p-1}(dx/y^p)$ or equivalently

$$y^{p-1}dx=[x(x-1)(x-\lambda)]^m dx$$

is exact for $m=(p-1)/2$. This form is exact if and only if the coefficient of x^{p-1} in $[x(x-1)(x-\lambda)]^m$ is zero, but this is also the coefficient of x^m in $[(x-1)(x-\lambda)]^m$ is zero. This coefficient is $G_p(\lambda)$ by (3.4). Now the proposition follows from (3.5).

These considerations can be used to count mod p the number of points on the elliptic curve E_λ for $\lambda\in\mathbb{F}_p$. For $m=(p-1)/2$ recall that $[x(x-1)(x-\lambda)]^m=+1$ or -1, an d$x(x-1)(x-\lambda)$ is a square, that is, of the form y^2, if and only if $[x(x-1)(x-\lambda)]^m=+1$. This leads to the formula

$$\#E_\lambda(\mathbb{F}_p)=\sum_{x\in\mathbb{F}_p}\{1+[x(x-1)(x-\lambda)]^m\}\mod p.$$

Separating out the contribution form $x=0$, and using the elementary character sum in $\mathbb{F}_p$.

$$\sum_{x \in \mathbb{F}_p} x^a = \begin{cases} 0 & \text{if } p-1 \text{ does not divide } a \\ -1 & \text{if } p-1 \text{ does divide } a, \end{cases}$$

we obtain another version of the above formula

$$\#E_\lambda(\mathbb{F}_p) = 1 - \{\text{coefficient of } x^{p-1} \text{ in } [x(x-1)(x-\lambda)]^m\}.$$

By (3.4) we have the following assertion.

(3.9) Proposition. *For $\lambda \in \mathbb{F}_p$ and $p > 3$ the number of rational points on the curve E_λ is given by*

$$\#E_\lambda(\mathbb{F}_p) \equiv 1 - G_p(\lambda) \mod p.$$

In particular, E_λ is supersingular if and only if $\#E_\lambda(\mathbb{F}_p) \equiv 1 \mod p$ in which case, we have $\#E_\lambda(\mathbb{F}_p) = p + 1$ by the Riemann hypothesis.

§4. Number of Supersingular Elliptic Curves

We already know from (3.7) that there are $(p-1)/2$ values such that the curve E_λ : $y^2 = x(x-1)(x-\lambda)$ is supersingular in characteristic $p \neq 2$ and in characteristic 2 there is just one supersingular curve $y^2 + y = x^3$ up to isomorphism. In order to describe the number of supersingular curves up to isomorphism, we count the j values of $j(\lambda)$ as λ ranges over the points on the λ-line such that E is supersingular. Recall the map from the λ-line to the j-line is given by

$$j(\lambda) = 2^8 \frac{(\lambda_2 - \lambda + 1)^3}{\lambda^2(\lambda - 1)^2}.$$

This is a map of degree 6 with ramification precisely over the points $\{\infty, 0, 12^3\}$.

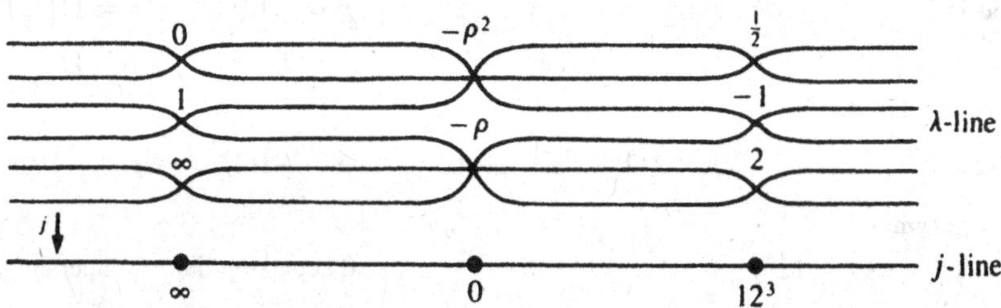

If a supersingular curve E_{λ_0} has a j-value $j_0 = j(\lambda_0)$, then all the points λ in $j^{-1}(j_0)$ correspond to supersingular curves E each isomorphic to E_{λ_0}. With two possible exceptions there are six such curves isomorphic to E_{λ_0}. Starting with the relation

$$\frac{p-1}{2} = \sum_{E_\lambda \text{ supersingular}} 1,$$

and dividing by 6, we obtain the relation

$$\frac{p-1}{12} = \left(\sum_{\substack{\text{supersingular } E \\ \text{up to isomorphism} \\ \text{with } j(E) \neq 0.1728}} 1 \right) + \frac{\alpha}{2} + \frac{\beta}{3},$$

where $\alpha = 0$ or depending on E with $j(E) = 12^3$, namely $\alpha = 1$ if and only if E is supersingular, and where $\beta = 0$ or 1 depending on E with $j(E) = 0$, namely $\beta = 1$ if and only if E is supersingular. Dividing further by 2 and interpretating the numbers in the denominators as orders of $\mathrm{Aut}(E)$, we obtain the following theorem.

(4.1) Theorem. *For a prime p the following sum taken over supersingular curves defined over* $\bar{\mathbb{F}}_p$ *up to isomorphism*

$$\frac{p-1}{24} = \sum_{\substack{E \text{ supersingular mod } p \\ \text{up to isomorphism}}} \frac{1}{\#\mathrm{Aut}(E)}$$

holds.

The above discussion proves the theorem for $p > 3$. The theorem predicts one supersingular curve for $p = 2$ with an automorphism group having 24 elements and one supersingular curve for $p = 3$ with an automorphism group having 12 elements. This is the case by (2.2) and 3(6.2) for $p = 2$ and by (2.3) and 3(5.2) for $p = 3$.

To make the above theorem more explicit, we decompose $(p-1)/24 = m/2 + \alpha/4 + \beta/6$, where α and β are 0 or 1 and $m = [p/12]$. Let $n(p)$ denote the number of supersingular curves in characteristic p.

Table 1. For supersingular curves in characteristic $p > 3$.

$p \pmod{12}$	$p = 1(12)$	$p = 5(12)$	$p = 7(12)$	$p = 11(12)$
$n(p) =$ number of supersingular curves	$\frac{p-1}{12}$	$\frac{p-5}{12} + 1$	$\frac{p-7}{12} + 1$	$\frac{p-11}{12} + 2$
$p \pmod 4$	$p = 1(4)$	$p = 1(4)$	$p = -1(4)$	$p = -1(4)$
Hasse invariant of $y^2 = x^3 - x$ $(j = 12^3)$	1	1	0 supersingular	0 supersingular
$p \pmod 3$	$p = 1(3)$	$p = -1(3)$	$p = 1(3)$	$p = -1(3)$
Hasse invariant of $y^2 = x^3 - 1$ $(j = 0)$	1	0 supersingular	1	0 supersingular

§5. Points of Order p and Supersingular Curves

Recall that the degree of an endomorphism $\lambda : E \to E'$ is $\deg(\lambda) = [k(E) : k(E')]$ coming from the induced embedding $k(E') \to k(E)$ factors into separable and purely inseparable parts

$$\deg(\lambda) = \deg(\lambda)_s \deg(\lambda)_i,$$

where $\deg(\lambda)_s = [k(E) : k(E')]_s$ and $\deg(\lambda)_i = [k(E) : k(E')]_i$. The purely inseperable degree is always a power of p, $\deg(\lambda)_i = p^h$, where h is called the height of λ. The kernel $\ker(\lambda)$ of λ is a finite subgroup of $E(k_s)$.

(5.1) Remark. The number of elements in the kernel is the separable degree. Let $k = k_s$. Given a finite subgroup G in $E(k_s)$ there exists a unique separable isogeny $\lambda_G : E \to E/G$ where $k(E/G) = k(E)^G$, the subfield invariant by the action of G on $D(E)$ induced by translations of G on E. The isogeny and the quotient elliptic curve are defined over an intermediate field k_1 between k and its separable algebraic closure k_s. Then $\ker(\lambda_G) = G$ and $\deg(\lambda_G) = \deg(\lambda_g)_s = \#G = [k(E) : k(E)^G]$ by Galois theory. This is an indication of how to construct all separable isogenies. We do not supply the details of the construction of E/G except to give $k(E/G)$ and the interested reader can consult Mumford, *Abelian Varieties*, Section 7 for further details.

(5.2) Definition. The iterated absolute Frobenius $\mathrm{Fr}_E^h : E \to E^{(p^h)}$ is defined by $\mathrm{Fr}_E^h(x, y) = (x^{p^h}, y^{p^h})$ where $E^{(p^h)}$ has Weierstrass equation with coefficients the p^hth power of the coefficients of E, i.e.,

$$y^2 + a_1^{p^h} xy + a_3^{p^h} y = x^3 + a_2^{p^h} x^2 + a_4^{p^h} x + a_6^{p^h}.$$

For k perfect Fr_E^h is purely inseparable of height h since $k(E^{(p^h)}) = k(E)^{p^h}$ in $k(E)$. Observe that E is defined over the finite field $\mathbb{F}_{p^h}$ if and only if E and $E^{(p^h)}$ are isomorphic and in this case $\mathrm{Fr}_E^h = \pi_E$.

(5.3) Lemma. *Let $K = k(x, y)$ be a separable algebraic extension of a purely transcendental extension $k(x)$ of the perfect field k. Then $[K : K^p] = p$.*

Proof. Since $K = k(x, y) = k(x, y^p)$, the element x generates K over $K^p = k^p(x^p, y^p) = k(x^p, y^p)$. Hence the degree is either p or 1. In the latter case x is in K^p and $x = t^p$ where t is both separable and purely inseparable over $k(x)$, which is impossible. Hence $[K : K^p] = p$.

(5.4) Proposition. *Let k be a perfect ground field, and let $\lambda : E \to E'$ be a purely inseparable isogeny of height h. Then there exists an isomorphism $u : E' \to E^{(p^h)}$ with $u\lambda = \mathrm{Fr}_E^h$.*

Proof. We have the inclusion of fields $k(E) \supset k(E') \supset k(E)^{p^h} = k(E^{(p^h)})$. By Lemma (5.3) we have $[k(E) : k(E')] = [k(E) : k(E)^{p^h}]$ and thus $k(E') = k(E)^{p^h}$. This defines the isomorphism $E' \to E^{(p^h)}$ with the desired factorization property.

Now we wish to apply this discussion to multiplication by p, denoted $p_E : E \to E$, on a curve E over a field k of positive characteristic p. The degree of p_E is p^2, and it factors as

$$p^2 = \deg(p_E) = \deg(p_E)_s \cdot \deg(p_E)_i.$$

Since p_E is not separable, it follows that the height h of p_E is 1 or 2, and $\#\ker(p_E)$ is p or 1, respectively, i.e., $\#_{p^i} E(\bar{k}) = (p^i)^{2-h}$.

(5.5) Proposition. *For an elliptic curve E over a perfect field k of positive characteristic with formal group Φ_E, the height h of p_E equals the height of the formal group Φ_E.*

Proof. Multiplication by p on Φ_E, denoted by $[p]_E(t) = c_1 t^p + c_2 t^{p^2} + \cdots$ is induced by p_E on E. The inseparable degree can be calculated in terms of the embedding $k[[t]] \to k[[t]]$ given by sustitution $f(t) \to f([p]_E(t))$. This is of degree p for $c_1 \neq 0$ and of degree p^2 otherwise.

(5.6) Theorem. *Let E be an elliptic curve over a perfect field k of characteristic $p > 0$ with formal group Φ_E. The following are equivalent:*

(1) ${}_pE(k_s) = 0$, *i.e., the curve has no points of order p.*
(2) p_E *is a purely inseparable isogeny.*
(3) *The formal group Φ_E has height 2.*
(4) *The invariant differential ω is exact, i.e., E is supersingular.*

Further, for a supersingular E we have an isomorphism $E \to E^{(p^2)}$ and E is defined over $\mathbb{F}_{p^2}$.

Proof. The equivalence of (1) and (2) follows from the formula for the order of the kernel of an isogeny in terms of its separable degree. The equivalence of (2) and (3) is (5.5) and of (3) and (4) is 6(6.7). The last assertion follows from (5.4) and the remarks preceding (5.3). This proves the theorem.

Since every supersingular elliptic curve is defined over $\mathbb{F}_{p^2}$. This gives proof that there are only finitely many up to isomorphism over $\bar{\mathbb{F}}_p$.

§6. The Endomorphism Algebra and Supersingular Curves

It might happen that the Frobenius endomorphism π_E of an elliptic curve E over $\mathbb{F}_q$ might have some power in the subring $\mathbb{Z}$ of $\mathrm{End}(E)$. Now we study the implications of this possibility. Recall π_E is in $\mathrm{End}_k(E)$ (resp. $\mathrm{End}_k^0(E)$) Which is a subring (resp. division subring) of $\mathrm{End}_{\bar{k}}(E)$ (resp. $\mathrm{End}_{\bar{k}}^0(E)$).

(6.1) Proposition. *If no power π_E^n of π_E is in $\mathbb{Z}$ contained in End(E), then $\mathrm{End}_k^0(E) = \mathrm{End}_{\bar{k}}^0(E) = \mathbb{Q}(\pi_E)$ is a purely imaginary quadratic field.*

Proof. An element φ in $\mathrm{End}_{\bar{k}}(E)$ is defined over some algebraic extension $\mathbb{F}_{q^n}$ of the ground field $k = \mathbb{F}_q$ and thus $\pi_E^n\varphi = \varphi\pi_E^n$. Thus φ is in $\mathbb{Q}(\pi_E^n)$ from commutation properties of algebras embedded in $M_2(Q)$, see Lang's *Algebra*. In fact, $\mathrm{End}_{\bar{k}}^0(E)$ is necessarily two-dimensional over $\mathbb{Q}$ and equals $\mathbb{Q}(\pi_E)$. The rest follows from (1.2).

(6.2) Proposition. *If some power of π_E is in $\mathbb{Z}$, then $\pi_E^n = p_E^m$ for some m, E is supersingular, and $\mathrm{End}_{\bar{k}}^0(E)$ is a quaternion algebra.*

Proof. Since π_E is purely inseparable, $\deg(\pi_E) = p^i$ and $\deg(\pi_E^n) = p^{ni}$. From the assumption that $\pi_E^n = c_E \in \mathbb{Z}$ and $\deg(\pi_E^n) = c_E^2$, we deduce that $\pi_E^n = p_E^m$, where $m = ni/2$.

Next, p_E is purely inseparable which by (5.5) means that E is supersingular. So it is among a finite set of isomorphism classes of supersingular curves. For a finite subgroup G of order prime to p, E/G is also supersingular as one can see by looking at the points of order p. Let S be any infinite set of primes not including p. Then there must exist a pair of distinct ℓ, ℓ' is S with E/G and E/G' isomorphic, where G and G' are cyclic subgroups of orders ℓ and ℓ', respectively. Consider the following commutative diagram which defined the isogeny φ of degree $\ell\ell'$, which is not a square:

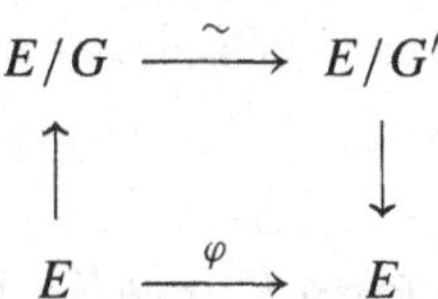

Since $\deg(n_E) = n^2$, we see that φ is not in $\mathbb{Q}$ which is always contained in $\mathrm{End}^0(E) = \mathrm{End}_{\bar{k}}^0(E)$. If $\mathrm{End}^0(E)$ were commutative, it would be an imaginary quadratic extension. In that case there would be an infinite set of primes S such that they and products of distinct pairs are not norms of elements from $\mathrm{End}^0(E)$, i.e., degrees of elements from $\mathrm{End}^0(E)$. The above construction shows that no such infinite set exists, and therefore we deduce that $\mathrm{End}^0(E)$ must be a quaternion algebra and is, in particular, noncommutative.

(6.3) Theorem. *For an elliptic curve E over a field k the following assertions are equivalent:*

(1) ${}_pE(k_2) = 0$, *i.e., the curve is supersingular.*
(2) $\mathrm{End}_{\bar{k}}(E)$ *is noncommutative.*
(3) *E is defined over a finite field and there exist strictly positive m and n with $\pi_E^m = p_E^n$.*

Proof. That assertion (1) implies (2) was given in the last paragraph of the previous proposition. Observe that in the argument only the condition that E is supersingular was used. Conversely, if ${}_pE(k_s)$ is nonzero, then $T_p(E)$ is a free $\mathbb{Z}_p$-module of rank 1, and the representation $\mathrm{End}(E) \to \mathrm{End}(\mathbb{Z}_p)$ is faithful into a commutative ring. This proves that (2) implies (1). When (1) holds E is defined over a finite field and now the same holds for (2) and of course (3).

By (6.2) we see that (3) implies (2) and (2) implies (3) follows from (6.1). This proves the theorem.

Finally, we are also in a position to derive Deuring's criterion for a curve to be defined over a finite field in terms of complex multiplications of the curve.

(6.4) Theorem (Deuring). *In characteristic $p > 0$ an elliptic curve E is defined over a finite field if and only if $\dim_{\mathbb{Q}} \mathrm{End}^0(E) > 1$, i.e., E has (nontrivial) complex multiplications.*

Proof. If E is defined over a finite field, then (6.1) and (6.2) tell us exactly what $\mathrm{End}^0(E)$ is, and it has the desired property.

Conversely, consider E_λ given by $y^2 + y + \lambda xy = x^3$ in characteristic 2 and by $y^2 = x(x-1)(x-\lambda)$ in characteristic different from 2. Assume that λ is transcendental over $\mathbb{F}_p$, and assume there exists u in $\mathrm{End}(E_\lambda)$ with $u^2 = N < 0$. For a prime ℓ not dividing pN we choose a nondivisible element x in $T_\ell(E_\lambda)$, and denote by G_n the cyclic group generated by the image of x in $z^m\ell$ under $T_\ell(E_\lambda) \to \ell^n E$. Then the separable quotient $p_n : E_\lambda/G_n = E_{\lambda(n)}$, where $\lambda(n)$ is transcendental over $\mathbb{F}_p$. If u_n in $\mathrm{End}(E_{\lambda(n)})$ corresponds to u under one of the two isomorphisms $E_\lambda \to E_{\lambda(n)}$, then $\pm u = p_n^{-1} u_n p_n$ and $u(\ker(p_n))$ is contained in $\ker(p_n) = G_n$ for all n. Hence $u(x) = ax$ is an eigenvector, and, therefore, u acts as a scalar on $T_\ell(E)$. This contradicts $u^2 = N < 0$. Hence if E is not defined over a finite field, then $\mathrm{End}(E) = \mathbb{Z}$, and this proves the theorem.

§7. Summary of Criteria for a Curve To Be Supersingular

Before giving the table summarizing the previous results (Table 2) we add one more criterion in terms of sheaf cohomology for a curve to be supersingular. The reader with insufficient background can skip this result.

(7.1) Proposition. *Let E be an elliptic curve over a field k in characteristic $p > 0$. Then E is supersingular if and only if*

$$\pi_E^* : H^1(E, \mathcal{O}_E) \to H^1(E, \mathcal{O}_E)$$

is nonzero, where π_E^* is induced by the Frobenius morphism $\pi_E E \to E$.

Proof. We have an commutative diagram using the fact that the ideal sheaf of E in $\mathbb{P}_2$ is isomorphic to $\mathcal{O}_{\mathbb{P}}(-3)$ and the Frobenius F' on $\mathbb{P} = \mathbb{P}_2$ maps $\mathcal{O}_E$ to $\mathcal{O}_E(p)$ where $E^{(p)}$ is the subscheme of $\mathbb{P}$ defined by $f^p = 0$.

$$\begin{array}{ccccccccc} 0 & \longrightarrow & \mathcal{O}_{\mathbb{P}}(-3p) & \longrightarrow & \mathcal{O}_{\mathbb{P}} & \longrightarrow & \mathcal{O}_E(p) & \longrightarrow & 0 \\ & & \downarrow{\scriptstyle f^{p-1}} & & \downarrow & & \downarrow & & \\ 0 & \longrightarrow & \mathcal{O}_{\mathbb{P}}(-3) & \longrightarrow & \mathcal{O}_{\mathbb{P}} & \longrightarrow & \mathcal{O}_E & \longrightarrow & 0. \end{array}$$

Table 2. Elliptic Curves in char p different from 0 (results mostly of Deuring and Hasse).

E can be defined over a finite field if and only if E has complex multiplication, i.e., $\dim_{\mathbb{Q}} \mathrm{End}^0(E) > 1$.

Elliptic curves E defined over a finite field divide into two classes:

ordinary (Hasse invariant $H = 1$; height = 1	Supersingular[†] (Hasse invariant $H = 0$; height = 2
Characterization in terms of p-division points	
1. p-Rank $(E) = 1$, i.e., the p division points are $\mathbb{Z}/p\mathbb{Z}$.	1. p-Rank $(E) = 0$, i.e., the p-division points are 0.
2. Height of the formal group at 0 is 1.	2. Height of the formal group at 0 is 2.
3. Frob: $E \to E^{p^2}$ does not factor through $p : E \to E$.	3. Frob: $E \to E^{p^2}$ factors through $p : E \to E$. Corollary: E can be defined over $\mathbb{F}_{p^2}$.
Characterization in terms of endomorphism rings	
4. $\mathrm{Frob}^n \neq p_E^m$ for all n, m.	4. $\mathrm{Frob}^n = p_E^m$ for some n, m.
5. $\dim_{\mathbb{Q}} \mathrm{End}^0(E) = 2^1\ (= 2)$. $\mathrm{End}^0(E)$ is an imaginary quadratic extension and $\mathrm{End}(E)$ is a maximal order of index prime to p.	5. $\dim_{\mathbb{Q}} \mathrm{End}^0(E) = 2^2\ (= 4)$. $\mathrm{End}^0(E) = h(p)$ is a quaternion algebra with $\mathrm{inv}_\ell = 0$ for $\ell \neq p, \infty$ and $\mathrm{inv}_\ell = 1/2$ otherwise. $\mathrm{End}(E)$ is a maximal order in the algebra.
Characterization in terms of f a cubic equation for E	
6. Coefficient of $(wxy)^{p-1}$ in $f(w, x, y)^{p-1}$ is $\neq 0$	6. Coefficient of $(wxy)^{p-1}$ in $f(w, x, y)^{p-1}$ is $= 0$
7. Frob on $H^1(\mathcal{O}_E)$ is an isomorphism.	7. Frob on $H^1(\mathcal{O}_E)$ is an zero.
Characterization in terms of the differential form $\omega = dx/y$	
8. $\omega = d \log \psi$ and is not exact.	8. $\omega = d\varphi$ and is exact.
9. $\omega \neq 0$ is of the first kind with $a_{p-1} \neq 0$.	9. $\omega \neq 0$ is of the first kind with $a_{p-1} = 0$.
$\omega = \sum_{0 \leq i} a_i t^i dt$	
Characterization in terms of number n_p of points for $q = p \leq 3$, i.e., only curves over $\mathbb{F}_p$	
10. $N_p \neq 1 + p$,	10. $N_p = 1 + p$.

[†] Supersingular elliptic curves in characteristic p are defined over either the prime field $\mathbb{F}_p$ or its quadratic extension $\mathbb{F}_{p^2}$.

This induces another commutative diagram in which the vertical morphisms are cohomolgy boundary morphisms and F is the Frobenius on E inducing F^* as the top vertical composite

$$\begin{array}{ccccc} H^1(E, \mathcal{O}_E) & \xrightarrow{F_1^*} & H^1(E^{(p)}, \mathcal{O}_E(p)) & \longrightarrow & H^1(E, \mathcal{O}_E) \\ \downarrow\wr & & \downarrow\wr & & \downarrow\wr \\ H^2(\mathbb{P}, \mathcal{O}_{\mathbb{P}}(-3)) & \xrightarrow{F_1^*} & H^2(\mathbb{P}, \mathcal{O}_{\mathbb{P}}(-3p)) & \xrightarrow{f^{p-1}} & H^2(\mathbb{P}, \mathcal{O}_{\mathbb{P}}(-3)). \end{array}$$

The image of

$$F_1^*((wxy)^{-1}) = (wxy)^{-p}$$

has as image $f^{p-1}(wxy)^{-p}$, and $H^2(\mathcal{O}_{\mathbb{P}}(-3))$ has as basis $(wxy)^{-1}$ times the coefficient of $(wxy)^{p-1}$ in f^{p-1}. This is the formulation of the Hasse invariant used in (3.1).

§8. Tate's Description of Homomorphisms

In 12(6.1) and 12(6.3), we saw that for each prime ℓ different from the characteristic of the ground field k, the induced homomorphisms

$$T_\ell : \mathrm{Hom}_k(E, E') \otimes Z_\ell \to \mathrm{Hom}_{\mathrm{Gal}(k_s/k)}(T_\ell(E), T_\ell(E')),$$
$$V_\ell : \mathrm{Hom}_k(E, E') \otimes Q_\ell \to \mathrm{Hom}_{\mathrm{Gal}(k_s/k)}(V_\ell(E), V_\ell(E')),$$

are injective. In the case of a finite field k which we have been considering in this chapter, the Galois group $\mathrm{Gal}(k_s/k)$ is topologically generated by Frobenius $\pi = \mathrm{Fr}^h$ where $\#k = p^h = q$, and the symbol $\mathrm{Hom}_{\mathrm{Gal}(k_s/k)}$ can be written $\mathrm{Hom}_{(\mathrm{Fr})}$, namely the module of homomorphisms commuting with the action of the Frobenius on the modules T_ℓ or V_ℓ.

(8.1) Theorem (Tate). *The homomorphisms T_ℓ and V_ℓ, defined above, are isomorphisms for ℓ different from p the ground field characteristic.*

This theorem was proved by Tate [1996] for an abelian varies over a finite field. The assertion that T_ℓ is an isomorphism is equivalent to the assertion that V_ℓ is an isomorphism since the cokernel of T_ℓ is torsion free by 12(6.3).

Recall that two elliptic curves E and E' are isogenous provided $\mathrm{Hom}(E, E')$ is nonzero. Just from the injectivity of T_ℓ and V_ℓ, we have the following elementary result using linear algebra.

(8.2) Proposition. *If two elliptic curves E and E′ are isogenous, then their characteristic polynomials $f_E = f_{E'}$ are equal. Moreover,*

$$\mathrm{rank}\ \mathrm{Hom}_{(\mathrm{Fr})}(V_\ell(E), V_\ell(E')) = \begin{cases} 2 & \textit{if E is ordinary,} \\ 4 & \textit{if E is supersingular.} \end{cases}$$

The zeros of f_E are distinct in the ordinary case and equal in the supersingular case.

Proof. In an extension of Q_ℓ diagonalize the action of π_E to $\begin{pmatrix} u & 0 \\ 0 & v \end{pmatrix}$ on $V_\ell(E)$ and the action of π_E to $\begin{pmatrix} u' & 0 \\ 0 & v' \end{pmatrix}$, where u, v are the conjugate roots of $T^2 f_E(1/T)$ and u', v' of $T^2 f_{E'}(1/T)$. For $\begin{pmatrix} a & b \\ c & d \end{pmatrix}$ to be in $\mathrm{Hom}_{(\mathrm{Fr})}(V_\ell(E), V_\ell(E'))$ it is necessary and sufficient that

$$\begin{pmatrix} au & bv \\ cu & cu \end{pmatrix} = \begin{pmatrix} a & b \\ c & d \end{pmatrix}\begin{pmatrix} u & 0 \\ 0 & v \end{pmatrix} = \begin{pmatrix} u' & 0 \\ 0 & u' \end{pmatrix}\begin{pmatrix} a & b \\ c & d \end{pmatrix} = \begin{pmatrix} au' & bu' \\ cv' & dv' \end{pmatrix}.$$

If one entry of $\begin{pmatrix} a & b \\ c & d \end{pmatrix}$ is nonzero, then one eigenvalue of π_E is equal to one eigenvalue of $\pi_{E'}$ and since they are conjugate pairs, the sets $\{u, v\}$ and $\{u', v'\}$ are equal. When $u = v$, the supersingular case, we see that $\mathrm{Hom}_{(\mathrm{Fr})}$ consists of all homomorphisms and has rank 4. When the eigenvalues are distinct, we cannot have both $a \neq 0$ and $c \neq 0$ otherwise $u = u'$ and $u = v'$ contradicting $u' = v'$ and, similarly, we cannot have both $b \neq 0$ and $d \neq 0$ or it would contradict $u \neq v$. Thus $\mathrm{Hom}_{(\mathrm{Fr})}(V_\ell(E), V_\ell(E'))$ consists of all $\begin{pmatrix} a & 0 \\ 0 & d \end{pmatrix}$ or all $\begin{pmatrix} 0 & b \\ c & 0 \end{pmatrix}$ and hence has rank 2. This proves the proposition.

(8.3) Remark. For two isogenies $u : E \to E'$ and $v : E' \to E''$ the composite $vu : E \to E''$ is an isogeny, in particular, nonzero. This means that the pairings

$$\mathrm{End}_k(E) \times \mathrm{Hom}_k(E, E') \to \mathrm{Hom}_k(E, E')$$

are nondegenerate. We know that rank $\mathrm{End}_k(E) = 2$ for E ordinary (not supersingular) and rank $\mathrm{End}_k(E) = 4$ for E supersingular. Hence we see that for $\mathrm{Hom}_k(E, E') \neq 0$, it follows that rank $\mathrm{Hom}_k(E, E') =$ rank $\mathrm{Hom}_{(\mathrm{Fr})}(T_\ell(E), T_\ell(E'))$, and this proves Theorem (8.1) in the case where $\mathrm{Hom}_k(E, E') \neq 0$. In other words,

$$T_\ell : \mathrm{Hom}_k(E, E') \otimes \mathbb{Z}_\ell \to \mathrm{Hom}_{(\mathrm{Fr})}(T_\ell(E), T_\ell(E'))$$

is a monomorphism between two modules of the same rank with torsion free cokernel, and, therefore, T_ℓ is an isomorphism.

(8.4) Theorem. *Let E and E′ be two elliptic curves over a finite field $k=\mathbb{F}_q$. Then the following are equivalent:*

(1) *E and E′ are isogenous over k.*
(2) *$V_\ell(E)$ and $V_\ell(E')$ are Gal(k/k)-isomorphic modules.*
(3) *$\#E(k) = \#E'(k)$, i.e., the two curves have the same number of elements over k.*
(4) *$\zeta_E(s) = \zeta_{E'}(s)$, the two curves have the same zeta functions.*

Proof. The equivalence of (1) and (2) follows by Tate's theorem, (8.1), since

$$\mathrm{Hom}(E, E') \otimes Q_\ell \quad \text{and} \quad \mathrm{Hom}_{\mathrm{Gal}(\bar{k}/k)}(V_\ell(E), V_\ell(E'))$$

are isomorphic. Since $\mathrm{Gal}(\bar{k}/k)$ acts semisimply on $V_\ell(E)$ by (8.2), the $\mathrm{Gal}(\bar{k}/k)$-module structure over Q_ℓ is determined by the trace of σ where $\sigma(a) = a^q$ or π, that is, by the cardinality of $E(k)$. This gives the equivalence of (1), (2), and (3). Finally, (3) and (4) are equivalent by (2.2) for N_1 determines all the N_m. This proves the theorem.

Consider integers a with $|a| \le 2\sqrt{q}$. We ask whether there is an elliptic curve $E/\mathbb{F}_q$ with $\#E(\mathbb{F}_q) = q + 1 - a$ or equivalently with $\mathrm{Tr}(\pi_E) = a$. The problem was solved more generally for abelian varieties by Honda [1968], see also Tate [1968]. We only state the result.

(8.5) Theorem. *Let a be an integer with $|a| \le 2\sqrt{q}$. If $a \equiv 0 \pmod q$, then $a^2 \equiv 0 \pmod q$. Then there exists an elliptic curve E over $\mathbb{F}_q$ with $\#E(\mathbb{F}_q) = 1 + q - a$ or equivalently* $\mathrm{Tr}(\pi_E) = a$.

(8.6) Example. For $p = 2$ and $N = \#E(\mathbb{F}_{2^a})$ we have for $s = 1, 2, 3$

$q = 2$, $q + 1 = 3$ possible $N = 1, 2, 3, 4, 5$.

$q = 4$, $q + 1 = 5$ possible $N = 1, 2, 3, 4, 5, 6, 7, 8, 9$.

$q = 8$, $q + 1 = 9$ possible $N = 4, 5, 6, 8, 9, 10, 12, 13, 14$.

Observe that $9 - 2 = 7$ and $9 + 2 = 11$ are missing values for $q = 8$.

§9. Division Polynomial

The division polynomial is associated with multiplication by N on an elliptic curve E over a field k. It is polynomial $\psi_N(x)$ in the ring generated by the coefficients of the equation for E and of degree $(N^2 - 1)/2$ when N is odd. We begin our sketch of the theory with the statement of a general result for an elliptic curve over any field giving a polynomial in two variables $\psi_N(x, y)$, called the Nth division polynomial.

(9.1) Proposition. *Let E be an elliptic curve over a field k with coefficients a_i, and let $N > 0$ be an odd number. There exist polynomials $\psi_N(x, y), \theta_N(x, y), \omega_N(x, y) \in \mathbb{Z}[x, y, a_i]$ such that multiplication by N on $(x, y) \in E(k) - \{0\}$ is given by*

$$[N](x, y) = \left(\frac{\theta_N(x, y)}{\psi_N(x, y)^2}, \frac{\omega_N(x, y)}{\psi_N(x, y)^3}\right).$$

The polynomials θ_N and ω_N are polynomials in ψ_N.

A general reference for this subject is Silverman [1994] GTM 151, for this assertion see pp. 105.

Over a field of characteristic $\neq 2, 3$ the situation simplifies to a single series of polynomials $\psi_N(x)$ describing $[N](x, y)$, for the dependence of $\psi_N(x, y)$ simplifies significantly. Here a reference is also Blake, Seroussi, and Smart [1999] LMS 265, pp. 39–42.

(9.2) Division polynomial over fields of characteristic unequal to 2 or 3. Let N be an odd number, and consider the curve $E = E_{a,b}$ defined by $y^2 = x^3 + ax + b$ over a field k of characteristic $p \neq 2$. We introduce the polynomial

$$\psi_N(x) = \prod_{P \in (E[N]-\{0\})/\{\pm 1\}} (x - x(P))$$

of degree $(N^2 - 1)/2$ which can be viewed as an element in $\mathbb{Z}[x, a, b]$. When N is even, y appears only with odd power in $\psi_N(x, y)$. When N is odd, y appears only with even power in $\psi_N(x, y)$. If N is odd, then on the curve we can eliminate y so that $\psi_N(x, y)$ becomes $\psi_N(x)$.

(9.3) Recurrence formulas. We have the following examples

$$\psi_0 = 0, \qquad \psi_1 = 1, \qquad \psi_2 = 2y, \qquad \psi_3 = 3x^4 + 6ax^2 + 12bx - a^2,$$
$$\psi_4 = 4y(x^6 + 5ax^4 + 20bx^3 - 5a^2x^2 - 4abx - 8b^2 - a^2),$$

and the following recurrence formulas

$$\psi_{2N+1} = \psi_{N+2}\psi_N^3 - \psi_{N-1}\psi_{N+1}, \qquad N \geq 2$$
$$\psi_{2N} = (\psi_{N+2}\psi_{N-1}^2 - \psi_{N-2}\psi_{N+1}^2)\frac{N}{2y}, \qquad N > 2.$$

(9.4) Formula for multiplication by N. In terms of the division polynomials we have

$$[N](x, y) = \left(x - \frac{\psi_{N-1}\psi_{N+1}}{\psi_N^2}, \frac{\psi_{N+2}\psi_{N-1}^2 - \psi_{N-2}\psi_{N+1}^2}{4y\psi_N^3}\right).$$

14

Elliptic Curves over Local Fields

We return to the ideas of Chapter 5 where the torsion in $E(\mathbb{Q})$ was studied using the reduction map $E(\mathbb{Q}) \to \bar{E}(\mathbb{F}_p)$ at a prime p. Now we study reduction in terms of $E(\mathbb{Q}) \to E(\mathbb{Q}_p)$. This leads to elliptic curves over any complete field K with a discrete valuation where the congruences of 5(4.3) are interpreted using the formal group introduced in Chapter 12 §7. We obtain a more precise version of 5(4.5).

For the reduction morphism we use an equation in normal form which is minimal as in 5(2.2), that is, such that the coefficients a_j are in the valuation ring R and the valuation of the discriminant $v(\Delta)$ is minimal among such equation. When the reduced curve $\bar{E}$ is nonsingular, it can be studied within the theory of the cubic equation, and the reduction map is a group homomorphism.

When the reduced curve $\bar{E}$ is singular, we study the possible singular behavior, and for this it is useful to introduce a second minimal model, the Néron minimal model $E^{\#}$. This concept depends on the notion of group scheme which we describe briefly. Following Tate [1975, LN 476], we enumerate the possible singular reductions and the corresponding Néron models rather than embarking on a general theory which would take us beyond the scope of this book. Included in the list of singular fibres are various numerical invariants, like the conductor, which are used in the study of the L-function of an elliptic curve over a global field.

Finally in this chapter we include an introduction to elliptic curves over R. This is a treatment that I learned from Don Zagier which had been worked out for his paper with B. Gross on the derivative of the L-function, see [1986].

The reader should refer to Silverman [1986], Chapter VII in this chapter.

§1. The Canonical p-Adic Filtration on the Points of an Elliptic Curve over a Local Field

Using the notations 5(1.1) and a fixed uniformizing parameter π, we recall that the is a filtration of the additive group of the local field K

$$K \supset R \supset R\pi \supset R\pi^2 \supset \cdots \supset R\pi^m \supset \cdots$$

and one of the multiplicative group

$$K^* \supset R^* \supset 1 + R\pi \supset 1 + R\pi^2 \supset \cdots \supset 1 + R\pi^m \supset \cdots .$$

There are isomorphisms

$$R\pi^n/R\pi^{n+1} \to k^+ \quad \text{for } n \geq 0, \qquad v : K^*/R^* \to \mathbb{Z}, \qquad r : R^*/1 + R\pi \to k^*,$$

and

$$1 + R\pi^m/1 + R\pi^{m+1} \to k^+ \qquad \text{for } m \geq 1.$$

In 5(4.5) there are analogous results for the kernel $E^{(1)}(K)$ of the reduction homomorphism. Using the formal group associated to E, we will be able to give a more precise picture of this p-adic filtration.

(1.1) Definition. Let E be an elliptic curve over K with reduction mapping $r : E(K) \to \bar{E}(k)$, and let $\bar{E}(k)_{\mathrm{ns}}$ the group of nonsingular points of $E(k)$. The canonical p-adic filtration of E is a sequence of subgroups of $E(K)$

$$E(K) \supset E^{(0)}(K) \supset E^{(1)}(K) \supset \cdots \supset E^{(n)}(K) \supset \cdots ,$$

where $E^{(0)}(K) = r^{-1}(E(k)_{\mathrm{ns}})$, $E^{(1)}(K) = r^{-1}(0)$, and for $n \geq 1$ $E^{(n)}(K)$ is the set of all (x, y) satisfying $v(x) \leq -2n$ and $v(y) \leq -3n$.

From the equation in normal form with $v(a_j) \geq 0$ we have $v(x) < 0$ if and only if $v(y) < 0$ and in this case $v(x) = -2n$ and $v(y) = -3n$ for some n. In particular, the two definitions of $E^{(1)}(K)$ are equivalent. It is also the case that these definitions of $E^{(m)}(K)$ are equivalent to those given in 5(4.1).

For the filtration $E(K) \supset E^{(0)}(K) \supset E^{(1)}(K)$ we have asserted that asserted that $E(K)/E^{(0)}(K)$ is finite and that $E^{(0)}(K)/E^{(1)}(K)$ is isomorphic to $\bar{E}(k)_{\mathrm{ns}}$ as in 5(3.4). An analysis of the structure of the quotients $E^{(1)}(K)$ was given in Chapter 5, §4 in terms of the structure of the quotients $E^{(n)}(K)/E^{(2n)}(K)$. Using the formal group law Φ_E of E where $\Phi_E(t_1, t_2)$ is in $R[[t_1, t_2]]$, under the assumption that K is complete, we can define a second group structure on $R\pi$ by $a +_E b = \Phi_E(a, b)$ in $R\pi$, and this modified version of $R\pi$ is isomorphic to $E^{(1)}(K)$. More precisely we have the following theorem which uses the considerations of Chapter 12, §7.

(1.2) Theorem. *Let E be an elliptic curve over a field K with a complete discrete valuation. The function $t(P) = -x(P)/y(P)$ is an isomorphism of $E^{(1)}(K)$ onto $R\pi$ where $R\pi$ has the group structure given by $a +_E b = \Phi_E(a, b)$ in terms of the formal group law Φ_E of E.*

Proof. The definition of $\Phi_E(t_1, t_2)$ in 12(7.2) is in terms of the group law on the elliptic curve E. Hence the function $t(P) = -x(P)/y(P)$ is a group homomorphism $E^{(1)}(K)/E^{(n)}(K) \to R\pi/R\pi^n$ for each n, and hence also in the limit $E^{(1)}(K) \to R\pi$. By 5(4.5) it induces a monomorphism $E^{(n)}(K)/E^{(2n)}(K) \to R\pi^n/R\pi^{2n}$ for each n, and this implies that $E^{(1)}(K) \to R\pi$ is a monomorphism. Finally the map is surjective, for given a value t in R, we can substitute into the power series for x and y in t, see 12(7.1), to obtain a point on $E^{(1)}(K)$ having the given t value. This proves the theorem.

(1.3) Corollary. *Let E be an elliptic curve over a field K with a complete discrete valuation. The group $E^{(1)}(K)$ is uniquely divisible by all integers m not divisible by the characteristic of k.*

Proof. For the formal group law $\Phi_E(t_1, t_2)$, the series $[m](x) = mx + \cdots$ has an inverse function in $R[[x]]$ since m is invertible in R. Hence the map, which is multiplication by m on $R\pi'$ with group structure $a +_E b$ is an isomorphism. By the theorem (1.2) it is an isomorphism on $E^{(1)}(K)$. This proves the corollary.

§2. The Néron Minimal Model

Now we touch on concepts which go beyond the scope of the methods of this book. The minimal model E over K is defined by an equation in normal form whose discriminant has a certain minimality property. There is another minimal model $E^\#$, due to A. Néron, which contains additional information about the reduced curve $\bar{E}$. This additional structure plays an important role in analyzing the entire p-adic filtration on $E(K)$ and in defining the conductor of an elliptic curve in the global theory. The basic references are Néron [1964], Ogg [1967], Serre–Tate [1968], and Tate [1975, LN 476].

(2.1) Definition. Let E be an elliptic curve over K. The Néron minimal model $E^\#$ associated with E is a smooth scheme over R together with an isomorphism of curves over K defined $\theta : E^\# \times_R K \to E$ such that for any smooth scheme X over $\mathrm{Spec}(R)$ this isomorphism induces an isomorphism

$$\mathrm{Hom}_R(X, E^\#) \to \mathrm{Hom}_K(X \times_R K, E)$$

given by $f \mapsto \theta(f \times_R K)$ for f in $\mathrm{Hom}_R(X, E^\#)$.

The minimal normal form of the cubic equation also defines a scheme $E^!$ over R which means that there is a map $E^! \to \mathrm{Spec}(R)$. Since $\mathrm{Spec}(R)$ consists of two points η and s, the general fibre $E^!_\eta$, i.e., the fibre of $E^!$ over η the open general point, is just the given elliptic curve E over K, and the special fibre $E^!_s$, i.e., the fibre $E^!$ over s the closed or special point, is just the reduced curve $\bar{E}$ over k. We also use the notation $E^!_\eta = E^! \times_R K$ and $E^!_s = E^! \times_R k$ for the general and special fibres. Observe that the scheme theoretical language allows us to view E over K and $\bar{E}$ over k as part of a single algebraic object.

(2.2) Remark. The two-dimensional scheme $E^!$ over $\mathrm{Spec}(R)$ is regular if and only if $\bar{E}$ is nonsingular. By resolving the possible singularities of $E^!$, we obtain a new scheme E^+ over $\mathrm{Spec}(R)$. If $E^\#$ denotes the subscheme of smooth points of the map $E^+ \to \mathrm{Spec}(R)$, then there is a map $E^\# \times E^+ \to E^+$ which restricts to a group structure on $E^\#$ over $\mathrm{Spec}(R)$. This $E^\#$ is the Néron model. In order to check the universal property of $E^\#$, we consider a morphism of the general fibre of a smooth scheme $g : X \times_R K = X_\eta \to E$ into E over K. The graph G_g

of g is a subscheme of the common general fibre $X_\eta \times E_\eta$ of all three schemes $X \times E^\# \hookrightarrow X \times E^+ \to X \times E^!$. The key observation is that the closure of G_g in $X \times E^+$ is in fact in $X \times E^\#$ and is the graph G_f of a morphism $f : X \to E^\#$ corresponding to g in the formulation of the universal property. For details the reader should see Néron [1964].

(2.3) Remarks. The special fibre $E_s^\#$ of the Néron model $E^\#$ of E is a richer object than the reduction $\bar{E}$ of E and the group $\bar{E}(k)_{\text{ns}} \subset \bar{E}(k)$. The reduction morphism for the group scheme $E^\#$ over Spec(R) defines an epimorphism for a complete R

$$E(K) = E^\#(K) \to E_s^\#(k)$$

with kernel $E^{(1)}(K)$. The original reduction $\bar{E}(k)_{\text{ns}}$ is $E_s^{\#0}(k)$, where $E_s^{\#0}$ is the connected component of the algebraic group $E_s^\#$. In particular, we have the isomorphisms

$$E(K)/E^{(1)}(K) \to E_s^\#(k) \quad \text{and} \quad E(K)/E^{(0)}(K) \to E_s^\#(k)/\bar{E}(k)_{\text{ns}},$$

where $E_s^\#(k)/\bar{E}(k)_{\text{ns}}$ is the finite group of connected components of $E_s^\#(k)$.

Now we come to a basic invariant of an elliptic curve over a local field K.

(2.4) Definition. The conductor $f(E) = \pi^f$ in R of an elliptic curve E over K is given by

$$f = v(\Delta) + 1 - n,$$

where n is the number of connected components of $E_s^\#$ over $\bar{k}$.

(2.5) Remark. The ordinal f of the conductor $f(E)$ is a sum

$$f = f' + d, \qquad \text{where } f' = \begin{cases} 0 & \text{if } E_{\text{ns}} \text{ is an elliptic curve,} \\ 1 & \text{if } E_{\text{ns}} \text{ is of multiplicative type, or} \\ 2 & \text{if } E_{\text{ns}} \text{ is of additive type,} \end{cases}$$

and d is zero unless the residue class field k has characteristic 2 or 3, and it is given in terms of the wild ramification of $E(K)$ as a Galois module, see Ogg [1967], where the relation for f is considered.

Tate [1975] in LN 476, Antwerp IV gives a detailed version of an algorithm for describing the special fibre of the Néron model. For the construction of the Néron model following (2.2) in terms of the valuations (or orders of zero) $v(a_i)$, $v(b_i)$, $v(c_i)$, and $v(\Delta)$ of the standard coefficients and the discriminant Δ of the minimal Weierstrass equation for E, see 5(2.2). In 5(7.2) we considered the three classes of good, multiplicative, and additive reduction. It is the curves with additive reduction that divide into seven different families when the special fibre of the Néron model is studied. We now give a version of this classification and a table of the possibilities.

(2.6) Special Fibre of the Néron Model. We preserve the notations of (2.3), and we denote by c the order of $\Gamma = E(K)/E^{(0)}(K)$ and n the number of connected components of $E_s^{\#}$ over $\bar{k}$. We divide the classification of the special fibers into three classes. The first is the good and multiplicative reduction cases, the second is three cases of additive reduction controlled by the quadratic part of the Weierstrass quation, and the third is four cases of additive reduction controlled by the cubic part of the Weierstrass equation.

Step I. We start with the coefficients a_i, b_i, and c_i from a minimal model in normal form.

(G) The curve E has good reduction which is equivalent to $v(\Delta) = 0$. In this case $E = E^{\#}$, and the special fibre is an elliptic curve $\bar{E}$ which is the reduction of E.

(M) The curve has multiplicative reduction which is equivalent to $v(\Delta) > 0$ and $v(b_2) = 0$. We change coordinates such that $v(a_3)$, $v(a_4)$, and $v(a_6) > 0$, i.e., so that the singularity of the reduced curve is at $(0, 0)$. Then $v(b_2) = 0$ is equivalent to $v(c_4) = 0$ since $v(b_4) > 0$, and we see that $v(j(E)) = -v(\Delta) = -n$. In Table 1 this is the Case I_n (or (b_n) in Néron's notation), where n is the number of connected components of $E_s^{\#}$ over $\bar{k}$. Let k' be the field of roots over k of the equation $T^2 + \bar{a}_1 T - \bar{a}_2$. There are two cases for the structure of the connected component $E_s^{\#0}$ of the identity.

(M1) If $k' = k$, then $f = 1$, $E_s^{\#0}(k) = k^*$, and Γ is cyclic of order $n = v(\Delta)$. In this case the curve is isomorphic to the Tate curve E_q, where q satisfies $v(q) = n$.

(M2) If $k' \neq k$, then $f = 1$, $E_s^{\#0}(k) = \ker(N_{k'/k})$, Γ is cyclic of order

$$c = \begin{cases} 1 & \text{if } v(\Delta) \text{ is odd,} \\ 2 & \text{if } v(\Delta) \text{ is even.} \end{cases}$$

(A) The curve has additive reduction. These are all the other cases and are treated in the next two steps.

Step II. In this and the next step assume that $v(\Delta) > 0$, $v(b_2) > 0$, and after change of variable $v(a_3)$, $v(a_4)$, and $v(a_6) > 0$. Recall that $b_2 = a_1^2 + 4a_2$ which is the discriminant of the quadratic $T^2 + a_1 T - a_2$. For local uniformizing parameter π there is a useful notation $a_{i,m} = \pi^{-m} a_i$, $x_m = \pi^{-m} x$, and $y_m = \pi^{-m} y$ for when the quantities have a suitably high valuation (or order of zero). We distinguish three cases assuming the above.

(A1) Assume $v(a_6) = 1$. This is the cusp, Case II (or (c1)) in Table 1 with $c = 1$, $n = 1$, and $f = v(\Delta) = 2$. In this case the Weierstrass model is the Néron model since $m = (\pi, x, y) = (x, y)$ and $A = R[x, y]_m$ is regular.

(A2) Assume $v(a_6) \geq 2$ and $v(b_8) = 2$. Observe that $v(a_6) \geq 2$ implies that $v(b_6), v(b_8) \geq 2$. This is Case III (or (c2)) in Table 1 with $c = 2$, $n = 2$, $f = 2$, and $v(\Delta) = 3$.

(A3) Assume $v(a_6) \geq 2$, $v(b_8) \geq 3$, and $v(b_6) = 2$. This is Case IV (or (c3)) in Table 1 with $n = 3$, $c = 1$ or 3, $f = 2$, and $v(\Delta) = 4$.

Under the assumptions of (A2) we can write the equation for the curve in the form

(a) $$y_1^2 + a_{1,0}x_1y_1 + a_{3,1}y_1 = \pi x_1^3 + a_{2,0}x_1^2 + a_{4,1}x_1 + a_{6,2}.$$

The singular point on the fibre, whose local ring is $A = R[x, y]_m$, blows up into the conic

(b) $$y^2 + a_{1,0}xy + a_{3,1}y = a_{2,0}x^2 + a_{4,1}x + a_{6,2}.$$

whose discriminant is $b_{8,2} = -\pi^{-2}b_8$ for (A2). Further, in (A3) the conic becomes

(c) $$t^2 + a_{3,1}t + a_{6,2} = 0,$$

where $t = y - ax$ is given by $(y-ax)^2 = y^2 + a_1xy - a_2x^2$. The discriminant of the quadratic (c) is $b_{6,2}$, and thus under the assumptions of (A3) the conic degenerates into two distinct lines. Dividing (a) by x_1^3 we obtain an equation $F(u, v) = \pi$ which modulo π factors into $L_1L_2L_3$. Since the local ring generated by any two factors L_1, L_2, or L_3 is regular, equation (a) gives a regular scheme over R with fibre the three lines $\bar{L}_i = 0$. Finally, c is the number lines rational over k. One line is rational over k and the others are given by $t^2 + a_{3,1}t + a_{6,2} = 0$. Hence the value $c = 1$ or 3, but it is 3 over a large enough k.

Step III. We assume that $v(a_6) \geq 2$, $v(b_8) \geq 3$, and $v(b_6) \geq 3$. With the singularity at $(0, 0)$ on $\bar{E}$ this leads to the relations $v(a_1) \geq 1$, $v(a_2) \geq 1$, $v(a_3) \geq 2$, $v(a_4) \geq 2$, and $v(a_6) \geq 3$. For the cubic polynomial $P(t) = t^3 + a_{2,1}t^2 + a_{4,2}t + a_{6,3}$, the equation of the curve becomes

$$y_2^2 + a_{1,1}x_1y_1 + a_{3,2}y_2 = P(x_1).$$

There are three cases depending on the multiplicity of the roots of $P(t)$. If there is a double root, there are two subcases depending on the multiplicity of the roots of a quadratic polynomials $y^2 + a_{3,3}y - a_{6,4}$. If there is a triple root, there are three subcases depending on the multiplicity of the roots of $y^2 + a_{3,2}y - a_{6,4}$ and the divisibility of a_4.

For details and a description of how Cases I_0^*, I_m^*, IV^*, III^*, and II^* come up, see Tate [1975, LN 476, pp. 50–52] and for a related analysis Ogg [1967]. We conclude with Table 1.

§3. Galois Criterion of Good Reduction of Néron–Ogg–Šafarevič

Let K be a local field with valuation ring R and residue class field $R \to R/R\pi = k$. There is a criterion in terms of the action of $\mathrm{Gal}(K_s/K)$ on $E(K_s)$ and the subgroups ${}_NE(K_s)$ for an elliptic curve E over K to have good reduction at k. The idea behind this condition is that the ${}_NE(K_s)$ is isomorphic to $(\mathbb{Z}/\mathbb{Z}N)^2$ for N prime to the characteristic of K, and the following are equivalent:

(a) ${}_N\bar{E}(k_s)_{\mathrm{ns}}$ is isomorphic to $(\mathbb{Z}/\mathbb{Z}N)^2$ for all N prime to char(k).
(b) $\bar{E}(k_s)_{\mathrm{ns}}$ is complete, i.e., $\bar{E} = \bar{E}_{\mathrm{ns}}$.
(c) E has good reduction.

For the multiplicative group we have ${}_N G_{\mathrm{m}}(k_{\mathrm{s}}) = \mu_N(1)(k_{\mathrm{s}})$ and for the additive group we have ${}_N G_{\mathrm{a}}(k_{\mathrm{s}}) = 0$. Thus it is exactly the case of good reduction where the reduction morphism $E(K_{\mathrm{s}}) \to \bar{E}(k_{\mathrm{s}})$ restricts to an isomorphism ${}_N E \to {}_N \bar{E}$, and it is the case of bad reduction when it has a nontrivial kernel. Unfortunately the above outline of ideas does not work for the Weierstrass model E over R, but we must use the Néron model $E^\#$ of E which has additional naturality properties.

The Galois group $\mathrm{Gal}(K_{\mathrm{s}}/K) \to \mathrm{Gal}(k_{\mathrm{s}}/k)$ maps surjectively and the kernel is called the inertia subgroup I of $\mathrm{Gal}(K_{\mathrm{s}}/K)$.

(3.1) Definition. A set S on which $\mathrm{Gal}(K_{\mathrm{s}}/K)$ acts is called unramified provided the inertia subgroup I acts as the identity on the set S.

In other words, in the unramified case, the action of $\mathrm{Gal}(K_{\mathrm{s}}/K)$ factors through an action of the Galois group $\mathrm{Gal}(k_{\mathrm{s}}/k)$ of the residue class field. This definition can apply to $S = {}_N E(K_{\mathrm{s}})$ or $T_\ell(E)$ associated with an elliptic curve E over K.

(3.2) Theorem (Criterion of Néron–Ogg–Šafarevič). *Let K be a local field with perfect residue class field k of characteristic p. Then the following assertions are equivalent for an elliptic curve E over K:*

(1) *The elliptic curve E has good reduction.*
(2) *The N-division points ${}_N E(K_{\mathrm{s}})$ are unramified for all N prime to the characteristic p of k.*
(2)′ *The N-division points ${}_N E(K_{\mathrm{s}})$ are unramified for infinitely many N prime to the characteristic p of k.*
(3) *The Tate module $T_\ell(E)$ is unramified for some prime ℓ unequal p.*

Now we sketch the proof given in Serre–Tate [1968]; see also Ogg [1967]. The line of argument that we give for elliptic curves works equally well for higher-dimensional abelian varieties.

Proof. Since $T_\ell(E)$ is unramified if and only if each ${}_{\ell^n} E(K_{\mathrm{s}})$ is unramified, it is clear that (2) implies (3) and (3) implies (2)′. In order to see that (1) implies (2) and (2)′ implies (1), we introduce the notation L for the fixed field of the inertial subgroup I of $\mathrm{Gal}(K_{\mathrm{s}}/K)$ and R_L for the ring of integers in L. Then $\bar{k}$ is the residue class field of both R_L and R_s of K_{s}. For the Néron model $E^\#$ we have two isomorphisms and a reduction morphism r in the sequence

$$E(K_{\mathrm{s}})^I \to E(L) \to E^\#(R_L) \xrightarrow{r} E^\#_s(\bar{k}).$$

The basic properties of r are derived from the facts that R_L is Henselian, i.e., Hensel's lemma applies to it, and that $E^\#$ is smooth. The morphism r is surjective and $\mathrm{Gal}(K_{\mathrm{s}}/K) \to \mathrm{Gal}(\bar{k}/k)$ equivariant, and the kernel $\ker(r)$ is uniquely divisible by any N prime to p.

If E has good reduction, i.e., (1) is satisfied, then $E^\#_s$ is an elliptic curve so that ${}_N E^\#_s(\bar{k})$ is isomorphic to $(\mathbb{Z}/N\mathbb{Z})^2$. Thus the same is true for ${}_N E(K_{\mathrm{s}})^I$ so that I acts trivially on ${}_N E(K_{\mathrm{s}})$ for all N prime to p. Thus (1) implies (2).

Table 1. Special Fibres.

Symbol of Kodaira/Néron	Cycle $E^{\#}$ (diagram)	E_{ns}	n	Γ	$v(\Delta)$ $(\delta = 0)$	Coefficients j $(p \neq 2, 3)$	$y^2 = 4x^3 - g_2x - g_3$
Case I_m (b_m)	$m = 1$; m arb.; 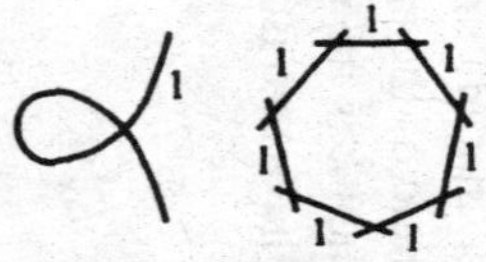 m components	$\mathbf{G}_m$	m	(m)	m	$v(j) = -m$	$v(g_2) = 0$, $m = -v(j)$
Case II (c1)		$\mathbf{G}_a$	1	(1)	2	$\tilde{j} = 0$	$v(g_2) \geq 1$, $v(g_3) = 1$
Case III (c2)		$\mathbf{G}_a$	2	(2)	3	$\tilde{j} = 12^3$	$v(g_2) = 1$, $v(g_3) \geq 2$
Case IV (c3)	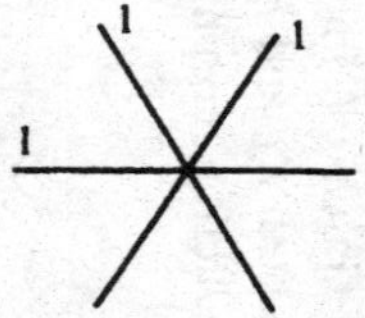	$\mathbf{G}_a$	3	(3)	4	$\tilde{j} = 0$	$v(g_2) \geq 2$, $v(g_3) = 2$

Case I_0^* (c4)	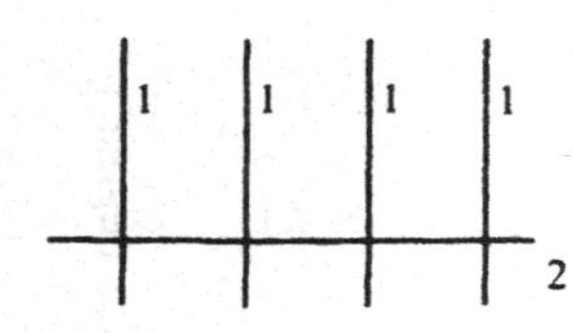	$\mathbf{G}_a$	5	(2×2)	6	$v(j) = 0$	$v(g_2) = 2$ $v(g_3) \geq 3$
Case I_m^* (c5m)	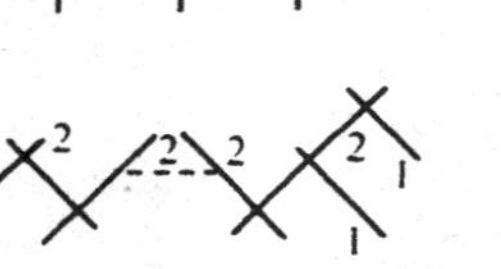 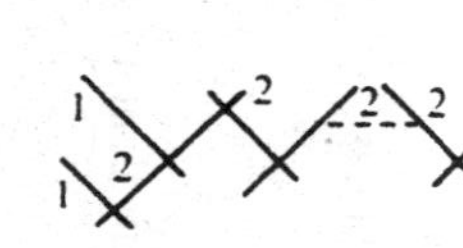 $m + 1$ double components	$\mathbf{G}_a$	$5 + m$	(4) m odd (2×2) m even	$6 + m$	$v(j) = -m$	$v(g_2) = 2$ $m = -v(j)$
Case IV* (c6)	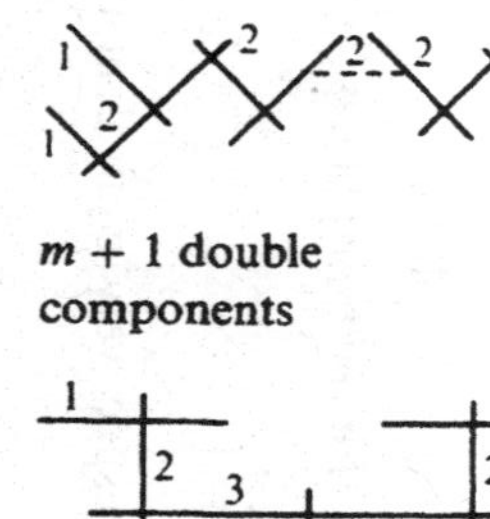	$\mathbf{G}_a$	7	(3)	8	$\tilde{j} = 0$	$v(g_2) \geq 3$ $v(g_3) = 4$
Case III (c7)	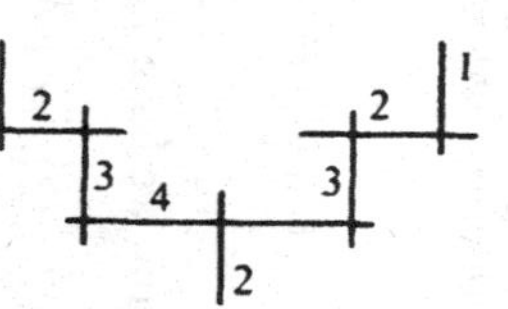 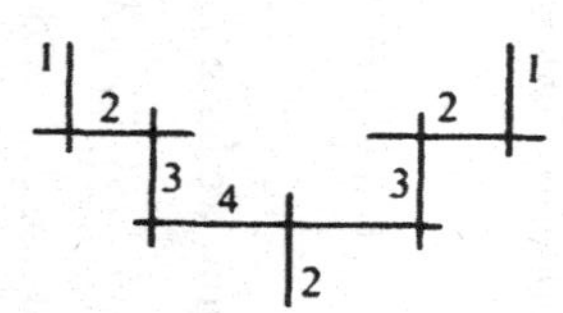	$\mathbf{G}_a$	8	(2)	9	$\tilde{j} = 12^3$	$v(g_2) = 3$ $v(g_3) \geq 5$
Case II* (c8)	 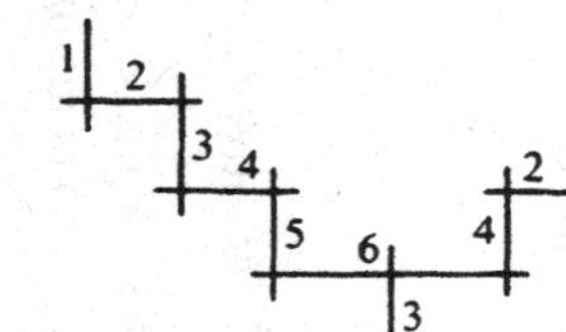	$\mathbf{G}_a$	9	(1)	10	$\tilde{j} = 0$	$v(g_2) \geq 4$ $v(g_3) = 5$

If $(2)'$ is satisfied, then for infinitely many N the fixed subgroup ${}_N E(K_s)^I = {}_N E(K_s)$. For N strictly bigger than the index $(E_s^\# : E_{s,0}^\#)$ there is a subgroup of $E_s^\#$ isomorphic to $(\mathbb{Z}/N\mathbb{Z})^2$, and hence, $E_{s,0}^\#$ has a subgroup isomorphic to $(\mathbb{Z}/N\mathbb{Z})^2$. In particular, $E_{s,0}^\#$ is an elliptic curve and therefore proper. It suffices to show that the scheme $E^\#$ is proper over $\mathrm{Spec}(R)$, and for this we use the next lemma which will complete the proof of the theorem.

(3.3) Lemma. *Let X be a smooth scheme over* $\mathrm{Spec}(R)$ *whose general fibre $X_\eta = X \times_R K$ is geometrically connected and whose special fibre $\bar{X}$ is proper. Then X is proper over* $\mathrm{Spec}(R)$, *and $\bar{X}$ is geometrically connected.*

For the proof of the lemma see Serre–Tate [1968, p. 496]. It follows from basic results on properness and correctedness for schemes.

Finally we consider the characteristic polynomial in the ramified case in a form to be used in the study of the L-function. Let K be a finite extension of $\mathbb{Q}_p$ with a valuation v extending ord_p on $\mathbb{Q}_p$ and with residue class field $k(v)$ and $Nv = \# k(v) = q = p^f$. Let I be the inertia subgroup of $G = \mathrm{Gal}(\bar{K}/K) = \mathrm{Gal}(\bar{\mathbb{Q}}_p/K)$ and Fr the arithmetic Frobenius which generates $G/I \xrightarrow{\sim} \hat{\mathbb{Z}}$.

For each $\ell \neq p$ the element Fr has a well-defined action on $T_\ell(E)^I$ independent of the choice of representative in the coset modulo I, and its characteristic polynomial

$$f_{E/K}(T) = \det\left(1 - \left(\mathrm{Fr} \mid T_\ell(E)^I\right) T\right)$$

has again integral coefficients independent of ℓ and is of degree ≤ 2. Its degree depends on whether the reduction is good, multiplicative, or additive.

(3.4) Remark. Let $\bar{E}$ denote the reduction of E over $k(v)$. Then the cardinality $\# E(k(v))_{\mathrm{ns}} = q f_{E/K}(1/q)$. For the three cases:

(1) Good reduction, we have $f_{E/K}(T) = 1 - \mathrm{Tr}(\pi)T + qT^2$.
(2) Multiplicative reduction, we have

$$f_{E/K}(T) = \begin{cases} 1 - T & \text{for the split case,} \\ 1 + T & \text{for the nonsplit case.} \end{cases}$$

(3) Additive reduction, we have $f_{E/K}(T) = 1$.

Observe the cases (1), (2), and (3) are respectively equivalent to the degree of $f_{E/K}(T)$ being of degree 2, 1, and 0.

§4. Elliptic Curves over the Real Numbers

We recall the notation for 3(8.5) for a parametrization of the family of elliptic curves over a field K. We will eventually specialize K to the real numbers $\mathbb{R}$.

(4.1) Notation. Let K be a field, and consider the curve $E\langle\alpha,\beta\rangle$: $y^2 = x^3 - 3\alpha x + 2\beta$. We have seen that $E\langle\alpha,\beta\rangle$ and $E\langle\alpha',\beta'\rangle$ are isomorphic if and only if there exists $\lambda \in K^*$ with $\lambda^4\alpha = \alpha'$ and $\lambda^6\beta = \beta'$. Since the J-value of $E\langle\alpha,\beta\rangle$ is $J(\alpha,\beta) = \alpha^3/\Delta(\alpha,\beta)$ where $\Delta(\alpha,\beta) = \alpha^3 - \beta^2$, we introduce the following two orbit spaces

$$\mathcal{E}\ell\ell(K) = K^2 - \{\alpha^3 = \beta^2\}/K^* \quad \text{and} \quad \mathcal{E}\ell\ell'(K) = K^2 - \{(0,0)\}/K^*.$$

In both cases the action of K^* on K^2 is by $\lambda \cdot (\alpha,\beta) = (\lambda^4\alpha, \lambda^6\beta)$. We have constructed a bijection from $\mathcal{E}\ell\ell(K) = K^2 - \{\alpha^3 = \beta^2\}/K^*$ to isomorphism classes of elliptic curves and a bijection from $\mathcal{E}\ell\ell'(K) = K^2 - \{(0,0)\}/K^*$ to isomorphism classes of possibly singular elliptic curves with at most a double point.

For K algebraically closed the J-function is a bijection $J : \mathcal{E}\ell\ell(K) \to K$ which extends to a bijection $J : \mathcal{E}\ell\ell'(K) \to K \cup \{\infty\}$ where $\mathbb{P}_1(K) = K \cup \{\infty\}$. The value $J = \infty$ corresponds to the curve with a double point $E\langle\lambda^2, \lambda^3\rangle$ at λ and third root at -2λ.

(4.2) The Upper Half Plane and the Cubic Curve. Recall the bijection

$$\Phi : \mathfrak{H}/\mathbb{SL}(2,\mathbb{Z}) \to \mathcal{E}\ell\ell(\mathbb{C})$$

where $\tau \in \mathfrak{H}$ is assigned to the curve $\Phi(\tau) = E\langle E_4(\tau), E_6(\tau)\rangle$ with cubic equation

$$y^2 = x^3 - 3E_4(\tau)x + 2E_6(\tau),$$

see 10(1.6). We diagram both the smooth and the singular cubics for both $\mathbb{C}$ and $\mathbb{R}$

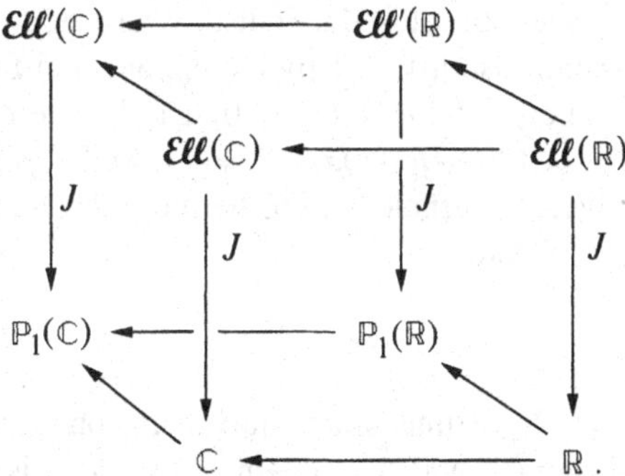

Recall the classical $j = 12^3 J$. Now we study which points in $\mathfrak{H}/\mathbb{SL}(2,\mathbb{Z})$, or equivalently, in the fundamental domain of $\mathfrak{H}$, map under Φ to $\mathcal{E}\ell\ell(\mathbb{R}) \subset \mathcal{E}\ell\ell(\mathbb{C})$. Then we describe the behavior of J on these domains.

(4.3) Eisenstein Series Coefficients. The cubic equation in the Weierstrass theory was originally derived in the form,

$$y^2 = 4x^3 - g_2(\tau)x - g_3(\tau),$$

see 9(4.1). In 10(1.4) other multiples of the basic Eisenstein series $G_k(L) = G_{2k}(\tau)$ for $L_\tau = \mathbb{Z}\tau + \mathbb{Z}$ were introduced where $g_2(\tau) = 60G_4(\tau)$ and $g_3(\tau) = 140G_6(\tau)$. Now we use another normalization with q-expansion having 1 as constant term, namely,

$$G_{2k}(\tau) = 2\zeta(2k)E_{2k}(\tau).$$

Then the two relevant functions of weights 4 and 6 are respectively

$$E_4 = 1 + 240\sum_{n\geq 1} \sigma_3(n)q^n \quad \text{and} \quad E_6 = 1 + 504\sum_{n\geq 1} \sigma_5(n)q^n$$

where $\sigma_s(n) \sum_{d|n} d^s$. After suitable renormalization we have the equation of the cubic as above $y^2 = x^3 - 3E_4(\tau)x + 2E_6(\tau)$.

(4.4) Fundamental Domain. The fundamental domain D of $\mathbb{SL}(2, \mathbb{Z})$ in $\mathfrak{H}$ is given by the inequalities $D : -1/2 \leq \mathbf{Re}(\tau) \leq 1/2$ and $|\tau| \geq 1$,

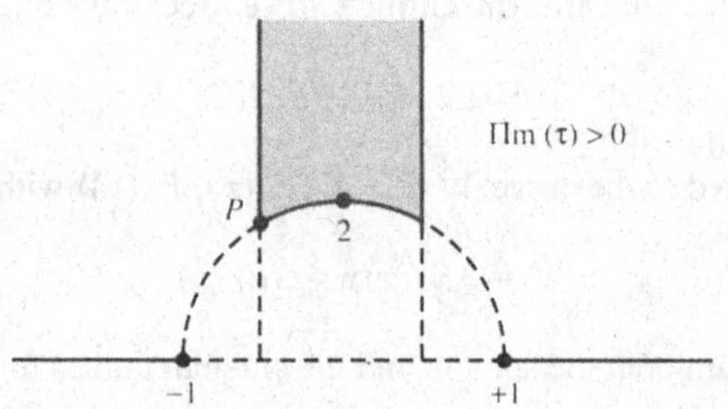

that is, the central strip outside the closed unit disc. When the strip $-1/2 \leq \mathbf{Re}(\tau) \leq 1/2$ is mapped into the unit disc $|q| < 1$ by the exponential function $q = q(t) = e^{2\pi i\tau}$ we see that q is real exactly for $\mathbf{Re}(\tau) = 0, \pm 1/2$. In terms of the equation of the cubic curve $E(\tau)$: $y^2 = x^3 - 3E_4(\tau)x + 2E_6(\tau)$, we know that E_4 and E_6 have expansions in $q(\tau)$ with real coefficients. Furthermore the function $j(\tau)$ of 9(4.8) becomes $j(\tau) = 12^3 J(\tau)$ where

$$J(\tau) = E_4(\tau)^3/\Delta(\tau) \quad \text{and} \quad \Delta(\tau) = E_4(\tau)^3 - E_6(\tau)^2.$$

Recalling 10(1.7), we have thus four equivalent conditions for the equation $E\langle E_4(\tau), E_6(\tau)\rangle$ to be defined with real coefficients: (1) q is real, (2) $\mathbf{Re}(\tau)$ is a half integer, (3) $\tau + \bar{\tau}$ is an integer, and (4) $\mathbb{Z}\tau + \mathbb{Z}$ is stable under complex conjugation.

(4.5) Notation. Let $\zeta_n = \exp(2\pi i/n)$, a special primitive nth root of unity. It satisfies the equation $1 + x + \cdots + x^{n-1} = 0$ like all nontrivial nth roots of unity.

The "corners" of the fundamental domain along the circle $\tau| = 1$ are respectively ζ_3 where $\mathbf{Re}(\zeta_3) = -1/2$, $\zeta_4 = i$ where $\mathbf{Re}(\zeta_4) = 0$, and ζ_6 where $\mathbf{Re}(\zeta_6) = 1/2$. These are the intersections with the lines given real elliptic curves. We will use the following basic modular transformations $S(\tau) = -1/\tau$ and $T(\tau) = -1/(\tau - 1)$. These two transformations have the following values on the corners.

(1) For S it preserves the line $\mathbf{Re}(\tau) = 0$ and $S(\zeta_4) = \zeta_4$. Also $S(\zeta_3) = \zeta_6$ and $S(\zeta_6) = \zeta_3$ interchanging the lines $\mathbf{Re}(\tau) = \pm 1$.

(2) For T we have $T(\zeta_6) = \zeta_6$, $T(\zeta_4) = (1/2)^{1/2}\zeta_8$, and $T(\zeta_3) = (1/3)^{1/2}\zeta_{12}$ carries the circle $|\tau| = 1$ to the line $\mathbf{Re}(\tau) = 1/2$ in the upper half plane.

(4.6) The Modular Forms E_4 and E_6. The transformation law

$$E_{2k}\left(\frac{a\tau + b}{c\tau + d}\right) = (c\tau + d)^{2k} E_{2k}(\tau) \qquad \text{for } \begin{pmatrix} a & b \\ c & d \end{pmatrix} \in \mathbb{SL}(2\mathbb{Z})$$

specializes for the two transformations $S(\tau)$ and $T(\tau)$ to the following: $E_{2k}(S(\tau)) = \tau^{2k} E_{2k}(\tau)$ and $E_{2k}(T(\tau)) = (\tau - 1)^{2k} E_{2k}(\tau)$. We have two cases for the lines $\mathbf{Re}(\tau) = 0$ and $1/2$.

(1) For $\mathbf{Re}(\tau) = 0$: $E_6(\zeta_4) = 0$ and $J(\zeta_4) = 1$. Moreover, $\tau = it, t > 0$, and $q \geq 0$ in this case, and the expansion $E_4 = 1 + 240\sum_{n\geq 1}\sigma_3(n)q^n$ shows that $E_4(it) \geq 0$ on this line. In particular $J(q) \geq 1$ on this line. The relation $S(it) = i/t$ shows that the two parts of the positive imaginary axis $i[1, \infty)$ in the fundamental domain and $i(0, 1]$ are interchanged. The coefficients of the curve $\Phi(it)$ are transformed by the modular relations to give

$$E_4\left(\frac{i}{t}\right) = t^4 E_4(it) \quad \text{and} \quad E_6\left(\frac{i}{t}\right) = -t^6 E_6(it).$$

(2) For $\mathbf{Re}(\tau) = 1/2$: We have three distinguished values:

$\tau =$	$T(\zeta_6) = \zeta_6$	$T(\zeta_4) = (1/2)^{1/2}\zeta_8$	$T(\zeta_3) = (1/3)^{1/2}\zeta_{12}$
$E_4(\tau) =$	0	$E_4(i) > 0$	0
$E_6(\tau) =$	$E_6(\zeta_6)$	0	$-27 \cdot E_6(\zeta_6)$
$J(t) =$	0	1	0

Here we use that $E_6(T(\zeta_3)) = (\zeta_3 - 1)^6 E_6(\zeta_6) = -27 \cdot E_6(\zeta_6)$ for the last entry in the table. For $\tau = 1/2 + it, t \geq 0$, we have $-1 < q \leq 0$ and $J(q)$ take values $(-\infty, 1]$, that is, the values not coming from $\mathbf{Re}(\tau) = 0$.

(4.7) Two Real Forms. The two equations

$$E\langle\alpha, \beta\rangle\colon\ y^2 = x^3 - 3\alpha x + 2\beta \quad \text{and} \quad E\langle t^4\alpha, -t^6\beta\rangle\colon\ y^2 = x^3 - 3\alpha t^4 x - 2t^6\beta$$

over the real numbers are nonisomorphic, but over the complex numbers become isomorphic under the substitution which carries (x, y) to $(-t^2 x, it^3 y)$ followed by dividing by $-t^6$. We can see how this can be carried out with the modular property of the coefficients.

(4.8) Notation. Let $C(ev)$ denote $i\mathbb{R}_+$ compactified with a point, denoted $0 = i\infty$, and $C(od)$ denote $(1/2) + i\mathbb{R}_+$ compactified with a point, denoted $1/2 = 1/2 + i\infty$.

The two curves $C(ev)$ and $C(od)$ are treated as disjoint spaces with the compactifying points also distinct. Returning to the diagram in (4.2), we describe $J : C(ev) \cup C(od) = \mathcal{E}\ell\ell'(\mathbb{R}) \to \mathbb{P}_1(\mathbb{R})$.

(4.9) Theorem. *The function $J : \mathcal{E}\ell\ell'(\mathbb{R}) \to \mathbb{P}_1(\mathbb{R})$ is 2:1 and the separate restrictions of J to $C(ev)$ and $C(od)$ have the form:*

(1) $J_{ev} = J|C(ev) : C(ev) \to [1, \infty] \subset \mathbb{P}_1(\mathbb{R})$ *and corresponds to* $\Delta(\tau) > 0$. *Over* $(\infty, 1)$ *the function* J_{ev} *is a smooth* 2: 1 *mapping with exactly on one point in each of the inverse images of* 1 *and* ∞.

(2) $J_{od} = J|C(od) : C(od) \to [\infty, 1] \subset \mathbb{P}_1(\mathbb{R})$ *and corresponds to* $\Delta(\tau) < 0$. *Over* $(\infty, 1)$ *the function* J_{od} *is a smooth* 2: 1 *mapping except over the point* $0 \in [\infty, 1]$. *There is exactly one point in each of the inverse images of* ∞ *and* 1.

Proof. The 2:1 character of the restrictions of J follow from (4.7) applied to $y^2 = x^3 - 3E_4(\tau)x + 2E_6(\tau)$. In the case (1) we have the relations $E_4(i/t) = t^4 E_4(it)$ and $E_6(i/t) = -t^6 E_6(it)$ for $\tau = it$ which up to rescaling give the two signs for E_6, and except for the two special points give a 2:1 mapping.

In the case (2) we have the two signs at $\zeta_6 = T(\zeta_6)$ and $T(\zeta_3)$ for $\tau = (1/2)+ti$. By continuity this sign difference extends over pairs of points $\tau' = T(\beta')$ and $\tau'' = T(\beta'')$ where β' and β'' are two boundary points on the fundamental domain giving the same elliptic curve over $\mathbb{C}$. since they are nonisomorphic at $\{\beta', \beta''\} = \{\zeta_3, \zeta_6\}$, they are nonisomorphic in general. This proves the theorem.

(4.10) Remark. At the two points over 0 where J_{od} is not smooth, we have vertical tangents.

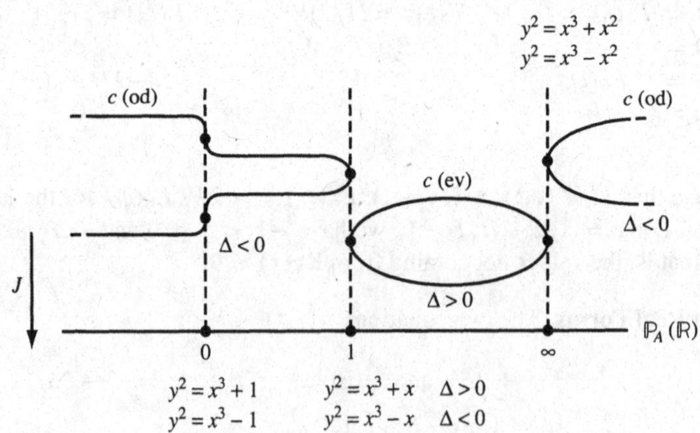

Table 2.
Table of Elliptic Curves. Sixteen Curves in Serre's *Inventiones* article [1972], pp. 309, 310, 315–316, 318–319.

		Semistable case		
5.5.1 $\sim \Gamma_1(11)$	$y^2 + y = x^3 - x^2$	$N = 11$	$\Delta = -11$	$j = -\frac{2^{12}}{11}$
5.5.2 $\sim \Gamma_0(11)$	$y^2 + y = x^3 - x^2 - 10x - 20$	$n = 11$	$\Delta = -11^5$	$j = \frac{-2^{12} \cdot 31^3}{11^5}$
5.5.6	$y^2 + y = x^3 - x$	$N = 37$	$\Delta = 37$	$j = \frac{2^{12} \cdot 3^3}{37}$
5.5.7	$y^2 + y = x^3 + x^2$	$N = 43$	$\Delta = -43$	$j = \frac{-2^{12}}{43}$
5.5.8	$y^2 + xy + y = x^3 - x^2$	$N = 53$	$\Delta = 53$	$j = \frac{-3^3 \cdot 5^3}{53}$
5.5.3	$y^2 + xy + y = x^3 - x$	$N = 2 \cdot 7$	$\Delta = -2^2 \cdot 7$	$j = \frac{-5^6}{2^2 \cdot 7}$
5.5.4	$y^2 + xy + y = x^3 - x^2 - 3x + 3$	$N = 2 \cdot 13$	$\Delta = -2^7 \cdot 13$	$j = \frac{-3^3 \cdot 43^3}{2^7 \cdot 13}$
5.5.5	$y^2 + xy + y = x^3 + x^2 - 4x + 5$	$N = 2 \cdot 3 \cdot 7$	$\Delta = -2^8 \cdot 3^2 \cdot 7$	$j = \frac{-193^3}{2^8 \cdot 3^2 \cdot 7}$
		Not semistable; j nonintegral		
5.7.1	$y^2 = x^3 + x^2 - x$	$N = 2^2 \cdot 5$	$\Delta = 2^4 \cdot 5$	$j = \frac{2^{14}}{5}$
5.7.2	$y^2 = x^3 - x^2 + x$	$N = 2^3 \cdot 3$	$\Delta = -2^4 \cdot 3$	$j = \frac{2^{11}}{3}$
5.7.3	$y^2 + xy = x^3 - x^2 - 5$	$N = 3^2 \cdot 3$	$\Delta = -3^7 \cdot 5$	$j = \frac{-1}{3 \cdot 5}$
5.7.4	$y^2 + xy + y = x^3 + x^2 - 3x + 1$	$N = 2 \cdot 5^2$	$\Delta = -2^5 \cdot 5^2$	$j = \frac{-5 \cdot 29^3}{2^5}$
		Not semistable; j integral		
5.9.1	$y^2 = x^3 - 2x^2 - x$	$N = 2^7$	$\Delta = 2^7$	$j = 2^5 \cdot 7^3$
5.9.2	$y^2 = x^3 + 6x - 2$	$N = 2^6 \cdot 3^3$	$\Delta = -2^6 \cdot 3^5$	$j = 2^9 \cdot 3$
5.9.3	$y^2 + xy = x^3 - x^2 - 2x - 1$	$N = 7^2$	$\Delta = -7^3$	$j = -3^3 \cdot 5^3$
5.9.4	$y^2 + xy = x^3 - x^2 - 2x - 7$	$N = 11^2$	$\Delta = -11^4$	$j = -11^2$

15

Elliptic Curves over Global Fields and ℓ-Adic Representations

In the previous two chapters we carried out the local study of elliptic curves, and a substantial part of the theory was related to how the fundamental symmetry, the Frobenius element, behaved on the curve modulo a prime. For an elliptic curve E over a number field K (or more generally any global field), we have for each prime a Frobenius element acting on the points of the curve. These Frobenius elements are in $\mathrm{Gal}(K_s/K)$, and this Galois group acts on the K_s-valued points $E(K_s)$, on the subgroup of N-division points ${}_NE = {}_NE(K_s)$, and on the limit Tate modules $T_\ell(E)$ where N and ℓ are prime to the characteristic of K. In Chapters 12 and 13, the action of $\mathrm{Gal}(K_s/K)$ on the endomorphisms $\mathrm{End}_{K_s}(E)$ and the automorphisms $\mathrm{Aut}_{K_s}(E)$ over K_s was considered in detail. As usual, K_s denotes a separable algebraic closure of K.

In this chapter we analyse in greater detail the action of $\mathrm{Gal}(K_s/K)$ on ${}_NE$, $T_\ell(E)$ and $V_\ell(E)$ to give a general perspective and to relate to the general concept of ℓ-adic representation. Taniyama [1957] first defined and investigated the notion of an ℓ-adic representation. General properties of these representations have been used by Faltings [1983] in the proof of the Mordell conjecture. Our aim is to give a brief introduction to the theory.

§1. Minimal Discriminant Normal Cubic Forms over a Dedekind Ring

Let R be a Dedekind ring with field of fraction K, and let v denote a finite place with valuation ring $R_{(v)} \subset K$ and $R \subset R_{(v)}$. Then R is the intersection of all $R_{(v)}$ as v runs over the places of R. Of special interest is $R = \mathbb{Z}$, $K = \mathbb{Q}$, and $R_{(v)} = \mathbb{Z}_{(p)}$ for any prime p.

Any elliptic curve E defined over K has a normal cubic equation with coefficients in R, hence also in each $R_{(v)}$. By the local theory considered in Chapter 10, we can choose, as described in 10(1.2), a minimal equation

$$F_v(x, y) = y^2 + a_{1,v}xy + a_{3,v}y - x^3 - a_{2,v}x^2 - a_{4,v}x - a_{6,v} = 0$$

for E with coefficients in $R_{(v)}$ having discriminant Δ_v, differential ω_r and conductor f_v. The discriminant Δ_v has the minimal property: $v(\Delta_v) \le v(\Delta')$, where Δ' is the discriminant of any other equation for E over $R_{(v)}$.

(1.1) Definition. With the above notations for an elliptic curve E over K, we define two divisors on R: $\mathfrak{D}_E = \sum_v v(\Delta_v)v$, called the discriminant of E, and $\mathfrak{f}_E = \sum_v v(f_v)v$, called the conductor of E. When R is a principal ideal ring, for example $\mathbb{Z}$, these divisors are principal: $\mathfrak{D}_E = (d_E)$ and $\mathfrak{f}_E = (N_E)$, where d_E in R is the discriminant of E and $N_E = N$ in R is the conductor. These elements of R are well-defined up to a unit in R, and in the case of the integers $R = \mathbb{Z}$, we choose the conductor always to be positive.

The conductor N_E can only be defined from local data, while for the discriminant d_E, we have a discriminant Δ_F for each equation F of E over R

$$F(x, y) = y^2 + a_1xy + a_3y - x^3 - a_2x^2 - a_4x - a_6 = 0$$

with which we can compare d_E. The curve E given by equation $F = 0$ also has an invariant differential ω_F. Recall that, if $F' = 0$ is a second normal cubic equation for E over R, then for some nonzero u we have $u^{12}\Delta_{F'} = \Delta_F$ and $u^{-1}\omega_{F'} = \omega_F$. From the relation $u^{-1}\omega_{F'} = \omega_F$, we can define the valuation of the quotient of two differentials by

$$v\left(\frac{\omega_F}{\omega_{F'}}\right) = v(u^{-1}) = -v(u).$$

This is related to the valuations of the corresponding discriminants by

$$12v(\omega_F/\omega_{F'}) = v(\Delta_{F'}) - v(\Delta_F).$$

There is an obvious transitivity condition for three curves $F = 0$, $F' = 0$, and $F'' = 0$

$$v\left(\frac{\omega_F}{\omega_{F'}}\right) + v\left(\frac{\omega_{F'}}{\omega_{F''}}\right) = v\left(\frac{\omega_F}{\omega_{F''}}\right).$$

For ω_F and each local ω_v, we form a divisor which relates $F = 0$ to the local minimal models

$$\mathfrak{A}_F = \sum_v v\left(\frac{\omega_F}{\omega_v}\right) v.$$

This satisfies $\mathfrak{A}_F + (u) = \mathfrak{A}_{F'}$ from $u^{-1}\omega_{F'} = \omega_F$ and the transitivity formula. Moreover, this and the relation

$$12v\left(\frac{\omega_F}{\omega_v}\right) = v(\Delta_v) - v(\Delta_F)$$

give the following result.

(1.2) Proposition. *The divisor $\mathfrak{A}_F$ is well-defined by E in the divisor class group $\mathrm{Cl}(R)$ of R, and as a divisor it satisfies the relation*

$$12\mathfrak{A}_F + (\Delta_F) = D_E = \sum_v v(\Delta_v)v$$

In the divisor class group D_E is $12\mathfrak{A}_F$.

If 12 kills the divisor class group $\mathrm{Cl}(R)$ of divisors modulo principal divisors, then D_E equals (d_E) the divisor of a function. The class group $\mathrm{Cl}(R)$ is zero if and only if R is a principal ideal ring, and in this case we have a globally minimal normal cubic equation for an elliptic curve. The following theorem was already seen from the discussion before 5(2.6), but it also follows from the discussion above.

(1.3) Theorem. *Let R be a principal ideal ring, and let E be an elliptic curve over the field of fractions K. There exists a normal cubic equation $F_E(x, y) = 0$ for E over K such that $F = F_E$ is minimal at all places v of R. Further, d_E equals Δ_F up to a unit in R.*

Proof. Since $\mathfrak{A}_F = (u)$ for some nonzero u in K, we can use u to define an isomorphism of the curve with equation $F = 0$ onto the curve with $\mathfrak{A}_F = 0$, that is, we can make a change of variable to the case $\mathfrak{A}_F = 0$. Then, for such an equation F over R, we have

$$(\Delta_f) = \mathfrak{D}_E = (d_E).$$

This proves the theorem.

We conclude this section with some notations concerning algebraic number fields.

(1.4) Notations. Let K denote an algebraic number field with ring of integers R and for each place v of K, we denote the completion of K at v by K_v. If v is Archimedean, then K_v is isomorphic to $\mathbb{R}$ or $\mathbb{C}$. For a non-Archimedean valuation v, let R_v (resp. $R_{(v)}$) denote the valuation ring of v in K_v (resp. K) consisting of all field elements a with $v(a) \geq 0$. Let π_v denote a local uniformizing parameter of v which can be chosen to be in K satisfying $v(\pi_v) = 1$, and let $k(v)$ denote the residue class field of v. Since $R_v\pi_v$ (resp. $R_{(v)}\pi_v$) is the unique maximal ideal of R_v (resp. $R_{(v)}$), the residue class field $k(v)$ is either $R_v/R_v\pi_v$ or $R_{(v)}/R_{(v)}\pi_v$. Finally, let Nv be the order of $k(v)$ and p_v the characteristic of $k(v)$ where $p_v^a = Nv$.

§2. Generalities on ℓ-Adic Representations

For a field k with separable algebraic closure k_s, the Galois group $\mathrm{Gal}(k_s/k)$ is given the Krull topology where a neighborhood base of the identity consists of subgroups fixing finite extensions K of k. With this topology, it is a compact and totally disconnected group.

(2.1) Remark. If V is an n-dimensional complex vector space, then a continuous homomorphism $\mathrm{Gal}(k_s/k) \to \mathbb{GL}(V) = \mathbb{GL}(n, \mathbb{C})$ factors through a finite quotient of the form $\mathrm{Gal}(k_s/k) \to \mathrm{Gal}(k'/k)$, where $k' \subset k_s$ and k'/k is a finite Galois extension. This follows from the fact that the only compact, totally disconnected subgroups of $\mathbb{GL}(n, \mathbb{C})$ are finite subgroups.

If we replace the complex numbers $\mathbb{C}$ by the ℓ-adic numbers $\mathbb{Q}_\ell$, then as usual finite dimensional vector spaces V and their endomorphism algebras $\mathrm{End}(V)$ have a natural product topology, and $\mathbb{GL}(V)$ has the subspace topology of $\mathrm{End}(V)$ making it into a locally compact, totally disconnected topological group. It is this fact which makes ℓ-adic representations more useful in studying the Galois group $\mathrm{Gal}(k_s/k)$ than complex representations.

(2.2) Definition. An n-dimensional ℓ-adic representation of the Galois group $\mathrm{Gal}(k_s/k)$, or of the field k, is a continuous homomorphism $\rho : \mathrm{Gal}(k_s/k) \to \mathbb{GL}(V) = \mathbb{GL}(n, \mathbb{Q}_\ell)$, where V is an n-dimensional $\mathbb{Q}_\ell$ vector space.

In this case, the image of ρ can be infinite as can be seen by the two basic examples of (2.3) and Tate modules.

(2.3) Example. For ℓ prime to the characteristic of k, $\mathbb{Q}_\ell(1)(k_s)$ is a one-dimensional ℓ-adic representation, and for any elliptic curve E, the rational Tate vector space $V_\ell(E)$ is a two-dimensional ℓ-adic representation (of $\mathrm{Gal}(k_s/k)$ or of k). For the details, see (5.1), (5.2), and (5.3).

The $\mathbb{Z}_\ell$-submodules of the ℓ-adic representation spaces in the examples $\mathbb{Z}_\ell(1)(k_s) \subset \mathbb{Q}_\ell(1)(k_s)$ and $T_\ell(E) \subset V_\ell(E)$ are stable under the action of $\mathrm{Gal}(k_s/k)$, and introducing a basis, we can factor the ℓ-adic representation through the subgroup of integral matrices

$$\mathrm{Gal}(k_s/k) \to \mathbb{GL}(n, \mathbb{Z}_\ell) \subset \mathbb{GL}(n, \mathbb{Q}_\ell),$$

where $n = 1$ in the first case and $n = 2$ in the case of $T_\ell(E) \subset V_\ell(E)$. More generally, a module T over Z_ℓ of a vector space V over $\mathbb{Q}_\ell$ is called a lattice provided the natural homomorphism $T \otimes_{Z_\ell} \mathbb{Z}_\ell \to V$ is an isomorphism. A basis of T is also a basis of V yielding compatible isomorphisms

$$\begin{array}{ccc} \mathbb{GL}(T) & \xrightarrow{\sim} & \mathbb{GL}(n, \mathbb{Z}_\ell) \\ \cap & & \cap \\ \mathbb{GL}(V) & \xrightarrow{\sim} & \mathbb{GL}(n, \mathbb{Q}_\ell) \end{array}$$

(2.4) Proposition. *If ρ is an ℓ-adic representation of* $\mathrm{Gal}(k_s/k)$ *on the vector space V over $\mathbb{Q}_\ell$, then there exists a lattice T in V with $\rho(s)T = T$ for all s in* $\mathrm{Gal}(k_s/k) = G$.

Proof. Let T' be a lattice in V with generators $x_1, \ldots, x_m$, and consider for $j = 1, \ldots, m$ the function which assigns to each s in G the coset $sx_j + T'$ in V/T'. This is a finite valued function with all sx_j in $\ell^{-N}T$ for some large N. The subgroup H of all s in G with $sx_j \in T'$, for all j, $sT' \subset T'$ is an open subgroup of finite index in G. Let T be the submodule over $\mathbb{Z}_\ell$ generated by sT' for $sH \in G/H$. Then $T' \subset T \subset \ell^{-N}T'$ and T is a lattice with $sT \subset T$ for all s in G. Hence $sT = T$ for all $s \in G$. This proves the proposition.

In the previous proof, we used the usual convention that sx denotes $\rho(s)x$ when ρ is clear from the context.

(2.5) Remarks. If $\rho : G \to \mathbb{GL}(V)$ is a representation of a group on a finite dimensional vector space V over a field F, then the character χ_ρ of ρ is the function $\chi_\rho(s) = \mathrm{Tr}(\rho(s))$. A particular value of χ_ρ is $\chi_\rho(1) = \dim_F V$. If F is of characteristic zero and if G is finite, then we know that two representations ρ and $\rho' : G \to \mathbb{GL}(V')$ are isomorphic if and only if $\chi_\rho = \chi_{\rho'}$.

There is a similar assertion for infinite groups and semisimple representations. This can be viewed in terms of the group algebra $A = F[G]$ of G over F consisting of (finite) linear combinations $\sum_{s \in G} a_s s$ with scalars $a_s \in F$ and algebra structure given by

$$\left(\sum_s a_s s\right)\left(\sum_t b_t t\right) = \sum_{s,t} (a_s b_t) st.$$

Then a representation $\rho : G \to \mathbb{GL}(V)$ defines an $A = F[G]$ module structure on V by $(\sum_s a_s s)v = \sum_s a_s \rho(s)v$ and, conversely, A-modules restrict to G-representations. Two A-modules V and V' are isomorphic if and only if the representations are isomorphic. The A-module V is simple if and only if the representation V is irreducible or simple. In either case, the term semisimple means a direct sum of simple modules or representations. The following proposition shows that χ_ρ determines ρ up to isomorphism in the case of finite dimension, semisimple representations over a field F of characteristic 0.

(2.6) Proposition. *Let F be a field of characteristic zero, let A be an algebra over F, and let M and N be two semisimple A-modules which are finite dimensional over F. If $\mathrm{Tr}_M(a) = \mathrm{Tr}_N(a)$ for all a in A, then M and N are isomorphic.*

Proof. Let I be the set of isomorphism classes of simple A-modules, and let $E(i)$ be a representative module in the class $i \in I$. Then M is isomorphic to $\sum_i E(i)^{m(i)}$ and N to $\sum_i E(i)^{n(i)}$ for some $m(i), n(i) \in N$. For each a in A and $j \in I$ there exist by the density theorem (see Lang, Algebra) elements $a(j) \in A$ with scalar action on $E(i)$ given by

$$a(j)_{E(i)} = \begin{cases} 0 & \text{for } i \neq j, \\ a_{E(j)} & \text{for } i = j. \end{cases}$$

The hypothesis $\mathrm{Tr}_M(a) = \mathrm{Tr}_N(a)$ for $a(j)$ becomes simply the relation

$$m(i)\mathrm{Tr}_{E(j)}(a) = n(i)\mathrm{Tr}_{E(j)}(a).$$

Specialize to $a = 1$, and we obtain the relation $m(i) = n(i)$ since $\mathrm{Tr}_{E(j)}(a) = \dim E(j) \neq 0$ in the field $F \subset A$. This shows that M and N are isomorphic and proves the proposition.

(2.7) Corollary. *Let G be a group, and let ρ and ρ' be two semisimple finite dimensional representations defined over a field F of characteristic zero. Then ρ and ρ' are isomorphic if and only if $\mathrm{Tr}(\rho(s)) = \mathrm{Tr}(\rho'(s))$ for all $s \in G$.*

The proof of the corollary of (2.6) follows from the discussion at the end of (2.5).

Later, we will also consider characteristic polynomials of $\rho(s)$ in $\mathbb{GL}(V)$ for a representation $\rho : G \to \mathbb{GL}(V)$, where V is finite dimensional over F. These polynomials are defined by

$$P_\rho(s)(T) = \det(1 - \rho(s)T) \in F[T].$$

Then $P_\rho(s)(0) = 1$ and the derivative at $T = 0$ is $P_\rho(s)'(0) = -\mathrm{Tr}(\rho(s))$ since $P_\rho(s)(T) = 1 - \mathrm{Tr}(\rho(s))T + \dots$. In fact, from the traces of elements, one can recover the characteristic polynomial.

The following formula comes up frequently in the context of ℓ-representations.

(2.8) Proposition. *Let $u : V \to V$ be a linear transformation of a finite dimensional vector space V over a field of characteristic zero. Then the characteristic polynomial satisfies*

$$\det(1 - uT)^{-1} = \exp\left(\sum_{n=1}^{\infty} \frac{\mathrm{Tr}(u^n)T^n}{n}\right).$$

Proof. We can choose a basis of V so that u is upper triangular. Then $\det(1 - uT) = (1 - c_1T)\dots(1 - c_mT)$ where the $c_1, \dots, c_m$, are the eigenvalues with multiplicities of u and $m = \dim_F V$. Now use the classical expansion

$$\log\frac{1}{1 - cT} = \sum_{n=1}^{\infty} \frac{c^nT^n}{n}$$

to prove the proposition.

§3. Galois Representations and the Néron–Ogg–Šafarevič Criterion in the Global Case

We keep the notations introduced in (1.4) for an algebraic number field K, in particular, for each place v of K, the valuation ring $R_{(v)}$ in K, the local field K_v with its valuation ring R_v, the residue class field $k(v)$ of either $R_{(v)}$ or R_v, and Nv equal to the cardinality of $k(v)$. We consider algebraic extensions L of K in a fixed algebraic closure $\bar{K} = \bar{\mathbb{Q}}$ and extensions w to L of places v of K. The fact that w extends v is denoted $w|v$.

(3.1) Notations. Let L be a Galois extension of K with Galois group $\mathrm{Gal}(L/K)$, and let w be a place of L extending a place v of K. The decomposition subgroup D_w of $\mathrm{Gal}(L/K)$ consists of all s in $\mathrm{Gal}(L/K)$ with $ws = w$. By reduction modulo the maximal ideal, we have an epimorphism of groups

$$D_w \to \mathrm{Gal}(k(w)/k(v))$$

with kernel the inertia subgroup I_w of the place w. There is also an isomorphism given by extension to the completed fields

$$D_w \to \mathrm{Gal}(L_w/K_v)$$

An element s of the decomposition group D_w belongs to the inertia subgroup I_w of D_w if and only if $w(s(a) - a) > 0$ for all a in $R_{(w)}$, or equivalently, in R_w. When the maximal ideal of $R_{(w)}$ or R_w is generated by an element π, then an element s in D_w is also in I_w if and only if $w(s(\pi) - \pi) > 0$.

Let E be an elliptic curve over K. Then $\mathrm{Gal}(\bar{K}/K)$ acts on $E(\bar{K})$, on ${}_N E(\bar{K})$, and on $T_\ell(E)$ for all rational primes ℓ. Further, for any place w of $\bar{K}$ extending a place v of K, the decomposition group D_w acts on $E(\bar{K})$, on ${}_N E(\bar{K})$, and on $T_\ell(E)$. Under the isomorphism $D_w \to \mathrm{Gal}(\bar{K}_w/K_v)$, the following natural homomorphisms are equivariant morphisms of groups

$$E(\bar{K}) \to E(\bar{K}_w), \quad {}_N E(\bar{K}) \to {}_N E(\bar{K}_w), \quad \text{and} \quad T_\ell(E/K) \to T_\ell(E/K_v),$$

where the second two are isomorphisms. With these remarks, we can translate the local Néron–Ogg–Šafarevič criterion 14(3.2) into a global assertion using the following definition.

(3.2) Definition. A set S on which $\mathrm{Gal}(L/K))$ acts is called unramified at a place v of K provided the inertia subgroups I_w act as the identity on the set S for all $w|v$, that is, places w of L dividing v.

(3.3) Theorem. *Let K be an algebraic number field, and let v be a place of K with residue class field of characteristic p_v. Then the following assertions are equivalent for an elliptic curve E over K:*

(1) *The elliptic curve E has good reduction at v.*
(2) *The N-division points ${}_N E(\bar{K})$ are unramified at v for all N prime to p_v.*
(2)′ *The N-division points ${}_N E(K)$ are unramified at v for infinitely many N prime to p_v.*
(3) *The Tate module $T_\ell(E)$ is unramified at v for some prime ℓ unequal to p_v.*
(3)′ *The Tate module $T_\ell(E)$ is unramified at v for all primes ℓ unequal to p_v.*

In addition, the vector space $V_\ell(E) = T_\ell(E) \otimes_{\mathbb{Z}_\ell} \mathbb{Q}_\ell$ could be used in conditions (3) *and* (3)′ *in place of $T_\ell(E)$.*

This theorem suggest that a systematic study of $\mathbb{Q}_\ell$ vector spaces with a $\mathrm{Gal}(\bar{K}/K)$ action should be undertaken. This is the subject of the next sections. It was first done by Taniyama [1957] and extended considerably by Serre.

(3.4) Corollary. *Let $u : E \to E'$ be an isogeny of elliptic curves over K. For a place v of K, the curve E has good reduction at v if and only if E' has good reduction at v.*

Proof. The isogeny u is a group homomorphism, and it induces an isomorphism $V_\ell(E) \to V_\ell(E')$ for all ℓ. Hence the result follows immediately from the previous theorem.

In the previous corollary, if ℓ is prime to the degree of u, then u already induces an isomorphism $T_\ell(E) \to T_\ell(E')$ as Galois modules.

§4. Ramification Properties of ℓ-Adic Representations of Number Fields: Čebotarev's Density Theorem

We begin by summarizing the notations of (1.4) and (3.1) relative to a number field K.

(4.1) Notations. Let K be a number field, and for each place v of K, we have the valuation ring $R_{(v)}$ in K, the local field K_v with its valuation ring R_v, and the residue class field $k(v)$. The cardinality Nv of $k(v)$ is of the form p_v^a. Let L be a Galois extension of K with Galois group $\mathrm{Gal}(L/K)$, and let w be a place of L extending a place v of K. The decomposition subgroup in $\mathrm{Gal}(L/K)$ of w is denoted D_w, and the inertia subgroup is denoted by I_w. The natural epimorphism $D_w \to \mathrm{Gal}(k(w)/k(v))$ induces an isomorphism $D_w/I_w \to \mathrm{Gal}(k(w)/k(v))$. The group $\mathrm{Gal}(k(w)/k(v))$ is generated by Fr_w, where $\mathrm{Fr}_w(a) = a^{Nv}$ is the Frobenius automorphism. If $k(w)/k(v)$ is an infinite extension, then Fr_w generates Gal(k(w)/k(v)) as a topological group, i.e., the powers of Fr_w are dense.

(4.2) Definition. Let K be a number field, and let $\rho : \mathrm{Gal}(\bar{K}/K) \to \mathbb{GL}(V)$ be an ℓ-adic representation. The representation ρ is unramified at a place v of K provided for each place w of $\bar{K}$ over K we have $\rho(I_w) = 1$, i.e., the inertia group I_w acts trivially on V.

(4.3) Examples. For the one-dimensional Galois representation $\mathbb{Q}_\ell(1)(\bar{K})$, the representation is unramified at v if and only if v does not divide ℓ, i.e., $v(\ell) = 0$. For an elliptic curve E over K, the two-dimensional Galois representation $V_\ell(E)$ is unramified at v if and only if E has good reduction at v, see 14(3.3).

(4.4) Remark. Let $H = \ker(\rho)$ where $\rho : \mathrm{Gal}(\bar{K}/K) \to \mathbb{GL}(V)$ is an ℓ-adic representation. Let L be the fixed subfield of $\bar{K}$ corresponding to H. Then ρ is unramified at v if and only if the extension L/K is unramified at the place v.

(4.5) Remark. If $\rho : G \to \mathbb{GL}(V)$ is unramified at v, then $\rho|D_w : D_w \to \mathbb{GL}(V)$ induces a morphism $D_w/I_w \to \mathbb{GL}(V)$ which, composed with the inverse of the natural isomorphism from (3.1), yields a canonical ℓ-adic representation of the field

$$\rho_w : \mathrm{Gal}(k(w)/k(v)) \to \mathbb{GL}(V).$$

The canonical (possibly topological) generator Fr_w has image denoted $\mathrm{Fr}_{w,\rho}$ in $\mathbb{GL}(V)$. Two places w and w' extending v are conjugate under the Galois group, and thus $\mathrm{Fr}_{w,\rho}$ and $\mathrm{Fr}_{w',\rho}$ are conjugate in $\mathbb{GL}(V)$. Hence they have the same characteristic polynomial which depends only on v and the representation ρ. We use the notation

$$P_{v,\rho}(T) = \det(1 - \mathrm{Fr}_{w,\rho}T)$$

for this polynomial. Then $P_{v,\rho}(T)$ is in $1 + TQ_\ell[T]$ and has degree equal to the dimension of V.

(4.6) Definition. The polynomials $P_{v,\rho}(T) = \det(1-\mathrm{Fr}_{w,\rho}T)$ are called the characteristic polynomials of the ℓ-adic representation ρ of the number field K. Following Deligne, the characteristic polynomials $P_{v,\rho}(T)$ can be defined at the ramified primes v too by

$$P_{v,\rho}(T) = \det(1 - \rho_w(\mathrm{Fr}_w)T),$$

where $\rho_w : \mathrm{Gal}(k(w)/k(v)) \to \mathbb{GL}(V^{\ell_w})$ is the ℓ-adic representation constructed as above on the fixed part V^{ℓ_w} of V under the inertia subgroup.

We will see in the next few sections that these polynomials are basic for the analysis of ℓ-adic representations. For this we need two types of results, namely that there are sufficiently many unramified places and that Frobenius elements are sufficiently numberous in the Galois group. This is the case for finite Galois extensions, i.e., ℓ-adic representations with kernel of finite index.

(4.7) Remark. If an ℓ-adic representation has a kernel of finite index in $\mathrm{Gal}(\bar{K}/K)$, then there are only finitely many ramified places. Those places which are ramified in the finite extension of K correspond to the kernel of the representation by 4.4. For a general ℓ-adic representation, it is not true that all but a finite number of places are unramified, although the examples $\mathbb{Z}_\ell(1)$ and $T_\ell(E)$ are the ones that we study in detail will have this property.

The question of which elements in the Galois group are Frobenius elements leads directly to one of the basic results in algebraic number theory, the Čebotarev density theorem, which we state after some preliminary definitions. Let Σ_K denote the set of places of the number field K. For each subset X of Σ_K, we denote by $N(t, X)$ the number of $v \in X$, $Nv \le t$.

(4.8) Definition. A subset X of Σ_K has a density provided the following limit exists:

$$d = \lim_{t\to+\infty} \frac{N(t, X)}{N(t, \Sigma_K)}$$

and the value d of the limit is called the density of the subset X.

The prime number theorem says that $N(t, \Sigma_K) \sim t/\log t$ so that a subset X has a density d if and only if

$$N(t, X) = d\frac{t}{\log t} + O\left(\frac{t}{\log t}\right)$$

Further, if X and X' are two subsets with $X - X'$ and $X' - X$ both finite sets, and if one has a density, then so does the other and the two densities are equal.

(4.9) Theorem (Čebotarov Density Theorem). *Let L/K be a finite Galois extension of number fields with $G = \mathrm{Gal}(L/K)$. For each subset C of G stable under conjugation, let X_C denote the set of places v of K unramified in L such that the Frobenius element Fr_w is in C for any $w|v$. Then X_C has a density and it equals*

$$\frac{\#C}{\#G}.$$

In particular, for each element s of $\mathrm{Gal}(L/K)$, there are infinitely many unramified places w of L such that $\mathrm{Fr}_w = s$.

(4.10) Corollary. *Let L be an algebraic Galois extension of a number field K which is unramified outside a finite number of places of K. Then the Frobenius elements of the unramified places of L are dense in $\mathrm{Gal}(L/K)$.*

Proof. By (4.9), the set of Frobenius elements maps subjectively onto every finite quotient of $\mathrm{Gal}(L/K)$, and , thus, every element of $\mathrm{Gal}(L/K)$ is arbitrarily close to a Frobenius element.

For the proof of the Čebotarov density theorem, see either Serre [1968; Appendix to Chapter 1], E. Artin, *Collected Works*, or S. Lang, *Algebraic Number Theory*, 1970, p. 169, where Lang records a simple proof due to M. Deuring.

In order to illustrate the use of these density results for the Frobenius elements, we have the following result for semisimple ℓ-adic representations.

(4.11) Theorem. *Let $\rho : G \to \mathbb{GL}(V)$ and $\rho' : G \to \mathbb{GL}(V')$ be two semisimple ℓ-adic representations of a number field K. Assume that ρ and ρ' are unramified outside a finite number of places S of K, and that $P_{v,\rho}(T) = P_{v,\rho'}(T)$ for all v outside S. Then ρ and ρ' are isomorphic.*

Proof. The equality of characteristic polynomials gives $\mathrm{Tr}(\rho(\mathrm{Fr}_w)) = \mathrm{Tr}(\rho'(\mathrm{Fr}_w))$ for all $w|p$, where v is a place outside of S. Since the elements Fr_w are dense by (4.9), and since $s \mapsto \mathrm{Tr}(\rho(s))$ is continuous on G, we deduce that $\mathrm{Tr}(\rho(s)) = \mathrm{Tr}(\rho'(s))$ for all s in G. Hence by (2.7), the representations ρ and ρ' are isomorphic. This proves the theorem.

§5. Rationality Properties of Frobenius Elements in ℓ-Adic Representations: Variation of ℓ

In this section we continue with the notations of (4.1) and for an ℓ-adic representation $\rho : \mathrm{Gal}(\bar{K}/K) \rightarrow \mathbb{GL}(V)$ the notation of (4.6) for the characteristic polynomials

$$P_{v,\rho}(T) = \det(1 - \rho_w(\mathrm{Fr}_w)T).$$

(5.1) Definition. An ℓ-adic representation $\rho : \mathrm{Gal}(\bar{K}/K) \rightarrow \mathbb{GL}(V)$ of a number field K is integral (resp. rational) provided there is a finite set of places S of K such that ρ is unramified at all v outside of S and such that $P_{v,\rho}(T) \in \mathbb{Z}[T]$ (resp. $\mathbb{Q}[T]$) for those v.

(5.2) Examples. For a number field K, the one-dimensional ℓ-adic representation $\mathbb{Q}_\ell(1)(\bar{K})$ is unramified at all v not dividing ℓ. The characteristic polynomial for v unramified is

$$P_{v,\rho} = 1 - (Nv)T$$

since $\mathrm{Fr}_w(z) = z^{Nv}$ on ℓth power roots of unity, in particular on the ℓth power roots of unity for $w|v$. Observe that $P_{v,\rho}(T)$ is in $1 + T\mathbb{Z}[T]$, so that the representation is integral and is independent of ℓ in this case.

For an elliptic curve E over a number field K, the two-dimensional ℓ-adic representation $V_\ell(E)$ is unramified at all v not dividing ℓ where E has good reduction. The characteristic polynomial for v unramified is

$$P_{v,\rho}(T) = 1 - a_v T + (Nv)T^2$$

where a_v is a rational integer with $|a_v| \le 2\sqrt{Nv}$ by the Riemann hypothesis. Here $1 - a_v + Nv = P_{v,\rho}(1)$ is the number of points on the reduced curve E_v with coordinates in $k(v)$. Observe that $P_{v,\rho}(T)$ is in $1 + T\mathbb{Z}[T]$, is independent of ℓ, and for the factorization $P_{v,\rho}(T) = (1 - \alpha_v T)(1 - \alpha'_v T)$, we have $|\alpha_v|, |\alpha'_v| \le (Nv)^{1/2}$. The representation $V_\ell(E)$ is in particular integral.

The importance of the notion of rational representations lies in the fact that it is possible to compare ℓ-adic representations for different rational primes ℓ using the fact that the $P_{v,\rho}(T)$ are independent of ℓ, see Examples (5.2).

(5.3) Definition. A rational ℓ-adic representation $\rho : \mathrm{Gal}(\bar{K}/K) \rightarrow \mathbb{GL}(V)$ is compatible with a rational ℓ'-adic representation $\rho' : \mathrm{Gal}(\bar{K}/K) \rightarrow \mathbb{GL}(V')$ provided there exists a finite set of places S such that ρ and ρ' are unramified outside S and $P_{v,\rho}(T) = P_{v,\rho'}(T)$ for all v outside S.

Examples of compatible pairs of representations are contained in (5.2) by using different rational primes. Compatibility is clearly an equivalence relation.

(5.4) Proposition. *Let $\rho : \mathrm{Gal}(\bar{K}/K) \rightarrow \mathbb{GL}(V)$ be a rational ℓ-adic representation and ℓ' a rational prime. There exists at most one semisimple rational ℓ'-adic representation ℓ' compatible with ρ up to isomorphism.*

Proof. This is an immediate corollary of Theorem (4.11).

The next definition describes Examples (5.2) where there is a representation for each rational prime. For a number field K, let $[\ell]$ denote the set of places of K dividing a rational prime ℓ, i.e., the set of v with $p_v = \ell$.

(5.5) Definition. Let K be an algebraic number field, and for each rational prime ℓ let $\rho_\ell : \mathrm{Gal}(\bar{K}/K) \to \mathbb{GL}(V_\ell)$ be a rational ℓ-adic representation of K. The system (ρ_ℓ) is called compatible provided ρ_ℓ and $\rho_{\ell'}$ are compatible for any pair of primes. The system (ρ_ℓ) is called strictly compatible provided there exists a finite set S of places of K such that ρ_ℓ is unramified outside $S \cup [\ell]$, $P_{v,\rho_\ell}(T)$ has rational coefficients, and

$$P_{v,\rho_\ell}(T) = P_{v,\rho_{\ell'}}(T) \qquad \text{for all } v \text{ outside } S \cup [\ell] \cup [\ell']$$

When a system (ρ_ℓ) is strictly compatible, there is a smallest finite set S having the strict compatibility properties of the previous definition. This S is called the exceptional set of the system.

(5.6) Remark. By (4.6), the characteristic polynomials $P_{v,\rho}(T)$ can be defined even for ramified places. In (5.1), (5.3), and (5.5), it is interesting to consider systems where all $P_{v,\rho}(T)$ have rational coefficients or integral coefficients as is the case in Examples (5.2).

Observe that for a characteristic polynomial $P_{v,\rho}(T)$ over $\mathbb{Q}$, the expression $P_{v,\rho(Nv)^{-s}}$ equals $P(p_v^{-s})$ for some polynomial $P(T) \in \mathbb{Q}[T]$.

Now we come to the notion of an Euler factor, which is an expression of the form for the representation ρ at the place v

$$\frac{1}{P_{v,\rho}((Nv)^{-s})} = \frac{1}{(1 - c_1 p_v^{-s}) \dots (1 - c_m p_v^{-s})}.$$

Here m is $\dim \rho$ times a where $Nv = p_v^a$.

(5.7) Definition. Let (ρ_ℓ) be a strictly compatible family of ℓ-adic representations of a number field K with exceptional set S. The formal Dirichlet series below is the L-function of (ρ_ℓ).

$$L_\rho(s) = \prod_{v \notin S} \frac{1}{P_{v,\rho}((Nv)^{-s})} \qquad \text{where choose } \ell, v|\ell.$$

This definition can be used for a single ℓ-adic representation as well without reference to compatibility. Also, the polynomials $P_{v,\rho}(T)$ are independent of which representation from the compatible family is used to determine them.

The question of the convergence of these Euler products was discussed in Chapter 11, §6, and the conditions considered in the next section lead to convergence properties.

§6. Weight Properties of Frobenius Elements in ℓ-Adic Representations: Faltings' Finiteness Theorem

(6.1) Definition. An n-dimensional integral ℓ-adic representation $\rho : G \to \mathbb{GL}(V)$ of an algebraic number field K has weight w provided there exists a finite set S of places of K such that, for all places v outside S, the integral polynomials

$$P_{v,\rho}(T) = \det(1 - \rho_w(\mathrm{Fr}_w)) = (1 - \alpha_1 T)\ldots(1 - \alpha_n T)$$

have reciprocal roots α_j with complex absolute value $|\alpha_j| = (Nv)^{w/2}$.

There is another convention were $|\alpha_j| = (Nv)^{-w/2}$.

(6.2) Examples. The examples in (5.2) are integral and they have weights. For $\mathbb{Q}_\ell(1)(\bar{K})$ we see from $P_{v,\rho}(T) = 1 - (Nv)T$ that this one-dimensinal representation has weight 2. For $V_\ell(E)$ where E is an elliptic curve over K, we see from

$$P_{v,\rho}(T) = 1 - a_v(T) + (Nv)T^2 = (1 - \alpha_v T)(1 - \alpha'_v T)$$

with $|\alpha_v|,|\alpha'_v| = (Nv)^{1/2}$ that this representation has weight 1.

There are some finiteness properties implicit in the weight condition which can be formulated as follows.

(6.3) Assertion. For given n, q, and w, the number of integral polynomials of the form

$$P(T) = 1 + a_1 T + \cdots + a_n T^n = (1 - c_1 T)\ldots(1 - c_n T)$$

with all c_j algebraic integers having absolute value $|c_j| = q^{w/2}$ is a finite set.

The absolute value conditions on the roots translate into boundedness conditions on the integral coefficients of the polynomials $P(T)$.

(6.4) Proposition. *Let K be an algebraic number field, let S be a finite set of places of K, and let m be an integer. There exists a finite Galois extension L of K such that every extension K' of K unramified outside S and with $[K' : K] \le m$ is isomorphic to a subextension of L.*

Proof. By 8(3.5) up to isomorphism the number of K' as above is finite. Let L be the Galois closure of the composite of these extensions. This proves the proposition.

This proposition is another version of 8(3.5) used to prove the weak Mordell theorem. Now we have one more preliminary result before Falting's finiteness theorem.

(6.5) Proposition (Nakayama's Lemma). *Let E be a $\mathbb{Z}_\ell$-algebra of finite rank, and let X be a subset of E whose image in $E/\ell E$ generates $E/\ell E$ as an algebra. Then X generates E as an algebra.*

Proof. Let E' be the subalgebra of E generated by X. Then $E = E + \ell E$ so that the finitely generated E/E' satisfies $\ell(E/E') = E/E'$. Thus from the structure theorem for modules over $\mathbb{Z}_\ell$ we have $E/E' = 0$ or $E = E'$. This proves the proposition.

(6.6) Theorem (Faltings). *Let ℓ be a prime. Let S be a finite set of places in an algebraic number field K, and let n and w be natural numbers. Then the set of isomorphism classes of n-dimensional, semisimple, integral ℓ-adic representations of weight w unramified outside S is finite.*

Proof. Extend S to $S(\ell)$ by adding the finite number of places dividing ℓ. Now apply (6.4) to $S(\ell)$ and $m = \ell^{2n^2}$ to obtain a finite Galois extension L of K with $G = \mathrm{Gal}(L/K)$. Let S^* be a finite set of places v outside of $S(\ell)$ such that G is the union of all Fr_w with $w|v$ and $v \in S^*$. This is possible by (4.9).

Assertion. Let $\rho : \mathrm{Gal}(\bar{K}/K) \to \mathbb{GL}(V)$ and $\rho' : \mathrm{Gal}(\bar{K}/K) \to \mathbb{GL}(V')$ be two n-dimensional, semisimple, integral ℓ-adic representations of weight w unramified outside S. If $\mathrm{Tr}(\rho(\mathrm{Fr}_v)) = \mathrm{Tr}(\rho'(\mathrm{Fr}_v))$ for all $v \in S^*$, then ρ and ρ' are isomorphic.

Proof of the assertion. By (2.4) we can choose Galois stable lattices of V and V' and further a basis of these lattices so that we can view ρ, ρ' as morphisms $\mathrm{Gal}(\bar{K}/K) \to \mathbb{GL}_n(\mathbb{Z}_\ell)$. For $s \in G$, let $\rho^*(S) = (\rho(s), \rho'(s))$, and let E be the subalgebra over $\mathbb{Z}_\ell$ of $M_n(\mathbb{Z}_\ell)^2$ generated by the elements $\rho^*(s), s \in G$. By (4.11), it suffices to show that for each $x = (x_1, x_2) \in E$,

$$\mathrm{Tr}(x_1) = \mathrm{Tr}(x_2)$$

since the representations are semisimple. By hypothesis, we know that $\mathrm{Tr}(\rho(\mathrm{Fr}_w)) = \mathrm{Tr}(\rho'(\mathrm{Fr}_w))$ for all $w|v$ and $v \in S^*$, and thus it suffices to show that these ρ^* generate E over $\mathbb{Z}_\ell$.

Let $\bar{\rho}^*$ denote ρ^* followed by reduction $\mathbb{GL}_n(\mathbb{Z}_\ell) \to \mathbb{GL}_n(\mathbb{F}_\ell)$. If we can show that the set of $\bar{\rho}^*(\mathrm{Fr}_w)$ for all $w|v$ and $v \in S^*$ generates $E/\ell E$, then by Nakayama's lemma (6.5) we have established the assertion. For this, observe that $\bar{\rho}^*$ factors through $\mathrm{Gal}(\bar{K}/K) = \mathrm{Gal}(L/K) = G$ and all elements of $\mathrm{Gal}(L/K)$ are of the form Fr_w with $w|v$ and $v \in S^*$. Hence $\mathrm{im}(\bar{\rho}^*)$ generates $E/\ell E$ and E is generated by $\rho^*(\mathrm{Fr}_w)$ where $w|v$ and $v \in S^*$. This proves the assertion.

Finally, returning to the proof of the theorem, we see by (6.3) that there are only finitely many choices for $P_{v,\rho}(T)$ at each $v \in S^*$. Since $P_{v,\rho}(T)$ determines $\mathrm{Tr}(\rho(\mathrm{Fr}_v))$, there are by the assertion only finitely many representations ρ up to isomorphism of the type in the statement of the theorem. This proves the finiteness theorem.

(6.7) Remark. It is this finiteness theorem which Faltings uses to deduce the Šafarevič conjecture, see (7.1) or Satz 5 in Faltings [1983], from the assertion of the Tate conjecture, see (7.2) or Satz 3 and Satz 4. The Galois modules $V_\ell(A)$ for a g-dimensional abelian variety over K satisfy the hypotheses of the finiteness theorem

(6.6) with $n = 2g$ and weight $w = 1$ with the case of elliptic curves being $g = 1$. The Tate conjecture allows one to reconstruct an abelian variety up to isogeny from its V_ℓ for suitable ℓ among the representations satisfying the hypotheses of the finiteness theorem.

The other application of weight properties arises with the convergence assertion of the L-function $L_\rho(s)$ associated with (ρ_ℓ) in (5.7).

(6.8) Remark. Let (ρ_ℓ) be a compatible family of ℓ-adic representations of a number field K with exceptional set S. If all ρ_ℓ have weight w, then the L-function $L_\rho(s)$ of (5.7) converges for $\mathrm{Re}(s) > 1 + w/2$ by 11(6.6).

§7. Tate's Conjecture, Šafarevič's Theorem, and Faltings' Proof

In Corollary (3.4) we saw how the Galois modules $T_\ell(A)$ and $V_\ell(A)$ could give information about isogenies, but actually this is only the beginning. In fact, the ideas are related with the Mordell conjecture, solved by Faltings [1983], and they are part of a broader story which was considered in part in Chapter 12, §6 and Chapter 13, §8. Recall the following commutative diagram for any field k and ℓ prime to char(k):

$$\begin{array}{ccc} \mathrm{Hom}_k(A, A') \otimes \mathbb{Z}_\ell & \xrightarrow{T_\ell} & \mathrm{Hom}_{\mathrm{Gal}(k_s/k)}(T_\ell(A), T_\ell(A')) \\ \cap & & \cap \\ \mathrm{Hom}_k(A, A') \otimes \mathbb{Q}_\ell & \xrightarrow{V_\ell} & \mathrm{Hom}_{\mathrm{Gal}(k_s/k)}(V_\ell(A), V_\ell(A')) \end{array}$$

where A and A' are elliptic curves, or more generally abelian varieties, over the field k and the endomorphisms are defined over k. We know that for any field k the maps T_ℓ and V_ℓ are injective and that T_ℓ has a torsion-free cokernel by 12(6.1) and 12(6.4). For the case of abelian varieties, see also Mumford [1970, Theorem 3, p. 176].

Tate's theorem, 13(8.1) says that T_ℓ and hence V_ℓ are isomorphisms for k a finite field, and, moreover, that the action of $\mathrm{Gal}(\bar{k}/k)$ is semisimple on $V_\ell(A)$. For the case of abelian varieties, see also Tate [1966].

In Tate [1974], the following conjecture appears as Conjecture 5, p. 200, for elliptic curves and it is a natural extension of Tate's work over finite fields.

(7.1) Conjecture of Tate. For an algebraic number field K, the homomorphisms T_ℓ and V_ℓ are isomorphisms for all abelian varieties A and A' over K. The action of $\mathrm{Gal}(\bar{K}/K)$ on $V_\ell(A)$ is semisimple.

As a special case, Serre [1968] showed, for elliptic curves over a number field, that if $V_\ell(E)$ and $V_\ell(E')$ are isomorphic Galois modules and if $j(E)$ is not an integer in K, then E and E' are isogenous over K. This was proved using the Tate model for the curve.

Faltings [1983] has shown that the Tate conjecture is true not just for elliptic curves but, in fact, also for all abelian varieties over a number field. This work of

Faltings is part of his proof of the Mordell conjecture, which he uses in the proofs of his basic finiteness theorems on abelian varieties and curves of higher genus. One such result that Faltings generalizes is the following.

(7.2) Theorem (Šafarevič). *Let S be a finite set of places of a number field K. The number of isomorphism classes of elliptic curves over K with good reduction at all v outside S is finite.*

We give an argument due to Tate which reduces the result to Siegel's theorem on the finiteness of integral points on curves which was mentioned in the Introduction. Faltings proves two generalizations to abelian varieties, one of which was conjectured by Šafarevič on the finiteness of isogeny classes of abelian varieties with good reduction outside a finite set S.

Returning to the reduction of the assertion of (7.2) to Siegel's theorem, we first extend S if necessary to include all places dividing 2 and 3 and so that R, the ring of S-integers in K, is principal. Recall that c in K is an element of R if and only if $v(c) = 0$ for all v outside S. For each E over K we can, by (1.3), choose a minimal model of the form

$$y^2 = 4x^3 - ux - v$$

with $\Delta_E = u^3 - 27v^2$ in R^*, the group of units in the ring R. If E has good reduction outside S, then $v(\Delta_E) = 0$ for all places outside of S. Now, for a given E, the class Δ_E in $R^*/(R^*)^{12}$ is well defined by the isomorphism class of E from 3(3.2). This group $R^*/(R^*)^{12}$ is finite by the Dirichlet unit theorem for number fields. Thus there is a finite set D in R^* such that any curve E over K with good reduction outside of S can be written in the form $y^2 = 4x^3 - ux - v$ with $d = u^3 - 27v^2$ in D.

For a given d, the affine elliptic curve

$$27v^2 = u^3 - d$$

has only a finite number of integral points (u, v) on it, i.e., points (u, v) in R^2. This is a generalization of a theorem of Siegel, see Lang [1962, Chapter VII].

We close with the statements of Faltings' two generalizations of (7.2) to abelian varieties. For the proofs, see his basic paper *Endlichkeitssätze für abelsche Varietäten über Zahlkörpern* [Faltings, 1983].

(7.3) Theorem (Satz 5). *Let S be a finite set of places in an algebraic number field K. In a given dimension, there are only finitely many isogeny classes of abelian varieties over K with good reduction outside S.*

(7.4) Theorem (Satz 6). *Let S be a finite set of places in an algebraic number field K and $d > 0$. In a given dimension n there are only finitely many isomorphism classes of d-fold polarized abelian varieties over K with good reduction outside S.*

§8. Image of ℓ-Adic Representations of Elliptic Curves: Serre's Open Image Theorem

In this section, let E denote an elliptic curve defined over a number field K. The action of $G = \text{Gal}(\bar{K}/K)$ on the N-division points of E yields a homomorphism

$$\Phi_n : G = \text{Gal}(\bar{K}/K) \to \mathbb{GL}({}_N E) \cong \text{GL}_2(\mathbb{Z}/N\mathbb{Z}).$$

and the inverse limit over all N is

$$\rho : G = \text{Gal}(\bar{K}/K) \to \varprojlim_N \mathbb{GL}({}_N E) = \mathbb{GL}(\widehat{\text{Tors}}\, E) \cong \prod_\ell \mathbb{GL}(T_\ell(E)).$$

This yields the following diagram of projections and quotients of ρ:

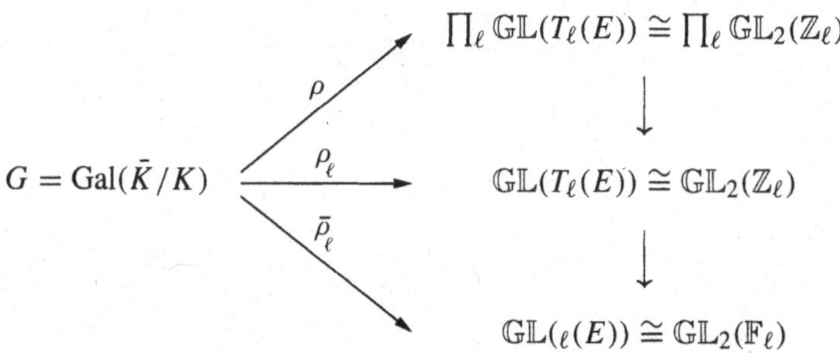

The main theorem of Serre is that $\text{im}(\rho)$ is an open subgroup of the product group when E does not have complex multiplication. Before discussing this result, we have the following remark on the contrasting case when E has complex multiplication.

(8.1) Remark. If the elliptic curve E has complex multiplication by an imaginary quadratic number field F, then F acts on each $V_\ell(E)$ and embeds as a subfield of $\text{End}(V_\ell(E))$ which commutes with the action of $\text{Gal}(\bar{K}/K)$. Hence $\text{im}(\rho_\ell)$ commutes with every element of $F^* \subset \mathbb{GL}(V_\ell(E)) = \mathbb{GL}_2(\mathbb{Q}_\ell)$. Since the image of F^* contains nonscalar elements, $\text{im}(\rho_\ell)$ will not be open in $\mathbb{GL}(T_\ell(E)) = \mathbb{GL}_2(\mathbb{Z}_\ell)$ and, hence, $\text{im}(\rho)$ will not be open in the product group. In the next chapter, we will study this action of the complex multiplication and its relation to the Galois action further.

The following theorem is the main result in the book by Serre [1968] and the *Inventiones* paper by Serre [1972].

(8.2) Theorem (Serre). *Let E be an elliptic curve without complex multiplication defined over a number field K. Then the image of $\rho : \text{Gal}(\bar{K}/K) \to \prod_\ell \mathbb{GL}(T_\ell(E))$ is an open subgroup.*

The proof of this theorem would take us beyond the scope of this book, and, in any event, it is difficult to improve on the exposition and methods of Serre. We will be content to simply observe that the following result from the book, see p. IV–19, is the criterion used by Serre to show the image is open.

(8.3) Proposition. *Let G be a closed subgroup of $\prod_\ell \mathbb{GL}_2(\mathbb{Z}_\ell)$ and denote by G_ℓ its projection to $\mathbb{GL}_2(\mathbb{Z}_\ell)$ and the reduction of G_ℓ into $\mathbb{GL}_2(\mathbb{F}_\ell)$ by $\bar{G}_\ell$. Then G is open in the product $\prod_\ell \mathbb{GL}_2(\mathbb{Z}_\ell)$ if the following three conditions hold:*

(a) *For all ℓ, the subgroup G_ℓ is open in $\mathbb{GL}_2(\mathbb{Z}_\ell)$.*
(b) *The image of G under* $\det : \prod_\ell \mathbb{GL}_2(\mathbb{Z}_\ell) \to \prod_\ell \mathbb{Z}_\ell^*$ *is open.*
(c) *For all but a finite number of ℓ, the subgroup G_ℓ contains the subgroup $\mathbb{SL}_2(\mathbb{Z}_\ell)$ of $\mathbb{GL}_2(\mathbb{Z}_\ell)$.*

16

L-Function of an Elliptic Curve and Its Analytic Continuation

We introduce the L-function of an elliptic curve E over a number field and derive its elementary convergence properties. L-functions of this type were first introduced by Hasse, and the concept was greatly extended by Weil. For this reason, they are frequently called the Hasse–Weil L-function.

There are two sets of conjectural properties of the L-function with the second depending on the first. The first conjecture, due to Hasse and Weil, asserts that the L-function L_E, defined as a Dirichlet series for $\mathbf{Re}(s) > 3/2$ has an analytic continuation to the complex plane and satisfies a functional equation under reflection s to $2 - s$. The second conjecture, due to Birch and Swinneton-Dyer, ties up the arithmetic of the curve E with the behavior of L_E at $s = 1$ which is the fixed point under s to $2 - s$. This conjecture clearly depends on the first and is discussed in the next chapter.

The Hasse–Weil conjecture has been verified for two general classes of elliptic curves. First, the L-function of an elliptic curve with complex multiplication is related to the L-function of Hecke Grossencharacters by a theorem of Deuring. In this case, the conjecture is deduced from the analytic continuation and functional equation of Hecke L-functions. Second, if an elliptic curve E over the rational numbers is the image of a map of curves $X_0(N) \rightarrow E$, then the L-function L_E is the Mellin transform of a modular form of weight 2 for $\Gamma_0(N)$. In this case, the conjecture follows from the functional equation of the modular form. The general case is considered in §8 and Chapter 18.

§1. Remarks on Analytic Methods in Arithmetic

In Chapters 9, 10, and 11, analytic methods were considered to describe the complex points on an elliptic curve. This led to elliptic functions, theta functions, and in order to describe all elliptic curves up to isomorphism, also modular functions.

In these last two chapters, we sketch some of the flavor of the use of analytic methods in the arithmetic of elliptic curves, that is, for elliptic curves over number fields. The idea is to assemble the modulo p information about the elliptic curve into an Euler factor and form the product of these Euler factors into a Dirichlet series. The study of Euler products was started in Chapter 11 where they arose in connection with modular forms.

A remarkable feature of the theory is that certain of these L-functions associated to modular forms by the Mellin transform, and which have Euler product decompositions, also arise as the L-functions of elliptic curves, which are defined as Euler products from the local arithmetic data of the elliptic curve defined over the rational numbers. This is the theory of Eichler–Shimura.

§2. Zeta Functions of Curves over $\mathbb{Q}$

In Chapter 13, §2, we discussed the zeta function ζ_C of a smooth curve over F_q. It took two forms, (2.2) and (2.5) respectively

$$\zeta_C(s) = \exp\left(\sum_{i\le m} \frac{N_m}{m} q^{-ms}\right),$$

where $N_m = \#C(F_{q^m})$ the number of points of C over F_{q^m}, and

$$\zeta_C(s) = \frac{P(q^{-s})}{(1-q^{-s})(1-q^{1-s})},$$

where $P(T) = (1-\alpha_1 T)\cdots(1-\alpha_{2g}T)$ and the reciprocal roots $\alpha_1, \ldots, \alpha_{2g}$ are paired off in such a way that $\overline{\alpha}_{2j-1} = \alpha_{2j}$ so that $|\alpha_i| = \sqrt{q}$ and $\alpha_{2j-1}\alpha_{2j} = q$. The numbers of points N_m and the reciprocal roots, which are all algebraic integers, are related by

$$N_m = q^m + 1 - \sum_{i=1}^{2q} \alpha_i^m.$$

The number g is the genus of C.

Since $q = p^a$, the terms $(1-q^{-s})$, $(1-q\cdot q^{-s})$ and $P(q^{-s})$ are the kinds of expressions used to make Euler products provided we are given these expressions for all primes. This comes up in the study of smooth curves C over a number field of genus g. Roughly, as in Chapter 5, we wish to choose equations of C over $\mathbb{Z}$ in such a way that the reduction C_p modulo p of C is a smooth curve of genus g for all primes p not in a finite set S. The curve C is said to have good reduction at $p \notin S$. More precisely, we extend C to a smooth scheme over $\mathbb{Z}[1/n]$ where $n = \prod_{p\notin S} p$. Then the curves C_p are the special fibres. As with elliptic curves, good reduction can be described in terms of a projective model over $\mathbb{Q}$.

(2.1) Definition. Let C be a curve over $\mathbb{Q}$ extending to a smooth scheme over $\mathbb{Z}[1/n]$ such that the reduction C_p (mod p) is smooth over F_p for $p \notin S$, where S is the set of prime divisors of n. The crude Hasse–Weil zeta function of C is

$$\zeta_C^* = \prod_{p \notin S} \zeta_{C_p}(s) = \prod_{p \notin S} \frac{P_p(p^{-s})}{(1-p^{-s})(1-p^{1-s})}$$

and the crude Hasse–Weil L-function of C is

$$L_C^*(s) = \prod_{p \notin S} P_p(p^{-s})^{-1}.$$

We know what the reduction mod p of an elliptic curve is in terms of the normal form. For a general curve C, one needs scheme theoretical techniques. For C over $\mathbb{Q}$, it has good reduction outside a finite set S.

The Hasse–Weil zeta and L-functions are the analytic objects which store the diophantine data of C related to congruences modulo p. In this definition, we have referred to the crude zeta and L-functions because the zeta function and L-function will have factors associated with $p \in S$. In this form, we can make some elementary assertions about convergence which also apply to the more precise versions of these functions.

(2.2) Remark. For each finite set of primes S, we can consider a modified Riemann zeta function $\zeta_S(s)$ defined by

$$\zeta_S(s) = \prod_{p \notin S} \frac{1}{(1-p^{-s})} = \prod_{p \in S}(1-p^{-s}) \cdot \zeta(s)$$
$$= \sum_{1 \le n,\ (S,n)=1} \frac{1}{n^s},$$

where $\zeta(s)$ is the ordinary Riemann zeta function corresponding to empty S. In terms of the modified zeta function ζ_S, we have the relation

$$L_C^*(s) \cdot \zeta_C^*(s) = \zeta_S(s) \cdot \zeta_S(s-1).$$

Since the Riemann zeta function, and therefore also the modified versions, are fairly well understood relative to convergence and functional equation, we consider the elementary study of the Hasse–Weil zeta function $\zeta_C^*(s)$ and L-function $L_C^*(s)$ as effectively equivalent. The L-function $L_C^*(s)$ is an Euler product

$$\prod_p \frac{1}{(1-\alpha_1(p)p^{-s}) \cdots (1-\alpha_{2g}(p)p^{-s})},$$

where we take all $\alpha_j(p) = 0$ for $p \in S$, $j = 1, \ldots, 2g$. For an elliptic curve E over $\mathbb{Q}$ as in the previous definition,

$$L_E^*(s) = \prod_p \frac{1}{(1-\alpha(p)p^{-s})(1-\beta(p)p^{-s})}$$

since $g = 1$. Now the considerations of 13(1.7) apply because the Riemann hypothesis asserts that

$$|\alpha(p)| = |\beta(p)| \le \sqrt{p}$$

holds for elliptic curves, 13(1.2) and 13(1.6) by Hasse, and also for general smooth curves

$$|\alpha_j(p)| \le \sqrt{p}$$

by Weil, see 13(2.5). Hence we have the following convergence result as a direct application of 11(6.6).

(2.3) Proposition. *The Hasse–Weil L-function $L_C^*(s)$ for a curve C over $\mathbb{Q}$ converges for* $\mathbf{Re}(s) > 1 + 1/2 = 3/2$.

(2.4) Remark. When $\mathbb{Q}$ is replaced by a number field K, we consider products over all but a finite set S of valuations (or places) v of K, and the finite field F_p is replaced by F_{q_v} where q_v is the number of elements in the residue class field of v. The crude Hasse–Weil L-function takes the form

$$L_C^*(s) = \prod_v \frac{1}{(1 - \alpha_1(p)q_v^{-s}) \ldots (1 - \alpha_{2g}(p)q_v^{-s})},$$

and there is a similar expression for the Hasse–Weil zeta function. This is an Euler product of degree $2g \cdot [K : \mathbb{Q}]$ since all v with $q_v = p^{n(v)}$ for given p combine to give the p-Euler factor in the product. Thus by the Riemann hypothesis, 13(1.2) and 13(2.5), we see that 11(6.6) applies to show that 2.3 is true for curves over a number field.

(2.5) Remark. For a discussion of the proof by Deligne of the Riemann hypothesis for smooth projective varieties, together with references, see Katz [1976]. In the last section of his paper, Katz considers the Hasse–Weil zeta function for varieties.

§3. Hasse–Weil *L*-Function and the Functional Equation

Now we concentrate on the case of an elliptic curve E over a number field k. For many considerations k can be taken to be the rational numbers. We know that for all valuations v of k such that $v(\Delta_E) = 0$, the curve E has an elliptic curve E_v for reduction over $k(v)$, the residue class field of v. In the previous section, we considered a quadratic polynomial $f_v(T)$ that we defined using E_v, giving an Euler factor in the L-function $L_E^*(s)$ of E. Next we introduce the Euler factors for the other primes v where E has bad reduction.

Let E_v denote the reduction of E at any v, a curve over $k(v) = F_{q_v}$, let $E_{v,0}(k(v))$ be the group of nonsingular points, and let $N_v = \#E_{v,0}(k(v))$ be the number of points in this group curve.

(3.1) Notations. For an elliptic curve E over a number field k and a non-Archimedean valuation v of k, we denote by $f_v(T)$ the polynomial:

(1) $f_v(T) = 1 - a_v T + q_v T^2$, where $a_v = q_v + 1 - N_v$, if E has good reduction at v.
(2) $f_v(T) = 1 - e_v T$, where

$$e_v = \begin{cases} -1 & \text{If } E \text{ has split multiplicative reduction at } v, \\ +1 & \text{if } E \text{ has nonsplit multiplicative reduction at } v, \\ 0 & \text{if } E \text{ has additive reduction at } v. \end{cases}$$

Observe that in all cases

$$N_v = q_v f_v \left(\frac{1}{q_v} \right).$$

The complete non-Archimedean part of the L-function is $\prod_v 1/L_v(s)$, where $L_v(s) = f_v(q_v^{-s})$.

(3.2) Definition. Let E/K be an elliptic curve over a number field K. The Hasse–Weil L-function of E over K is

$$L_{E/K}(s) = \prod_v \frac{1}{L_v(s)},$$

where the local factor $L_v(s)$ is given in the previous section (3.1). Other notations for $L_{E/K}(s)$ in current use are $L_E(s)$, $L(s, E/K)$, and $L(E/K, s)$.

The question of the local factors at infinity is taken up by J.-P. Serre in *Facteurs locaux des Fonctions Zêta des Variétés algébriques: Séminaire Delange-Pisot-Poitou, 11 mai 1970*. Serre in this 1970 DPP seminar gives a general definition for any smooth projective variety over a global field.

(3.3) Definition. Let E/K be an elliptic curve over a number field K. The modified Hasse–Weil L-function of E over K is

$$\Lambda_{E/K}(s) = A^{s/2}\Gamma_K(s)L_{E/K}(s),$$

where $A = A_{E/K}$ and $\Gamma_K(s)$ are defined as follows:

(1) The constant $A_{E/K}$ is given by

$$A_{E/K} = N_{K/\mathbb{Q}}(f_{E/K}) \cdot d_{K/\mathbb{Q}}^2,$$

where $d_{K/\mathbb{Q}}$ is the absolute discriminant of K and $f_{E/K}$ is the conductor of E over K, see 14(2.4) and Serre, DPP, pp. 19–12.

(2) The gamma factor for the field K where $n = [K : \mathbb{Q}]$ is

$$\Gamma_K(s) = [(2\pi)^{-s}\Gamma(s)]^n.$$

The functions $L_{E/K}(s)$ and $\Lambda_{E/K}(s)$ are defined as Dirichlet series and Dirichlet series multiplied with elementary factors which converge to holomorphic functions on the half plane

$$\mathbf{Re}(s) > \frac{3}{2}.$$

This leads to one of the main conjectures of the subject.

(3.4) Conjecture (Hasse–Weil). *Let E be an elliptic curve over a number field K. The modified Hasse–Weil L-function $\Lambda_{E/K}(s)$ has an analytic continuation to the entire complex plane as an analytic function, and it satisfies the functional equation*

$$\Lambda_{E/K}(2-s) = w\Lambda_{E/K}(s)$$

with $w = \pm 1$.

If $\Lambda_{E/K}$ has an analytic continuation to the complex plane, then so will $L_{E/K}(s)$ and further from the gamma factor $\Gamma(s)^n$ and its poles we see that $L_{E/K}(s)$ will have zeros of order at least n at all the negative integers. The order of the possible zero of $\Lambda_{E/K}(s)$ at $s = 1$ is related to the sign $w = 1$ or -1 in the functional equation. If $w = -1$, then $\Lambda_{E/K}(1) = -\Lambda_{E/K}(1)$ and $\Lambda_{E/K}(s)$ has a zero at $s = 1$. In fact, the parity of $\mathrm{ord}_{s=1}\Lambda_{E/K}(s)$ is even for $w = 1$ and odd for $w = -1$.

(3.5) Remark. Elliptic curves over $\mathbb{Q}$ will be of special interest, and in the case of $E/\mathbb{Q}$, the modified Hasse–Weil L-function becomes

$$\Lambda_E(s) = N^{s/2}(2\pi)^{-s}\Gamma(s)L_E(s).$$

This conjecture has been established in two cases. First, if E/K has complex multiplication, then the Hasse–Weil L-function is related to the L-functions of type $L(s, \chi)$, where χ is a Hecke Grossencharacter. The functional equation for L-functions with Hecke Grossencharacters implies the functional equation for elliptic curves. The object of this chapter is primarily to explain this result. Second, if E is an elliptic curve over the rational numbers which is the image of a certain modular curve related to its conductor, then $L_E(s)$ is the Mellin transform of a modular from and the functional equation follows from this fact. These topics are taken up in the last two sections.

In the next section we review the functional equation of the classical zeta and L-functions and lead into the theory of algebraic Hecke Grössencharakters, see Hecke, *Mathematische Werke*, No. 14. We define algebraic Hecke characters as extensions of algebraic group characters over $\mathbb{Q}$ to algebraic valued characters on an idèle class group. The Grössencharakters in the sense of Hecke, which we are interested in, are Archimedean modifications defined on the idèle class group. From this perspective, there are also ℓ-adic modifications related to ℓ-adic representations and the Hecke characters coming from elliptic curves with complex multiplication.

§4. Classical Abelian L-Functions and Their Functional Equations

The classical zeta and L-functions were defined and studied by Riemann, Dirichlet, Dedekind, and Hecke. The Dedekind zeta function of an algebraic number field is the generalization of the Riemann zeta function for the rational numbers, and Hecke L-functions for a number field are generalizations of the Dirichlet L-functions of characters on the rational integers. We give the definition of each of these functions as a Dirichlet series, give the Euler product expansion, and state the functional equation. Of special interest is the Hecke L-function with "Grossencharakter" for it is this L-function which is related to the Hasse–Weil L-function of an elliptic curve with complex multiplication.

(4.1) Riemann Zeta Function. The Dirichlet series expansion of the Riemann zeta function is for $\mathbf{Re}(s) > 1$

$$\zeta(s) = \sum_{1\le n} \frac{1}{n^s}.$$

The Euler product, which is the analytic statement that natural numbers have a unique factorization into primes, is given by

$$\zeta(s) = \prod_p \frac{1}{1-p^{-s}} \quad \text{for} \quad \mathbf{Re}(s) > 1.$$

The Riemann zeta function has a meromorphic continuation to the entire complex plane, and it satisfies a functional equation which is most easily described by introducing the function $\xi(s) = \pi^{-s/2}\Gamma(s/2)\zeta(s)$ related to the zeta function. The functional equation for the zeta function is simply

$$\xi(s) = \xi(1-s).$$

One proof of the functional equation results from using the elementary theta function $\theta(t) = \sum_n e^{-\pi n^2 t}$ and the integral for the Γ-function . Then

$$\xi(s) = \int_0^\infty x^{x/2}\theta(x)\frac{dx}{x} = \frac{1}{s} + \frac{1}{1-s} + \int_1^\infty \left(x^{(1-s)/2} + x^{s/2}\right)\theta(x)\frac{dx}{x}.$$

This follows by dividing the first integral into two integrals at $x = 1$. The above representation holds because the theta function satisfies $\theta(1/t) = \sqrt{t}\theta(t)$, which is a modular type relation. The subject of modular forms was considered in Chapter 11. The theta-function relation is proved by applying the Poisson summation formula to $e^{-\pi t x^2}$ and using the fact that $e^{-\pi x^2}$ is its own Fourier transform.

In the rational integers $\mathbb{Z}$, each nonzero ideal $\mathfrak{a}$ is principal $\mathfrak{a} = (n)$ where $n \ge 1$, the absolute norm $N\mathfrak{a} = n$, and $\mathfrak{a}$ is a prime ideal if and only if $\mathfrak{a} = (p)$ for a prime number p. These remarks lead to Dedekind's generalization of $\zeta(s)$ to a number field K using the ideals $\mathfrak{a}$ in $\mathcal{O}$, the ring of algebraic integers in K, and the absolute norm $N\mathfrak{a} = \#(\mathcal{O}/\mathfrak{a})$, the cardinality of the finite ring $\mathcal{O}/\mathfrak{a}$.

(4.2) Dedekind Zeta Function. The Dirichlet series expansion of the Dedekind zeta function for the number field K is

$$\zeta_K(s) = \sum_{\mathfrak{a} \neq 0} \frac{1}{N\mathfrak{a}^s} \quad \text{for } \mathbf{Re}(s) > 1.$$

If the degree $[K : \mathbb{Q}] = n$, then there are at most n ideals $\mathfrak{a}$ with $N\mathfrak{a}$ equal to a given prime p, and this is the reason why the Dirichlet series for $\zeta_K(s)$ has the same half plane of convergence os the Riemann zeta function. The Euler product, which is the analytic statement that ideals in the ring of integers in a number field have unique factorization into prime ideals, is given by

$$\zeta_K(s) = \sum_{\mathfrak{p}} \frac{1}{N\mathfrak{p}^s} \quad \text{for } \mathbf{Re}(s) > 1.$$

The Dedekind zeta function has a meromorphic continuation to the entire complex plane, and it satisfies a functional equation. For the functional equation we need $d = d_K$ the absolute discriminant of K and the decomposition of the degree $[K : \mathbb{Q}] = n = r_1 + 2r_2$, where r_1 is the number of real places of K and r_2 is the number of conjugate pairs of complex places of K. In 1917, Hecke proved for

$$\xi_K(s) = \left(\frac{\sqrt{|d|}}{2^{r_2}\pi^{n/2}}\right)^s \Gamma\left(\frac{s}{2}\right)^{r_1} \Gamma(s)^{r_2} \xi_K(s)$$

the functional equation

$$\xi_K(s) = \xi_K(1-s).$$

The zeta functions contain information about the distribution of prime numbers and prime ideals. In an L-function this information is further organized by weighting the primes via character values. On a finite abelian group, a character is just a homomorphism into $\mathbb{C}^*$. If χ is a character of the finite abelian group $(\mathbb{Z}/m\mathbb{Z})^*$ of units in the ring $\mathbb{Z}/m\mathbb{Z}$, then we extend the complex valued function χ to be zero on $\mathbb{Z}/m\mathbb{Z}$ outside of $(\mathbb{Z}/m\mathbb{Z})^*$, and we compose it with the quotient map $\mathbb{Z} \to \mathbb{Z}/m\mathbb{Z}$ to define a function on $\mathbb{Z}$ also denoted χ. On $\mathbb{Z}$ we have $\chi(nn') = \chi(n)\chi(n')$ and $\chi(n) = 0$ if and only if the greatest common divisor $(n, m) > 1$.

(4.3) Dirichlet L-Function. The Dirichlet series expansion of the Dirichlet L-function associated with a character $\chi \mod m$ is

$$L(s,\chi) = \sum_{1 \leq n} \frac{\chi(n)}{n^s}.$$

The Euler product is given by

$$L(s,\chi) = \prod_p \frac{1}{1 - \chi(p)p^{-s}}$$

and both the Dirichlet series and the Euler product converge absolutely for $\mathbf{Re}(s) > 1$. For the trivial character $\chi = 1$, $L(s, 1) = \zeta(s)$ the Riemann zeta function.

As with the zeta function, there is a functional equation for the Dirichlet L-function, but for this we need the notion of the conductor of a character. It is a natural number.

(4.4) Definition. For a character χ modulo m, consider factorizations of $\chi = \chi'\pi$ where

$$(\mathbb{Z}/m\mathbb{Z})^* \xrightarrow{\pi} (\mathbb{Z}/f\mathbb{Z})^* \xrightarrow{\chi'} \mathbb{C}^*.$$

f divides m, χ' is a character modulo f, and π is reduction modulo f. The character χ is imprimitive provided such a factorization exists with f unequal to m, otherwise χ is called primitive. The conductor f of χ is the unique natural number f such that there is a factorization $\chi = \chi'\pi$ and χ' is primitive.

In particular, if χ is a primitive character modulo m, then m is the conductor of χ. If $\chi = \chi'\pi$ as above, then the L-functions are related by

$$L(s, \chi') = L(s, \chi) \prod_{p|m} \left(1 - \frac{\chi'(p)}{p^s}\right)$$

since $L(s, \chi')$ and $L(s, \chi)$ have the same Euler factors at all p not dividing m. Hence $L(s, \chi') = L(s, \chi)$ if and only if every prime dividing m also divides f.

(4.5) Functional Equation of the Dirichlet L-Function. Let χ be a primitive character with conductor f, and let

$$\Lambda(s, \chi) = (f/\pi)^{s/2}\Gamma\left(\frac{s+a}{2}\right) L(s, \chi),$$

where $a = 0$ for $\chi(-1) = 1$ and $a = 1$ for $\chi(-1) = -1$. Then $\Lambda(x, \chi)$ and $L(s, \chi)$ have meromorphic continuations to the entire plane $\mathbb{C}$ with $L(s, \chi)$ holomorphic on $\mathbb{C}$ for $f > 1$. Furthermore $\Lambda(s, \chi)$ satisfies a functional equation

$$\Lambda(s, \chi) = W(\chi)\Lambda(1 - s, \overline{\chi}),$$

where $W(\chi) = (-1)^a[g(\chi)/\sqrt{f}]$ and $g(\chi) = \sum_{n \bmod f} \chi(n)\exp(2\pi i/f)^n$, a Gauss sum. We also have $|W(\chi)| = 1$.

(4.6) Hecke L-functions. Let K be a number field. The Hecke L-function $L_K(s, \chi)$ for K and a character combine a sum over nonzero ideals of $N\mathfrak{a}^{-s}$, as in the Dedekind zeta function, with character values on ideals $\chi(\mathfrak{a})$ as in the Dirichlet L-functions. In all cases they have the form

$$L_K(s, \chi) = \sum_{\mathfrak{a} \neq 0} \frac{\chi(\mathfrak{a})}{N\mathfrak{a}^s} = \prod_{\mathfrak{p}} \frac{1}{1 - \chi(\mathfrak{p})N\mathfrak{p}^s}.$$

As in the case of a Dirichlet character , the character χ in a Hecke L-function has a modulus $\mathfrak{m}$ where χ is defined modulo $\mathfrak{m}$ and $\chi(\mathfrak{a}) = 0$ for $(\mathfrak{a}, \mathfrak{m}) \neq 1$. A primitive character is one which cannot be factored nontrivially as in (4.4). The modulus is denoted by $\mathfrak{f}$ and called the conductor in this case.

(4.7) Functional Equation of the Hecke L-Function. Let K be a number field with discriminant d and degree $n = [K : \mathbb{Q}] = r_1 + 2r_2$ with real monomorphisms $u_1, \ldots, u_{r_1} : K \to \mathbb{R}$. Let χ be a primitive character with conductor $\mathfrak{f}$ and $N(\mathfrak{f}) = f$. Let

$$\Lambda_K(s, \chi) = A_K(\chi)^s \Gamma_K(s, \chi) L_K(s, \chi),$$

where $A_K(\chi) = \sqrt{|d| f}/2^{r_2}\pi^{n/2}$ and the gamma factor has the form

$$\Gamma_K(s, \chi) = \prod_{j=1}^{r_1} \Gamma\left(\frac{s + a(j)}{2}\right) \Gamma(s)^{r_2}$$

for numbers $a(1), \ldots, a(r_1)$ equal to 0 or 1 such that

$$\chi((x)) = (\mathrm{sgn} u_1(x))^{a(1)} \ldots (\mathrm{sgn} u_{r'}(x))^{a(r_1)}$$

for any $x \equiv 1 \pmod{f}$. Then $\Lambda_K(s, \chi)$ and $L_K(s, \chi)$ have a meromorphic continuation to the entire plane $\mathbb{C}$ with $L_K(s, \chi)$ holomorphic on $\mathbb{C}$ for $\chi \neq 1$. Furthermore, $\Lambda_K(s, \chi)$ satisfies a functional equation

$$\Lambda_K(s, \chi) = W(\chi) \Lambda_K(1 - s, \overline{\chi}),$$

where $W(\chi)$ is given by Gauss sums and has absolute value 1.

§5. Grössencharacters and Hecke L-Functions

Hecke [1918, 1920] extended the notion of characters which give rise to L-functions, and he called these Grössencharacters. Later, Iwasawa and Tate reformulated the notion of Grössencharacter, and, using Fourier analysis on locally compact groups, they derived the functional equation from the Poisson summation formula. Weil [1947] discussed Grössencharacters with algebraic values. These characters, also referred to as type (A_0) Grössencharacters, are the ones with an arithmetic meaning in terms of infinite abelian extensions of a number field, and, for us, in terms of elliptic curves with complex multiplication. We introduce a version of these which are called algebraic Grössencharacters.

(5.1) Definition. The idèle group of a number field K is an inductive limit over finite subsets S of places of K, that is

$$I_K = \varinjlim_S \prod_{v \notin S} \mathcal{O}_v^* \times \prod_{v \in S} K_v^*.$$

In this discussion, $\mathcal{O}_v = K_v$ for all Archimedean places v. An element of I_K has the form (a_v), where a_v is a unit in $\mathcal{O}_v$ for all but a finite set of v. An idèle $(a_v) \in I_K$ is principal provided there exists $a \in K^*$ with $a_v = a$ for all v. We decompose $I_K = I_K^f \times I_K^\infty$ where $I_K^\infty = \prod_{v \text{ arch.}} K_v^*$ and is the group I_K^f of finite idèles formed from the finite places as with I_K.

We embed $K^* \subset I_K$ as the subgroup of principal idèles. It is a discrete subgroup. Grössencharacters are certain continuous homomorphisms $I_K/K^* \to \mathbb{C}^*$. In order to relate Grössencharacters to the characters considered in previous sections, we need some definitions.

(5.2) Definition. Let K and F be two extension fields of k. An algebraic group homomorphism $\beta : K^* \to F^*$ over k is a function which assigns to each algebra A over k a group homomorphism $\beta_A : (K \otimes_k A)^* \to (F \otimes_k A)^*$ which is functorial in A, that is, such that for all morphisms $f : A \to A'$ of algebras over f the following diagram is commutative:

$$\begin{array}{ccc} (K \otimes_k A)^* & \xrightarrow{\beta_A} & (F \otimes_k A)^* \\ {\scriptstyle (K\otimes_k f)^*}\downarrow & & \downarrow{\scriptstyle (F\otimes_k f)^*} \\ (K \otimes_k A')^* & \xrightarrow{\beta_{A'}} & (F \otimes_k A')^* \end{array}$$

(5.3) Remarks. When K and F are algebraic extensions of k, the algebraic group homomorphisms $\beta : K^* \to F^*$ can be described as a group homomorphism β such that for each pair of embeddings $\varphi : F \to \bar{k}$, an algebraic closure of k, and for each embedding $\sigma : K \to \bar{k}$, there exists an integer $n(\sigma, \varphi)$ with

$$\varphi\beta(x) = \prod \sigma(x)^{n(\sigma,\varphi)} \quad \text{for all} \quad x \in K^*$$

where the product is taken over all embeddings $\sigma : K \to \bar{k}$. The fact that β is defined over k is the extra condition that $n(\tau\sigma, \tau\varphi) = n(\sigma, \varphi)$ for all automorphisms $\tau \in \mathrm{Gal}(\bar{k}/k)$.

(5.4) Examples. The zero (or one) homomorphism $K^* \to F^*$ is algebraic where all integers $n(\sigma, \varphi) = 0$. In the case where K is a finite extension of F, the norm $N_{K/F} : K^* \to F^*$ is algebraic with all $n(\sigma, \varphi) = 1$.

(5.5) Notations. For two algebraic number fields K and F, we consider algebraic group homomorphisms $\beta : K^* \to F^*$ over $\mathbb{Q}$. For A, as in (5.2), equal to $\mathbb{Q}$, $\mathbb{Q}_\ell$, or $\mathbb{R}$, we denote $K_\ell = \mathbb{Q}_\ell \otimes_{\mathbb{Q}} K$, $F_\ell = \mathbb{Q}_\ell \otimes_{\mathbb{Q}} F$, $K_\infty = \mathbb{R} \otimes_{\mathbb{Q}} K$, and $F_\infty = \mathbb{R} \otimes_{\mathbb{Q}} F$, and the corresponding group homomorphisms as $\beta_0 : K^* \to F^*$, $\beta_\ell : K_\ell^* \to F_\ell^*$, and $\beta_\infty : K_\infty^* \to F_\infty^*$. We can embed $K_\ell^* \subset I_K$ and $K_\infty^* \subset I_K$, and, in fact, $K_\infty^* = I_K^\infty$. The subgroup of finite idèles I_K^f is contained in $\prod_\ell K_\ell^*$ where the product is taken over all rational primes ℓ. For an element $x \in I_K$, we denote by x_ℓ the ℓ-component of x and by x_∞ the ∞-component of x.

(5.6) Definition. Let $\beta : K^* \to F^*$ be an algebraic group homomorphism over $\mathbb{Q}$ of number fields. An algebraic Hecke character of type β is a homomorphism of groups $\psi : I_K \to F^*$ such that the restriction $\psi|K^* = \beta_0$ and ψ is continuous in the sense that $\ker(\psi)$ is an open subgroup of I_K.

Let H be an open subgroup of I_K, for example $H = \ker(\psi)$ for an algebraic Hecke character ψ. Then $H_\infty = H \cap I_K^\infty$ is an open subgroup of I_K of index 2^r, where r is less than r_1, the number of real places of K, and $H_v = H \cap K_v$ is of the form $\mathcal{O}_v^*$ for all finite places v outside a finite set S and of the form $(1 + m_v)^{k(v)}$ for $v \in S$ with $h(v) \geq 1$. We define $h(v) = 0$ for $v \notin S$. The product group $\prod_v H_v \subset H$ has finite index in the original group H. Now suppose that $H = \ker(\psi)$. With these notations, we make the following definition.

(5.7) Definition. The conductor $\mathfrak{f}_\psi$ of an algebraic Hecke character ψ of type β is $\mathfrak{f}_\psi = \prod_v \mathfrak{m}_v^{h(v)}$. The ℓ-twist $\psi_\ell : I_K \to F_\ell^*$ and the ∞-twist $\psi_\infty : I_K \to F_\infty^*$ are defined by the relations

$$\psi_\ell(x) = \psi(x)\beta_\ell(x_\ell)^{-1} \quad \text{and} \quad \psi_\infty(x) = \psi(x)\beta_\infty(x_\infty)^{-1}.$$

The twists satisfy $\psi_\ell(K^*) = 1$ and $\psi_\infty(K^*) = 1$, and thus they can be defined on the group $C_K = I_K/K^*$ of idèle classes. This follows from the relation $\psi(x)\beta_0(x)^{-1} = 1$ for $x \in K^*$.

(5.8) Remarks. Let ψ be algebraic Hecke character $I_K \to F^*$. The ℓ-twist $\psi_\ell : C_K \to F_\ell^*$ defined on the idèle class group $C_K = I_K/K^*$ carries the connected component C_K^0 of the identity in C_K to 1 in F_ℓ^* since F_ℓ^* is totally discontinuous. The quotient character also denoted $\psi_\ell : C_K/C_K^0 \to F_\ell^*$. This can be composed with the reciprocity homomorphism of global class field theory $\eta : \mathrm{Gal}(\overline{K}/K) \to C_K/C_K^0$. In this way, the family of all ℓ-twists ψ_ℓ of an algebraic Hecke character defines a family (ψ, η) of abelian ℓ-adic representations of the Galois group $\mathrm{Gal}(\overline{K}/K)$ of dimension $[F : \mathbb{Q}]$.

(5.9) Remarks. Let ψ be an algebraic Hecke character $I_K \to F^*$, and let $\varphi : F \to \mathbb{C}$ be a complex embedding of F. Then the twist ψ_∞ composed with $\varphi : F_\infty^* \to \mathbb{C}^*$ defines a Grössencharacter of the type (A_0) considered by Weil [1955]. Now definition (4.6) applies to these compositions $\psi_\varphi = \varphi\psi_\infty = \chi$. We have only considered Grössencharacters, that is, characters $\chi : \mathbb{C}_K \to \mathbb{C}^*$ which come from algebraic Hecke characters. The continuity of the character implies that $\chi(\mathcal{O}_v^*) = 1$ for all v outside a finite set S, and hence $\chi(\pi_v)$ is well defined also called $\chi(\mathfrak{p}_v)$ giving the linear Euler factor.

$$\frac{1}{1 - \chi(\mathfrak{p}_v)(N\mathfrak{p}_v)^{-s}}$$

Again, π_v is a local uniformizing element in $\mathcal{O}_v$. A version of (4.7) for s to $2 - s$ also holds for this L-function of a Grössencharacter.

In the next section, the field F will be an imaginary quadratic number field, and there are two embeddings into $\mathbb{C}$ which differ by complex conjugation so we will have two complex valued Grössencharacters from the algebraic Hecke character of the form χ and $\overline{\chi}$.

§6. Deuring's Theorem on the L-Function of an Elliptic Curve with Complex Multiplication

For an elliptic curve E over a number field, the L-factor $L_v(s)$ at a place v has the form $L_v(s) = f_v(q_v^{-s})$ where the quadratic polynomial factors

$$f_u(T) = 1 - a_v T + q_v T^2 = (1 - \alpha_v T)(1 - \overline{\alpha}_v T).$$

Let E over K be a curve with complex multiplication by an imaginary quadratic field F. Deuring's theorem asserts roughly that the function $v \mapsto \alpha_v$ can be interpreted as an algebraic Hecke character $\chi_{E/K}$ on K with values in F, and the L-function $L_{E/K}(s)$ of E factors as a product of two Hecke L-functions

$$L_{E/K}(s) = L(s, \chi_{E/K})L(s, \overline{\chi}_{E/K}).$$

The essential point is not that the α_v are all in a fixed quadratic field F, indeed, they are always imaginary quadratic numbers by the Riemann hypothesis. Complex multiplication allows us to make a choice α_v among two conjugate numbers at each prime v, namely the one corresponding to Frobenius as we see in the next proposition. These are the α_v which are assembled together continuously to an algebraic Hecke character.

(6.1) Proposition. *Let E be an elliptic curve over a number field K such that E has good reduction E_v at a place v of K with residue class field $k(v)$. Then the reduction homomorphisms*

$$\mathrm{End}_K(E) \xrightarrow{\ r\ } \mathrm{End}_{k(v)}(E_v) \quad \textit{and} \quad \mathrm{End}_K(E) \otimes \mathbb{Q} \xrightarrow{\ r_v\ } \mathrm{End}_{k(v)}(E_v) \otimes \mathbb{Q}$$

are injective. If $\mathrm{End}_K(E) \otimes \mathbb{Q} = F$, a quadratic imaginary field, then there exists an element α_v in F with $r_v(\alpha_v) = \pi_v$, the Frobenius endomorphism of E_v over the field $k(v)$.

Proof. The injectivity of the reduction homomorphism r follows from the isomorphism $V_\ell(E) \to V_\ell(E_v)$ for ℓ a prime different from the characteristic of $k(v)$. Since the endomorphism algebras act faithfully on the V_ℓ spaces by 12(6.1) and 12(6.2), the map r is an injection.

For the second statement, there are two cases. Either r is an isomorphism between quadratic imaginary fields where the statement is clear, or r is an injection of a quadratic imaginary field into a quaternion algebra. In this case by 13(6.3) the Frobenius endomorphism is in the center of the algebra, hence in the image of r. The first case is ordinary reduction E_v, and the second is supersingular reduction. This proves the proposition.

(6.2) Remark. The quadratic imaginary field $F = \mathrm{End}_K(E) \otimes \mathbb{Q}$, where E has complex multiplication over K, has a natural embedding $\theta : F \to K$ in K given by the action of F on an invariant differential ω. For a in F we have $a^*(\omega) = \theta(a)\omega$. Corresponding to each place v where E has good reduction, we have an element $a_v \in F \subset K$ corresponding to the Frobenius element of the reduced curve $E_{(v)}$ over the residue class field $k(v)$.

In the next theorem, essentially due to Deuring, we state a result which describes how all the elements α_v fit together globally. For a proof, see Shimura [1971].

(6.3) Theorem. *Let E be an elliptic curve defined over a number field K with complex multiplication by $F = \mathrm{End}_K(E) \otimes \mathbb{Q}$. There exists a unique algebraic Grössencharacter $\chi_E : I_K \to F^*$ such that for all idèles $a = (a_v)$ with $a_v \equiv 1$ at v where E has bad reduction or v Archimedean*

$$\chi_E(a) = \prod_{v \text{ good}} a_v^{\mathrm{ord}_v(a_v)}.$$

The conductor $f_{E/K}$ of χ_E is concentrated at places of bad reduction. This character has the properties:

(1) *The ℓ-twists $\chi_{E,(\ell)}$ composed with the reciprocity epimorphism ν are the ℓ-adic representations of* $\mathrm{Gal}(\overline{K}/K)$ *acting on the F_ℓ-module $V_\ell(E)$ of rank 1.*
(2) *The ∞-twists define L-functions and the Hasse–Weil L-function of E over K can be recovered by the formula*

$$L_{E/K}(s) = L(s, \chi_{E,(\infty)})L(s, \overline{\chi}_{E,(\infty)}).$$

(6.4) Corollary. *Let E be an elliptic curve defined over a number field K with complex multiplication. The L-function $L_{E/K}(s)$ has an analytic continuation to the complex plane and satisfies a functional equation by* (5.5).

For examples of $L_E(s)$ see the book by Koblitz [1984], Chapter 2. For further properties of Grössencharacters and the L-functions associated with them, see S. Lang, *Algebraic Number Theory*.

§7. Eichler–Shimura Theory

The complex points on the modular curve $X_0(N)$ were discussed in Chapter 11, see (3.3).

(7.1) Remarks. The algebraic curve $X_0(N)$ can be defined over $\mathbb{Q}$ using the fact that it is a moduli space. By Igusa [1959] the curve $X_0(N)$ has good reduction at all primes p not dividing N, and for such p the smooth reduced curve $X_0(N)_p$ has a Frobenius correspondence Fr_p of the curve to itself. The Hecke operator on modular forms, defined in 11(7.1), also defines a correspondence of the reduced curve $X_0(N)_p$ to itself.

Recall that correspondences of one algebraic curve to another are defined as divisors on the product of the two curves, i.e., in terms of their graphs. By inverting the order of the factors, every correspondence has a transpose, and it is possible to add correspondences by adding divisors.

(7.2) Theorem (Eichler–Shimura). *In the group of correspondences of $X_0(N)_p$, the following relation holds*

$$T_p = \mathrm{Fr}_p + \mathrm{Fr}'_p$$

where Fr'_p *is the transpose of Frobenius* Fr_p.

For a proof of this result, see Shimura [1971, §7.2]. This book has a complete discussion of the ideas centering around this result.

The meaning of this fundamental congruence relation of Eichler–Shimura is that the eigenvalues of T_p are the same as the eigenvalues of Fr_p together with their conjugates coming from Fr'_p. Now we use the eigenvalues of Fr_p to construct L-functions of elliptic curves as Euler products, and the eigenvalues of T_p appear in the Euler product decompositions of the L-function of certain modular forms. This leads naturally to the following result of Eichler–Shimura.

(7.3) Theorem. *Let E be an elliptic curve over $\mathbb{Q}$ for which there is a surjective map over $\mathbb{Q}$ defined by $X_0(N) \to E$. Then there is a cusp form f for $\Gamma_0(N)$, which is an eigenfunction for all T_p where p does not divide N, such that the p factor in the Mellin transform of f is equal to the p Euler factor in the L-function $L_{E/\mathbb{Q}}(s)$ of the elliptic curve for all p not dividing N.*

In fact, the papers of Langlands and Deligne in the Antwerp conference 1972, see LN 349, show that the factors for p dividing N can also be handled. The result is dependent on representation theory and brings in the relation of the theory of elliptic curves to the Langlands Ansatz.

(7.4) Theorem. *Let E be an elliptic curve over $\mathbb{Q}$ for which there is a surjective map over $\mathbb{Q}$ defined by $X_0(N) \to E$. Then there is a normalized cusp form f for $\Gamma_0(N)$ of weight 2 whose L-function $L_f(s) = L_{E/\mathbb{Q}}(s)$, the L-function of the elliptic curve E over $\mathbb{Q}$.*

As in the case of complex multiplication where the L-function of an elliptic curve is identified with another L-function having analytic continuation and a functional equation, we have the following result. First, we introduce a definition suggested by the previous results.

(7.5) Definition. A modular parameterization at N of an elliptic curve E over $\mathbb{Q}$ is a surjective map $X_0(N) \to E$ defined over $\mathbb{Q}$. Such a curve E is called a modular elliptic curve.

(7.6) Theorem. *The L-function $L_{E/\mathbb{Q}}(s)$ of a modular elliptic curve E over $\mathbb{Q}$ satisfies the Hasse–Weil conjecture, that is, it has an analytic continuation to the plane with a functional equation.*

§8. The Modular Curve Conjecture

The modular curve conjecture goes under the name of the Taniyama–Weil conjecture or the Shimura–Taniyama–Weil conjecture in various places in the mathematical literature.

(8.1) Conjecture (Shimura–Taniyama–Weil). *Let E be an elliptic curve over the rational numbers with conductor N and L-function $L_{E/\mathbb{Q}}(s) = \sum_{1\le n} a_n n^{-s}$.*

(1) *The function $f(z) = \sum_{1\le n} a_n z^n$ is a cusp form of weight 2 for the group $\Gamma_0(N)$. It is an eigenfunction for all Hecke operators T_p with p not dividing N. Moreover,*

$$f\left(-\frac{1}{Nz}\right) = -wf(z),$$

where w is the sign in the conjectural functional equation (3.4).

(2) *There is a rational map over $\mathbb{Q}$ giving a modular parameterization $X_0(N) \to E$ such that the canonical differential on E pulls back to a constant times the differential form $f(z)\,dz$ on $X_0(N)$ defined by the cusp form $f(z)$ of weight 2.*

It is difficult to point to a place in the literature where this conjecture is stated as a conjecture. It seems to have matured in the minds of people around the time of the International Symposium on Algebraic Number Theory in Tokyo, September 1955. It is alluded to in the paper by Weil [1967] where he discusses an important extension of Hecke's work on modular forms or where the Dirichlet series which are Mellin transforms of modular forms are characterized. Hecke did the case of $SL_2(\mathbb{Z})$, and Weil extended it to $\Gamma_0(N)$. Using these results of Weil, one can see the following.

(8.2) Remark. Conjectures (8.1) and (3.4) are equivalent for elliptic curves over the rational numbers.

Henceforth, we call this conjecture the modular curve conjecture.

(8.3) Remark. Shimura [1971] showed that elliptic curves defined over the rational numbers with complex multiplication have a modular parameterization as in (8.1).

The modular curve conjecture is taken up further with the work of Andrew Wiles in Chapter 18.

17

Remarks on the Birch and Swinnerton–Dyer Conjecture

Let E be an elliptic curve over a number field K, and assume that E satisfies the Taniyama-Weil conjecture for L_E. The first assertion in the Birch and Swinnerton–Dyer conjecture is that $L_E(s)$ has a zero of order $r = \mathrm{rk}(E(K))$ at $s = 1$. The rank of the Mordell-Weil group was the invariant of E that was completely inaccessible by elementary methods unlike, for example, the torsion subgroup of $E(K)$. In the original papers where the conjecture was made, Birch and Swinnerton–Dyer checked the statement for a large family of curves of the form $y^2 = x^3 - Dx$ and $y^2 = x^3 - D$, which, being curves with complex multiplication, have an L-function with analytic continuation.

The conjecture goes further to describe the leading term

$$\lim_{s\to 1}(s-1)^{-r}L_E(s)$$

in terms of the determinant of the height pairing, the conductor, and the order of the group Ш_E introduced in Chapter 8, §3. This expression in the full conjecture is formed by analogy with a similar formula of Dirichlet in algebraic number theory. In 1986, there was not a single curve E known for which Ш_E is proven finite, so that the conjecture was not verified for a single case, (see §8). Now the the situation has changed completely.

In this chapter, we only state the main results and give a guide to the recent literature on the subject. The subject has grown very far in the last fifteen years.

§1. The Conjecture Relating Rank and Order of Zero

In the previous chapter, the L-function $L_{E/K}(s)$ of an elliptic curve E over a number field K was considered in terms of its analytic properties. No explicit reference to invariants of the curve were involved in the discussion. Now we take up the discovery of Birch and Swinnerton–Dyer that the L-function plays a basic role in the arithmetic of E over K as given by the group $E(K)$.

(1.1) First Birch and Swinnerton–Dyer Conjecture. Let E be an elliptic curve over a number field K with $L_{E/K}(s)$ having an analytic continuation to the complex plane. Let $g_{E/K} = g$ denote the rank of $E(K)$. Then g is equal to the order of the zero of $L_{E/K}(s)$ at the point $s = 1$.

There is extensive numerical evidence in favor of this conjecture which is recorded in the papers of Birch and Swinnerton–Dyer, for example. J. Coates has a weaker formulation of the conjecture which brings in the question of the structure of $Ш_{E/K}$. We will sometimes refer to (1.1) as the rank conjecture.

(1.2) Remark. Coates conjectures in his H. Weyl lectures at the IAS,

$$\mathrm{ord}_{s=1} L_{E/K}(s) \geq g_{E/K}$$

with equality if and only if $Ш_{E/K}$ is finite. It is conjectured that in fact $Ш_{E/K}$ is finite for all E. As mentioned above, there is not a single example of a curve E for which this is known to be true before 1986. This is still one of the very difficult problems in diophantine geometry.

(1.3) Remark. Recall that the Dedekind zeta function $\zeta_K(s)$ of a number field K of degree $r_1 + 2r_2 = [K : \mathbb{Q}]$ has zeros at negative integers of the following orders:
(1) $r_1 + r_2 - 1$ at zero,
(2) r_2 at negative odd integers,
(3) $r_1 + r_2$ at strictly negative even integers.

The unit theorem implies that the rank of the group $\mathcal{O}_K^*$ of units in the ring $\mathcal{O}_K$ of integers of K equals the order of the zero of $\zeta_K(s)$ at $s = 0$. Lichtenbaum conjectures that the orders of zeros at $1 - 2m$ and $2m$ are to be related to ranks of the Quillen algebraic K groups of $\mathcal{O}_K$.

In fact, at $s = 0$, we have

$$\zeta_K(s) = -\frac{h_K R_K}{w_K} s^{r_1+r_2-1} + \mathcal{O}(s^{r_1+r_2})$$

where h_K is the class number of K, w_K is the number of roots of unity, and R_K is the regulator of K. This is the Dirichlet class formula.

§2. Rank Conjecture for Curves with Complex Multiplication I, by Coates and Wiles

The first general result related to the conjecture of Birch and Swinnerton–Dyer is the following assertion due to Coates and Wiles [1977].

(2.1) Theorem. *Let E be an elliptic curve over K with complex multiplication. Assume that K is either the rational numbers or the field of complex multiplication. If $g_{E/K}$ is nonzero, then the L-function of E over K has a zero at $s = 1$, that is, $L_{E/K}(1) = 0$.*

First observe that $L_{E/K}(s)$ has an analytic continuation to the entire plane by the theorem of Deruing, and, in particular, the value at $s = 1$ has a meaning.

The idea of the proof is that one can factor from $L_{E/K}(1)$ a well-determined algebraic number $L^*_{E/K}(1)$. Then, the strategy is to show that under the hypotheses of the theorem, $L^*_{E/K}(1)$ is divisible by infinitely many primes. Like so much progress recently in the theory of elliptic curves, there is an analogy with results in the theory of cyclotomic fields and many constructions are motivated by those in Iwasawa theory. The result in question is the following: let U'_0 be the group of local units in $\mathbb{Q}_p(\mu_p)$ which are $\equiv 1$ and have norm 1 in the p-adic numbers $\mathbb{Q}_p$, and let C_0 be the class group of cyclotomic units in $\mathbb{Q}(\mu_p)$ which are $\equiv 1$ modulo the unique prime $\mathfrak{p}_0$ dividing p. Let χ be the canonical character on $G_0 = \mathrm{Gal}(\mathbb{Q}_p(\mu_p)/\mathbb{Q}_p)$ giving the action of G_0 on the pth roots of unity μ_p. Coates and Wiles use an analogue for elliptic units of the following classical result: for each even integer k with $1 < K < p-1$, the χ^kth eigenspace for the action of G_0 on $U'_0/\overline{C}_0$ is nontrivial if and only if p divides $(1/2\pi i)\zeta(k)$.

(2.2) Remark. Arthaud [1978] extended the Coates–Wiles result of (2.1) to fields K which are abelian extensions of F and satisfy some extra assumptions.

(2.3) Remark. K. Rubin refined the result (2.1) to give more precise information about the order of vanishing of $L_{E/K}(s)$ at $s = 1$, see Rubin [1981].

§3. Rank Conjecture for Curves with Complex Multiplication II, by Greenberg and Rohrlich

The following theorem of Ralph Greenberg is a partial converse of the Coates and Wiles theorem. Here Greenberg shows really the nontriviality of the Selmer group.

(3.1) Theorem. *Let E be an elliptic curve over the rational numbers with complex multiplication. If $L_{E/\mathbb{Q}}(s)$ has a zero of odd multiplicity at $s = 1$, then either $g_{E/\mathbb{Q}} > 0$ or the Tate–Šafarevič group $Ш_E$ contains a subgroup isomorphic to $\mathbb{Q}_p/\mathbb{Z}_p$ for each prime p strictly bigger than 3 where E has good ordinary reduction.*

The results of Greenberg underscore the interest in showing that the Tate–Šafarevič group is finite. It is also related to the modification (1.2) of the rank conjecture (1.1). Rohrlich [1984], following along the lines of Greenberg's work, considers more general towers of field extensions and an infinite set X of Hecke characters transferred to characters of $\mathrm{Gal}(\overline{K}/K)$ by class field theory. For each such character χ, there is a sign $W(\chi)$ in the functional equation of the related L-function $L(s, \chi)$.

(3.2) Theorem (Rohrlich). *For all but finitely many χ is the set X, we have*

$$\mathrm{ord}_{s=1} L(s,\chi) = \begin{cases} 0 & \text{if } W(\chi) = 1, \\ 1 & \text{if } W(\chi) = -1. \end{cases}$$

§4. Rank Conjecture for Modular Curves by Gross and Zagier

Modular elliptic curves (so defined over $\mathbb{Q}$) and elliptic curves over a number field with complex multiplication are the two families of elliptic curves for which $L_{E/K}$ is known to have an analytic continuation. Shimura [1971] showed that elliptic curves defined over the rational numbers with complex multiplication are modular elliptic curves. The following assertion is similar in nature to Greenberg's result.

(4.1) Theorem. *Let E be a modular elliptic curve defined over the rational numbers. If $L_E(s)$ has a simple zero at $s = 1$, then the group $E(\mathbb{Q})$ is infinite, i.e., the rank g_E is nonzero.*

This assertion says there is a rational point of infinite order which is the trace of a Heegner point. This is the first result of this nature.

The proof of this result is part of the entire discussion of the work of Gross and Zagier. We will say a little about it when their formula for the derivative of the L-function at $s = 1$ is considered in §7. We just state another result that also comes out of their work starting with Heegner points.

(4.2) Example. There exists a modular elliptic curve $E/\mathbb{Q}$ with rank 3 over $\mathbb{Q}$ and $L_{E/\mathbb{Q}}(s)$ has a zero of odd order ≥ 3 at $s = 1$. The curve comes from $X_0(37)$ with a twist related to 139. The equation is $-139y^2 = x^3 + 10x^2 - 20x + 8$ and conductor $37 \cdot 139^2 = 714877$.

§5. Goldfeld's Work on the Class Number Problem and Its Relation to the Birch and Swinnerton–Dyer Conjecture

In a remarkable paper Goldfeld [1976] discovered a relation between an effective lower bound of Siegel for the class number h_F of an imaginary quadratic field F in terms of the discriminant d_F and the existence of an L-function $L(s)$ with a zero of order ≥ 3 at $s = 1$. In other words, Goldfeld reduced the effective bound question to the existence of such a function which should be the L-function of an elliptic curve over $\mathbb{Q}$ of rank 3. The curve in (4.2) discovered by Gross and Zagier yields such an L-function.

The classical formula of Dirichlet on the class number was a starting point for Goldfeld and is related to the formula for the derivative of the L-function considered in the next two sections due to Gross and Zagier.

(5.1) Dirichlet's Class Number Formula. Form the Dirichlet L-function $L(s, \chi) = \sum_{1\leq n} \chi(n)n^{-s}$ for a real primitive character χ mod d, and denote by $F_\chi = \mathbb{Q}(\sqrt{\chi(-1)d})$ a quadratic number field with class number h, w roots of unity, and $\varepsilon_0 > 1$ the fundamental unit in the real quadratic case. The value of the Dirichlet L-function at $s = 1$ is given by the formula

$$L(1,\chi) = \begin{cases} \dfrac{2\pi h}{w\sqrt{d}} & \text{if } \chi(-1) = -1, \\[2ex] \dfrac{2h \log \varepsilon_0}{\sqrt{d}} & \text{if } \chi(-1) = +1. \end{cases}$$

The Dirichlet formula transferred the problem of a lower bound for h_F in terms of d_F into a problem of a lower bound for $L(1, \chi)$ in terms of d_{F_χ}. In terms of $L(1, \chi)$, Siegel [1935] or Goldfeld [1974] showed that for all $0 < \varepsilon$ there was a constant c_ε with $L(1, \chi) \geq c_\varepsilon d_F^{\varepsilon - 1/2}$. Unfortunately, if one wanted to make a list of all $\mathbb{Q}(\sqrt{-d}) = F$ with, for example, $h_F \geq 5$, this result would not help.

It was Dorian Goldfeld who made the spectacular connection between this problem of an effective constant c_ε and elliptic curves. He proved the following theorem.

(5.2) Theorem. *For all real numbers $0 < \varepsilon < 1$ there exists an effective $c_\varepsilon > 0$ such that for all imaginary quadratic fields F*

$$h_F \geq c_\varepsilon (\log |d_F|)^\varepsilon$$

if there exists an elliptic curve E over $\mathbb{Q}$ with $g_E = 3$ and $L(s, \chi)$ having a zero of order at least 3 at $s = 1$.

Putting together the theorem of Goldfeld (5.2) and one of the by-products of the work of Gross–Zagier (4.2), we can assert the following theorem.

(5.3) Theorem. *There exists an effective lower bound for the class numbers h_F of imaginary quadratic fields F in terms of an arbitrary positive power of* $\log |d_F|$ *and an effective positive constant.*

See also the report of Oesterlé in the Séminaire Bourbaki, 1983–84, exposé 631.

§6. The Conjecture of Birch and Swinnerton–Dyer on the Leading Term

If the L-function of an elliptic curve has a zero of order g at $s = 1$, then there is a power series expansion

$$L_{E/K}(s) = c_g (s-1)^g + \ldots$$

assuming as usual for these questions that $L_{E/K}$ has an analytic continuation to the plane, or at least to a neighborhood of $s = 1$. The second part of the Birch and Swinnerton–Dyer conjecture gives a formula for the value of c_g. There are several factors in the formula reflecting different aspects of the arithmetic of the curve.

(1) The order of the Tate–Šafarevič group Ш_E, which itself was only conjecturally finite.

(2) In Chapter 6, in the proof of the Mordell theorem, the canonical height was used, see (7.2), along with the associated bilinear pairing $\langle P, Q\rangle$. We introduce the following number $V_\infty(E)$ by choosing $P_1, \ldots, P_g$ independent elements of the group $E(\mathbb{Q})$ of rank g:

$$V_\infty(E) = \begin{cases} \dfrac{1}{\#E(\mathbb{Q}} & \text{for } g = 0, \\[2ex] \dfrac{\det\langle P_i, P_j\rangle}{(E(\mathbb{Q}) : B)^2} & \text{otherwise,} \end{cases}$$

where B is the subgroup of finite index in $E(\mathbb{Q})$ generated by $P_1, \ldots, P_g$.

(3a) Let $c_p = (E(\mathbb{Q}_p) : E_0(\mathbb{Q}_p))$, where E_0 is defined to be the subgroup of points mapping to nonsingular points mod p, see (1.1) where the notation was $E^{(0)}$ for E_0. It is also the number of components of multiplicity 1 rational over F_p on the special fibre of the Néron's minimal model for E at p.

(3b) Let $c_\infty = \int_{E(\mathbb{R})} |\omega|$ where ω is the differential form associated with a global minimal model for E. It is unique up to sign, i.e., units in $\mathbb{Z}$.

The number c_∞ is the number of components of $E(\mathbb{R})$ times the positive real period of ω. The numbers $c_p = 1$, except possibly for p, where E has bad reduction, see Table 1 in Chapter 14. The above definitions were formulated for $K = \mathbb{Q}$. For the general case see, for example, the article by Bloch [1980] where there is a general formulation and the conjecture is interpreted as a conjecture about Tamagawa numbers. Now we can state the full conjecture over $\mathbb{Q}$.

(6.1) Birch and Swinnerton–Dyer Conjecture Over $\mathbb{Q}$. Let E be an elliptic curve over $\mathbb{Q}$ such that $L_{E/\mathbb{Q}}(s)$ has an analytic continuation to the entire plane and $Ш_E$ is finite. Then the order of $L_{E/\mathbb{Q}}(s)$ at $s = 1$ is g, the rank of $E(\mathbb{Q})$, and the coefficient of $(s-1)^g$ in the Taylor expansion is given by

$$\#(Ш_E) \cdot V_\infty(E) c_\infty \prod_p c_p.$$

For numerical and theoretical evidence see Birch [1971], Birch and Swinnerton–Dyer [1965], and Swinnerton–Dyer [1967]. For a modular curve it is not too hard to show that the leading coefficient is c_∞ times a rational number.

§7. Heegner Points and the Derivative of the L-function at $s = 1$, after Gross and Zagier

Let E be a modular elliptic curve over $\mathbb{Q}$ such that $L_E(1) = 0$ where, for example, the sign in the functional equation is -1. In (5.1) there is a conjecture for $L'_E(1)$ when the order of the zero at $s = 1$ is 1. Gross and Zagier proved a formula for $L'_E(1)$ involving the height of a Heegner point on $X_0(N)$ associated with E. This formula is related to (6.1) , but it avoids the question of whether or not $Ш_E$ is finite.

The proof of (4.1) requires the construction or the proof of the existence of a point of infinite order. The method is to construct points on the modular curves $X_0(N)$ and with the modular parameterization map these to points on E. These special points on certain $X_0(N)$ were introduced by Heegner [1952], and they were first studied intensively by Birch. He defined them as pairs of N-isogenous elliptic curves E and E' with the ring $\text{End}(E) = \text{End}(E')$ and having complex multiplication. If $d < 0$ is the discriminant of the complex multiplication field, then the Heegner points are images of roots $z \in \mathfrak{H}$ of quadratic equations

$$\alpha z^2 + \beta z + \gamma = 0$$

where $\beta^2 - 4\alpha\gamma = d, \alpha \equiv 0 \pmod{N}$, and $\beta \equiv \delta \pmod{2N}$, where δ is a square root of $d \pmod{4N}$.

For a point y on the curve $X_0(N)$, we can consider y^σ, the conjugate under an automorphism of $\mathbb{C}$ and $T_m(y^\sigma)$ its image under a Hecke correspondence T_m. The divisors $T_m(y^\sigma) - (\infty)$ were studied by Gross [1985], and the local heights of these points were calculated. The main results, see Gross and Zagier [1986], are formulas for the derivative of an L-function at $s = 1$ where the L-function is made from a cusp form of weight 2, so corresponding to a certain elliptic curve, and data from a Heegner point. In their paper, Theorem 6.1 is a formula for $L'(1)$ in terms of an inner product of the cusp form with a cusp form of weight 2 built out of data from the $T_m(y^\sigma)$ terms mentioned above. The second formula, Theorem 6.3 in their paper [Gross and Zagier, 1986], is a formula for $L'(1)$ in terms of the inner product of the original cusp form with itself times the canonical height of the Heegner point together with certain general constants. The proof involves an understanding of about a dozen height terms with corresponding terms related to modular expressions.

§8. Remarks On Postscript: October 1986

In the first edition of *Elliptic Curves*, we had the good luck to be in a position to report on three recent developments in the theory of elliptic curves. They were results by K. Rubin, by N. Elkies, and by G. Frey and K. Ribet.

The work of G. Frey and K. Ribet is considered in Chapter 18, §1 and §2. As for the results of Elkies, we reported the following:

(8.1) Infinitude of Primes with Supersingular Reduction by N. Elkies. From the results of Table 1, p. 264, we know that the curve defined by $y^2 = x^3 - x$ has supersingular reduction for primes $p \equiv 7, 11 \pmod{12}$ and the curve defined by $y^2 = x^3 - 1$ has supersingular reduction for primes $p \equiv 5, 11 \pmod{12}$. This means that each curve reduces to a supersingular curve at infinitely many primes. Further analysis shows that any elliptic curve with complex multiplication has supersingular reduction at infinitely many primes, and it was conjectured to be true for all elliptic curves over the rational numbers, or more generally over a number field. In a letter to B. Mazur in July 1986, N. Elkies proved this result for all elliptic curves over $\mathbb{Q}$. He assumes that there is supersingular reduction at primes $p_1 < \cdots < p_r$, and using the result for curves with complex multiplication, he deduces that there is a prime $p_{r+1} > p_r$ with supersingular reduction.

The reference to the article which has since appeared is Elkies [1987].

(8.2) Progress on the Finiteness of Ш(E) by K. Rubin. This was published in the same issue of Invent. math. *89* (1987) in two articles. These articles were only the beginning, for they came at the same time as the methods of Kolyvagin on Euler Systems. There are four very useful surveys of Kolyvagin's work by Gross [1991], McCallum [1991], Perrin-Riou [1990], and Rubin [1989].

Later, we have the book by Karl Rubin, *Euler Systems*, Annals of Math. Studies [2000] based on his H. Weyl lectures. Here the general notion of Euler systems is studied and applied to bounds on Galois cohomology which in turn lead to finiteness properties of Ш(E) and relations between cases where $\mathrm{rk}E(\mathbb{Q}) = 0, 1$ and where order of the pole at $s = 1$ of $L(s, E/\mathbb{Q})$ is equal to $0, 1$.

The subject of the Birch and Swinnerton–Dyer conjecture merits a very extended treatment which lies outside the purpose of this book. It is a subject for a separate book.

18

Remarks on the Modular Elliptic Curves Conjecture and Fermat's Last Theorem

A modular elliptic curve is an elliptic curve over the rational numbers which has a finite covering by a modular curve of the form $X_0(N)$. Using the theory of molular functions for $\Gamma_0(N)$, we deduce that the Hasse–Weil Zeta function of a modular elliptic curve has two basic properties: it has an analytic continuation to the complex plane, and it satisfies a functional equation for the interchange between s and $2-s$, see 16(3.4).

The modular elliptic curve conjecture, or simply modular curve conjecture, is a conjecture growing out of the work of Shimura and Taniyama in the period 1950–1965. It asserts that every elliptic curve over $\mathbb{Q}$ is modular. In Weil [1967] conceptual evidence for the conjecture was given, and the conjecture became widely known from this paper. For this reason the modular curve conjecture goes under the name of the Shimura–Taniyama–Weil conjecture or the earlier name of the Taniyama–Weil conjecture. If an elliptic curve over $\mathbb{Q}$ is modular, then also any elliptic curve with the same j-value is modular. Such j-values are thus called modular. Before Wiles [1995] only a finite number of j-values over $\mathbb{Q}$ were known to be modular.

Ten years earlier Frey made the remarkable observation that the modularity conjecture should imply Fermat's Last Theorem. The program of Frey, relating the two conjectures, was reformulated by Serre as the ε-conjecture, and as mentioned in the first edition of this book, it was proved by Ribet in the summer of 1986, see Ribet [1990]. In fact, Ribet's result required only that the modular curve conjecture be true for semistable curves in order to deduce Fermat's Last Theorem.

In his 1995 Annals paper Wiles was able to prove the modular curve conjecture for sufficiently many curves to establish Fermat's Last Theorem by completing the above program. One basic point in Wiles' proof was carried out jointly with Richard Taylor in Taylor and Wiles [1995]. The proof was completed in the Fall of 1994.

As outlined in the last section §9, the entire modular curve conjecture was proved at the end of 1999 by Breuil, Conrad, Diamond, and Taylor.

In this chapter we begin by reviewing some of the material in earlier chapters which have a direct bearing on the modular curve conjecture starting with the role of the Tate module of an elliptic curve. In terms of the Tate module with its structure as an ℓ-adic Galois module, we are able to give an alternate formulation of the modular curve conjecture. We illustrate the main idea with the following diagram of the basic sets of objects connected by two functions α and β together with reduction mod ℓ of a sublattice of a vector space representation.

$$\begin{Bmatrix}\text{elliptic curves}\\ \text{over } \mathbb{Q} \text{ with}\\ \text{conducteur}\end{Bmatrix} \xrightarrow{\alpha} \begin{Bmatrix}\ell\text{-adic}\\ \text{representations}\\ \text{of } \mathrm{Gal}(\bar{\mathbb{Q}}/\mathbb{Q})\end{Bmatrix} \xleftarrow{\beta} \begin{Bmatrix}\text{modular forms of}\\ \text{weight 2 for } \Gamma_0(N)\\ \text{which are Hecke}\\ \text{eigenfunctions}\end{Bmatrix}$$

$$\Big\downarrow \text{reduction mod } \ell$$

$$\begin{Bmatrix}\text{mod } \ell\\ \text{representations}\\ \text{of } \mathrm{Gal}(\bar{\mathbb{Q}}/\mathbb{Q})\end{Bmatrix}.$$

By a result of Faltings [1984] the function α, which assigns to an elliptic curve E over $\mathbb{Q}$ the ℓ-adic representation on the Tate module $T_\ell(E)$, is injective up to isogeny. In other words, if $T_\ell(E')$ and $T_\ell(E'')$ are isomorphic Galois representations, then E' and E'' are isogenous over $\mathbb{Q}$. For general abelian varieties this was conjectured by Šafarevič and proved by Faltings.

The function β assigns to a modular form f associated with $\Gamma_0(N)$, which is an eigenform for the Hecke algebra, an ℓ-adic representation $\beta(f)$ with the property that the eigenvalue λ_p for the Hecke operator $T_p(f) = \lambda_p f$ associated to a modular form f has a Frobenius trace $\mathrm{tr}(F_p)$ at p on $\beta(f)$ equal to λ_p. In fact to each weight 2 modular Hecke eigenform such that the eigenvalues are rational numbers there is an elliptic curve E with conductor $|N$ and a map of the modular curve $X_0(N) \to E$ onto E. These are the modular elliptic curves, and the function β is injective. The modular elliptic curve conjecture is equivalent to the observation that every ℓ-adic representation coming from Tate modules of $E/\mathbb{Q}$ has this form.

The strategy of Wiles. Let V' and V'' be two irreducible 2-dimensional ℓ-adic representations of $\mathrm{Gal}(\bar{\mathbb{Q}}/\mathbb{Q})$ which are isomorphic modulo ℓ. Under suitable conditions if either V' or V'' is a modular representation, then so is the other of weight two and expected level.

In the application of this strategy a modular form of weight one will be the starting point coming from Langlands and Tunnell. We go on to outline some of the main results and ideas in the proof.

We conclude with a guide to the recent literature on the subject.

§1. Semistable Curves and Tate Modules

We begin by recalling the notions of good reduction and semistable reduction. In 14(3.2) we introduced the criterion of Néron–Ogg–Šafarevič for an elliptic curve E over a local field to have good reduction. The global version of the theorem is stated in 15(3.3).

Besides good reduction, we will also need to consider the weaker notion of semistable reduction.

(1.1) Definition. Let E be an elliptic curve over $\mathbb{Q}$. The curve E has good reduction (resp. semistable reduction) at a prime p provided there exists a cubic equation over $\mathbb{Z}$ for E with the reduction mod p a nonsingular cubic (resp. a cubic with at most a double point as singular point). An elliptic curve E over $\mathbb{Q}$ is semistable provided it has either good or semistable reduction at each prime number p.

(1.2) Example. At the end of chapter 14, Table 2 we have given a list of 16 elliptic curves coming from an article of Serre. It is indicated which curves are semistable. It is the case that all complex multiplication curves are not semistable.

In the semistable case the reduced curve will be either an elliptic curve over $\mathbb{F}_p$ or a curve whose smooth points over a finite extension of $\mathbb{F}_p$ form the multiplicative group.

(1.3) Remark. Among the cubic equations over $\mathbb{Z}$ for an elliptic curve E over $\mathbb{Q}$ there is one with minimal discriminant $D \in \mathbb{Z}$. There is also the conductor N of E which is a natural number. The curve E has good reduction at a prime p if and only if $p \nmid D$, or equivalently if $p \nmid N$. The curve E has semistable reduction at p if and only if $p^2 \nmid N$, and hence, the semistable curves are exactly those with square free conductor N. Recall that the conductor N is defined locally in 14(2.4) and globally in 15(1.1), and the minimal discriminant in 5(2.6) and 15(1.3).

In 12(3.3) the definition of isogeny was given. We say that E and E' are isogenous provided there exists an isogeny $E \to E'$. This is seen to be an equivalence relation. Two isogenous elliptic curves have the same conductor, but the discriminants can be different. In the next theorem the direct implication is easy to show, but the striking converse is due to Faltings [1984, §5].

(1.4) Theorem. *Two elliptic curves E and E' over $\mathbb{Q}$ are isogenous if and only if E and E' have the same good primes and $\#E(F_p) = \#E'(F_p)$ for each good prime p.*

(1.5) Remark. Let G denote the absolute Galois group $\mathrm{Gal}(\bar{\mathbb{Q}}/\mathbb{Q})$ over $\mathbb{Q}$. Then G acts on the K-valued points $E(K)$ of E for any Galois extension K of $\mathbb{Q}$ contained in the algebraic closure $\bar{\mathbb{Q}}$ of the rational numbers $\mathbb{Q}$. As was described in 12(5.2), G acts also on the Tate module $T_\ell(E)$ giving a two dimensional Z_ℓ-representation. It has as a quotient $E[\ell] = {}_\ell E(\bar{\mathbb{Q}})$, the ℓ-division points of E, giving a two dimensional F_ℓ-representation.

§2. The Frey Curve and the Reduction of Fermat Equation to Modular Elliptic Curves over $\mathbb{Q}$

Although the cubic curve $y^2 = x(x - a^p)(x - c^p)$ was studied by Hellyovarch in the mid 70's, its possible significance for Fermat's Last Theorem was pointed out by Gerhard Frey in an article in *Annales Universitatis Saraviensis*, Vol. 1, No. 1, 1986. For this reason the next definition has come into common usage.

(2.1) Definition. The Frey curve $E(a, c; p)$ is defined for an odd prime p and for two distinct integers a and c by the cubic equation: $y^2 = x(x - a^p)(x - c^p)$.

Frey had the idea that if a and c are related to third integer b satisfying the Fermat equation $a^p + b^p = c^p$, then we would have an elliptic curve $E(a, c; p) = E(a, b, c; p)$ over $\mathbb{Q}$ which was not modular. This would contradict the modular curve conjecture.

(2.2) Remark. Note that $E(a, b, c; p)$ has good reduction at a prime q if and only if q does not divide the product abc. In particular it always has bad reduction at 2. On the other hand this curve is semistable for relatively prime a, b, c having a suitable order. In any case the early methods of Wiles could handle this curve even when it is not semistable at 2.

(2.3) Fundamental Property of the Frey Curve. For $G = \mathrm{Gal}(\bar{\mathbb{Q}}/\mathbb{Q})$ and each prime $\ell > 2$ the modular represention of G on $E[\ell]$ would have a very special property. It would be unramified at $\ell \neq p$. At primes ℓ where $E(a, b, c : p)$ has good reduction, by the Néron–Ogg–Šafarevič criterion, but it also holds for all odd primes dividing abc from the special structure of the equation. At $\ell = p$ it is crystalline. This concept will be considered later.

(2.4) Basic Assertion. The curve $E(a, b, c : p)$ with these properties can not be modular. This is the content of the article of Ribet [1990]. It is a careful study of raising and lowering the levels of modular Galois representations.

In other words the Fermat conjecture is reduced to the modular elliptic curve conjecture for semistable curves.

§3. Modular Elliptic Curves and the Hecke Algebra

In chapter 11, §4, we introduced modular forms $f(\tau)$ for subgroups of $\mathbb{SL}_2(\mathbb{Z})$ and their q-expansions at cusps. In 11(5.2) the Mellin transform of a modular form was analytically continued to the entire plane, and in 11(5.3) for $\mathbb{SL}_2(\mathbb{Z})$ and in 11(5.7) for $\Gamma_0(N)$, and the functional equation of the L-function which is the Mellin transform up to a factor was explained. In 11(8.1) the Hecke operators T_p were introduced acting on modular forms for the group $\mathbb{SL}_2(\mathbb{Z})$, and their role with respect to Euler expansions was indicated in 11(8.2). We need these operators on $S(N)$ the space of cusp forms of weight 2 for $\Gamma_0(N)$.

(3.1) Definition. The Hecke operators

$$T_p : S(N) \to S(N) \quad \text{for } p \nmid N \quad \text{and} \quad U_p : S(N) \to S(N) \quad \text{for } p \mid N$$

are defined by giving the q-expansion coefficients in terms of the q-expansion coefficients $a_n(f)$ of $f(\tau)$ as follows

$$a_n(T_p f) = a_{np}(f) + p a_{n/p}(f) \quad \text{and} \quad a_n(U_p f) = a_{np}(f).$$

The Hecke algebra $\mathbb{T}(N)$ is the commutative algebra generated by the Hecke operators acting on $S(N)$, the space of cusp forms for $\Gamma_0(N)$ of weight 2.

Here $a_r(f) = 0$ if $r \in \mathbb{Q} - \mathbb{Z}$.

(3.2) Definition. An eigenform $f(\tau)$ is a cusp form for $\Gamma_0(N)$ of weight 2 such that f is a common eigenform for all Hecke operators normalized so that $a_1(f) = 1$.

(3.3) Remark. If f is an eigenform, then $a_p(F) = a_1(T_p f)$ for $p \nmid N$ and $a_p(f) = a_1(U_p f)$ for $p|N$, or in other words the q-expansion coefficient $a_p(F)$ is the eigenvalue of the related Hecke operator. The other coefficients $a_n(f)$ are determined recursivelly, so that we see that eigenform f is determined by the prime indexed set of eigenvalues $a_p(F)$ of f.

Conversely a prime indexed set of numbers (a_p) determines a q-expansion $f(\tau) = \sum_{1\le n} a_n e^{2\pi i \tau}$ for a possible eigenform using the recursion relations arising from the fact it is an eigenform for T_p when $p \nmid N$ and U_p when $p|N$.

(3.4) Definition. An L-prime indexed set of numbers a_p is a prime indexed set of numbers for which the related q-expansion $f(\tau) = \sum_{1\le n} a_n e^{2\pi i \tau}$ and any Direchlet character χ mod N has a related L-function

$$L_f(s, \chi) = \sum_{1\le n} a_n \chi(n) n^{-s}.$$

which has an analytic continuation to the entire s-plane bounded in vertical strips and which satisfies a functional equation as in 11(5.7).

(3.5) Theorem (A. Weil [1968]). *A prime indexed set of numbers a_p corresponds to a modular form $f(\tau)$ as in (3.3) if and only if it is an L-prime indexed set.*

(3.6) Assertion. An eigenform f for N or equivalently an L-indexed set of numbers a_p, which are all rational numbers, determines an isogeny class of elliptic curves defined over $\mathbb{Q}$ with conductor N and good reduction at all $p \nmid N$. Also at p where there is good reduction, i.e., $p \nmid N$, the number $\#E(\mathbb{F}_p)$ of $\mathbb{F}_p$ rational points of the reduced curve is $\#E(\mathbb{F}_p) = p + 1 - a_p$ where E is in the isogeny class. The class contains a minimal elliptic curve E_f with respect to the degree of the map from $X_0(N) \to E_f$. The elliptic curves which arise in this way are exactly the elliptic curves E over $\mathbb{Q}$ for which there is a non-constant map $X_0(N) \to E$.

Now we can return to the modular curve conjecture in this context.

(3.7) Modular Curve Conjecture. If E is an elliptic curve over $\mathbb{Q}$ with the reduction properties given in (3.5) and with conductor N, then there is an eigenform $f(\tau)$ with rational eigenvalues such that E and E_f are isogenous. In particular we have

$$a_p(F) = p + 1 - \#E(\mathbb{F}_p).$$

§4. Hecke Algebras and Tate Modules of Modular Elliptic Curves

In general there are Galois representations associated with eigenforms $f(\tau)$. We are interested in eigenforms with rational coefficients or with rational eigenvalues.

(4.1) Notation. Now we return to the Hecke algebra $\mathbb{T}(N)$. It is a finitely generated $\mathbb{Z}$-module. Consider a prime ℓ and a suitable non-Eisenstein maximal ideal $\mathfrak{M}(\ell)$ such that the residue field $\mathbb{T}/\mathfrak{M}(\ell)$ has characcteristic ℓ. Form the completion $\hat{\mathbb{T}}$ at this ideal of $\mathbb{T}$, that is, $\mathbb{T} = \varprojlim_q \mathbb{T}/\mathfrak{M}(\ell)^q$. For background, see B. Mazur, *Modular curves and the Eisenstein ideal*, IHES no 47 (1977), especially p. 37 and Chap II. §9, p. 95.

(4.2) Assertion. There is a basic two dimensional Galois representation $\rho = \rho_{\mathfrak{M}(\ell)}$: $\mathrm{Gal}(\bar{\mathbb{Q}}/\mathbb{Q}) \to \mathbb{GL}_2(\hat{\mathbb{T}})$ with the following two properties:

(1) ρ is unramified at $p \nmid Nl$; ρ is crystalline at $p = \ell$ if $\ell \nmid N$.

Unramified means that it is trivial when restricted to the inertial group at p. By 14(3.2) unramified Galois action on Tate modules is related to good reduction of elliptic curves. A crystalline representation is more difficult to define, but for the Tate module it is again related to good reduction.

(2) For $p \nmid Nl$ we have $\mathrm{trace}(\rho(\mathrm{Frob}_p)) = T_p$ and $\det(\rho(\mathrm{Frob}_p)) = p$.

Eigenforms are related to this basic representation in the following way.

(4.3) Assertion. An eigenform $f(\tau)$ with rational eigenvalues defines a morphism $\theta_f : \mathbb{T} \to \mathbb{Z}_\ell$, and if θ_f factors as $\theta_f : \hat{\mathbb{T}} \to \mathbb{Z}_\ell$, then applying the functor $\mathbb{GL}_2$ to θ_f and composing with $\rho_{\mathfrak{M}(\ell)}$ we obtain a Galois module $\mathbb{GL}_2(\theta_f)\rho_{\mathfrak{M}(\ell)}$ which is isomorphic as a Galois module to the Tate module $T_\ell(E_f)$ of the associated elliptic curve E_f.

In this way we describe the action of the absolute Galois group on the ℓ^m-division points of E_f using the eigenform f.

(4.4) Assertion. Let E be an elliptic curve over $\mathbb{Q}$ such that $E[\ell]$ is an irreducible Galois representation. The elliptic curve E is modular if and only if the associated Tate module can be constructed by such a morphism θ_f and representation $\rho_{\mathfrak{M}(\ell)}$ as in the following diagrams where the functor $\mathbb{GL}_2$ is applied to θ_f

$$\begin{array}{ccc} \hat{T} & \longrightarrow & T/\mathfrak{M} \\ \uparrow & & \uparrow \\ \mathbb{Z}_\ell & \longrightarrow & \mathbb{F}_\ell \end{array} \qquad \begin{array}{c} \hat{T} \\ \big\downarrow{\scriptstyle \theta_f} \\ \mathbb{Z}_\ell \end{array} \qquad \begin{array}{ccc} \mathrm{Gal}(\bar{\mathbb{Q}}/\mathbb{Q}) & \xrightarrow{\rho_{\mathfrak{M}(\ell)}} & \mathbb{GL}_2(\hat{\mathbb{T}}) \\ {\scriptstyle \rho_{\ell,E}}\big\downarrow & & \big\downarrow{\scriptstyle \mathbb{GL}_2(\theta_f)=\theta_f} \\ \mathrm{Aut}(T_\ell(E_f)) & = & \mathbb{GL}_2(\mathbb{Z}_\ell). \end{array}$$

§5. Special Properties of mod 3 Representations

Starting with the three division points $E[3]$ of an elliptic curve E. We obtain a Galois representation $\mathrm{Gal}(\bar{\mathbb{Q}}/\mathbb{Q}) \to \mathbb{GL}_2(\mathbb{F}_3)$. The group $\mathbb{GL}_2(\mathbb{F}_3)$ has as quotient $\mathbb{PGL}_2(\mathbb{F}_3)$ which is isomorphic to S_4, the symmetric group on the four points of the line $\mathbb{P}_1(\mathbb{F}_3)$.

(5.1) Assertion. We assume that $\mathrm{Gal}(\bar{\mathbb{Q}}/\mathbb{Q}) \to \mathbb{GL}_2(\mathbb{F}_3)$ is surjective and that E has good reduction at 3. When this is not the case, there is a related argument with $E[5]$ and the corresponding Galois representation $\mathrm{Gal}(\bar{\mathbb{Q}}/\mathbb{Q}) \to \mathbb{GL}_2(\mathbb{F}_5)$. For the three division points the quotient symmetric group $S_4 = \mathbb{PSL}_2(\mathbb{F}_3)$ is solvable, and lifting theorems of Langlands and Tunnell apply to show that the representation is modular. The deep theory of lifting of Langlands yields a weight one cusp form $g(\tau)$ for some $\Gamma_1(N)$, which is an eigenfuction for all the ralted Hecke operators, such that for almost all primes p the coefficient b_p is related to the trace of Frob_p on the Galois representation.

Next, this cusp form is multiplied by the following weight one modular form E for $\Gamma_1(3)$ defined with cubic character

$$\chi(d) = 0, \pm 1 \quad \text{where } d \equiv 0, \pm 1 \pmod 3$$

and having as q-expansion

$$E(\tau) = 1 + 6 \sum_{n \geq 1} \sum_{d|n} \chi(d) e^{2\pi i n\tau}.$$

The product $g(\tau)E(\tau)$ is a weight two cusp for $\Gamma_0(N)$, and its coefficients are congruent mod p to b_n for any prime $\mathfrak{p}$ in $\bar{\mathbb{Q}}$ containing $1 + 2^{1/2}$. By the Deligne–Serre [1974, lemma 6.11] it is possible to modify $g(\tau)E(\tau)$ to an eigenform for $\Gamma_1(N)$ such that its coefficients are congruent mod $\mathfrak{p}$ to b_n. Hence the representation $E[3]$ on the 3 division points is described by a modular form for $\Gamma_0(3N)$. Using the theory in Ribet [1990] and Carayol, we modify the conductor to obtain a form in $\Gamma_0(N_{\bar{\rho}}^-)$ where $N_{\bar{\rho}}^-$ is the minimal level for which the representation can arise.

For further details see Wiles [1995, Chapter 5].

(5.2) Remark. Starting with the Galois representation on $E[3]$, we can think of it as a quotient of $T_3(E)$ the 3-adic Tate module. The core of the proof is the assertion that the universal lifting of $E[3]$ is modular. This is considered in the next section.

§6. Deformation Theory and ℓ-Adic Representations

(6.1) Deformation Data. Let D denote a set of local conditions at all primes $p \neq \ell$ on liftings of $\bar{\rho}$ such that outside Σ a finite set of primes this condition is simply being unramified, but at the primes of Σ there are conditions of being ordinary or flat. For details see Wiles [1995, p. 458].

There is a minimal such set of conditions $D(\min)$ consisting of properties of liftings which are as unramified as possible.

The universal deformation ring $\mathcal{R}(D)$, depending on deformation data D, is constructed for anly prime ℓ, but the main application is at the prime 3.

(6.2) Universal Galois Representation. There is an algebra $\mathcal{R}(D)$ over the ℓ-adic numbers $\mathbb{Z}_\ell$ and a Galois representation $\rho : \mathrm{Gal}(\bar{\mathbb{Q}}/\mathbb{Q}) \to \mathbb{GL}_2(\mathcal{R}(D))$ satisfying the following properties:

(1) as a $\mathbb{Z}_\ell$-algebra $\mathcal{R}(D)$ has the form $\mathcal{R} = \mathbb{Z}_\ell[[T_1, \ldots, T_r]]/I$ where I is an ideal in the formal series algebra and $\rho_\ell \bmod(\ell, T_1, \ldots, T_r)$ is isomorphic to the representation $E[\ell]$,
(2) ρ is crystalline at ℓ (which corresponds to the good reduction of E at ℓ),
(3) ρ has local properties D at all primes $p \neq \ell$,
(4) $\det(\rho(\mathrm{Frob}_p)) = p$ for $p \nmid N$, and
(5) (universal property) any other representation $\mathrm{Gal}(\bar{\mathbb{Q}}/\mathbb{Q}) \to \mathbb{GL}_2(\mathcal{A})$ satisfying (1)–(4) is conjugate to the representation $\mathbb{GL}_2(\theta)\rho$ for a unique algebra morphism $\theta : \mathcal{R}(D) \to \mathcal{A}$.

(6.3) Construction of $\mathcal{R}(D)$. Starting with a finite set of generators of the Galois group, we take 4 formal variables over $\mathbb{Z}_\ell$ for each generator to make the ρ-image 2×2 matrix. Next, we divide the resulting formal series algebra by the minimal ideal such that properties (1)–(4) of (6.2) are satisfied together with a $\mathbb{GL}(2)$ adjoint action quotient.

(6.4) Remark. For each suitable class of deformation data D we can choose a level N and the maximal ideal $\mathfrak{M}(\ell, D)$ such that

$$\rho_{\mathfrak{M}(\ell,D)} : \mathrm{Gal}(\bar{\mathbb{Q}}/\mathbb{Q}) \to \mathbb{GL}_2(\hat{\mathbb{T}})$$

is the maximal modular deformation f satisfying D in (5.1)(5). We will write $\hat{\mathbb{T}}(D)$ for the completion of the Hecke algebra $\mathfrak{M}(\ell, D)$. Thus we have a commutative diagram

$$\begin{array}{ccccc} \mathcal{R}(D) & \longrightarrow & \hat{T}(D) & \longrightarrow & T/\mathfrak{M} \\ & \nwarrow & \uparrow & & \uparrow \\ & & \mathbb{Z}_\ell & \longrightarrow & \mathbb{F}_\ell . \end{array}$$

We remark that (5.1) tells us that if $\ell = 3$ then $\mathbb{T}(D(\min)) \neq 0$.

(6.5) Key Idea of Wiles. Show that the map $\mathcal{R}(D) \to \hat{T}(D)$ coming from the universal property of $\mathcal{R}(D)$ for such a D is an isomorphism.

For suitably large D there is a map $\mathcal{R}(D) \to \mathbb{Z}_3$ giving rise to the Galois representation $T_3(E)$. Thus we deduce that for such data D there is a morphism $\hat{T}(D) \to \mathbb{Z}_3$ giving rise to $T_3(E)$, that is, E is modular.

§7. Properties of the Universal Deformation Ring

We continue with the notation D for a choice of local data.

(7.1) Assumptions. For simplicity we will assume that there is a morphism

$$\hat{\mathbb{T}}(D(\min)) \to \mathbb{Z}_\ell.$$

In general there will be a morphism $\hat{\mathbb{T}}(D(\min)) \to \mathcal{O}$ for $\mathcal{O}$ the ring of integers in some finite extension of $\mathbb{Q}_\ell$, and one can apply a similar argument. Let

$$\rho_n : \mathrm{Gal}(\mathbb{Q}/\mathbb{Q}) \to \mathbb{GL}_2(\mathbb{Z}/\ell^n)$$

be the Galois representation corresponding to this morphism $\hat{\mathbb{T}}(D(\min)) \to \mathbb{Z}_\ell$. Let W_n denote the adjoint representation in $\mathbb{SL}_2(\mathbb{Z}/\ell^n)$ on $\mathbb{GL}_2(\mathbb{Z}/\ell^n)$ given by conjugation.

(7.2) Notation. Let $H^1_{f,D}(\mathbb{Q}, W_n)$ denote the cohomology group with the local conditions related to (6.2)(2) and (3). The definition consists of elements which are zero when restricted to subgroups of $\mathrm{Gal}(\bar{\mathbb{Q}}/\mathbb{Q})$ which are Galois groups of $\bar{\mathbb{Q}}_p$ over maximal unramified extensions of $\mathbb{Q}_p$. For the definitions see Wiles [1995, pp. 460–462].

(7.3) Remark. This is a form of a Selmer group similar to those considered in 8(3.2).

(7.4) Assertion (M. Flach). The orders of the Selmer groups $H^1_{f,D}(\mathbb{Q}, W_n)$ are uniformally bounded in n.

This order comes about from following numerical criterion for the morphism $\mathcal{R}(D) \to \hat{T}(D)$ to be an isomorphism.

(7.5) Assertion. The ring $\hat{T} = T(D)$ is Gorenstein, and this means that the dual $\mathrm{Hom}(\hat{T}, \mathbb{Z}_\ell)$ over $\mathbb{Z}_\ell$ is a free $\hat{T}$-module. Thus the surjection $\hat{T} \to \mathbb{Z}_\ell$ has an adjoint $\mathbb{Z}_\ell \to \hat{T}$, and the composition of these morphisms is multipliccation by an element $\eta_D \in \mathbb{Z}_\ell$. This element η_D is nonzero and is well defined up to a unit.

(7.6) Criterion for $\mathcal{R}(D) \to \hat{T}(D)$ to Be an Isomorphism. Let $\mathfrak{p}_D$ denote the kernel of the composite surjection $\mathcal{R}(D) \to \hat{T}(D) \to \mathbb{Z}_\ell$. We have the following inequalities of orders $\#(\mathfrak{p}_D/\mathfrak{p}_D^2) \geq \#(\mathbb{Z}_\ell/\eta_D\mathbb{Z}_\ell)$ with equality if and only if $\mathcal{R}(D) \to \hat{T}(D)$ is an isomorphism.

(7.7) Remark. The first attempt to prove this equality was with the Euler systems of Kolyvagin. Only the fact that η_D annihilated $\mathfrak{p}_D/\mathfrak{p}_D^2$ could be shown, and this is the content of theorem of M. Flach. At the time it seems to be very difficult to construct higher Euler systems.

(7.8) Remark. The problem is reducing to a problem in the commutative algebra properties of the two algebras $\mathcal{R}(D)$ and $\hat{T}(D)$.

§8. Remarks on the Proof of the Opposite Inequality

In the final stages of the proof of Wiles, subtle points in commutative algebra play a very special role.

We continue with the notation D for a choice of local data for the minimal case see Wiles [1995, p. 513]. Now there are two steps:

(1) $\mathcal{R}(D(\min)) \to \hat{T}(D(\min))$ is an isomorphism and this is discussed in (8.1),
(2) $\mathcal{R}(D) \to \hat{T}(D)$ is an isomorphism and this is discussed in (8.2) where we reduce to the minimal case.

(8.1) Remark. In the minimal case all primes of bad reduction occur already mod 3 and not only mod 3^i for $i > 1$. This is the content of Wiles [1995, Chapter 3]. Here he usees the estimates on the size of Selmer group, see especially Theorem 3.1 on p. 518. An important step in completing the proof was to use that $\mathbb{T}$ is not just a Gorenstein algebra, but also a complete intersection, see Taylor and Wiles [1995]

(8.2) Reduction to the Minimal Case. Now we know that

$$\#(\mathfrak{p}_{D(\min)}/\mathfrak{p}^2_{D(\min)}) = \#(\mathbb{Z}_\ell/\eta_{D(\min)}\mathbb{Z}_\ell),$$

and we must deduce, as in (7.6), that

$$\#(\mathfrak{p}_D/\mathfrak{p}_D^2) \le \#(\mathbb{Z}_\ell/\eta_D\mathbb{Z}_\ell).$$

(1) For the left hand side one notes that

$$\#(\mathfrak{p}_D/\mathfrak{p}_D^2) = \#H^1_{f,D}(\mathbb{Q}, W_n)$$

for $n \gg 0$, and one can estimate (from above) the quotient

$$H^1_{f,D}(\mathbb{Q}, W_1^*)/H^1_{f,D(n)}(\mathbb{Q}, W_1^*).$$

(2) The right hand side measures congruences between modular forms. Here a method of Ribet and Ihara allows one to estimate $\#(\mathfrak{p}_D/\mathfrak{p}_D^2)/\#(\mathfrak{p}_{d(\min)}/\mathfrak{p}^2_{D(\min)})$ from below.

These bounds are the same as proving $\#(\mathfrak{p}_D/\mathfrak{p}_D^2) \le \#(\mathbb{Z}_\ell/\eta_D\mathbb{Z}_\ell)$. Hence $\mathcal{R}(D) \to \hat{T}(D)$ is an isomorphism, and this proves the theorem.

§9. Survey of the Nonsemistable Case of the Modular Curve Conjecture

After the semistable case was finished in the main paper of Wiles and the Taylor–Wiles article, progress on the general case of elliptic curves over $\mathbb{Q}$ continued with various types of refinements along the lines of Wiles' program by especially F. Diamond between 1994 and 1997.

In the paper of B. Conrad, F. Diamond, and R. Taylor [1999] the modular curve conjecture was established expect for curves with very singular behavior at 3, that is, any curve with conducctor not divisible by 3^3. Finally in C. Breuil, B. Conrad, F. Diamond, and R. Taylor [2001] with preprint in 1999 the special problems at the prime 3 were resolved, and the modular curve conjecture is established for all elliptic curves over the rational numbers.

19

Higher Dimensional Analogs of Elliptic Curves: Calabi–Yau Varieties

In any discussion of higher dimensional analogs of elliptic curves, the first point which has to be emphasized is that there are compact complex manifolds, which are not algebraic. In the absence of an algebraic structure, a complex Kähler metric becomes important. Deformations of these manifolds in a family can mean deformations of either the complex structure or the Kähler metric. In particular, complex differential geometry plays a crucial role in higher complex dimensions. The first examples begin with tori $\mathbb{C}^g/\Gamma$, where Γ is a discrete subgroup of rank $2g$. Complex tori and abelian varieties, which are algebraic tori, constitute the first obvious extension of elliptic curves, and this subject is highly developed, see the books of A. Weil and D. Mumford.

Our aim is to give an introduction to the study of another class of varieties X/k with similar properties to elliptic curves. These varieties are smooth, projective, and having zero canonical class $K_X = 0$. The last condition can be stated that the canonical line bundle $\omega_X = \Omega_X^n$ is trivial for $\dim X = n$. Such smooth varieties with trivial canonical line bundle exist in all dimensions. For example, abelian varieties and smooth hypersurfaces in $\mathbb{P}^{n+1}$ of degree $n+2$. Recall that in dimension one, an elliptic curve is both a one dimensional abelian variety and can be represented as a smooth cubic plane curve.

In dimension two, the smooth hypersurface of degree four in $\mathbb{P}^3$ is not an abelian surface. For it is simply connected, and abelian varieties are never simply connected. The smooth surface of degree four in projective three space is an example of a K3-surface which is algebraic. A K3-surface has an everywhere nonzero holomorphic 2-form, like an abelian surface, but it is simply connected. K3-surfaces were named after Kummer, Kähler, and Kodaira.

This condition points the way to the Calabi–Yau generalization of elliptic curves as compact n-dimensional varieties with an everywhere nonzero holomorphic n-form or trivial canonical bundle with the added condition that the fundamental group is finite. In dimension three, the compact varieties with trivial ω_X and with finite fundamental group are called Calabi–Yau 3-folds, and they include smooth quintic hypersurfaces in projective four space. Some complete intersections in toric varieties are also important examples of Calabi–Yau varieties.

A Calabi–Yau manifold takes its name from a conjecture about Kähler manifolds by Calabi [1955] which was proved by Yau [1978]. This is a result in complex analytic geometry which is formulated in terms of a Ricci flat Kähler metric or in terms of a Levi–Civita connection with $SU(n)$ holonomy. In the context of complex geometry these conditions are equivalent to a trivial canonical line bundle.

To survey this part of the theory, we introduce briefly some basic differential geometry, some discussion of Calabi–Yau manifolds as Kähler manifolds, and then move to the algebraic theory. The aim here is to give a relatively self-contained account of basic Kähler geometry in order to show clearly the basic concept of Calabi–Yau manifold as a higher dimensional generalization of an elliptic curve. Here the role of a conjecture of Calabi and the proof of the conjecture by Yau is taken up in the context of Kähler geometry. This means we can use a Kähler structure to formulate geometric flatness type conditions for a smooth projective algebraic variety over the complex numbers to be a Calabi–Yau manifold. These conditions are considered in the context of algebraic geometry where they have a purely algebraic formulation for smooth algebraic varieties. This extends even to varieties with mild singularities. Some of the preliminary results on Calabi–Yau manifolds are mentioned, the main source of examples from toric geometry is introduced, and a guide to the literature is given.

As mentioned above, we make a quick survey of relevant topics in complex differential geometry. The following remarks are a background to the point of view used in these first sections.

Differential geometry is formulated in terms of smooth vector bundles where the main examples are the tangent bundle and exterior powers of the cotangent bundle. As was made precise by Serre and Swan, a vector bundle over M can be viewed as a bundle $p : E \to M$ with the fibres $E_x = p^{-1}(x), x \in M$, having smoothly varying vector space structures. There is a principal bundle of frames in the background, and also a finitely generated projective module over the algebra $C^\infty(M)$ of smooth functions on the manifold M. This related module is just $\Gamma^\infty(M, E)$ the vector space of smooth cross sections of E with the $C^\infty(M)$-module structure coming from multiplication of a scalar valued smooth function with a vector field. In the case of the tangent bundle $T(M)$ on M, it is the module of vector fields $\mathrm{Vec}(M)$ on M.

We formulate everything in terms of vector fields and differential forms with only a passing word on the fibre bundle approach. The reader is invited to carry out this approach and even fill in the discussion with coordinates and the resulting flood of indices.

In the final four sections of this chapter, we survey the Enriques classification of surfaces where the existence of an elliptic fibration of a surface plays a basic role. The place of the two dimensional Calabi–Yau variety, namely the K3 surface, in this classification is pointed out, and the chapter concludes with a general discussion of K3 surfaces.

§1. Smooth Manifolds: Real Differential Geometry

(1.1) Notation. Let M be a smooth real manifold, and let $C^\infty_\mathbb{R}(M)$ denote the algebra of smooth real valued functions on M. The extension of scalars $\mathbb{C} \otimes_\mathbb{R} C^\infty_\mathbb{R}(M)$ is the algebra $C^\infty(M)$ of smooth complex functions on M. A coordinate chart is a pair of open sets $U \subset V$ in M, with the closure $\overline{U} \subset V$, and a system of real valued functions $x_1, \ldots, x_n \in C^\infty_\mathbb{R}(M)$ equal to zero outside V and having the property that every $f \in C^\infty_\mathbb{R}(M)$ can be represented as a composite function $f = \phi(x_1, \ldots, x_n)$ on U by a unique smooth function ϕ defined on the open subset $(x_1, \ldots, x_n) \subset \mathbb{R}^n$.

We will describe the tangent bundle $T(M)$, the cotangent bundle $T^*(M)$, and the pth exterior powers $\Lambda^p T^*(M)$ in terms of their smooth cross sections $\mathrm{Vect}(M)$ of smooth vector fields for $T(M)$ and $A^p(M)$ of smooth p-forms for $\Lambda^p T^*(M)$. Since each germ of a smooth function, of a smooth vector field, or of a p-form is represented by a global section, we can recover the vector bundles from these $C^\infty_\mathbb{R}(M)$ modules. The same formulation will apply to complex structures and Riemannian, Hermitian, and Kähler metrics.

(1.2) Definition. A real (resp. complex) vector field on a smooth real manifold M is a derivation $\xi : C^\infty_\mathbb{R}(M) \to C^\infty_\mathbb{R}(M)$ (resp. $\xi : C^\infty(M) \to C^\infty(M)$). Let $\mathrm{Vec}_\mathbb{R}(M)$ (resp. $\mathrm{Vec}(M)$) denote the real (resp. complex) vector space of vector fields on M where the formula $(a\xi + b\eta)(f) = a\xi(f) + b\eta(f)$ defines the vector space structure over the scalars $\mathbb{R}$ (resp. $\mathbb{C}$).

The derivation property for ξ means that ξ is linear over the scalars and for two functions f, g we have

$$\xi(fg) = f\xi(g) + g\xi(f)$$

(1.3) Remark. To derive assertions about real vector fields from assertions about complex vector fields, we use the complex conjugate vector field $\overline{\xi}$ defined by $\overline{\xi}(\overline{f}) = \overline{\xi(f)}$. Just as a function $f \in C^\infty(M)$ has the property that $f \in C^\infty_\mathbb{R}(M)$ if and only if $f = \overline{f}$, we have $f \in \mathrm{Vec}_\mathbb{R}(M)$ if and only if $\xi = \overline{\xi}$. From now on, we state everything for complex valued functions and vector fields.

There are two additional operations on vector fields.

(1.4) Remark. Let $h \in C^\infty(M)$ and $\xi, \eta \in \mathrm{Vec}(M)$. The $C^\infty(M)$-module structure on $\mathrm{Vec}(M)$ is given by the formula $(h\xi)(f) = h(\xi(f))$, and the Lie bracket structure on $\mathrm{Vec}(M)$ is given by the formula

$$[\xi, \eta](f) = \xi(\eta(f)) - \eta(\xi(f))$$

We leave it to the reader to check that the resulting maps are derivations in each case.

(1.5) Remark. The complex vector space $\mathrm{Vec}(M)$ is a module over C^∞ with the operation $h\xi$ and a Lie algebra over $\mathbb{C}$ with the Lie bracket as in (1.4). It is not a Lie algebra over $C^\infty(M)$. Indeed, for $\xi, \eta \in \mathrm{Vec}(M)$ and $f, g \in C^\infty(M)$,

$$[f\xi, g\eta] = fg[\xi, \eta] + \xi(g)\eta - \eta(f)\xi.$$

(1.6) Definition. A p-form on a smooth manifold is a C^∞-multilinear map $\omega : \mathrm{Vec}(M)^p \to C^\infty(M)$. Let $A^p(M)$ denote the $C^\infty(M)$-module of p-forms. The module $A^1(M)$ is the $C^\infty(M)$-dual of $\mathrm{Vec}(M)$ of cotangent vector fields or 1-forms, and each function f defines a 1-form df by $df(\xi) = \xi(f)$.

The resulting morphism $d : A^0(M) = C(M) \to A^1(M)$ also satisfies the derivation property $d(fg) = fd(g) + gd(f)$.

(1.7) Remark. For a smooth function f on a manifold, the support of f, denoted $\mathrm{supp}(f)$, is the closure of all points $x \in M$ where $f(x) \neq 0$. For a vector field ξ, we have $\mathrm{supp}(\xi(f)) \subset \mathrm{supp}(f)$. In terms of local coordinates $x_1, \ldots, x_n$ on a coordinate chart $U \subset V \subset M$ with corresponding partial derivatives $\partial_1, \ldots, \partial_n$ on U, we have $\xi = \xi(x_1)\partial_1 + \cdots + \xi(x_n)\partial_n$. Thus the partial derivatives are local cross sections of the tangent bundle $T(M)$ of M, and the local coordinate of ξ is the n-tuple $(\xi(x_1), \ldots, \xi(x_n))$. The 1-forms $dx_1, \ldots, dx_n$ are local cross sections of the cotangent bundle $T^*(M)$ of 1-forms on M, and the exterior products $dx_{i(1)} \wedge \cdots \wedge dx_{i(p)}$, denoted $dx_{i(*)}$ for all $i(*) : 1 \le i(1) < \cdots < i(p) \le n$, are a basis of local cross sections of the bundle $\wedge^p T^*(M)$ of p-forms. In particular, a p-form ω can be written in local coordinates as $\omega = \sum_{i(*)} \omega_{i(*)} dx_{i(*)}$ where the $\omega_{i(*)}$ are smooth functions.

(1.8) Definition. In terms of local coordinates, we can define a map $d : A^p(M) \to A^{p+1}(M)$ by the formula

$$d\omega = \sum_{i(*)} d\omega_{i(*)} \wedge dx_{i(*)}.$$

Here we have used the exterior multiplication of forms which is defined $A^p(M) \times A^q(M) \to A^{p+q}(M)$. We leave it to the reader to check that the map is independent of local coordinates and satisfies $d(\omega' \wedge \omega'') = d\omega' \wedge \omega'' + (-1)^p \omega' \wedge \omega''$ for $\omega' \in A^p$, $\omega'' \in A^q$, and $d(d(\omega)) = 0$.

Metric properties of manifolds come from the following concepts.

(1.9) Definition. A pseudo-Riemannian metric on M is a $C^\infty_{\mathbb{R}}(M)$-bilinear form $g : \mathrm{Vec}_{\mathbb{R}}(M) \times \mathrm{Vec}_{\mathbb{R}}(M) \to C^\infty_{\mathbb{R}}(M)$ such that

(1) $g(\xi, \eta) = g(\eta, \xi)$ for all $\xi, \eta \in C^\infty_{\mathbb{R}}(M)$, and
(2) for each 1-form θ, there exists a unique vector field $\eta \in \mathrm{Vec}_{\mathbb{R}}(M)$ with $g(\xi, \eta) = \theta(\xi)$ for all $\xi \in \mathrm{Vec}_{\mathbb{R}}(M)$.

Also, g is Riemannian provided $g(\xi, \xi) \ge 0$ for ξ real, or $\xi = \overline{\xi}$.

(1.10) Remark. The function which assigns to $\eta \in \mathrm{Vec}_{\mathbb{R}}(M)$, then element $\theta(\xi) = g(\xi, \eta)$ with $\theta \in A^1_{\mathbb{R}}(M)$ is an isomorphism

$$\mathrm{Vec}_{\mathbb{R}}(M) \longrightarrow A^1_{\mathbb{R}}(M) \quad \text{of } C^\infty_{\mathbb{R}}(M)\text{-modules.}$$

A pseudo-Riemannian or Riemannian metric g defines by extension of scalars a $C^\infty(M)$-bilinear form $g : \mathrm{Vec}(M) \times \mathrm{Vec}(M) \to C^\infty(M)$.

(1.11) Remark. Returning to the local coordinates $x_1, \ldots, x_n$ of (1.7) on the coordinate chart $U \subset V \subset M$, we can write a pseudo-Riemannian metric $g(\xi, \eta)$ in local coordinates using the local cross sections $\partial_1, \ldots, \partial_n$ by decomposing

$$\begin{aligned} g(\xi, \eta) &= g(\xi(x_1)\partial_1 + \cdots + \xi(x_n)\partial_n, \eta(x_1)\partial_1 + \cdots + \eta(x_n)\partial_n) \\ &= \sum_{1 \le i,j \le n} g(\partial_i, \partial_j)\xi(x_i)\eta(x_j) \\ &= \sum_{1 \le i,j \le n} g(\partial_i, \partial_j)(dx_1 \otimes dx_j)(\xi, \eta) \end{aligned}$$

where $\xi = \xi(x_1)\partial_1 + \cdots + \xi(x_n)\partial_n$ and $\eta = \eta(x_1)\partial_1 + \cdots + \eta(x_n)\partial_n$. Thus in local coordinates we have

$$g = \sum_{1 \le i,j \le n} g(\partial_i, \partial_j)\, dx_i \otimes dx_j \quad \text{or simply} \quad g = \sum_{1 \le i,j \le n} g_{i,j}\, dx_i\, dx_j$$

where $g_{i,j} = g(\partial_i, \partial_j) = g_{j,i}$ from the symmetry of g.

§2. Complex Analytic Manifolds: Complex Differential Geometry

Complex differential geometry begins with the concept of a complex structure on a real manifold which generalizes the complex structure on $\mathbb{R}^{2n}$ making the complex Euclidean space $\mathbb{C}^n$. A real manifold with a complex structure is a complex analytic manifold. A complex structure on X can be defined by returning to coordinate charts $U \subset X$ with complex valued coordinate maps

$$(z_1, \ldots, z_n) : U \to U' \subset \mathbb{C}^n$$

where the change of coordinates is by holomorphic (or complex analytic) functions in n variables.

A second approach is to speak of an almost complex structure on a real manifold of even dimension $2n$.

(2.1) Definition. An almost complex structure on a manifold X is a $C^\infty_{\mathbb{R}}(X)$-morphism $J_{\mathbb{R}} : \mathrm{Vec}_{\mathbb{R}}(X) \to \mathrm{Vec}_{\mathbb{R}}(X)$ which satisfies $J^2_{\mathbb{R}} = -1$. This morphism $J = \mathbb{C} \otimes_{\mathbb{R}} J_{\mathbb{R}}$ induces an eigenspace decomposition which splits the complex vector fields for eigenvalues $+i$ and $-i$.

$$\mathbb{C} \otimes_{\mathbb{R}} \mathrm{Vec}_{\mathbb{R}}(X) = \mathrm{Vec}(X) = \mathrm{Vec}(X)^{(1,0)} \oplus \mathrm{Vec}(X)^{(0,1)}$$

called respectively the holomorphic and antiholomorphic parts. The elements of $\mathrm{Vec}_X^{(1,0)}$ have the form $v - iJ(v)$ and those of $\mathrm{Vec}(X)^{(0,1)}$ have the form $v + iJ(v)$ for arbitrary $v \in \mathrm{Vec}(X)$.

(2.2) Example. Let X be a holomorphic manifold with local complex coordinate charts with complex coordinate functions $(z_1, \ldots, z_n)$. Let $(x_1, \ldots, x_n, y_1, \ldots, y_n)$ be the corresponding real coordinate functions associated with the complex coordinate functions $z_1 = x_1 + iy_1, \ldots, z_n = x_n + iy_n$. Then

$$z_1, \ldots, z_n, \bar{z}_1, \ldots, \bar{z}_n$$

can also be used formally as local coordinates. The related almost complex structure J has the following values on the local partial derivatives as complex vector fields:

$$J(\partial_{x_j}) = \partial_{y_j}, \quad J(\partial_{y_j}) = -\partial_{x_j}, \quad J(\partial_{z_j}) = i\partial_{z_j}, \quad J(\partial_{\bar{z}_j}) = -i\partial_{\bar{z}_j}$$

for $j = 1, \ldots, n$. For this we use the following relations between real and complex coordinate functions in a coordinate chart:

$$x = \frac{1}{2}(z_j + \bar{z}_j) \quad \text{and} \quad y_j = \frac{1}{2i}(z_j - \bar{z}_j) \quad \text{for } j = 1, \ldots, n.$$

(2.3) Definition. An almost complex structure on a manifold is a complex structure when it comes from a holomorphic structure as in example (2.2).

The splitting of $\mathrm{Vec}(X)$ under a complex structure induces a dual splitting on the complex differential forms $A^1(X)$ and generally $A^p(X)$.

(2.4) Definition. Let X be an almost complex manifold with splitting $\mathrm{Vec}(X) = \mathrm{Vec}(X)^{(1,0)} \oplus \mathrm{Vec}(X)^{(0,1)}$. The dual splitting is $A^1(X) = A^{1,0}(X) \oplus A^{0,1}(X)$ where
(1) $\omega \in A^{1,0}(X)$ if and only if $\omega(\mathrm{Vec}(X)^{(0,1)}) = 0$ and
(2) $\omega \in A^{0,1}(X)$ if and only if $\omega(\mathrm{Vec}(X)^{(1,0)}) = 0$

for $\omega \in A^1(X)$.

This splitting can be carried through to all forms by

$$A^{p,q}(X) = \Lambda^p A^{1,0}(X) \otimes \Lambda^q A^{0,1}(X).$$

For local coordinate functions $z_1, \ldots, z_n, \bar{z}_1, \ldots, \bar{z}_n$, as a module over the smooth functions the forms $A^{p,q}(X)$ has a basis

$$dz_{i(*)} \wedge d\bar{z}_{j(*)} = dz_{i(*)} \otimes d\bar{z}_{j(*)}$$

for all $i(*) : 1 \le i(1) < \cdots < i(p) \le n$, and $j(*) : 1 \le j(1) < \cdots < j(q) \le n$. When $p = 0$, the basis consists of $d\bar{z}_{j(*)}$, and when $q = 0$, the basis consists of $dz_{i(*)}$.

(2.5) Remark. The operation cc of complex conjugation $cc(f) = \overline{f}$ on complex valued functions defines a complex conjugation $cc : \mathrm{Vec}(X) \to \mathrm{Vec}(X)$ and $cc : A^r(X) \to A^r(X)$ which restricts to isomorphisms $\mathrm{Vec}(X)^{(1,0)} \to \mathrm{Vec}(X)^{(0,1)}$ and $A^{p,q}(X) \to A^{q,p}(X)$.

(2.6) Remark. The exterior differential on a complex manifold is a sum $d = d' + d''$ also denoted $\partial = d'$ and $\overline{\partial} = d''$. They satisfy the relations $(d')^2 = 0$, $(d'')^2 = 0$, and $d'd'' + d''d' = 0$. Moreover, $d' : A^{p,q}(X) \to A^{p+1,q}(X)$ and $d'' : A^{p,q}(X) \to A^{p,q+1}(X)$ explain the fact that d' has bidegree $(+1, 0)$ and d'' has bidegree $(0, +1)$.

For an almost complex structure to come from a holomorphic complex structure, there is an integrability condition. We state the result, see also Kobayashi and Nomizu [1969].

(2.7) Theorem. *Let X be an almost complex manifold. Then the following are equivalent:*

(1) *For* $\xi, \eta \in \mathrm{Vec}(X)^{(1,0)}$ *the bracket* $[\xi, \eta] \in \mathrm{Vec}(X)^{(1,0)}$.
(2) *For* $\xi, \eta \in \mathrm{Vec}(X)^{(0,1)}$ *the bracket* $[\xi, \eta] \in \mathrm{Vec}(X)^{(0,1)}$.
(3) *For the exterior differential on 1-forms we have*

$$d(A^{1,0}(X)) \subset A^{2,0}(X) \oplus A^{1,1}(X) \quad \textit{and} \quad d(A^{0,1}(X)) \subset A^{1,1}(X) \oplus A^{0,2}(X).$$

(4) *For the exterior differential on* (p, q)*-forms, we have*

$$d(A^{p,q}(X)) \subset A^{p+1,q}(X) \oplus A^{p,q+1}(X).$$

(5) *The almost complex structure comes from a holomorphic structure on the manifold.*

We close this section by explaining the cohomological implications of the fact that exterior differentiation d satisfies $dd = 0$ and the relations in (2.7) for a complex manifold.

(2.8) Definition. The de Rham cohomology $H^*_{\mathrm{DR}}(M)$ is defined in degree m to be the following quotient

$$H^m_{\mathrm{DR}}(M) = \frac{\ker(d : A^m(M) \to A^{m+1}(M))}{\mathrm{im}(d : A^{m-1}(M) \to A^m(M))}.$$

There are two versions of de Rham cohomology, namely $H^*_{\mathrm{DR}}(M, \mathbb{R})$ and $H^*_{\mathrm{DR}}(M, \mathbb{C})$ using real or complex valued differential forms.

(2.9) Remark. For a general smooth manifold M, the de Rham cohomology $H^*_{\mathrm{DR}}(M)$ is naturally isomorphic to $H^*_{\mathrm{sing}}(M)$, the singular cohomology. This is basically a theorem in sheaf theory.

(2.10) Remark. When M is a compact oriented manifold with a Riemannian matrix, then there exists a prehilbert space structure on $A^m(M)$ with inner product

$$(\alpha|\beta) = \int_X \alpha \wedge \overline{\beta}^*.$$

There is an adjoint $d^* : A^*(M) \to A^*(M)$ to the operator d. It has degree -1. The combination $\Delta = dd^* + d^*d$ is a Laplace operator generalizing the usual Laplace operator as one sees in a local computation on Euclidean space. Using the norm property

$$|(\Delta\alpha|\alpha)| = \|d\alpha\|^2 + \|d^*\alpha\|^2,$$

we can define a from $\alpha \in A^m(M)$ to be harmonic provided $\Delta\alpha = 0$ or equivalently $d\alpha = 0$ and $d^*\alpha = 0$, that is α is harmonic if and only if it is closed and coclosed.

A de Rham cohomology class $c \in H^m_{\mathrm{DR}}(M)$ is a coset

$$c = \omega + dA^{m-1}(M)$$

determined by a closed form ω. A fundamental property of compact Riemannian manifolds is that every cohomology class c is represented by a unique harmonic form α, i.e. $c = \alpha + dA^{m-1}(M)$.

(2.11) Remark. For a complex manifold X, the two exterior operators d' and d'' define a double complex which has an associated spectral sequence. The E_1-term of this spectral sequence has the following form $E_1^{p,q} = H^q(X, \Omega_X^p)$ converging to de Rham cohomology. Here Ω_X^p is the sheaf of holomorphic p-forms on X. This provides H^*_{DR} with an additional structure $F^p H^*_{\mathrm{DR}}(X)$ called the Hodge filtration. We will return to this later for compact Kähler manifolds. As a reference, see Griffths and Harris [1978].

§3. Kähler Manifolds

Consider a complex valued $C^\infty_{\mathbb{R}}(X)$-bilinear map

$$h : \mathrm{Vec}_{\mathbb{R}}(X) \times \mathrm{Vec}_{\mathbb{R}}(X) \to C^\infty(X) \supset C^\infty_{\mathbb{R}}(X)$$

with $h = g - i\omega$ where

$$g : \mathrm{Vec}_{\mathbb{R}}(X) \times \mathrm{Vec}_{\mathbb{R}}(X) \to C^\infty_{\mathbb{R}}(X) \quad \text{and} \quad \omega : \mathrm{Vec}_{\mathbb{R}}(X) \times \mathrm{Vec}_{\mathbb{R}}(X) \to C^\infty_{\mathbb{R}}(X)$$

are $C^\infty_{\mathbb{R}}(X)$-bilinear forms. A complex version of a Riemannian metric is formulated using the complex structure J on the manifold X.

(3.1) Definition. Let X be a complex manifold with complex structure J. A Hermitian metric h on (X, J) is a

$$C^\infty_{\mathbb{R}}\text{-bilinear map} \quad h : \mathrm{Vec}_{\mathbb{R}}(X) \times \mathrm{Vec}_{\mathbb{R}}(X) \to C(X)$$

with real and imaginary parts $h = g - i\omega$ satisfying either of the following two equivalent conditions and the positivity axiom:

(1) $h(\eta, \xi) = \overline{h(\xi, \eta)}$, the condition of Hermitian symmetry, and $h(J(\xi), \eta) = ih(\xi, \eta)$ the condition of $\mathbb{C}$-linearity in the first variable of h. The two conditions imply $h(\xi, J(\eta)) = -ih(\xi, \eta)$ which is antilinearity over $\mathbb{C}$ in the second variable of h.

(2) $\omega(\eta, \xi) = -\omega(\xi, \eta)$ is $C^\infty_{\mathbb{R}}(X)$-bilinear antisymmetric and $g(\xi, \eta) = \omega(\xi, J(\eta))$ is $C^\infty_{\mathbb{R}}(X)$-bilinear symmetric and pseudo-Riemannian.

Positivity: The metric g is Riemannian or equivalently $h(\xi, \xi) \geq 0$ for all $\xi \in \mathrm{Vec}_{\mathbb{R}}(X)$.

To see that (1) gives (2), we write out

$$\overline{h(\xi, \eta)} = g(\xi, \eta) + i\omega(\xi, \eta) = g(\eta, \xi) - i\omega(\eta, \xi) = h(\eta, \xi)$$

and

$$h(\xi, J(\eta)) = g(\xi, J(\eta)) - i\omega(\xi, J(\eta)) = (-i)(\omega(\xi, \eta) - ig(\xi, J(\eta)) = -ih(\xi, \eta).$$

In turn, these expressions also show that condition (2) gives (1).

(3.2) Remark. For a Hermitian metric $h = g - i\omega$ on a complex manifold (X, J), the condition of J-invariance of ω, that is, $\omega(J(\xi), J(\eta)) = \omega(\xi, \eta)$ is equivalent to $\omega \in A^{1,1}(X)$. Also, for $\xi \in \mathrm{Vec}(X)^{(1,0)}$ and $\eta \in \mathrm{Vec}(X)^{(0,1)}$, we have $h(\xi, \overline{\eta}) = 0$ since

$$h(\xi, \overline{\eta}) = h(J(\xi), J(\overline{\eta})) = h(i\xi, i\overline{\eta}) = -h(\xi, \overline{\eta}).$$

(3.3) Remark (Hermitian Form in Local Coordinates). Locally on the almost complex manifold X, the holomorphic vector fields $\mathrm{Vec}(X)^{(1,0)}$ have a basis $\partial/\partial z_1, \ldots, \partial/\partial z_n$ and the holomorphic 1-forms $A^{1,0}(X)$ have the related dual basis denoted $dz_1, \ldots, dz_n$. The coefficients of the Hermitian form in these local coordinates are

$$h_{j,k}(z, \overline{z}) = h\left(\frac{\partial}{\partial z_i}, \frac{\partial}{\partial z_j}\right),$$

and the differential form $\omega = -2i \sum_{j,k=1}^{n} h_{j,k}(z, \overline{z})\, dz_j \wedge d\overline{z}_k$.

(3.4) Definition. A Kähler metric h on a complex manifold X is a Hermitian metric h on X with $h = g - i\omega$ where the related 2-form ω is J-invariant and closed. The form ω associated with the Kähler metric h is called the Kähler form.

Observe that the Kähler form $\omega(\xi, \eta)$ and J determine the Kähler metric by the relation $h(\xi, \eta) = \omega(\xi, J(\eta)) - i\omega(\xi, \eta)$.

(3.5) Remark. The condition that the Kähler form ω is closed is equivalent to the complex structure J being parallel with respect to the connection ∇ associated with g, see §4.

(3.6) Remark. To each Kähler metric h and Kähler form ω, there is a Kähler cohomology class $[\omega] \in H^2(X, \mathbb{R})$. Under the natural inclusion $H^2(X, \mathbb{R}) \subset H^2(X, \mathbb{C}) = H^{2,0}(X) \oplus H^{1,1}(X) \oplus H^{0,2}(X)$, the Kähler class $[\omega] \in H^{1,1}(X)$. The set $\mathcal{K}$ of all Kähler classes $[\omega]$ is a cone viewed in either real vector space $H^2(X, \mathbb{R})$ or $H^{1,1}(X)$.

(3.7) Remark. The Kähler condition $d\omega = 0$ reduces to the differential equations because $(d' + d'')\omega = 0$ is locally

$$\frac{\partial}{\partial z_\ell} h_{j,k} = \frac{\partial}{\partial z_j} h_{\ell,k} \quad \text{where } j, k, \ell = 1, \ldots, n.$$

Further, locally there exists function K called the Kähler potential satisfying $\omega = id'd''K$.

Kähler potentials defined on an open covering can lead to the Kähler form immediately as in the case of complex projective space $\mathbb{P}_N(\mathbb{C})$ with homogeneous coordinates $z_0 : \cdots : z_N$.

(3.8) Complex Projective Space. As a complex manifold $\mathbb{P}_N(\mathbb{C})$ has $N + 1$ open sets U_j with $z_j \neq 0$ which are coordinate charts with coordinate functions taking values in $\mathbb{C}^N$ denoted either w_i or

$$z_i(j) = \begin{cases} z_{i-1}/z_j & \text{for } i \leq j, \\ z_i/z_j & \text{for } i > j. \end{cases}$$

To see that $\mathbb{P}_N(\mathbb{C})$ has a natural Kähler manifold structure, we introduce the following functions which turn out to be Kähler potentials

$$K_j = \log\left(1 + \sum_{k \neq j} |z_k/z_j|^2\right) = \log\left(\sum_{k=0}^{N} |z_k|^2\right) - \log|z_j|^2$$

On the intersection of two coordinate charts $U_j \cap U_k$, the difference

$$K_j - K_k = \log|z_k|^2 - \log|z_j|^2 = \log\left(z_k/z_j\right) - \log\left(\overline{z}_k/\overline{z}_j\right)$$

satisfies the equation $d'd''(K_j - K_k) = 0$, and hence there exists a global form ω on $\mathbb{P}_N(\mathbb{C})$ with $\omega|U_j = id'd''K_j$.

To see that this form ω is the form associated with a Kähler structure, we consider its coefficients $h_{k,\ell}$ in local coordinates where

$$\omega|U_j = id'd''K_j = i\sum_{k,\ell} h_{k,\ell}\, dw_k \wedge d\overline{w}_\ell$$

and $w_k = z_k(j) = z_k/z_j$.

(3.9) Assertion. The coefficient matrix $(h_{k,\ell})$ is positive definite and Hermitian symmetric. For this we calculate with

$$K_j = \log\left(1 + \sum_{k\neq j} |z_k/z_j|^2\right) = \log\left(1 + \sum_{k=1}^{N} |w_k|^2\right)$$

the differentials

$$d''K_j = \left(1 + \sum_{k=1}^{N} |w_k|^2\right)^{-1} \cdot \sum_{k=1}^{N} w_k d\overline{w}_k$$

and

$$\begin{aligned} d'd''K_j = {} & \left(1 + \sum_{k=1}^{N} |w_k|^2\right)^{-1} \cdot \sum_{k=1}^{N} dw_k \wedge d\overline{w}_k \\ & - \left(1 + \sum_{k=1}^{N} |w_k|^2\right)^{-2} \cdot \sum_{k,\ell=1}^{N} \overline{w}_k dw_k \wedge w_l d\overline{w}_\ell \\ = {} & \left(1 + \sum_{k=1}^{N} |w_k|^2\right) \cdot \sum_{k,\ell=1}^{N} \left[\delta_{k,\ell}\left(1 + \sum_{k=1}^{N} |w_k|^2\right) - \overline{w}_k w_\ell\right] dw_k \wedge d\overline{w}_\ell. \end{aligned}$$

For a complex vector $\zeta = (\zeta_1, \ldots, \zeta_N) \in \mathbb{C}^N$, we study the positivity properties of the following expression using the Hermitian inner product $(\omega|\zeta) = \sum_{k=1}^{N} w_k \overline{\zeta}_k$ to obtain the inequality

$$\begin{aligned} & \sum_{k,\ell=1}^{N} \left\{\delta_{k,\ell}\left(1 + \sum_{k=1}^{N} |w_k|^2\right) - \bar{w}_k w_\ell\right\} \zeta_k \bar{\zeta}_\ell \\ & \quad = (\zeta \mid \zeta)^2(1 + (w \mid w)^2) - |(w \mid \zeta)|^2 > 0 \qquad \text{for } \zeta \neq 0. \end{aligned}$$

for $\zeta \neq 0$. Here we have used the Schwarz inequality $|(w|\zeta)|^2 \leq (\zeta|\zeta)^2(w|w)^2$.

(3.10) Remark. A complex submanifold of a Kähler manifold is a Kähler manifold so that all smooth projective algebraic varieties are Kähler manifolds. Conversely, we have the following theorem.

Theorem of Chow. *All closed submanifolds of $\mathbb{P}_N(\mathbb{C})$ are the zero locus of a finite number of homogeneous polynomial equations, that is, they are smooth projective algebraic varieties.*

In section (2.11) we gave a short indication of the fact that for a smooth, complex analytic manifold the de Rham cohomology has a natural filtration, called the Hodge filtration. For this we use the sheaf Ω_X^p of holomorphic p-forms on X.

(3.11) Betti Numbers and Hodge Numbers of Varieties. Let X be a smooth, compact complex manifold. The singular cohomology groups are known to be finitely generated, and over any field k with $\mathbb{Q} \subset k \subset \mathbb{C}$, the dimension $b_i = \dim_k H^i(X, k)$ is independent of the field k and is called the ith Betti number of X. The cohomology groups of X with values in these sheaves Ω_X^p of forms are finite dimensional over $\mathbb{C}$, and the dimension $h^{p,q} = \dim H^q(X, \Omega_X^p)$ is called the (p, q)-Hodge number. This arose because the differential forms defining the cohomology have a bigraded decomposition.

In (2.10) a Riemannian structure resulted in a unique harmonic form associated to a cohomology class in a given degree. Now, the Kähler condition allows us to use harmonic forms to give a bigraded decomposition of cohomology classes. The result is the following which we only state and refer to Griffths and Haris [1978].

(3.12) Theorem. *Let X be a compact Kähler manifold. The Hodge spectral sequence has the property that $E_1 = E_\infty$, and the Hodge filtration splits so that $H^m_{\mathrm{DR}}(X, \mathbb{C}) = \oplus_{p+q=m} H^q(X, \Omega_X^p)$. Each cohomology class in degree m is represented by a harmonic form α, and it decomposes $\alpha = \sum_{m=p+q} \alpha_{p,q}$ where $\alpha_{p,q} \in A^{p,q}(X)$ is a harmonic (p, q)-form. Hence we have a second decomposition $H^m_{\mathrm{DR}}(X, \mathbb{C}) = \oplus_{p+q=m} H^{p,q}$ where $H^{p,q}$ is the subspace of $H^m_{\mathrm{DR}}(X, \mathbb{C})$ represented by harmonic (p, q)-forms. Also, $H^{p,q} = H^q(X, \Omega_X^p)$ as subspaces of $H^{p+q}_{\mathrm{DR}}(X, \mathbb{C})$.*

We return to this subject in section 6, where the low dimensional examples of curves, surfaces, and threefolds are considered especially in the context of Calabi–Yau manifolds. The Hodge numbers of a Calabi–Yau manifold are one of the first basic invariants of the manifold.

§4. Connections, Curvature, and Holonomy

We consider the theory of affine connections and curvature for complex vector bundles over both smooth and complex analytic manifolds. Characteristic classes defined by connections and curvature are introduced. Connections associated to a Hermitian metric are considered, especially in the case where the metric is Kähler. The metric, curvature tensor, and Chern classes yield the following types of differential forms:

(1) Bilinear forms associated with a Kähler metric and the trace of the curvature tensor, called the Ricci tensor. These are of type $(2, 0)$.

(2) Differential $(1, 1)$-forms defining the first Chern class of a complex vector bundle and as the anti-symmetrization of the forms associated with a Kähler metric.

It is the interplay between these forms which is central in the Calabi–Yau considerations.

Recall that exterior differentiation $d : A^m(M) \to A^{m+1}(M)$ is defined for smooth manifolds M and is split as a sum $d = d' + d''$ for a complex analytic manifold X where

$$d' : A^{p,q}(X) \to A^{p+1,q}(X) \quad \text{and} \quad d'' : A^{p,q}(X) \to A^{p,q+1}(X).$$

Connections are extensions of exterior differentiation to sections $A^0(M, E)$ of a bundle vector E, and we denote by $A^m(M, E) = A^0(M, \Lambda^m(M) \otimes E)$ the vector valued forms in E.

(4.1) Definition. Let M be a smooth manifold. A connection on a real or complex vector bundle E over M is a morphism $\nabla : A^0(M, E) \to A^1(M, E)$ satisfying the Leibniz rule

$$\nabla(f\sigma) = f\nabla(\sigma) + df \otimes \sigma \quad \text{for } f \in A^0(M) \text{ and } \sigma \in A^0(M, E).$$

A connection on a complex vector bundle E over a complex analytic manifold X is a splitting of a smooth connection into $\nabla' : A^0(X, E) \to A^{1,0}(X, E)$ and $\nabla'' : A^0(X, E) \to A^{0,1}(X, E)$ with $\nabla = \nabla' + \nabla''$ each satisfying the Leibniz rule $\nabla'(f\sigma) = f\nabla'(\sigma) + d'f \otimes \sigma$ and $\nabla''(f\sigma) = f\nabla''(\sigma) + d''f \otimes \sigma$ for $\sigma \in A^0(M, E)$.

There is a second equivalent approach to the notation of a connection ∇ as a bilinear map $\mathrm{Vect}(X) \times A^0(M, E) \to A^0(M, E)$ denoted $\nabla_\xi \eta$. It is defined by pairing $\xi \in \mathrm{Vect}(X) = A(M, T(M))$ with the form component of $A^1(M, E) = A(M, T^*(M) \otimes E)$ giving $C^\infty(M)$-linear morphism, also denoted $\xi : A^1(M, E) \to A^0(M, E)$. With this notation we define $\nabla_\xi(\eta) = \xi(\nabla(\eta))$. This function of two variables $\nabla_\xi \eta$ is biadditive, and it satisfies

$$\nabla_{f\xi}(\eta) = f\nabla_\xi(\eta)$$

and the Leibniz rule which becomes

$$\nabla_\xi(f\eta) = f\nabla_\xi \eta + \xi(f)\eta$$

for $\xi \in \mathrm{Vect}(M)$, $\eta \in A^0(M, E)$, and $f \in A^0(M) = C^\infty(M)$.

(4.2) Example. If E is a complex analytic bundle on a complex manifold X, then $d'' : A^0(X, E) \to A^{0,1}(X, E) \subset A^1(X, E)$ is a well defined connection ∇ with $\nabla^{1,0} = 0$ and $\nabla^{0,1} = d''$. In effect, consider a frame $\mathfrak{s} = (s_1, \ldots, s_N) \in A^0(E)^N$ is a local trivialization of E on an open set U as in (1.11). Any $\sigma \in \Gamma(U, E)$ has an expansion $\sigma = \sum_{i=1}^N f_i s_i$, and the formula $d''(\sigma) = \sum_{i=1}^N d''(f_i) \otimes s_i$ defines the connection.

To show it is well defined, consider a second frame $\mathfrak{s} = g\mathfrak{s}'$ or $s_i = \sum_i g_{i,j} s'_j$ and the representation of

$$\sigma = \sum_{i=1}^N f_i s_i = \sum_{i,j=1}^N f_i g_{i,j} s'_j = \sum_{j=1}^N f'_j s'_j.$$

So

$$d''(\sigma) = \sum_{i=1}^N d''(f_i) \otimes s_i = \sum_{i,j=1}^N d''(f_i g_{i,j} \otimes s'_j) = \sum_{i,j=1}^N d''(f_i) g_{i,j} \otimes s'_j$$
$$= \sum_{i=1}^N d''(f'_i) \otimes s.$$

(4.3) Definition. A connection ∇ on a complex vector bundle E over a complex analytic manifold X is compatible with the complex structure on E provided $\nabla^{0,1} = d''$. It is also called a holomorphic connection.

Moreover, if ∇ is compatible with the complex structure, then $\nabla_\xi \eta$ is holomorphic for $\xi, \eta \in \mathrm{Vec}(X)^{1,0}$.

(4.4) Remark. A connection $\nabla : A^0(M, E) \to A^1(M, E)$ on a bundle over a smooth manifold extends to a map $\nabla : A^m(M, E) \to A^{m+1}(M, E)$ satisfying the extended Leibniz rule

$$\nabla(\alpha \otimes \sigma) = d\alpha \wedge \sigma + (-1)^m \alpha \wedge \nabla(\sigma) \quad \text{for } \sigma \in A^0(M, E) \text{ and } \alpha \in A^m(M).$$

In the case of a complex manifold X and a connection ∇ on a complex bundle E, the splitting $\nabla = \nabla' + \nabla''$

$$\nabla' : A^0(X, E) \to A^{1,0}(X, E) \quad \text{and} \quad \nabla'' : A^0(X, E) \to A^{0,1}(X, E)$$

there are extensions on the (p, q)-forms defined as follows

$$\nabla' : A^{p,q}(X, E) \to A^{p+1,q}(X, E) \quad \text{and} \quad \nabla'' : A^{p,q}(X, E) \to A^{p,q+1}(X, E)$$

satisfying the extended Leibniz rules

$$\nabla'(\sigma\theta) = \nabla'(\sigma) \wedge \theta + (-1)^p \sigma \wedge d'\theta \quad \text{and} \quad \nabla''(\sigma\theta) = \nabla''(\sigma) \wedge \theta + (-1)^q \sigma \wedge d''\theta$$

for $\sigma \in A^0(M, E)$ and $\theta \in A^{p,q}(M)$.

This extension of the differential calculus to forms with coefficients in a bundle with connection leads to the notion of curvature.

(4.5) Proposition. *The square $\nabla\nabla : A^0(M, E) \to A^2(M, E)$ is $A^0(M)$-linear, and there is $\Omega \in A^2(M, \mathrm{End}(E))$ such that $\nabla^2(\sigma) = \Omega\sigma$ for all $\sigma \in A^0(M, E)$. In the complex analytic case, the decomposition*

$$\nabla\nabla = \nabla'\nabla' + (\nabla'\nabla'' + \nabla''\nabla') + \nabla''\nabla''$$

induces a splitting of the curvature form Ω as a sum

$$\Omega = \Omega^{(2,0)} + \Omega^{(1,1)} + \Omega^{(0,2)} \quad \textit{where } \Omega^{(p,q)} \in A^{p,q}(M, \mathrm{End}(E)).$$

If ∇ is compatible with the complex structure, then $d''d'' = 0$ and $\Omega^{(0,2)} = 0$.

Proof. We calculate

$$\nabla^2(f\sigma) = \nabla(f\nabla\sigma + df \otimes \sigma) = f\nabla^2(\sigma) + df \wedge \nabla\sigma - df \wedge \nabla\sigma = f\nabla^2(\sigma).$$

The second assertion is a property of $A^0(M)$-linear operators which is seen locally. This proves the proposition.

For a local study of curvature consider a local trivialization of E given by a frame $\mathfrak{s} = (s_1, \ldots, s_N) \in A^0(E)^N$ over a suitable open set as (1.11). Applying the connection to the frame, we obtain $\nabla s_i = \sum_{i=1}^{N} \omega_{i,j} \otimes s_j$ where $\omega_{i,j} \in A^1(E)$ and $\omega_{i,j} \in A^{1,0}(E)$ in the case of ∇ compatible with a complex structure. In matrix notation, $\nabla\mathfrak{s} = \omega\mathfrak{s}$ where $\mathfrak{s}$ and $\nabla\mathfrak{s}$ are column matrices and ω is a square matrix of 1-forms. Now we calculate the curvature form Ω locally

$$\nabla^2\mathfrak{s} = \nabla(\omega \otimes \mathfrak{s}) = d\omega\mathfrak{s} - \omega \wedge \omega\mathfrak{s} = (d\omega - \omega \wedge \omega)\mathfrak{s} = \Omega\mathfrak{s},$$

hence $\Omega = d\omega - \omega \wedge \omega$. The exterior derivative has the following form

$$d\Omega = -d\omega \wedge \omega + \omega \wedge d\omega = \omega \wedge \Omega - \Omega \wedge \omega = [\omega, \Omega].$$

(4.6) Remark. Except in the abelian case of line bundles the curvature is not a closed form, but suitable traces of it are closed forms from the formula $d\Omega = [\omega, \Omega]$. Under change of frame of the form $s'_i = \sum_{j=1}^{N} a_{i,j}s_j$ or $\mathfrak{s}' = A\mathfrak{s}$ in matrix notation, we have two matrix 1-forms ω and ω' where $\nabla\mathfrak{s} = \omega \otimes \mathfrak{s}$ and $\nabla\mathfrak{s}' = \omega' \otimes \mathfrak{s}'$ and two related curvature matrix 2-forms $\Omega = d\omega - \omega \wedge \omega$ and $\Omega' = d\omega' - \omega' \wedge \omega'$. In the *Fibre Bundles*, pp. 284–285 we show that the two pairs of matrices of forms are related $\omega' = A\omega A^{-1} + (dA)A^{-1}$ and $\Omega' = A\Omega A^{-1}$. This is used to define characteristic classes, see §5, in de Rham cohomology.

Associated with any Riemannian metric is a unique connection, called the Levi-Cita connection, such that parallel transport preserves the metric.

(4.7) Theorem. *Let g be a pseudo-Riemannian metric on $T_{\mathbb{R}}(M)$ where M is a smooth manifold. Then there exists a connection on $T_{\mathbb{R}}(M)$ with the property that*

$$\begin{aligned}2g(\nabla_\xi\eta, \zeta) = \xi(g(\eta, \zeta)) + \eta(g(\xi, \zeta)) - \zeta(g(\xi, \eta))\\ + g([\xi, \eta], \zeta) + g([\zeta, \xi], \eta) + g(\xi, [\zeta, \eta]).\end{aligned}$$

The torsion of $\nabla_\xi\eta$ given by the tensor $T(\xi, \eta) = \nabla_\xi\eta - \nabla_\eta\xi - [\xi, \eta]$ is zero, and this connection is unique among connections leaving g invariant and having zero torsion. Moreover, the curvature of the connection $\nabla_\xi\eta$ is given by $R(\xi, \eta) = [\nabla_\xi, \nabla_\eta] - \nabla_{[\xi,\eta]}$.

Associated with any Hermitian metric on a complex vector bundle there is a unique natural connection preserving the metric.

(4.8) Theorem. *Let E be a Hermitian vector bundle on a complex manifold X. There exists a unique connection ∇ which is compatible with the complex structure and compatible with the metric $h(\xi, \eta)$ in the sense that*

$$dh(\xi, \eta) = h(\nabla\xi, \eta) + h(\xi, \nabla\eta).$$

Proof. Let $\mathfrak{s} = (s_1, \ldots, s_N) \in A^0(E)^N$ be a local frame of E over a suitable open set as in (1.11). Let $h_{i,j} = h(s_i, s_j)$. If ∇ exists, then its matrix ω with respect to $\mathfrak{s}$ must be of type $(1,0)$ forms, and hence $dh_{i,j} = dh(s_i, s_j) = \sum_k \omega_{i,k} h_{k,j} + \sum_k \overline{\omega}_{j,k} h_{i,k}$ decomposes into forms of type $(1,0)$ and $(0,1)$ as follows

$$d'h_{i,j} = \sum_k \omega_{i,k} h_{k,j} \quad \text{and} \quad d''h_{i,j} = \sum_k \overline{\omega}_{j,k} h_{i,k}.$$

In matrix notation these relations reduce simply to $d'h = \omega h$ and $d''h = h^t\overline{\omega}$. Then $\omega = (d'h)h^{-1}$ is the unique solution to both equations, and since ω is well determined by compatibility under change of frame, it is globally defined. This proves the theorem.

(4.9) Remark. If the local frame of a Hermitian vector bundle E in the previous theorem is unitary, that is, if $h(s_i, s_j) = \delta_{i,j}$, then we have $0 = dh(s_i, s_j) = \omega_{i,j} + \overline{\omega}_{j,i}$, and thus the matrix associated with a unitary frame is skew-Hermitian. In terms of covariant differentiation parallel transport takes the following form.

(4.10) Parallel Transport. Let M be a smooth manifold with a connection ∇ on a vector bundle $q : E \to M$. A vector field v along a curve $c : [a, b] \to M$ is a lifting $v : [a, b] \to E$ with $qv = c$. Let $c' : [a, b] \to T(M)$ denote the tangent vector lifting to c. The vector field v is a parallel transport (with respect to ∇) provided

$$\nabla_{c'(t)} v = 0 \quad \text{for all } t \in [a, b].$$

In local coordinates, this is a first order differential equation, and, as such, it has a unique solution for given initial data. This initial data is a vector $v_a \in E_{c(a)}$, and the solution is a vector $v(t) \in E_{c(t)}$ depending smoothly on $t \in [a, b]$.

The parallel transport defined by these curves c is the linear transformation $T_c : E_{c(a)} \to E_{c(b)}$ assigning to a vector $w \in E_{c(a)}$ firstly the solution $v(t)$ to the parallel transport equation with $v(a) = w$ and then the value $v(b) = T_c(v(a)) = T_c(w) \in E_{c(b)}$.

We must remark that the theory of connections and vector fields was developed only globally, and here we have used it for tangent vectors along a curve. The pointwise properties are left to the reader to work out. For the product of two parameterized curves $c.d$, that is, c followed by d, we have also the transitivity property $T_{c.d} = T_d T_c : E_{(c.d)(a)} \to E_{(c.d)(b)}$ so that the product of curves gives composition of parallel transport. Also, the parallel transport of the inverse path is the inverse linear map.

(4.11) Holonomy Groups. Let M be a smooth manifold with a connection ∇ on a vector bundle $q : E \to M$. The holonomy group for (E, ∇) at $x \in M$ is the subgroup of all parallel transport elements $T_c \in \mathrm{GL}(E_x)$ for loops c at x. From the composition of parallel transport property we know that T_c is an automorphism and the set of these elements is closed under multiplication of linear automorphisms of E_x. This is a closed subgroup of the linear group.

(4.12) Complex Holonomy Groups. Let X be a complex analytic manifold with a holomorphic connection ∇ on a complex analytic vector bundle $q : E \to X$

preserving a Hermitian product on E. Then the holonomy group, which is contained in $\mathrm{GL}(N, \mathbb{C})$, the complex linear group for $N = \dim E$. It is in the subgroup $U(N) \subset \mathrm{GL}(N, \mathbb{C})$ in the case of a Kähler manifold where the complex analytic tangent bundle is $T(X)^{(1,0)}$.

The holonomy group, being a closed subgroup of $\mathrm{GL}(N, \mathbb{C})$, has a Lie algebra which is a subalgebra of $\mathrm{LieGL}(N, \mathbb{C}) = M_N(\mathbb{C})$. The restriction of $\exp : M_N(\mathbb{C}) \to \mathrm{GL}(N, \mathbb{C})$ to this Lie algebra maps into the holonomy group. Ambrose and Singer describe this sub-Lie algebra with curvature.

(4.13) Curvature Generates the Lie Algebra of the Holonomy Group. An early reference in this direction is the paper of Ambrose and Singer, TAMS, **75** (1953).

(4.14) Remark. The connection associated with a metric has the property that parallel transport preserves the metric. A Hermitian metric is Kähler if and only if parallel transport commutes with multiplication by i.

(4.15) SU(*N*)-holonomy and Ricci Curvature. A fundamental property of

$$\exp : M_N(\mathbb{C}) \to \mathrm{GL}(N, \mathbb{C})$$

is that $\exp(\mathrm{tr}(A)) = \det(\exp(A))$. In particular, $\exp : \mathfrak{su}(N) \to \mathrm{SU}(N)$ is a local surjection which restricts from $\exp : \mathfrak{u}(N) \to U(N)$.

Here is where the Calabi–Yau theory starts from the complex differential geometry perspective in the next sections.

§5. Projective Spaces, Characteristic Classes, and Curvature

For additional details to this sketch of the theory of Chern classes see the last chapters of Husemöller, *Fibre Bundles*. The theory of characteristic classes begins with the first Chern class of a line bundle. This is a topological theory starting with complex projective space, and the fact that many considerations in complex geometry, Kähler geometry, and algebraic geometry start with projective space suggests that Chern classes are basic to many areas of geometry. This is the case.

(5.1) Remark. The finite and infinite complex projective spaces $\mathbb{P}_N(\mathbb{C}) \subset P_\infty(\mathbb{C})$ have four aspects.

(1) The second cohomology group is infinite cyclic with a canonical generator $H^2(\mathbb{P}_N(\mathbb{C}), \mathbb{Z}) = \mathbb{Z}\iota_N$ such that ι_∞ restricts to ι_N for all N. For the reader with a little background in topology, this can be seen from the isomorphism

$$\begin{aligned} H^2(\mathbb{P}_N(\mathbb{C}), \mathbb{Z}) &= \mathrm{Hom}(H_2(\mathbb{P}_N(\mathbb{C}), \mathbb{Z}), \mathbb{Z}) \\ &\to \mathrm{Hom}(\pi_2(\mathbb{P}_N(\mathbb{Z})), \mathbb{Z}) = \mathrm{Hom}(\mathbb{Z}, \mathbb{Z}) \end{aligned}$$

which is the universal coefficient theorem followed by the dual of the Hurewicz isomorphism. Then ι_N maps to the identity on $\mathbb{Z}$.

(2) The homotopy groups of the infinite projective space are

$$\pi_i(\mathbb{P}_\infty(\mathbb{C})) = \begin{cases} \mathbb{Z} & \text{for } i = 2, \\ 0 & \text{for } i \neq 2. \end{cases}$$

Such spaces with exactly one nonzero homotopy group play a basic role in classifying cohomology, and we say that such a space $\mathbb{P}_\infty(\mathbb{C})$ is a $K(\mathbb{Z}, 2)$-space. For each cohomology class $c \in H^2(X, \mathbb{Z})$ there exists a map $f : X \to \mathbb{P}_\infty(\mathbb{C})$ such that $c = f^*(\iota_\infty)$, and f is unique up to homotopy with this property.

If X is finite dimensional, for example a manifold or if X is compact, then every map $f : X \to \mathbb{P}_\infty(\mathbb{C})$ factors by some inclusion $\mathbb{P}_N(\mathbb{C}) \to \mathbb{P}_\infty(\mathbb{C})$ giving a map $g : X \to \mathbb{P}_N(\mathbb{C})$ with the property that $c = g^*(\iota_N)$.

(3) The projective space $\mathbb{P}_N(\mathbb{C})$ with homogeneous coordinates $z_0 : \cdots : z_N$ has an open covering of $N+1$ open coordinate domains given by $z_j \neq 0$. On the domain $z_j \neq 0$, the coordinate function is

$$z_k(j) = z_k/z_j \quad \text{for } k \neq j, 0 \leq k \leq N.$$

We have seen in (3.8) that the finite projective space $\mathbb{P}_N(\mathbb{C})$ is a Kähler manifold of complex dimension N.

(4) The projective spaces have canonical line bundles $L(N) \to \mathbb{P}_N(\mathbb{C})$ where $L(\infty)$ restricts to $L(\infty)|\mathbb{P}_N(\mathbb{C}) = L(N)$. The space $L(N)$ is the subspace of $\mathbb{P}_N(\mathbb{C}) \times \mathbb{C}^{N+1}$ consisting of $(z, \lambda z)$ for $\lambda \in \mathbb{C}$. For every topological $\mathbb{C}$-line bundle $L \to X$ over a space X there exists a map $f : X \to \mathbb{P}_\infty(\mathbb{C})$ such that L and $f^*(L(\infty))$ are isomorphic, and f is unique up to homotopy. If X is compact or finite dimensional, then we can choose $f : X \to \mathbb{P}_N(\mathbb{C})$.

(5.2) First Chern Class of a Line Bundle. This is defined by observing that the set of homotopy classes $[X, \mathbb{P}_\infty(\mathbb{C})]$ of mappings $x \to \mathbb{P}_\infty(\mathbb{C})$ classify isomorphism classes of topological complex line bundles $L \to X$ and also elements of $H^2(X, \mathbb{Z})$ by (5.1) (1). The cohomology class $c_1(L) \in H^2(X, \mathbb{Z})$ associated to the line bundle L is called the first Chern class of L. The function

$$c_1 : \{\text{isomorphism classes of complex line bundles}/X\} \to H^2(X, \mathbb{Z})$$

is a bijection satisfying the multiplication property

$$c_1(L' \otimes L'') = c_1(L') + c_1(L'').$$

The higher Chern classes $c_j(E)$ of a complex vector bundle E have an axiomatic characterization, due to Hirzebruch, based on the first Chern classes $c_1(L)$ of a complex line bundle.

(5.3) Axioms for Chern Classes. The Chern class of a complex vector bundle E over X is an element $c(E) \in H^{\text{ev}}(X, \mathbb{Z}) = \bigoplus_{i \geq 0} H^{2i}(X, \mathbb{Z})$ satisfying the following properties:

(1) The Chern class is a sum $c(E) = 1 + c_1(E) + \cdots + c_n(E)$ where $c_0(E) = 1$, $c_i(E) = 0$ for $i > \dim(E)$ and $c_i(E) \in H^{2i}(X, \mathbb{Z})$.
(2) If $f : X' \to X$ is a map, and if $E \to X$ is a complex vector bundle over X, then $c(f^*(E)) = f^*(c(E))$ in $H^{\mathrm{ev}}(X', \mathbb{Z})$ where $f^*(E)$ is the induced bundle on X'.
(3) The Chern class of the Whitney sum $E' \oplus E''$ is the cup product

$$c(E' \oplus E'') = c(E')c(E'')$$

in $H^{\mathrm{ev}}(X, \mathbb{Z})$.
(4) For a line bundle L, the Chern class $c(L) = 1 + c_1(L)$ where $c_1(L)$ is defined in (5.2).

(5.4) Remark. As Grothendieck observed in algebraic geometry, the existence and uniqueness of the Chern classes $E \to X$ can be established by working with the related bundle $\mathbb{P}(E) \to X$ of projective spaces and the standard line bundle $L_E \to \mathbb{P}(E)$ reducing to the canonical line bundle on each fibre. The class $c = c_1(L_E)$ is in $H^{\mathrm{ev}}(\mathbb{P}(E), \mathbb{Z})$ and $1, c, c^2, \ldots, c^{n-1}$ is a basis $H^{\mathrm{ev}}(\mathbb{P}(E), \mathbb{Z})$ as a free $H^{\mathrm{ev}}(X, \mathbb{Z})$ module under the cup product preserving morphism $H^{\mathrm{ev}}(X, \mathbb{Z}) \to H^{\mathrm{ev}}(\mathbb{P}(E), \mathbb{Z})$ induced by the projection $\mathbb{P}(E) \to X$. The Chern classes $c_i(E)$ are the coefficients of the equation

$$c^n - c_1(E)c^{n-1} + \cdots + (-1)^{n-1}c_{n-1}(E)c + (-1)^n c_n(E) = 0.$$

For further details, see *Fibre Bundles*, pp. 249–252.

In chapter 19 of *Fibre Bundles* we described how to define the Chern classes of a complex vector bundle using a connection and its curvature form. We sketch this theory and extend it to holomorphic connections on complex bundles on complex manifolds.

(5.5) Elementary Symmetric Functions. The elementary symmetric functions $\sigma_q(x_1, \ldots, x_n) = \sum_{i(1)<\cdots<i(q)} x_{i(1)} \cdots x_{i(q)}$ are also encoded in the expression

$$Q_x(t) = \prod_{1\le j\le n} (1 + x_j t) = \sum_{0\le q\le n} \sigma_q(x_1, \ldots, x_n)t^q.$$

This leads to other polynomials e_q which as polynomials of the symmetric function $\sigma_q(x_1, \ldots, x_n)$ have the form

$$-t\frac{d}{dt} \log Q_x(t) = \sum_{q\ge 1} e_q(\sigma_1, \ldots, \sigma_n)(-t)^q.$$

The ring of symmetric functions is the subalgebra of $k[x_1, \ldots, x_n]$ invariant under the action of the symmetric group on the variables $x_1, \ldots, x_n$. It is itself a polynomial ring on $\sigma_1, \ldots, \sigma_n$ and it contains the functions $e_1, \ldots, e_n$. If k is of characteristic zero, then $k[\sigma_1, \ldots, \sigma_n] = k[e_1, \ldots, e_n]$.

Let k be a field of characteristic zero in the next parts.

(5.6) Conjugation Invariant Polynomials in Matrix Elements. The ring of $\mathrm{GL}(n,k)$-conjugation invariant polynomials in the polynomial ring $k[x_{1,1},\dots,x_{n,n}]$ on n^2-variables $x_{i,j}$ is $k[c_1(x),\dots,c_n(x)]$ where the polynomial $c_q(x)$ is defined by the following $\mathrm{GL}(n,k)$-invariant formula

$$R_x(t) = \det(I + [x_{i,j}]t) = \sum_{0\le q\le n} c_q(x)t^q.$$

In the case $[x_{i,j}]$ is diagonal with $x_{i,j} = \delta_{i,j}\lambda_i$, we see that c_q is an elementary symmetric function

$$c_q(x_{1,1},\dots,x_{n,n}) = \sum_{i(1)<\cdots<i(q)} \lambda_{i(1)}\dots\lambda_{i(q)} = s_q(\lambda_1,\dots,\lambda_n).$$

Again we consider the logarithmic derivative

$$-t\frac{d}{dt}\log(\det(I + [x_{i,j}]t)) = \sum_{q\ge 1} \mathrm{Tr}(X^q)(-t)^q.$$

Thus as in the case of symmetric functions, the subalgebra of $\mathrm{GL}(n,k)$-conjugation invariant functions has two forms

$$k[c_1(x_{i,j}),\dots,c_n(x_{i,j})] = k[\mathrm{Tr}(X),\mathrm{Tr}(X^2),\dots,\mathrm{Tr}(X^n)]$$

where X denotes the matrix $[x_{i,j}]$. The intersection of this subspace and the homogeneous polynomials of degree q is denoted by $\mathrm{Inv}_q(n)$.

Let $\mathrm{Inv}(n)$ denote the direct sum over the homogeneous $\bigoplus_q \mathrm{Inv}_q(n)$.

(5.7) Definition. Let (E,∇) be a pair consisting of a complex vector bundle E over a smooth manifold M with connection ∇ having a curvature form Ω locally well defined up to the inner automorphism, see (4.6). For any $\phi \in \mathrm{Inv}(n)$, we have a well defined form $\phi(\Omega)$ independent of the local coordinates of E. In particular, we define the Chern forms $c_q(E,\nabla) = \frac{1}{(2\pi i)_q} c_q(\Omega)$ using the invariant polynomial c_q introduced in (5.6). These Chern forms depend on the connection ∇ and the complex vector bundle E.

(5.8) Remark. We can see that these forms $\phi(\Omega)$ are closed by using the Bianchi identity, see (4.6). In effect, $d\Omega = [\omega,\Omega]$ where ω is the corresponding connection form related to ∇ for a local trivialization of E. Since it suffices to check that $\phi(\Omega)$ is closed on generators of

$$k[c_1(x_{i,j}),\dots,c_n(x_{i,j})] = k[\mathrm{Tr}(X),\mathrm{Tr}(X^2),\dots,\mathrm{Tr}(X^n)]$$

we calculate

$$d\mathrm{Tr}(\Omega^q) = \sum_{i+j=q-1} \mathrm{Tr}(\Omega^i(d\Omega)\Omega^j) = \sum_{i+j=q-1} \mathrm{Tr}(\Omega^i[\omega,\Omega]\Omega^j) = \mathrm{Tr}([\omega,\Omega^q]) = 0.$$

(5.9) Curvature Forms for the Kähler Case. For a complex vector bundle $E \to X$ over a Kähler manifold X, the curvature form is given locally by a matrix of the form

$$\Omega = \left(\sum_{k,\ell} \Omega_{i,j,k,\ell} dz_k \wedge d\bar{z}_\ell \right) \in M_N(A^{1,1}(X)).$$

(5.10) First Chern Class. Returning to the calculation in (5.5), we see that $c_1(\Omega) = e_1(\Omega) = \mathrm{Tr}(\Omega)$. This trace is an other well known curvature form, the Ricci curvature form,

$$\mathrm{Ric} = \sum_{j,k,\ell} \Omega_{j,j,k,\ell} dz_k \wedge d\bar{z}_\ell \in A^{1,1}(X).$$

In particular, the vanishing of the Ricci curvature form implies that the first Chern class is trivial. If $\Omega' = \frac{1}{2\pi i}\Omega$, then we have also the following formula for the total Chern class

$$c(X) = 1 + \sum_j c_j(X) = \det(1 + \Omega') = 1 + \mathrm{tr}(\Omega') + \mathrm{tr}(\Omega' \wedge \Omega' - 2(\mathrm{tr}(\Omega'))^2) + \dots.$$

In each case the conjugation invariance leads to intrinsic quantities independent of framings. Now we consider some special features of line bundles and their first Chern class.

(5.11) Remark. The group of line bundles up to isomorphism, with multiplication given by tensor product, can be described as the Čech cohomology group $\check{H}^1(X, \mathcal{O}_X^*)$. Here, a Čech cocycle

$$g = (g_{a,b}) \in \check{Z}^1(\mathcal{U}, \mathcal{O}_X^*)$$

defines a line bundle L on X by gluing trivial bundles on U_a where $\mathcal{U} = (U_a)$ with the invertible $g_{a,b}$ on the intersections $U_a \cap U_b$. For $g_{a,b} = g_{b,a}^{-1}$ we see that the lifting $\sigma_{a,b} = \frac{1}{2\pi i}\log(g_{a,b}) \in C^1(X, \mathcal{O}_X)$ has a coboundary

$$(\delta\sigma)_{a,b,c} = \frac{1}{2\pi i}\left(\log(g_{b,c}) + \log(g_{c,a}) + \log(g_{a,b})\right) \in \check{Z}^2(\mathcal{U}, \mathbb{Z}).$$

This is a Čech cocycle for $c_1(L)$. We return to this subject in (7.4).

(5.12) Remark. For the local calculation in (4.8), we saw that a Hermitian metric h on a complex analytic line bundle leads to a holomorphic connection ∇_h with local connection matrix $\omega_a = (d'h_a)h_a^{-1}$ where h_a is the value of h on the frame over U_a as in the previous paragraph (5.11). The curvature of a line bundle is the exterior derivative of the connection form

$$\Omega_a = d\omega_a = d'(h_a) \wedge d''(h_a^{-1}) = d'd'' \log(h_a).$$

Hence the first Chern class is given by the following de Rham cohomology class:

$$c_1(L, h) = \frac{1}{2\pi i} d'd'' \log(h_a).$$

On $U_a \cap U_b$, the metric and change of frames functions are related by $h_a = |g_{b,a}|^2 h_b$ or $\log(h_a) = \log(h_b) + \log(g_{b,a}) + \log(\overline{g}_{b,a})$. For the connection form $\omega_a = (d'h_a)h_a^{-1} = d'\log(h_a)$, we calculate its Čech coboundary

$$(\delta\omega)_{a,b} = \omega_b - \omega_a = \frac{1}{2\pi i}d(\log(h_b/h_a)) = \frac{1}{2\pi i}d(\log(g_{a,b}\overline{g}_{a,b}))$$
$$= \frac{1}{2\pi i}d(\log(g_{a,b}) = d\sigma_{a,b}$$

Hence the closed two form associated with the cocycle $\delta\sigma$ is the first Chern class form derived from the Hermitian metric h, namely

$$c_1(L,h) = \frac{1}{2\pi i}d'd''\log(h_a).$$

§6. Characterizations of Calabi–Yau Manifolds: First Examples

(6.1) Equivalent Definitions of Calabi-Yau Manifolds. Let X be a compact n dimensional complex manifold. Then X is a Calabi–Yau manifold provided it satisfies any of the following equivalent conditions:

(1) X is a Kähler manifold with a vanishing first Chern class.
(2) X admits a Levi–Civita connection with $\mathrm{SU}(n)$ homology.
(3) X has a Ricci flat Kähler metric.
(4) X is a Kähler manifold with a nowhere vanishing holomorphic n-form.
(5) X is a Kähler manifold with a trivial canonical line bundle ω_X, that is, ω_X is isomorphic to $\mathcal{O}_X$.

(6.2) Remark. The first Chern class is represented by the trace of the curvature 2-form, that is, the Ricci tensor. Hence (3) implies (1). The Calabi conjecture and proven by Yau is the converse implication.

(6.3) Yau's Theorem. If X is a complex Kähler manifold with Kähler form ω and vanishing first Chern class, then there exists a unique Ricci-flat metric on X whose Kähler form is in the same cohomology class as ω.

An early reference on Calabi–Yau 3-folds is Hirzebruch, *Gesammelte Abhandlungen*, T. II, no. 75, pp. 757-770, "Some examples of threefolds with trivial canonical bundle."

The first indication of the importance of the Calabi–Yau variety concept is seen in the many equivalent versions of the definition. Now for basic examples together with their Euler numbers in low dimensions starting with curves of genus one.

(6.4) Example (dimension one). A Calabi–Yau in the extended sense in dimension one can be either a smooth cubic curve in $\mathbb{P}_2$ or the complete intersection of two smooth quadrics in $\mathbb{P}_3$. The Euler number is 0. Note that the fundamental group is infinite.

(6.5) Example (dimension two). Either a smooth quartic surface in $\mathbb{P}$ or the complete intersection of a smooth quadric with a smooth cubic hypersurface in $\mathbb{P}_4$ is a Calabi–Yau in dimension two. The Euler number is 24. There are two new considerations not seen in dimension one.

(1) These examples are special cases of K3 surfaces, for unlike smooth cubic curves in $\mathbb{P}_2$ giving all smooth curves of genus one, not every K3 surface is of this form.
(2) There are K3 surfaces over the complex numbers which are not algebraic surfaces. On the other hand, the Euler number is always 24.

These facts will be partly explained in the next two sections when the theory of K3 surfaces is put in the general Enriques classification theory of surfaces.

(6.6) Example (threefolds). There are five cases of complete intersections of smooth hypersurfaces in $\mathbb{P}_{3+m}$ in general position with the resulting variety is a Calabi–Yau

(1) A quintic in $\mathbb{P}_4$, and the Euler number is -200.
(2) The intersection of a quartic and quadratic in $\mathbb{P}_5$, and the Euler number is -176. The intersection of two cubics in $\mathbb{P}_5$, and the Euler number is -144.
(3) The intersection of a cubic and two quadrics in $\mathbb{P}_6$, and the Euler number is -144.
(4) The intersection of four quadrics in $\mathbb{P}_7$, and the Euler number is -128.

(6.7) Remark. More generally, the m dimensional complete intersection X of k smooth hypersurfaces of degree $d_1, \ldots, d_k$ in $\mathbb{P}_{m+k} = \mathbb{P}$ is a variety with trivial first Chern class zero if and only if $d_1 + \cdots + d_k = m + k + 1$. This follows from the following exact sequence for the tangent sheaf

$$0 \to T_X \to T_{\mathbb{P}}|X \to \bigoplus_{i=1}^{k} \mathcal{O}_X(d_i) \to 0$$

giving the canonical sheaf $K_X = K_{\mathbb{P}}|X\left(\sum_i d_i\right) = (\mathcal{O}_{\mathbb{P}}(-m-k-1)|X)\left(\sum_i d_i\right)$ so that $K_X = \mathcal{O}_X\left(-m-k-1+\sum_i d_i\right)$.

Cohomology provides the first invariants of varieties and Calabi–Yau manifolds in particular. For this we look closer at the Hodge to de Rham spectral sequence considered before in (2.11), (3.11), and (3.12) especially in low dimensions.

(6.8) The Hodge to de Rham Spectral Sequence. The Betti numbers and Hodge numbers are related by $b_i = \sum_{p+q=1} h^{p,q}$ for the cohomology $H^i(X, \mathbb{C})$ has a decreasing filtration $F^p H^i(X, \mathbb{C})$ with $H^q(X, \Omega_X^p)$ isomorphic to $F^p H^{p+q}(X, \mathbb{C})/F^{p+1} H^{p+q}(X, \mathbb{C})$. This is all related to a Hodge to de Rham spectral sequence where the first differential

$$d_1 : E_1^{i,j} = H^j(X, \Omega_X^i) \to E_1^{i+1,j} = H^j(X, \Omega_X^{i+1})$$

is induced by the holomorphic differential

$$0 \longrightarrow \mathcal{O}_X \xrightarrow{d'} \Omega^1_X \xrightarrow{d'} \cdots \xrightarrow{d'} \Omega^i_X \xrightarrow{d'} \cdots$$

The spectral sequence $E_r^{i,j}$ is defined for all $r \geq 1$ with differentials

$$d_r : E_r^{i,j} \to E_r^{i+r,j-r+1}$$

such that $H(E_r, d_r) = E_{r+1}$.

In dimension 1, we have the differential $d_1 : E_1^{p,q} \to E_1^{p+1,q}$

$$H^0(X, \mathcal{O}_X) \longrightarrow H^1(X, \Omega^1_X)$$

$$H^0(X, \mathcal{O}_X) \longrightarrow H^1(X, \Omega^1_X)$$

In dimension 2, we have the differential $d_1 : E_1^{p,q} \to E_1^{p+1,q}$

$$H^2(X, \mathcal{O}_X) \longrightarrow H^2(X, \Omega^1_X) \longrightarrow H^2(X, \Omega^2_X)$$

$$H^1(X, \mathcal{O}_X) \longrightarrow H^1(X, \Omega^1_X) \longrightarrow H^1(X, \Omega^2_X)$$

$$H^0(X, \mathcal{O}_X) \longrightarrow H^0(X, \Omega^1_X) \longrightarrow H^0(X, \Omega^2_X).$$

In dimension 3, we have the differential $d_1 : E_1^{p,q} \to E_1^{p+1,q}$

$$H^3(X, \mathcal{O}_X) \longrightarrow H^3(X, \Omega^1_X) \longrightarrow H^3(X, \Omega^2_X) \longrightarrow H^3(X, \Omega^3_X)$$

$$H^2(X, \mathcal{O}_X) \longrightarrow H^2(X, \Omega^1_X) \longrightarrow H^2(X, \Omega^2_X) \longrightarrow H^2(X, \Omega^3_X)$$

$$H^1(X, \mathcal{O}_X) \longrightarrow H^1(X, \Omega^1_X) \longrightarrow H^1(X, \Omega^2_X) \longrightarrow H^1(X, \Omega^3_X)$$

$$H^0(X, \mathcal{O}_X) \longrightarrow H^0(X, \Omega^1_X) \longrightarrow H^0(X, \Omega^2_X) \longrightarrow H^0(X, \Omega^3_X).$$

The differentials d_r with $r \geq 1$ are all zero for smooth algebraic varieties and Kähler manifolds.

(6.9) Poincaré Duality and Serre Duality. For a complex manifold X of complex dimension n we have a nondegenerate pairing

$$H^i_{\mathrm{DR}}(X) \times H^{2n-i}_{\mathrm{DR}}(X) \to \mathbb{C}$$

so that the Betti numbers satisfy the symmetry relation

$$b_i = \dim H^i_{\rm DR}(X) = \dim H^{2n-i}_{\rm DR}(X) = b_{2n-i}.$$

Serre duality is a nondegenerate pairing

$$H^i(X, \mathcal{L}) \times H^{n-i}(X, \omega_X \otimes \mathcal{L}^{(-1)\otimes}) \to k$$

for a line bundle $\mathcal{L}$ where ω_X is the dualizing sheaf. For smooth manifolds, we have $\omega_X = \Omega^n_X$. Hence the dimensions satisfy the symmetry

$$h^i(\mathcal{L}) = \dim H^i(X, \mathcal{L}) = \dim H^{n-i}(X, \omega_X \otimes \mathcal{L}^{(-1)\otimes}) = h^{n-i}(\omega_X \otimes \mathcal{L}^{(-1)\otimes}).$$

(6.10) Cohomology Properties of Calabi-Yau Manifolds. The fifth Calabi–Yau condition that X is a Kähler manifold with a trivial canonical line bundle ω_X implies that

$$H^0(X, \Omega^i_X) = \begin{cases} 0 & \text{for } 0 < i < n, \\ k = \mathbb{C} & \text{for } i = 0, n. \end{cases}$$

This leads to vanishing of the E_r terms. For example, the three dimensional diagram becomes for the differential d_1 and ground field k

$$\begin{array}{ccccccc}
k = H^3(X, \mathcal{O}_X) & \longrightarrow & 0 & \longrightarrow & 0 & \longrightarrow & H^3(X, \Omega^3_X) = k \\
0 & \longrightarrow & H^2(X, \Omega^1_X) & \longrightarrow & H^2(X, \Omega^2_X) & \longrightarrow & 0 \\
0 & \longrightarrow & H^1(X, \Omega^1_X) & \longrightarrow & H^1(X, \Omega^2_X) & \longrightarrow & 0 \\
k = H^0(X, \mathcal{O}_X) & \longrightarrow & 0 & \longrightarrow & 0 & \longrightarrow & H^0(X, \Omega^3_X) = k
\end{array}$$

The corresponding nonzero Hodge numbers are either one or $h^{1,1}$, $h^{2,1} = h^{1,2}$, and $h^{2,2} = h^{1,1}$ by Poincarè duality.

(6.11) Recommended Reading. We recommend that the reader consult now the books of Cox and Katz [1999], Joyce [1998], and Voisin [1996].

§7. Examples of Calabi–Yau Varieties from Toric Geometry

The first examples of Calabi–Yau varieties were hypersurfaces or intersections of hypersurfaces in projective space. For these examples the concrete description of the canonical divisor is used. Projective spaces are special cases of weighted projective spaces, and weighted projective spaces are special cases of toric varieties. In each case hypersurfaces and complete intersections of hypersurfaces have a canonical divisor which is described in terms the combinatorial data of the toric variety. In many cases it is trivial,and this gives many more examples of Calabi–Yau varieties.

(7.1) Projective Spaces. The first examples of Calabi–Yau varieties in dimension m were the complete intersections of multiple degree $(d_1, \dots, d_k)$ in projective space $\mathbb{P}_{m+k}$ satisfying

$$(CY)(d_1, \dots, d_k; m) : \quad m + k + 1 = d_1 + \cdots + d_k$$

which is the condition for a trivial canonical bundle. When $m = 3$, for example, we have the following solutions of $4 + k = d_1 + \cdots + d_k$ with $d_i > 1$:

$$5 = 5, \quad 6 = 4 + 2 \quad \text{or} \quad 6 = 3 + 3, \quad \text{and} \quad 7 = 3 + 2 + 2$$

In looking for further three dimensional examples, we consider complete intersections in weighted projective spaces.

(7.2) Weighted Projective Spaces. A weighted projective space $\mathbb{P}_N(w)$ is a generalization of $\mathbb{P}_N$. both are quotients of $\mathbb{C}^{N+1} - \{0\}$ by an action of $\mathbb{C}^* = \mathbb{C} - \{0\}$. The weights w are sequences of natural numbers $w = (w(0), \dots, w(N)) \in \mathbb{N}^{N+1}$ and the action of $\lambda \in \mathbb{C}^*$ on $(z_0), \dots, z_N) \in \mathbb{C}^{N+1} - \{0\}$ is given by the formula

$$\lambda \cdot (z_0, \dots, z_n) = (\lambda^{w(0)} z_0, \dots, \lambda^{w(N)} z_N).$$

We assume that the greatest common divisor of the $w(i)$ is 1.

For each subset $S \subset \{w(0), \dots, w(N)\}$ we denote by $q(S)$ the greatest common divisor of the $w(i)$ with $i \in S$. Let $H(S)$ denote the subset of all $(z_j) \in \mathbb{P}_N(w)$ with $z_i = 0$ for $i \notin S$. The points in $H(S)$ are cyclic quotient singularities for the group $\mathbb{Z}/q(S)\mathbb{Z}$.

A general reference on weighted projective spaces is I. Dolgachev [1982, SLN 956].

The equations of hypersurfaces in the weighted projective space $\mathbb{P}_N(w)$ of degree d are given by polynomial equations $f(z_0, \dots, z_n) = 0$ where

$$f(\lambda^{w(0)} z_0, \dots, \lambda^{w(N)} z_N) = \lambda^d f(z_0, \dots, z_N).$$

(7.3) Complete Intersections in Weighted Projective Spaces. The complete intersections of multiple degree $(d_1, \dots, d_k)$ in the weighted projective space $\mathbb{P}_{m+k}(w)$ with trivial canonical bundle are those satisfying the following condition

$$(CY)(d_1, \dots, d_k; w) : \quad w(0) + \cdots + w(m + k) = d_1 + \cdots + d_k.$$

This condition reduces to $(CY)(d_1, \dots, d_k)$ in (7.1) for a projective space $\mathbb{P}_{m+k}$.

The existence of singularities in a weighted projective space, which were not present in the standard projective space, leads to examples with these quotient singularities. Of special interest are the hypersurfaces transverse to the singularities.

There is a complete classification of Calabi–Yau varieties arising from transverse hypersurfaces in $\mathbb{P}_4(w)$, see A. Klemm and R. Schimmrigk [1994] and M. Kreuzer and H. Skarke [1992], where there are 7555 cases.

In dimension one elliptic curves arise as cubic curves in the plane or as complete intersections of two quadrics in 3-space. Elliptic curves also arise either as quartic curves in $\mathbb{P}_2(1,1,2)$ and sextic curves in $\mathbb{P}_2(1,2,3)$. We can take for example $w^4 + x^4 + y^2 = 0$ and $w^6 + x^3 + y^2 = 0$ respectively.

(7.4) General Toric Varieties. Toric varieties are varieties with an action of a torus such that there is a dense orbit. They include weighted projective spaces, hence also projective spaces. The affine pieces of a toric variety are defined by monomial equations, and the affine pieces are organized by a combinatorial configuration relating the toric actions on the affine open sets. For a general reference we recommend the book of Fulton [1993]. We give a guide to some of the sections in Fulton.

From the combinatorial description it is possible to know when a toric variety is proper and nonsingular. For this, see p. 39 and p. 29 respectively of Fulton [1933] and in section 2.6 the resolution of singularities of a toric variety can be prescribed from the combinatorial data needed to prescribe a toric structure.

(7.5) Complete Intersections in Toric Varieties. In Fulton, chapter 3, divisors and line bundles on a toric variety are studied. Of special importance are the T-invariant divisors on a tori variety X where T is the torus acting on X. In section 4.3, the canonical bundle is described using the T-invariant divisors. Then the complete intersections with trivial canonical bundle can be determined. Hence there is a combinatorial description of which complete intersections are Calabi–Yau manifolds.

(7.6) Remark. For applications to string theory there is the notion of the mirror Calabi–Yau manifold, and it can be very concretely determined in cases where the Calabi–Yau is a complete intersection in a toric variety. A reference for this is Voisin [1996, chapter 4].

§8. Line Bundles and Divisors: Picard and Néron–Severi Groups

In (5.2) we introduced the first Chern class of a line bundle in the setting of homotopy classes $[X, \mathbb{P}_\infty(\mathbb{C})]$ of mappings $X \to \mathbb{P}_\infty(\mathbb{C})$. Basic to this is the double interpretation of $\mathbb{P}_\infty(\mathbb{C})$ leading to the bijection

$$c_1 : \{\text{isomorphism classes of complex line bundles}/X\} \to H^2(X, \mathbb{Z}) = [X, \mathbb{P}_\infty(\mathbb{C})]$$

carrying the tensor product to the sum in the cohomology groups

$$c_1(L' \otimes L'') = c_1(L') + c_1(L'').$$

(8.1) Analytic/Algebraic Line Bundles and Divisors. Let X/k be a proper scheme. Let $\mathcal{M}_X$ denote the sheaf of total rings of fractions of $\mathcal{O}_X$ on the scheme X. The multiplicative structure of sheaves of rings leads to the following two diagrams relating line bundles to closed subschemes of codimension one.

$$\begin{array}{ccccccccc} 1 & \longrightarrow & \mathcal{O}_X^* & \longrightarrow & \mathcal{M}_X^* & \longrightarrow & \mathfrak{D}_X = \mathcal{M}_X^*/\mathcal{O}_X^* & \longrightarrow & 1 \\ & & \downarrow \mathrm{id} & & \downarrow \mathrm{id} & & \downarrow \mathrm{ord} & & \\ 1 & \longrightarrow & \mathcal{O}_X^* & \longrightarrow & \mathcal{M}_X^* & \xrightarrow{\mathrm{ord}} & \coprod_{\mathrm{codim}(x)=1} \mathbb{Z}_x & \longrightarrow & 0 \end{array}$$

and for positive divisors

$$\begin{array}{ccccccccc} 1 & \longrightarrow & \mathcal{O}_X^* & \longrightarrow & \mathcal{O}_X - \{0\} & \longrightarrow & \mathfrak{D}_X^+ & \longrightarrow & 1 \\ & & \downarrow \mathrm{id} & & \downarrow \mathrm{id} & & \downarrow \mathrm{ord} & & \\ 1 & \longrightarrow & \mathcal{O}_X^* & \longrightarrow & \mathcal{O}_X - \{0\}^* & \xrightarrow{\mathrm{ord}} & \coprod_{\mathrm{codim}(x)=1} \mathbb{N}_x & \longrightarrow & 0 \end{array}$$

where ord is the function which assigns to a germ the order of zero or pole.

For the first diagram we extract the following exact sequence of low dimensional cohomology groups for X over k

$$1 \to H^0(\mathcal{O}_X^*) = k^* \to H^0(\mathcal{M}_X^*) = k(X)^*$$
$$\to H^0(\mathfrak{D}_X) = \mathrm{Div}(X) \to H^1(\mathcal{O}_X^*) = \mathrm{Pic}(X) \to \dots$$

This leads to the exact sequence

$$1 \to k(X)^*/k^* = \mathrm{Div}_p(X) \to \mathrm{Div}(X) \to H^1(\mathcal{O}_x^*) = \mathrm{Pic}(X) \to H^1(\mathcal{M}_X^*) \to \dots$$

with the first arrow mapping a germ of the nonzero function f to the principal divisor (f).

(8.2) Groups of Divisors and of Line Bundles. In terms of sheaf cohomology we define divisors as elements of $\mathrm{Div}(X) = H^0(\mathfrak{D}_X)$. Line bundles up to isomorphism are described by elements of

$$H^1(\mathcal{O}_X^*) = \mathrm{Pic}(X)$$

Finally the group of divisor classes is the quotient group $\mathrm{Div}(X)/\mathrm{Div}_p(X)$ which maps by an injection $\mathrm{Div}(X)/\mathrm{Div}_p(X) \to \mathrm{Pic}(X)$ into the Picard group $\mathrm{Pic}(X)$.

In the algebraic case $H^1(\mathcal{M}_X^*) = 0$ and we have an isomorphism

$$\mathrm{Div}(X)/\mathrm{Div}_p(X) \to \mathrm{Pic}(X)$$

from the group of divisors classes to the group of isomorphism classes of line bundles.

(8.3) The Line Bundle of a Positive Divisor. For a divisor $D \geq 0$ viewed as a closed subscheme $D \to X$, we have the exact sequence

$$0 \to \mathcal{J}_D = \mathcal{O}(-D) \to \mathcal{O}_X \to \mathcal{O}_D \to 0$$

of structure sheaves on X and D together with the ideal sheaf $\mathcal{J}_D$ of the locus D in X.

The group $\mathrm{Pic}(X)$ is related to H^2 by the exponential function.

(8.4) First Chern Class of Analytic/Algebraic Line Bundles. Consider the exponential sequences

$$0 \longrightarrow \mathbb{Z} \longrightarrow \mathcal{O}_Z \xrightarrow{e} \mathcal{O}_X^* \longrightarrow 1$$

where $e(f) = \exp(2\pi i f)$. The boundary morphism in the cohomology exact sequence

$$\ldots \to H^1(X, \mathcal{O}_X) \to H^1(X, \mathcal{O}_X^*) = \mathrm{Pic}(X) \xrightarrow{c_1} H^2(X, \mathbb{Z}) \to H^2(X, \mathcal{O}_X) \to \ldots$$

is the first Chern class $c_1 : \mathrm{Pic}(X) \to H^2(X, \mathbb{Z})$ of line bundle classes. From the exact sequence we see that this algebraic or analytic Chern class is an isomorphism if $H^1(X, \mathcal{O}_X) = H^2(X, \mathcal{O}_X) = 0$.

(8.5) Definition. Let

$$\mathrm{Pic}^0(X) = \ker(c_1 : \mathrm{Pic}(X) \to H^2(X, \mathbb{Z}))$$

contained in $\mathrm{Pic}(X)$. The Néron–Severi group of X is the quotient $\mathrm{NS}(X) = \mathrm{Pic}(X)/\mathrm{Pic}^0(X)$.

With the intersection form we will give another description of the Néron–Severi group $\mathrm{NS}(X)$ as a quotient of $\mathrm{Pic}(X)$ in (7.7).

In the algebraic case there is a purely algebraic first Chern class in étale cohomology using the Kummer sequence instead of the exponential sequence.

(8.6) Degrees and Intersection Properties of Divisors. We consider the theory for curves, surfaces, and threefolds.

(1) For a curve X, the deg : $\mathrm{Div}(X) \to \mathbb{Z}$ is a function which defines on the quotient deg : $\mathrm{Pic}(X) = \mathrm{Div}(X)/\mathrm{Div}_p(X) \to \mathbb{Z}$ by the first Chern class evaluated on the top class $[X] \in H_2(X, \mathbb{Z})$ in homology

$$\deg(\varphi) = c_1(\varphi)[X] \quad \text{or} \quad \deg(D) = c_1((D))[X].$$

It defines a quotient morphism on $\mathrm{Pic}(X)$ since $\deg((f)) = 0$, that is, the number of zeros equals the number of poles of a function f.

(2) For a surface X, deg is replaced by the intersection pairing defined by the first Chern class cup product and evaluated on the top class

$$\mathcal{L}_1 \cdot \mathcal{L}_2 = c_1(\mathcal{L}_1)c_1(\mathcal{L}_2)[X] \quad \text{or} \quad D_1 \cdot D_2 = c_1(D_1)c_1(D_2)[X].$$

The intersection pairings on surfaces are defined

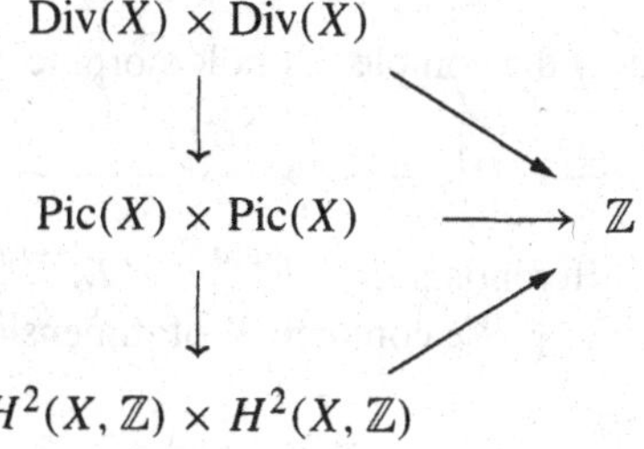

(3) For a threefold X the deg is replaced by a trilinear form on $\mathrm{Pic}(X)$

$$\mathcal{L}_1 \cdot \mathcal{L}_2 \cdot \mathcal{L}_3 = c_1(\mathcal{L}_1)c_1(\mathcal{L}_2)c_1(\mathcal{L}_3)[X]$$

for line bundles, and

$$D_1 \cdot D_2 \cdot D_3 = c_1(D_1)c_1(D_2)c_1(D_3)[X]$$

for divisors.

Returning to the Néron–Severi group $\mathrm{NS}(X)$, we consider the following relation between divisors.

(8.7) Definition. Two divisors D' and D'' are numerically equivalent provided $D' \cdot E = D'' \cdot E$ for all divisors E on a surface X.

(8.8) Proposition. *Let X be a surface such that the intersection form on* $\mathrm{Pic}(X)$ *is nondegenerate. Then the Néron–Severi group* $\mathrm{NS}(X)$ *is the quotient* $\mathrm{Div}(X)/\mathrm{Div}_n(X)$ *where* $\mathrm{Div}_n(X)$ *is the subgroup of* $\mathrm{Div}(X)$ *consisting of divisors numerically equivalent to zero.*

By passing to the quotient, the intersection form $D' \cdot D''$ is defined on NS(X) with values in the integers.

Since every divisor of a function is numerically equivalent to zero, that is, $\mathrm{Div}_p(X) \subset \mathrm{Div}_n(X)$, we have a quotient mapping from the divisor classes to $\mathrm{NS}(X) = \mathrm{Pic}(X)/\mathrm{Pic}^0(X) = \mathrm{Div}(X)/\mathrm{Div}_n(X)$ preserving the intersection form.

(8.9) Remark. The group $\mathrm{NS}(X)$ is finitely generated with rank $\rho(X)$ which is called the Picard number of the surface X. The intersection form extends to a bilinear pairing on the extension of scalars $\mathbb{Q} \otimes \mathrm{NS}(X)$ and $\mathbb{R} \otimes \mathrm{NS}(X)$. These are vector spaces of dimension equal the Picard number $\rho(X)$.

§9. Numerical Invariants of Surfaces

In this section we consider the invariants used in the Enriques classification of surfaces with a special emphasis on the surfaces satisfying the Calabi–Yau property. This classification begins with cohomological invariants which extend the concept of the genus of a curve to surfaces. Let k denote the algebraically closed ground field.

(9.1) The Hodge to de Rham Spectral Sequence. It has the form

$$E_1^{p,q} = H^q(X, \Omega_X^p)$$

with differentials induced by the complex of holomorphic differential forms

$$0 \longrightarrow \mathcal{O}_X \xrightarrow{d'} \Omega_X^1 \xrightarrow{d'} \cdots \xrightarrow{d'} \Omega_X^i \xrightarrow{d'} \cdots .$$

The terms $E_r^{p,q}$ and the differentials $d_r : E_r^{p,q} \to E_r^{p+r,q-r+1}$ are defined for all $r \geq 1$, and $H(E_r, d_r) = E_{r+1}$. We compare X of dimension 1 and X of dimension 2.

(1) Let X be a curve. There is a differential $d_1 : E^{p,q} \to E_1^{p+1,q}$ and the possible nonzero groups are:

$$\begin{array}{ccc} H^1(X, \mathcal{O}_X) & \longrightarrow & H^1(X, \Omega^1_X) = k \\ & & \\ k = H^1(X, \mathcal{O}_X) & \longrightarrow & H^1(X, \Omega^1_X) \end{array}$$

Here, $g(X) = g = h^{1,0} = h^{0,1}$ is the genus of the curve X.

(2) Let X be a surface. There is a differential $d_1 : E_1^{p,q} \to E_1^{p+1,q}$ and the possible nonzero groups are:

$$\begin{array}{ccccc} H^2(X, \mathcal{O}_X) & \longrightarrow & H^2(X, \Omega^1_X) & \longrightarrow & H^2(X, \Omega^2_X) = k \\ & & & & \\ H^1(X, \mathcal{O}_X) & \longrightarrow & H^1(X, \Omega^1_X) & \longrightarrow & H^1(X, \Omega^2_X) \\ & & & & \\ k = H^0(X, \mathcal{O}_X) & \longrightarrow & H^0(X, \Omega^1_X) & \longrightarrow & H^0(X, \Omega^2_X) \end{array}$$

Here $q = \dim H^1(X, \mathcal{O}_x)$ is the irregularity of X, and $p_g = \dim H^2(X, \mathcal{O}_X)$ is the geometric genus of the surface X.

As usual, the Betti numbers and Hodge numbers are related by $b_i = \sum_{p+q=i} h^{p,q}$ for the cohomology $H^i(X, \mathbb{C})$ has a decreasing filtration

$$F^p H^i(X, \mathbb{C}) \quad \text{with } H^q(X, \Omega^p_X)$$

isomorphic to the quotient $F^p H^{p+q}(X, \mathbb{C}) / F^{p+1} H^{p+q}(X, \mathbb{C}$.

(9.2) Riemann–Roch. Riemann–Roch theorems have to do with the calculation of the Euler–Poincaré characteristic

$$\chi(\mathcal{L}) = \sum_{i=0}^{n} (-1)^i \dim_k H^i(X, \mathcal{L}).$$

There are two parts, firstly, a relation between $\chi(\mathcal{L})$ or $\chi(\mathcal{O}(D))$ for a divisor D and $\chi(\mathcal{O}_X)$ and, secondly, a relation between $\chi(\mathcal{O}_X)$ and Chern classes of the tangent bundle.

For curves: $\chi(\mathcal{L}) = \chi(\mathcal{O}_X) + \deg(\mathcal{L})$ for a line bundle $\mathcal{L}$, and

$$\chi(\mathcal{O}(D)) = \deg(D) + \chi(\mathcal{O}_X) = \deg(D) + 1 - g(X)$$

for a divisor D on the curve X. Here $g = g(X) = \dim H^1(X, \mathcal{O}_X)$ is the genus of the curve X.

Moreover, $\chi(\mathcal{O}_X) = (1/2)c_1 = 1 - g(X)$ where $c_1 = c_1(T_X)[X]$. This implies the integrality assertion: c_1 is always an even integer.

For surfaces: $\chi(\mathcal{L}) = \chi(\mathcal{O}_X) + (1/2)\mathcal{L} \cdot \left(\mathcal{L} \otimes \omega_X^{(-1)\otimes}\right)$ for a line bundle $\mathcal{L}$, and

$$\chi(\mathcal{O}(D)) = \frac{1}{2}D \cdot (D - K_X) + \chi(\mathcal{O}_X) = \frac{1}{2}D \cdot (D - K_X) + 1 - q + p_g$$

for a divisor D on the surface X. Here $q = \dim H^1(x, \mathcal{O}_X)$ is the irregularity and $p_g = \dim H^2(X, \mathcal{O}_X)$ is the geometric genus of the surface X.

Moreover, $\chi(\mathcal{O}_X) = (1/12)(c_1^2 + c_2) = 1 - q + p_g$ where $c_i = c_i(T_X)[X]$ for $i = 1, 2$. This implies the integrality assertion: $c_1^2 + c_2$ is always an integer divisible by 12.

(9.3) Role of Serre Duality. Recall that Serre duality is a nondegenerate pairing

$$H^i(X, \mathcal{L}) \times H^{n-i}\left(X, \omega_X \otimes \mathcal{L}^{(-1)\otimes}\right) \to k$$

In particular, the two cohomology vector spaces have the same dimension. For divisors, this takes the form

$$\dim H^i(X, \mathcal{O}(D)) = \dim H^{n-i}(X, \mathcal{O}(K - D)).$$

In the special case of curves, this leads directly to

$$\deg(K) = 2g - 2 \quad \text{and} \quad g = \dim H^1(\mathcal{O}_X) = \dim H^0(\omega_X).$$

As an application of the Riemann–Roch formula, Grothendieck gave an algebraic proof of the following theorem.

(9.4) Theorem (Algebraic Index Theorem). *On* $\mathbb{R} \otimes \mathrm{NS}(X)$, *the intersection form is of signature* $(1, \rho(X) - 1)$.

(9.5) Index Theorem for Complex Surfaces. Let $\sigma = b^+ - b^-$ be the signature or index of the cup product quadratic form on $H^2(X, \mathbb{R})$.

$$\sigma = b^+ - b^- = \frac{1}{3}(c_1^2 - 2c_2) = \frac{1}{3}(K^2 - 2e).$$

For nonalgebraic complex surfaces this and the Riemann–Roch formula are proved by applying the Atiyah–Singer index formula.

(9.6) Genus Formula. Let $D > 0$ be a divisor, and let $\mathcal{O}_D$ be the structure sheaf on the scheme D defined by the ideal sheaf $\mathcal{J}_D = \mathcal{O}(-D)$. From the exact sequence

$$0 \to \mathcal{J}_D = \mathcal{O}(-D) \to \mathcal{O}_X \to \mathcal{O}_D \to 0.$$

we have the Euler–Poincaré characteristic relation

$$\chi(\mathcal{O}_D) = \chi(\mathcal{O}_X) - \chi(\mathcal{O}(-D))$$

which by Riemann–Roch is the genus formula

$$\chi(\mathcal{O}_D) = -\frac{1}{2}D \cdot (D + K).$$

For an irreducible curve C we recover the usual genus formula

$$2g(C) - 2 = C \cdot K + C^2.$$

(9.7) Relation Between Numerical Invariants. From Poincaré duality the topological Euler number $e(X) = \Delta \cdot \Delta$ on $X \times X$ is

$$c_2 = e(X) = b_0 - b_1 + b_2 - b_3 + b_4 = 2 - 2b_1 + b_2 = 2 - 2b_1 + b^+ + b^-.$$

Combining the topological Euler number calculation with the Riemann–Roch formula $12\chi(\mathcal{O}_X) = c_1^2 + c_2$ we obtain the relation

$$12 - 12q + 12p_g = K^2 + 2 - 2b_1 + b_2$$

which we rewrite in the form

$$10 - 8q + 12p_g = K^2 + b_2 + 2\Delta$$

where $\Delta = 2q - b_1$.

(9.8) Remark. For complex surfaces either b_1 is even when $\Delta = 0$ or b_1 is odd when $\Delta = 1$. Furthermore, $\dim H^0(\Omega^1) = q - \Delta$ and $b^+ = 2p_g + 1 - \Delta$. he following two assertions are used to prove this statement.

Firstly, every holomorphic 1-form is closed. This is false in dimensions strictly greater than two and in characteristic $p > 0$. Secondly, if $\omega_1, \ldots, \omega_r$ is a basis for $H^0(\Omega^1)$, then the set of forms $\omega_1, \ldots, \omega_r, \overline{\omega}_1, \ldots, \overline{\omega}_r$ is linearly independent in $H^1(\mathbb{C})$ and thus we have the inequality $2h^{1,0} \leq b_1$.

For algebraic surfaces there is a quantity Δ in characteristic $p > 0$ with properties as in the previous paragraph.

§10. Enriques Classification for Surfaces

(10.1) Exceptional Curves and Minimal Surfaces. The Enriques classification theorem is for surfaces without exceptional curves, that is, without rational curves E with $E, E = -1$. These curves on a surface X are the result of blowing up a smooth point P on a another surface Y. This blowing up is a map $X \to Y$ such that the restriction $X - E \to Y - P$ is an isomorphism. A surface is called minimal provided there are no exceptional curves. Every surface can be mapped onto a minimal surface with only exceptional curves mapping to points, and the process is unique with one type of exception related to $\mathbb{P}_2$ and $\mathbb{P}_1 \times \mathbb{P}_1$.

(10.2) Nature of the Classification of Minimal Surfaces. We can classify very roughly curves into three classes by the genus g where $g = 0$, $g = 1$, and $g \geq 2$. In this classification, the subject of the book is in the middle $g = 1$. The Enriques classification comes in four parts where the canonical divisor K with $\mathcal{O}(K) = \Omega_X^2$ again plays a basic role.

Step 1. For curves the first step is to identify genus $g = 0$ as $\mathbb{P}_1$ with a global meromorphic function with one simple pole. For surfaces, the corresponding class of minimal surfaces X consists of those with a curve C satisfying $K \cdot C < 0$. These are further classified by the irregularity q:

(a) If $q = 0$, then either X is isomorphic to $\mathbb{P}_2$ so $K \cdot K = 9$, or X is a ruled surface over $\mathbb{P}_1$, i.e., a $\mathbb{P}_1$-bundle over $\mathbb{P}_1$.
(b) If $q > 0$, then X is a ruled surface over a curve of genus q with $K \cdot K = 8(1-q)$.

Step 2. The analog of curves of genus $g > 1$ for surfaces are surfaces of general type, that is, surfaces such that

(1) $K \cdot D \geq 0$ for all divisors $D \geq 0$ and
(2) $K \cdot K > 0$.

As with curves, these surfaces can be birational embedding in projective space using sections from $\Gamma(X, K^{m\otimes})$ for large enough m.

Step 3. There are two broad classes of minimal surfaces which are the analogue of genus $g = 1$ for curves, that is, elliptic curves. In both cases, $K \cdot D \geq 0$ for all divisors $D \geq 0$ and $K \cdot K = 0$. The first class consists of those surfaces with $12K = 0$ and the second class with $12K \neq 0$.

The second class where $12K \neq 0$ are the properly elliptic surfaces. These surfaces are fibred over a curve of genus $q > 1$ by elliptic curves with only a finite number of exceptional fibres.

The first class where $12K = 0$ contains the surfaces with $K = 0$. These are the natural generalizations of elliptic curves, that is, the class of two dimensional Calabi–Yau manifolds.

Step 4. The classification of minimal surfaces with $12K = 0$ and $K \cdot D \geq 0$ for all divisor $D \geq 0$:

	b_2	b_1	$e(X)$	q	p_g	$\chi(\mathcal{O})$
K3 surfaces	22	0	24	0	1	2
Enriques surfaces	10	0	12	0	0	1
Abelian surfaces	6	4	0	2	1	0
Hyperelliptic surfaces	2	2	0	1	0	0

(10.3) Remark. Our main interest is in surfaces X with $K_X = 0$ or equivalently with $\Omega_X^2 = \mathcal{O}_X$. Only K3 surfaces and abelian surfaces satisfy this condition. Enriques surfaces satisfy $K_X \neq 0$, but $2K_X = 0$, and $12K_X = 0$ for hyperelliptic surfaces. There are hyperelliptic surfaces with $nK_X = 0$ for all divisors n of 12, but $n'K_X \neq 0$ for a proper divisor n' of n.

(10.4) Remark. Both complex K3 surfaces and complex tori can be nonalgebraic, but if a surface has an embedding in a projective space, then by Chow's lemma, it is algebraic. The term abelian surface is usually reserved for algebraic tori.

§11. Introduction to K3 Surfaces

We begin by collecting some elementary data about K3 surfaces. Firstly, we start with a definition which has a meaning for compact complex surfaces and for algebraic surfaces.

(11.1) Definition. A K3 surface is either a compact complex surface or an algebraic surface over an algebraically closed field k with

(1) $K_X = 0$ or equivalently $\Omega^2_{X/k}$ and $\mathcal{O}_X$ are isomorphic as $\mathcal{O}_X$-sheaves.
(2) $b_1(X) = 0$ or in the algebraic case X is regular, that is, $q = \dim_k H^1(X, \mathcal{O}_X) = 0$.

(11.2) Numerical Invariants from Cohomology. The Betti numbers are $b_0 = b_4 = 1, b_1 = b_3 = 0, b_2 = 22$ in the sense of singular cohomology for $k = \mathbb{C}$ or in étale cohomology in general. Therefore,

$$24 = e(X) = 12\chi(\mathcal{O}_X) = 12(h^{0,0} - h^{0,1} + h^{0,2})$$

so that the other Hodge numbers are $h^{2,0} = h^{0,2} = 1$ and $h^{1,1} = 22$. In the Hodge to de Rham spectral sequence $E_1 = E_\infty$ where as usual $E_1^{p,q} = H^q(X, \Omega^p_{X/k})$ converges to $H^{p+q}_{\mathrm{DR}}(X/k)$. It is also the case that a K3 surfaces is simply connected.

(11.3) Tangent and Cotangent Sheaves. The tangent sheaf $\mathcal{T}_{X/k} = (\Omega^1_{X/k})\check{}$ is by definition the dual of the cotangent sheaf $\Omega^1_{X/k}$. Since $(\mathcal{O})_X$ and $\Omega^2_{X/k}$ are isomorphic as $\mathcal{O}_X$-sheaves, we deduce that $\mathcal{T}_{X/k}$ and $\Omega_{X/k}$ are isomorphic.

The main assertion about the tangent sheaf on a K3 surface is that there are no nonzero tangent vector fields, or equivalently

$$H^0(X, \mathcal{T}_{X/k}) = 0.$$

There are two references for this result.

(a) Rudakov and Shafarevitch, n^0 6, Akad. Sc. SSSR *40* (1976), pp. 1264–1307.
(b) Nygaard, Annals *110* (1979), pp. 515–528.

Since $H^0(X, \Omega_{X/k}) = 0$ by the above discussion, Serre duality gives

$$H^2(X, \Omega^1_{X/k}) = 0,$$

an isomorphism between $H^2(X, \mathcal{O}_X)$ and $H^0(X, \Omega^2_{X/k}) = 0$ which is just k, and an isomorphism between $H^2(X, \Omega^2_{X/k}) = 0$ and $H^0(X, \mathcal{O}_X)$ which is just k also. The possible nonzero groups are:

$$H^2(X, \mathcal{O}_X) = k \quad H^2(X, \Omega^1_X) = 0 \quad H^2(X, \Omega^2_X) = k$$

$$H^1(X, \mathcal{O}_X) = 0 \quad H^1(X, \Omega^1_X) = k^{20} \quad H^1(X, \Omega^2_X) = 0.$$

$$H^0(X, \mathcal{O}_X) = k \quad H^0(X, \Omega^1_X) = 0 \quad H^0(X, \Omega^2_X) = k$$

Here $q = \dim H^1(X, \mathcal{O}_X)$ is the irregularity of X, and $p_g = \dim H^2(X, \mathcal{O}_X)$ is the geometric genus of the surface X, and $q = 0$ and $p_q = 1$.

(11.4) Signature and Intersection Form. The intersection form on $H_2(X, \mathbb{Z})$ or its dual the cup product form on $H^2(X, \mathbb{Z})$ is integral on this twenty-two dimensional lattice. Since $K_X = 0$, by (9.4) the signature

$$\sigma = (1/3)(K_X \cdot K_X - 2e(X)) = (-2/3)(e(X)) = (-2/3) \cdot 24 = -16$$

There is just one possibility for such an integral form, and it is

$$3\begin{pmatrix} 0 & 1 \\ 1 & 0 \end{pmatrix} \oplus (-2)E_8 = \Lambda(3, 19)$$

where $\begin{pmatrix} 0 & 1 \\ 1 & 0 \end{pmatrix}$ denotes the rank two hyperbolic form and E_8 is the unique even form of rank 8.

(11.5) Remark. We denote the symmetry group by $\Gamma = \Gamma(3, 19) = \mathrm{Aut}(\Lambda(3, 19)) \subset \mathrm{SO}(3, 19)$. Consider $H = H^2 = H_2$ the lattice of rank 22 with the cup product or intersection form respectively. Then the set $\mathrm{Isom}(\Lambda(3, 19), H)$ is a right principal homogeneous Γ-set with action given by right composition.

(11.6) Real Structure on $H^2(X,\mathbb{C})$. Consider the Hodge structure on $H^2(X, \mathbb{C}) = H^{2,0} \oplus H^{1,1} \oplus H^{0,2}$. Complex conjugation $c : H^2(X, \mathbb{C}) \to H^2(X, \mathbb{C})$ satisfies $c^2 = id$ and it interchanges the two summands $H^{2,0}$ and $H^{0,2}$. The intersection form restricts to a positive definite form on the two dimensional subspace

$$P(X) = (H^{2,0} \oplus H^{0,2}) \cap H^2(X, \mathbb{R}) \subset H^2(X, \mathbb{R}).$$

This implies that the restriction of the intersection form to $H_{\mathbb{R}}^{1,1} = H^{1,1} \cap H^2(X, \mathbb{R})$ has signature (1, 19). In $H_{\mathbb{R}}^{1,1}$, we denote by $V(X)$ the subspace of elements of strictly positive norm. Since the signature is (1, 19), $V(X)$, the subset of all $x \in H_{\mathbb{R}}^{1,1}$ with $(x|x) > 0$, has two disjoint components $V(X) = V^+(X) \cup V^-(X)$ where $V^+(X)$ denotes the component with Kähler class.

In (7.2) and (7.8) we introduced the general theory of the Picard lattice of line bundles and the Néron–Severi group. The kernel of $c_1 : \mathrm{Pic}(X) \to H^2(X)$ was denoted $\mathrm{Pic}^0(X)$ and the quotient $\mathrm{Pic}(X)/\mathrm{Pic}^0(X) = \mathrm{NS}(X)$ is the Néron–Severi group. Since $H^1(X, \mathcal{O}_X)$ maps onto $\mathrm{Pic}^0(X)$ and $H^1(X, \mathcal{O}_X) = 0$ for a K3 surface, we see that the natural map $\mathrm{Pic}(X) \to \mathrm{NS}(X)$ is an isomorphism. We can in fact bound the size of the Néron–Severi group using the real cohomology.

(11.7) Picard Lattice or Néron–Severi Group. The first Chern class monomorphism $c_1 : \mathrm{Pic}(X) \to H^2(X)$ restricts to an isomorphism defined $c_1 : \mathrm{Pic}(X) \to H^2(X, \mathbb{Z}) \cap H_{\mathbb{R}}^{1,1} \subset H^2(X, \mathbb{Z}) \cap H^{1,1}$. The assertion that c_1 is an isomorphism in this case is a theorem of Lefschetz. The Picard number or rank $\rho(X)$ of $\mathrm{Pic}(X)$ satisfies $\rho(X) \leq 20$ and the Hodge algebraic index theorem (8.6) says that the signature is $(1, \rho(X) - 1)$.

(11.8) Remark. For algebraic K3 surfaces $1 \leq \rho(X) \leq 20$. In the case of the Fermat surface $0 = y_0^4 + y_1^4 + y_2^4 + y_3^4$ we have $\rho = 20$ while $\rho = 1$ for the generic quadric surface. For generic complex K3 surfaces we have $\rho = 0$. The algebraic K3 surfaces have hyperplane sections, hence there exist curves on the surface, but they are all homologous.

(11.9) Remark. Every complex K3 surface has a Kähler form. The Hodge data and the Kähler form can be used to parameterize complex K3 surfaces. This is the Torelli theorem, and for further details see the book by Barth, Peters, and van de Ven [1980], *Compact complex surfaces*, Springer-Verlag, Ergebnisse der Mathematik und ihrer Grenzgebiete, 3 Folge, Band 4.

20

Families of Elliptic Curves

The purpose of this chapter is to return to the concept of families of elliptic curves in the context of scheme theory and to point out some of the many areas of mathematics, and now also even of physics, in which families of elliptic curves play a role. The idea of considering an elliptic curve over two related fields, for example, $\mathbb{Q}$ and $\mathbb{F}_p$ arose in chapter 5 when studying torsion in $E(\mathbb{Q})$ for an elliptic curve E over $\mathbb{Q}$, and in chapter 14 elliptic curves over a local field K were considered as objects over the ring $\mathcal{O}$ of integers in K and over the residue class field k of K. Both of these cases are included in the general concept of a morphism of schemes $\pi : E \to B$ having the property that the fibres of π are elliptic curves, and we refer to π as a family of elliptic curves.

To analyze the concept further, we could start with a morphism of schemes $\pi : X \to B$ where all the fibres are curves, or with an eye towards elliptic curves, all fibres are curves of genus one. For a family of genus one curves to be a family of elliptic curves, we must be given a zero point on each fibre with the property that it varies algebraically over the base scheme, that is, it is a section of the morphism π. Since each fibre has a group law with the value of this section as zero, there must be a morphism $X \times_B X \to X$ inducing the group law on each fibre. Such a general family is not of much use without some combination of conditions on π, that is, flat, smooth, and proper. The Néron models in chapter 14, §2, is a first illustration of such families; it is smooth, but not in general proper. In other contexts we require the family to be flat and proper, but it is not necessary for it to be smooth.

The next direction where we restrict the problem is with the base B. The classical literature on families of curves is very extensive for one dimensional B, for example a curve over a field k or $B = \mathrm{Spec}(R)$ where R is a discrete valuation ring or more generally any Dedekind ring. This means that X is a surface over the field k in the first case, or X is an arithmetic surface over the Dedekind ring R. Chapters 14 and 15 can be thought of as introductions to arithmetic surfaces.

We saw in the previous chapter that the classification of surfaces X over a field k is closely related to the existence of morphisms $\pi : E \to B$ onto a curve B with given fibres. The two cases of fibres either of rational curves or elliptic curves are especially basic. These morphisms are called fibrations when π satisfies suitable conditions. In the classification of surfaces, the role of the existence of elliptic fibrations is central. The problem of whether or not an elliptic curve on a surface X over k can be the fibre of some fibration $\pi : X \to B$ by curves of genus one or elliptic curves is an important starting point in the theory, and we examine this problem further for K3 surfaces.

Finally passing to three dimensional varieties X, we can, by extension of the study of elliptic fibrations on surfaces, consider fibrations by K3 surfaces and abelian surfaces. This we do only for three dimensional Calabi–Yau varieties where criterions for the existence of fibrations by elliptic curves and/or by K3 surfaces are given.

There are many questions in this direction that were raised by physicists working on string theory. For this, see the appendix by S. Theisen.

§1. Algebraic and Analytic Geometry

In the previous chapters on elliptic curves we saw that there is an algebraic theory and a closely related analytic theory. The concept used to include both theories is the following.

(1.1) Definition. A local ringed space X is a pair $(X, \mathcal{O}_X)$ consisting of a topological space X together with a sheaf of rings $\mathcal{O} = \mathcal{O}_X$ on X such that the stalks $\mathcal{O}_x$ are local rings for each $x \in X$.

For each $x \in X$, we have the unique maximal ideal $\mathfrak{m}_x \subset \mathcal{O}_x$ and the residue class field $\mathcal{K}(x) = \mathcal{O}_x/\mathfrak{m}_x$. For a cross-section $s \in \Gamma(U, \mathcal{O})$ over an open set $U \subset X$, the value of s at $x \in U$ is a germ denoted $s_x \in \mathcal{O}_x$ and the value of s at $x \in U$ in the residue class field is the image $s(x) \in \mathcal{K}(x)$ of s_x under the natural quotient morphism $\mathcal{O}_x \to \mathcal{K}(x)$.

(1.2) Remark. Since the projection of a sheaf $\mathfrak{F}$ to its base space is a local homeomorphism, the set $x \in U$ where two cross sections $s', s'' \in \Gamma(U, \mathfrak{F})$ are $s'(x) \neq s''(x)$ is a closed set. For a section $s \in \Gamma(U, \mathcal{O})$, the set of points where $s_x \neq 0$ is a closed set, while on the other hand, the set of points where $s(x) \neq 0$ is an open set. This is a special property of local ringed spaces for $s(x) \neq 0$ means that s_x is a unit in $\mathcal{O}_x$. Thus there exists an open set $U(x)$ containing x and $t \in \Gamma(U(x), \mathcal{O}_X)$ with $s_x t_x = 1$. Since s and t are sections of a sheaf, there exist an open set $U'(x) \subset U(x)$ still containing x such that $s_y t_y = 1$ for all $y \in U'(x)$. This implies that $s(y) \neq 0$ for $y \in U'(x)$.

(1.3) Definition. A morphism $f : (X, \mathcal{O}_X) \to (Y, \mathcal{O}_Y)$ of local ringed spaces is a pair consisting of a continuous function $f : X \to Y$ together with a morphism of sheaves of local rings $f^* : (\mathcal{O}_Y \to \mathcal{O}_X)$, so it preserves the maximal ideals in the stalks.

The category of local ringed spaces is denoted (l/rg/sp), and its objects are those defined in (1.1) and its morphisms are those defined in (1.2) with composition being given by composition of maps and suitable sheaf morphisms. We denote the category of (commutative) rings by (rg) with morphisms of rings always preserving the unit. The cross section functor is a contravariant functor with the following universal property.

(1.4) Assertion. For each ring R in (rg), there is a local ringed space, denoted $\mathrm{Spec}(R)$, such that there is a natural isomorphism between the morphism sets

$$\mathrm{Hom}_{(\mathrm{rg})}(T, \Gamma(X, \mathcal{O}_X)) \to \mathrm{Hom}_{(\mathrm{l/rg/sp})}(X, \mathrm{Spec}(R)).$$

Moreover, the local ringed space $\mathrm{Spec}(R)$ has the set of prime ideals in R as the underlying set.

Under the above bijection, a ring morphism $\phi : R \to \Gamma(X, \mathcal{O}_X)$ corresponds to the morphism $f : X \to \mathrm{Spec}(R)$ given by the following construction. We compose the ring morphism ϕ with its evaluation on the fibre $\mathcal{O}_x$ to define a ring morphism $\phi'_x : R \to \mathcal{O}_x$ and compose further with the reduction morphism π_x to define a second ring morphism $\phi_x : R \to \mathcal{K}(x)$. Since $\phi_x : R \to \mathcal{K}(x)$ is a ring morphism into a field, the kernel $\mathfrak{p}(x)$ is a prime ideal, and the corresponding function $f : X \to \mathrm{Spec}(R)$ is defined by the formula $f(x) = \mathfrak{p}(x)$. The various morphisms under consideration are displayed as follows

$$\phi_x = \pi_x \phi'_x \quad \text{and} \quad \begin{array}{ccccc} \mathfrak{p}(x) = \ker(\phi_x) \subset R & \xrightarrow{\phi} & \Gamma(X, \mathcal{O}_X) & & \\ \downarrow & \searrow & \downarrow & & \\ R_{\mathfrak{p}(x)} & \xrightarrow[\phi_x^*]{} & \mathcal{O}_x & \xrightarrow{\pi_x} & \mathcal{K}(x). \end{array}$$

The closed sets $Z(E)$ of $\mathrm{Spec}(R)$ are given by subsets $E \subset R$ where $Z(E)$ consists of all $\mathfrak{p} \in \mathrm{Spec}(R)$ with $E \subset \mathfrak{p}$. Thus the open sets of $Y = \mathrm{Spec}(R)$ have a basis consisting of set Y_f for $f \in R$ where Y_f equals the set of all prime ideals $\mathfrak{p} \in \mathrm{Spec}(R) = Y$ with $f \notin \mathfrak{p}$. The fibre of the structure sheaf $\mathcal{O}_{\mathrm{Spec}(R)}$ at $\mathfrak{p} \in \mathrm{Spec}(R)$ is the local ring which is the localization $R_{\mathfrak{p}}$ at the prime ideal $\mathfrak{p}$. Finally, the morphism $f^* : \mathcal{O}_Y \to \mathcal{O}_X$ on the germs at $x \in X$ is the morphism $\phi_x^* : R_{\mathfrak{p}(x)} \to \mathcal{O}_x$ in the above diagram.

Since with sheaf theory we can localize on open sets and glue local ringed space over open sets, we can describe the basic structures in geometry as local ringed spaces which are locally of certain type.

(1.5) Definition. An affine scheme X is a local ringed space isomorphic to some $\mathrm{Spec}(R)$. A scheme X is a local ringed space locally isomorphic at each point to an affine scheme, that is, for $x \in X$ there exists an open set U with $x \in U$ and $(U, \mathcal{O}_X|U)$ is an affine scheme. The schemes define a full subcategory (sch) of (l/rg/sp) called the category of schemes.

Smooth manifolds and complex analytic manifolds are local ringed spaces locally isomorphic to the local ringed spaces of germs of smooth functions on $\mathbb{R}^n$ and analytic functions on open subset of $\mathbb{C}^n$ respectively.

(1.6) Remark. If A is an algebra over a ring R, then the morphism $R \to A$ defines a morphism $\mathrm{Spec}(A) \to \mathrm{Spec}(R)$. By definition, a scheme X over a ring R is a scheme together with a morphism $X \to \mathrm{Spec}(R)$. A scheme X over R is sometimes denoted X/R, and in this case, every ring $\Gamma(U, \mathcal{O}_X)$ will have a given structure as an R-algebra. For schemes over R, the sheaf $\mathcal{O}_X$ and the rings $\Gamma(U, \mathcal{O}_X)$ of sections are R-algebras.

(1.7) Remark. The natural extension of the ideas in the previous remark is to consider the category (Sch/S) of schemes X with a morphism $X \to S$ to a fixed scheme S. This is a general categorical construction also denoted $(\mathrm{Sch})_{/S}$ where the morphisms $X \to Y$ are morphisms in (Sch) giving a commutative triangle with the two structure morphisms over S. For an object $X \to S$ the ring $\Gamma(X, \mathcal{O}_X)$ is an algebra over $\Gamma(S, \mathcal{O}_S)$. This category has the fundamental property that the product, called the fibre product and denoted $X \times_S Y$ exists for two objects $X \to S$ and $Y \to S$ in (Sch/S). For affine schemes, we have $\mathrm{Spec}(A) \times_{\mathrm{Spec}(R)} \mathrm{Spec}(B) = \mathrm{Spec}(A \otimes_R B)$, and this leads to the general existence theorem for products in (Sch/S) which are fibre products of schemes.

For two fields F and K, the scheme $\mathrm{Spec}(F)$ has only one point, and a morphism $\mathrm{Spec}(K) \to \mathrm{Spec}(F)$ is equivalent to a morphism of fields $F \to K$; it is always injective. The morphisms $\mathrm{Spec}(F) \to X$ are given by point $x \in X$ together with a morphism $\mathcal{O}_x \to F$ which factors as $\mathcal{K}(x) \to F$. The geometric points of a scheme are by definition the points $x \in X$ where there is an algebraically closed field F and a morphism $\mathrm{Spec}(F) \to X$ with image x.

(1.8) Remark. In remark (1.7), we pointed out that a morphism $p : X \to B$ of schemes as giving a $\Gamma(V, \mathcal{O}_B)$-algebra structure on the rings $\Gamma(p^{-1}(V), \mathcal{O}_X)$ in a functorial manner. But there is another basic interpretation of $p : X \to B$ as a family of schemes $X_b = \{b\} \times_B X$ given by the fibres of the morphism $p : X \to B$. This is the picture of $p : X \to B$ as the family of the schemes X_b for $b \in B$.

One way to study special types of varieties or schemes, like curves, surfaces, 3-folds, or abelian varieties, is to analyze how they form families. Before doing this, we consider morphisms given by line bundles and divisors in the next section.

We conclude with a few further remarks about schemes. One very clear advantage of working in the concept of scheme theory is that questions of families are well defined. For example, the moduli space of elliptic curves should be a family $p : X \to B$ where each elliptic curve under consideration should be isomorphic to exactly one fibre X_b in the family. The isomorphism classes of elliptic curves are no longer a discrete set, but they have the structure of a scheme on a set of representatives of the isomorphism classes. In previous chapters various families of elliptic curves had this structure of a scheme given in terms of coefficients of the equations defining the curve.

(1.9) Remark. Most of the ideas and results in commutative algebra have an analogue in scheme theory through local calculations in affine open neighborhoods. For example, a point $x \in X$ is a prime ideal in any affine open $\mathrm{Spec}(R)$ containing x. Prime ideals have invariants of dimension, height, and depth, which carry over to the points on schemes. Ideal sheaves determine closed subschemes as an ideal $I \subset R$ determines $\mathrm{Spec}(R/I)$ a closed subset of $\mathrm{Spec}(R)$. There are Noetherian conditions and the concepts of irreducibility and Krull dimension.

For basic (commutative) ring theory, we recommend the following book: H. Matsumura, *Commutative Algebra*, Second Edition, Benjamin, 1980. It covers those topics most relavent to local scheme theory.

For general scheme theory in English we prefer: D. Mumford, *Red Book*, SLN 1358. We will make reference to this book though this chapter.

There is still just one very basic reference of scheme theory, namely EGA, Grothendieck, A. and J. Dieudonné, *Eléments de Géométrie Algébrique*, Publ. Math. I.H.E.S., 4, 8, 11, 17, 20, 24, 28, 32 (1961–1967), and Springer-Verlag, Berlin, 1971. The serious student can only look forward to many hours of encounter with this book.

§2. Morphisms Into Projective Spaces Determined by Line Bundles, Divisors, and Linear Systems

In the previous section, we have characterized the morphisms $\mathrm{Hom}_{(\mathrm{l/rg/sp})}(X, \mathrm{Spec}(R))$ from a local ringed space X into an affine scheme $\mathrm{Spec}(R)$ with a natural isomorphism

$$\mathrm{Hom}_{(\mathrm{rg})}(R, \Gamma(X, \mathcal{O}_X)) \to \mathrm{Hom}_{(\mathrm{l/rg/sp})}(X, \mathrm{Spec}(R)).$$

Now, we try to characterize the set of morphisms $\mathrm{Hom}_{(\mathrm{Sh/R})}(X/R, \mathbb{P}^n_R)$ from a scheme X over R to the scheme given by n-dimensional projective space $\mathbb{P}^n_R$ also defined over a ring R.

(2.1) Remark. By using local isomorphisms with $\mathcal{O}^n_X$, we can extend the assertion of (1.2) to a locally free $\mathcal{O}_X$-sheaf $\mathcal{E}$ of finite rank, that is, a vector bundle. For a section $s \in \Gamma(X, \mathcal{E})$, the set X_s of points $x \in X$ where $s(x) \neq 0$ is an open set which in turn is a subset of the closed set of points $x \in X$ where $s_x \neq 0$. Consider the case where $\mathcal{E} = \mathcal{L}$ a line bundle, that is, an $\mathcal{O}_X$-sheaf locally isomorphic to $\mathcal{O}_X$ and $s \in \Gamma(X, \mathcal{L})$. Then, for each $t \in \Gamma(X_s, \mathcal{L})$, we can form $t/s \in \Gamma(X_s, \mathcal{O}_X)$ which is uniquely described by the relation $(t/s).s = t$ in $\Gamma(X_s, \mathcal{L})$.

Returning to the description of $\mathrm{Hom}_{(\mathrm{Sh/R})}(X/R, \mathbb{P}^n_R)$, we will define $\mathbb{P}^n_R = \mathrm{Proj}(R[y_0, \ldots, y_n])$ and its canonical line bundle $\mathcal{O}(1)$ which is generated by sections $y_0, \ldots, y_n \in \Gamma(\mathbb{P}^n_R, \mathcal{O}(1))$ as follows. The scheme $\mathbb{P}^n_R$ is covered by $n+1$ open sets W_i where $y_i(x) \neq 0$ and as affine subschemes of $\mathbb{P}^n_R$ they are given by

$$W_i = \mathrm{Spec}(R[y_0/y_i, \ldots, y_n/y_i]).$$

The covering condition $\mathbb{P}^n_R = W_0 \cup \cdots \cup W_n$ implies that the sections generate the line bundle, which for the line bundle $\mathcal{O}(1)$ means that at least one section is nonzero at each x of the space.

(2.2) Proposition. *For a local ringed space, let $\Phi(X)$ denote the set of pairs $(\mathcal{L}; s_0, \ldots, s_n)$ where $\mathcal{L}$ is a line bundle on X and $s_0, \ldots, s_n$ is a set of sections generating $\mathcal{L}$, all up to isomorphism. The function $\mathrm{Hom}_{(Sh/R)}(X/R, \mathbb{P}^n_R) \to \Phi(X)$ which assigns to a morphism $f : X/R \to \mathbb{P}^n_{R_*}$ the pair of the line bundle $f^*(\mathcal{O}(1))$ and the generating sections $f^*(y_0), \ldots, f^*(y_n)$ is a bijection.*

Proof. We prove the result by constructing an inverse to this function by starting with $\mathcal{L}$ a line bundle on X with generating sections $s_0, \ldots, s_n$. Denote by X_i the open set of $x \in X$ where $s_i(x) \neq 0$. The related f is the result of gluing the compatible morphism $f_i : X_i \to W_i = \mathrm{Spec}(R[y_0/y_i, \ldots, y_n/y_i])$ defined by $f_i^*(y_k/y_i) = s_k/s_i$ using the adjunction in (1.4). The rest is easily checked and left to the reader.

(2.3) Remark. For a section $s \in \Gamma(X, \mathcal{L})$ of a line bundle on X the closed subset $Z(s)$ of zeros of s. It is defined by $x \in Z(s)$ provided $s(x) = 0$ or equivalently the germ $s_x \in \mathfrak{m}_x \mathcal{L}_x$. In fact, $Z(s)$ is a closed subscheme of X with ideal sheaf $\mathcal{J}$ generated by one element. Locally at each $x \in X$ the line bundle $\mathcal{L}|U$ is isomorphic to $\mathcal{O}_X|U$ and the section s of $\mathcal{L}$ corresponds to $f \in \Gamma(U, \mathcal{O}_X)$ where the germ f_x generates the ideal $\mathcal{J}_x \subset \mathcal{O}_x$ defining the closed subscheme $Z(s)$.

For a nonzero section $s \in \Gamma(X, \mathcal{L})$ the closed subscheme $Z(s)$ has codimension one and is called a divisor. Now we recall some elements of the theory of divisors using the sheaf of germs of total rings of fractions $\mathcal{K}_X$ on a scheme $(X, \mathcal{O}_X)$ which is the general setting for zero and poles.

(2.4) Definition. Let $(X, \mathcal{O}_X)$ denote a local ringed space. The sheaf of rational functions $\mathcal{K}_X$ on X is the sheafification of the presheaf on X which assigns to each open set U the total ring of fractions of the ring $\Gamma(U, \mathcal{O}_X)$.

Observe that $\mathcal{K}_X$ is a sheaf of $\mathcal{O}_X$-algebras.

(2.5) Remark. For $x \in X$, the $\mathcal{O}_x$-algebra $\mathcal{K}_x$ is the total ring of fractions of the ring $\mathcal{O}_x$ since the operation of forming the total ring of fractions commutes with direct limits. For a Noetherian scheme X and an affine open subscheme U the ring $\Gamma(U, \mathcal{K}_U)$ is the total ring of fractions of the ring $\Gamma(U, \mathcal{O}_U)$.

(2.6) Notation. Let $\mathcal{O}_X^*$ denote the sheaf of invertible elements in $\mathcal{O}_X$, $\mathcal{K}_X^*$ the sheaf of invertible elements in $\mathcal{K}_X$, and $\mathcal{D}_X = \mathcal{K}_X^*/\mathcal{O}_X^*$. We have a short exact sequence of sheaves of abelian groups containing a short exact sequence of sheaves of monoids

$$\begin{array}{ccccccccc} 1 & \longrightarrow & \mathcal{O}_X^* & \longrightarrow & \mathcal{K}_X^* & \longrightarrow & \mathcal{D}_X & \longrightarrow & 1 \\ & & \| & & \cup & & \cup & & \\ 1 & \longrightarrow & \mathcal{O}_X^* & \longrightarrow & \mathcal{K}_X^* \cap \mathcal{O}_X & \longrightarrow & \mathcal{D}_X^* & \longrightarrow & 1. \end{array}$$

The sheaf $\mathcal{D}_X$ of abelian groups is called the sheaf of germs of divisors and the subsheaf $\mathcal{D}_X^*$ is called the sheaf of germs of positive divisors. For a germ $D_x \in \mathcal{D}_x$ a local equation is $f \in \Gamma(U, \mathcal{K}_x)$ where U is a neighborhood of x and $D_x = f_x$ mod $\mathcal{O}_{X,x}^*$ in the quotient. Compare with 19(8.1).

(2.7) Remark. The group of divisors $\Gamma(X, \mathcal{D}_X)$ is written additively so that if f', f'' are local equations of divisors D', D'', at $x \in X$ respectively, then the product $f' \cdot f''^{\pm 1}$ is a local equation of $D' \pm D''$. Also, if $D'_x = D''_x$, then $f'/f'' \in \operatorname{im}(\Gamma(U, \mathcal{O}_X^*) \to \Gamma(U, \mathcal{K}_X^*))$ in some open neighborhood U of x.

The short exact sequence

$$1 \longrightarrow \mathcal{O}_X^* \longrightarrow \mathcal{K}_X^* \longrightarrow \mathcal{D}_X \longrightarrow 1$$

leads to a long exact sequence of cohomology starting with the terms

$$0 \to \Gamma(X, \mathcal{O}_X^*) \to \Gamma(X, \mathcal{K}_X^*) \to \Gamma(X, \mathcal{D}_X^*) \to H^1(X, \mathcal{O}_X^*) \to H^1(X, \mathcal{K}_X^*) \to \dots .$$

(2.8) Definition. Associated to each rational function $f \in \Gamma(X, \mathcal{K}_X^*)$ is a divisor (f) whose local equation at any $x \in X$ is f. The divisor class $[D]$ of $D \in \Gamma(X, \mathcal{K}_X^*)$ is its image in $H^1(X, \mathcal{O}_X^*)$.

Hence in terms of the above additive notation we have $[D] = [D']$ if and only if $D = D' + (f)$ for some $f \in \Gamma(X, \mathcal{K}_X^*)$.

(2.9) Definition. To a divisor $D \in \Gamma(X, \mathcal{D}_X)$, we associate an $\mathcal{O}_X$-subsheaf $\mathcal{O}_X(D) \subset \mathcal{K}_X$ by the condition that $\mathcal{O}_X(D)_x = \mathcal{O}_x[f^{-1}]_x$ where f is any local equation of D on an open set U with $x \in U$.

Observe that $\mathcal{O}_X(D) \otimes \mathcal{O}_X(D')$ and $\mathcal{O}_X(D + D')$ are isomorphic, and in terms of the homomorphism sheaf we have an isomorphism

$$\mathcal{O}_X(-D) = \mathrm{Hom}\check{\mathcal{O}_X}(D) = \mathrm{Hom}(\mathcal{O}_X(D), \mathcal{O}_X).$$

(2.10) Proposition. *For two divisors D and D', the divisor classes $[D] = [D']$ are equal if and only if $\mathcal{O}_X(D)$ and $\mathcal{O}_X(D')$ are isomorphic as $\mathcal{O}_X$-sheaves.*

This is another way of looking at the second morphism in the following exact sequence $\Gamma(X, \mathcal{K}_X^*) \to \Gamma(X, \mathcal{D}_X^*) \to H^1(X, \mathcal{O}_X^*)$, namely, to the divisor D we assign the isomorphism class of the line bundle $\mathcal{O}_X(D)$.

A divisor D is positive (or effective) provided $\mathcal{O}_X \subset \mathcal{O}_X(D) \subset \mathcal{K}_X$, and this is equivalent to $\mathcal{O}_X(-D)$ is an ideal sheaf in $\mathcal{O}_X$.

(2.11) Remark. For an effective divisor D with structure sheaf $\mathcal{O}_D$ for the related closed subscheme D we have the exact sequence $0 \to \mathcal{O}_X(-D) \to \mathcal{O}_X \to \mathcal{O}_D \to 0$ of sheaves. The closed subscheme D has $\mathcal{O}_X(-D)$ as defining sheaf of ideals, and the local equation f of D at x has the property that $\mathcal{O}_X(-D)_x = f \cdot \mathcal{O}_x$. In particular, the divisor D is determined by a codimension 1 closed subscheme D. For an effective divisor D, the image s of 1 under the natural morphism $\Gamma(X, \mathcal{O}_X) \to \Gamma(X, \mathcal{O}_X(D)) \subset \Gamma(X, \mathcal{K}_X)$ is like a global equation for D and $Z(s) = D$.

(2.12) Proposition. *Let $\mathcal{L}$ be a line bundle. The sections $s \in \Gamma(X, \mathcal{L})$, which are nonzero divisors, up to multiplication by elements of $\Gamma(X, \mathcal{O}_X^*)$ are in bijective correspondence with effective divisors D such that $\mathcal{O}_D$ is isomorphic to $\mathcal{L}$. If $\mathcal{O}_X(D)$ and $\mathcal{O}_X(D')$ are isomorphic, then $D = D' + (f)$ where $f \in \Gamma(X, \mathcal{K}_X^*)$.*

(2.13) Remark. Let X be a scheme over a field k, and consider a line bundle $\mathcal{L}$ such that $\Gamma(X, \mathcal{L})$ is a k-vector space of dimension $n+1$. For any basis $s_0, \ldots, s_n \in \Gamma(X, \mathcal{L})$ we consider the open set U of $x \in X$ where at least one $s_i(x) \neq 0$. Applying (2.2), we have a morphism $X \supset U \to \mathbb{P}^n(k)$ denoted $\phi_{\mathcal{L}}$. This morphism is independent of basis, and we can denote it simply as

$$\phi_{\mathcal{L}} : U \to \mathbb{P}\Gamma(X, \mathcal{L})$$

Here $\mathbb{P}(V)$ denotes the projective space on the vector space V, and if $\mathcal{L} = \mathcal{O}_X(D)$, then the projective space $\mathbb{P}\Gamma(X, \mathcal{L})$ is also denoted $|D|$, and the mapping is denoted $\phi_D : U \to |D| = \mathbb{P}^n$. The projective space $|D|$ is called the (complete) linear system associated with the divisor in the classical terminology, and $n = \dim|D|$ is its dimension.

The aim in the next sections is to construct morphisms $\phi_D : X \to P$ a projective space with image B, and then further the morphism $\phi : X \to B$ whose fibres are identified from properties of $\mathcal{L}$ or D in the total space X.

§3. Fibrations Especially Surfaces Over Curves

(3.1) Definition. A fibering or fibration $p : X \to B$ is a proper, flat morphism of finite presentation. A fibration of curves of genus g (resp. K3-surfaces) is a fibration $p : X \to B$ such that every fibre X_b is a curve of genus g (resp. a K3 surface).

For definitions of properties of morphisms, see Mumford, Red Book, SLN 1358, pp. 121 and 215.

This brings up the first question related to the smoothness of a fibration p, for where p is not smooth the fibre can have a singular point. In the case of genus 1 curves we understand these singular curves, but for $K3$ surfaces the situation is more complicated. In these two cases, we can "enrich" the fibration with additional structures.

(3.2) Definition. An elliptic fibration $p : X \to B$ is a fibration of genus one curves together with a section $e : B \to X$ of p such that the geometric fibres are irreducible reduced curves. A polarized fibration of K3 surfaces is K3 fibration $p : X \to B$ is a fibration of surfaces such that the geometric fibres of p are K3 surfaces together with a class $\xi \in \mathrm{Pic}(X)$ such that its restriction to each fibre $\xi_b \in \mathrm{Pic}(X_b)$ is the class of an ample line bundle.

It is possible that all fibres over geometric points are singular rational curves of genus 1 in a genus one fibration over a curve. This happens only in characteristic $p = 2$ and 3, but in general, only a finite number of fibres are singular.

(3.3) General Fibre. If $f : X \to B$ is a fibration with general fibre F, then except for a finite set of $b \in B$, the fibre X_b of f over b is an irreducible curve. We can apply the genus formula (8.4) to obtain the genus $q(F)$ as $2q(F) - 2 = K_x \cdot F$ since $F \cdot F = 0$. We have an elliptic fibration if and only if $K_X \cdot F = 0$.

In the case of a surface over a curve, the fibres are curves or more generally one dimensional schemes of the form $D = \sum_i n_i E_i$ where the E_i are irreducible curves. Hence the fibre is this divisor D. Now we study the intersection properties of D and E_i with the following elementary result from linear algebra.

(3.4) Proposition. *Let V be an inner product space over $\mathbb{Q}$ generated by vectors e_i for $i \in I$ with $(e_i|e_j) \geq 0$ for all $i \neq j$. If there exists a vector $z = \sum_i a_i e_i$ with all $a_i > 0$ such that $(z|e_j) = 0$ for all j, then we have $(z|x) = 0$ and $(x|x) \leq 0$ for all $x \in V$.*

Proof. Since any x is a linear combination of the e_j and $(z|e_j) = 0$ for all indices j, it follows that $(z|x) = 0$. for the negative definite statement, we write any $x \in V$ as $x = \sum_i c_i a_i e_i$ where $c_i \in \mathbb{Q}$ and calculate

$$
\begin{aligned}
(x|x) - \sum_i c_i^2 a_i^2 (e_i|e_i) &= \sum_{i \neq j} c_i c_j (a_i e_i | a_j e_j) \leq \sum_{i \neq j} \frac{1}{2}(c_i^2 + c_j^2)(a_i e_i | a_j e_j) \\
&= \sum_i \frac{1}{2} c_i^2 (a_i e_i | z - a_i e_i) + \sum_j \frac{1}{2} c_j^2 (a_j e_j | z - a_j e_j) \\
&= -\sum_i c_i^2 a_i^2 (e_i|e_i).
\end{aligned}
$$

Thus the sequence of inequalities gives $(x|x) \leq 0$. This proves the proposition.

(3.5) Remark. In the previous inequality (3.3) we can ask when is it true that $(x|x) = 0$ as we already know one example $(z|z) = 0$. The inequality that we used above is $uv \leq (1/2)(u^2 + v^2)$ which comes from the positivity of squares $0 \leq (u - v)^2$. Clearly this inequality is an equality if and only if $u = v$. This means that $0 = (x|x)$ if and only if for all $i \neq j$ we have either $(e_i|e_j) = 0$ or $c_i = c_j$. Let I be the set of indices, and define two indices i and j to be connected provided there exists a sequence of indices $i = i(0), \ldots, i(q) = j$ with $(e_{i(\ell-1)}|e_{i(\ell)}) > 0$ for all $\ell = 1, \ldots, q$. This puts an equivalence relation of connectedness on I, and the equivalence classes are called connected components of I. For a connected component K of I, we form the part $z_K = \sum_{i \in K} a_i e_i$ of z, and observe that $z = \sum_K z_K$ where the sum is over the connected components of z. With these notations, we have the following assertion.

(3.6) Assertion. For $x \in V$, we have $0 = (x|x)$ if and only if we can write $x = \sum_K c_K z_K$ where the sum again is over the connected components of I and each $c_K \in \mathbb{Q}$. This sum formula for x is just the condition that for $i \neq j$ we have either $(e_i|e_j) = 0$ or $c_i = c_j$. In the special case where I is connected, this becomes $0 = (x|x)$ if and only if $x = cz$ for some $c \in \mathbb{Q}$.

(3.7) Application I. Let $f : X \to B$ be a morphism of a surface onto a curve, and consider a fibre $D = \sum_i n_i E_i$ where the E_i are distinct irreducible curves and $n_i > 0$. Then $D \cdot E_i = 0$ because D is algebraically equivalent to any near by fibre, and $E_i \cdot E_j \geq 0$ for $i \neq j$. Form the $\mathbb{Q}$-vector space generated by the curves E_i with the intersection form and apply (3.3) to obtain $E_i \cdot E_i \leq 0$. This is a necessary condition for any divisor to be the fibre of a fibration to a curve.

(3.8) Application II. Let X be a surface with $K_X \cdot K_X = 0$ and $K_X \cdot D \geq 0$ for all divisors $D \geq 0$. In the linear system of mK_X for $m > 0$ consider $D \in |mK_X|$ with $D \neq 0$. Decompose $D = \sum_i n_i E_i$ with $n_i > 0$, and hence, we have $0 = (mK_X) \cdot K_X = D \cdot K_X = \sum_i n_i (E_i \cdot K_X)$. Since each $E_i \cdot K_X \geq 0$, we see that $E_i \cdot K_X = 0$ and $E_i \cdot D = 0$ for all i. Apply (3.3) this time to the $\mathbb{Q}$-vector space generated by these E_i's, and we deduce that $E_i \cdot E_i \leq 0$ for $D = \sum_i n_i E_i \in |mK_X|$ on such a surface. This is a useful property of surfaces whose canonical divisor has zero self intersection.

A general reference for this section is Barth, Peters, and van de Ven [1980].

§4. Generalities on Elliptic Fibrations of Surfaces Over Curves

In the carrying out of the Enriques classification of surfaces, an understanding of the canonical divisor on an elliptic surface plays a basic role.

(4.1) Remark. For many considerations concerning a family $f : X \to B$ we can reduce to the case of $f_*(\mathcal{O}_X) = \mathcal{O}_B$ by Stein factorization. Any proper morphism $f : X \to Y$ between Noetherian schemes has a factorization

$$X \xrightarrow{f'} B \xrightarrow{q} Y$$

where f' is proper and q is finite with B Spec$(f_*(\mathcal{O}_X))$. In this factorization, $f'_*(\mathcal{O}_X) = \mathcal{O}_B$, and under a separability hypothesis on f, the morphism q is étale. In general, a standard modification of a family of curves or surfaces over B is to induce it by an étale map $B' \to B$ to a family over B'.

(4.2) Remark. Let $f : X \to B$ be an elliptic fibration with $f_*(\mathcal{O}_X) = \mathcal{O}_B$. Then $\mathbb{R}^1 f_*(\mathcal{O}_X) = \mathcal{L} \oplus T$ where $\mathcal{L}$ is a line bundle and T is a torsion sheaf. This includes quasi-elliptic fibrations which are those nonsmooth fibrations arising only in characteristic 2 and 3 mentioned in (3.2).

(4.3) Definition. With the previous notation for an elliptic fibration $f : X \to B$ with $f_*(\mathcal{O}_X) = \mathcal{O}_B$ the exceptional points of f are $b \in$ Supp(T). A fibre $D = f^{-1}(b)$ is exceptional provided it satisfies either of the following two equivalent conditions:

(1) b is an exceptional point, i.e. $b \in$ Supp(T).
(2) $\dim H^0(\mathcal{O}_D) \geq 2$.

Recall that a curve C is of the first kind provided C is rational and $C \cdot C = -1$, and a surface is minimal provided there are no curves of the first kind.

(4.4) Definition. A fibration $f : X \to B$ of a surface over a curve B is a relatively minimal fibration provided there are no curves of the first kind contained in any fibre.

(4.5) Theorem. *Let $f : X \to B$ be a relatively minimal elliptic or quasi-elliptic fibration of a surface over a curve B satisfying $f_*(\mathcal{O}_X) = \mathcal{O}_B$. There is an enumeration of the multiple fibres $f^{-1}(b_i) = m_i P_i$ such that the P_i are indecomposable. With the notation $\mathbb{R}^1 f_*(\mathcal{O}_X) = \mathcal{L} \oplus T$ we have the following formula for the canonical line bundle on X*

$$\omega_X = f^*\left(\mathcal{L}^{(-1)\otimes} \otimes \omega_B\right) \otimes \mathcal{O}\left(\sum_i a_i P_i\right)$$

where

(i) $0 \le a_i \le m_i - 1$, *and*
(ii) $a_i = m_i - 1$ *if P_i is not exceptional.*

Hence the length of T, denoted $l(T)$ is greater than or equal to the number of $a_i < m_i - 1$. Moreover,

$$\deg\left(L^{(-1)} \otimes \omega_B\right) = 2g(B) - 2\chi(\mathcal{O}_X) + l(T)$$

or

$$\deg(\mathcal{L}) \doteq -\chi(\mathcal{O}_X) - l(T).$$

Proof (Sketch of parts of the proof).

(1) We show that the line bundle $\mathcal{O}(K_X)$ equals $f^*(\mathcal{M}) \otimes \mathcal{O}\left(\sum_i a_i P_i\right)$ where $0 \le a_i < m_i$ for some line bundle $\mathcal{M}$ on B. In the language of divisors, we show that K_X is linearly equivalent to $f^{-1}(M) + \sum_i a_i P_i$ for some divisor M on B. For this, consider r nonsingular fibres $C(i)$ of f. Then the structure sheaves $\mathcal{O}_{C(i)}$ on $C(i)$ are isomorphic to $\mathcal{O}_{C(i)} \otimes \mathcal{O}(K_X + C(i))$ because $C(i)$ is an indecomposable divisor of canonical type, see (4.5), and in fact they are curves with $p(C(i)) = 1$.

(4.6) Definition. An effective divisor $D = \sum_i n_i C_i$ on a surface X is of canonical type provided $K \cdot C_i = D \cdot C_i = 0$ for all i. In addition, D is indecomposable of canonical type provided D is connected and the greatest common divisor of the n_i equals 1.

Tensoring the following exact sequence with $\mathcal{O}(K_X)$

$$0 \to \mathcal{O} \to \mathcal{O}\left(\textstyle\sum_i C(i)\right) \to \coprod_i \mathcal{O}_{C(i)} \otimes \mathcal{O}(C(i)) \to 0,$$

we obtain the exact sequence

$$0 \to \mathcal{O}(K_X) = \omega_X \to \mathcal{O}\left(K + \textstyle\sum_i C(i)\right) \to \coprod_i \mathcal{O}_{C(i)} \to 0.$$

The cohomology sequence is the following exact sequence

$$\begin{array}{ccc} 0 \to H^0(\mathcal{O}(K_X)) \longrightarrow & H^0(\mathcal{O}\left(K_X + \sum_i C(i)\right) & \\ & \downarrow & \\ & \coprod_i H^0(C(i), \mathcal{O}_{C(i)}) & \longrightarrow H^1(\mathcal{O}(K_X)) \to \dots \end{array}$$

and from this we derive the inequality

$$\dim H^0\left(\mathcal{O}\left(K_X + \sum_i C(i)\right)\right) \geq r + p_g - q = r - 1 + \chi(\mathcal{O}_X)$$

or

$$\dim \left| K_x + \sum_i C(i) \right| \geq r - 2 + \chi(\mathcal{O}_x).$$

When the number of fibres is large enough, we can thus find $D \in |K_X + \sum_i C(i)|$.

From (3.3) we see that $D \cdot F = 0$ for each fibre F of f, and hence, D is supported in a union of fibres. This means that D can be decomposed as a sum

$$D = \Delta + \sum_j m_j \Gamma_j + \sum_i a_i P_i$$

where $0 \leq a_i < m_i$ and where Δ does not contain any fibres or multiples of P_i.

Each component Δ' of Δ satisfies $\Delta' \cdot C = 0$ and $\Delta' \cdot \Delta' = 0 = C \cdot C$ for each component C of a fibre. By (3.3), this means that Δ' is a rational multiple of a fibre containing it again, and therefore, it is zero. Thus we have

$$K_X + \sum_i C(i) = \sum_j m_j \Gamma_j + \sum_i a_i P_i,$$

and this proves the first assertion.

(2) The duality theorem for a map says that

$$\mathcal{M} = f_*(\omega_X) = \mathrm{Hom}(R^1 f_*(\mathcal{O}_X), \mathcal{O}_B) = \mathcal{L}^{(-1)\otimes} \otimes \omega_B,$$

because the dual of the torsion part is zero. This formula can be found in Deligne–Rapoport [1973, LM 349] see pp. 19–20, (2.2.3). This means that

$$\omega_X = f^*\left(\mathcal{L}^{(-1)\otimes} \otimes \omega_B\right) \otimes \mathcal{O}\left(\sum_i a_i P_i\right)$$

as asserted.

(3) From the spectral sequence of the map f we deduce

$$\begin{aligned}\chi(\mathcal{O}_X) &= \chi(\mathcal{O}_B - \chi(\mathbb{R}^1 f_*(\mathcal{O}_X)) \\ &= 1 - p(B) - \chi(\mathcal{L}) - \mathrm{length}(T) \\ &= 1 - p(B) - (\deg(\mathcal{L}) + 1 - p(B)) - \mathrm{length}(T)\end{aligned}$$

Hence we deduce the formula $-\deg(\mathcal{L}) = \chi(\mathcal{O}_X) + \mathrm{length}(T)$ the last formula asserted above.

§5. Elliptic K3 Surfaces

The object of this section is to prove that a K3 surface with Picard number at least 5 has an elliptic fibration. Recall that a K3 surface X satisfies two conditions: $\mathcal{O}_X$ and Ω_X^2 are isomorphic (so $K_X = 0$) and $H^1(X, \mathcal{O}_X) = 0$. We use two notations:

$$h(D) = \dim H^0(X, \mathcal{O}_X(D)) \quad \text{and} \quad l(D) = \dim H^0(X, \mathcal{O}_X(D)) - 1.$$

(5.1) Remark. For a K3 surface X, the general Riemann–Roch formula $\chi(\mathcal{O}_D) = (1/2)D \cdot (D - K) + \chi(\mathcal{O}_X)$ implies the following inequality

$$l(D) + l(-D) \geq \frac{1}{2}(D^2)$$

The general genus formula $\chi(\mathcal{O}_D) = -(1/2)D \cdot (D + K)$ yields a formula for the genus of an irreducible curve $C \subset X$ of the following form

$$p_a(C) = \frac{1}{2}(C^2) + 1.$$

If g denotes the genus of the normalization of C, then $p_a(C) = g + \delta$ where $\delta \geq 0$ is an invariant depending on the singularities of C with $\delta = 0$ if and only if C is smooth.

(5.2) Proposition. *Let X be a K3 surface.*

(1) *If C is an irreducible curve on X, then either $(C^2) \geq 0$ or $(C^2) = -2$, in which case C is a rational smooth curve.*

(2) *If D is a divisor on X with $(D^2) = -2$, then either $-D$ or D is equivalent to an effective divisor (which may be reducible).*

(3) *If D is a divisor on X with $(D^2) = 0$, then either $-D$ or D is equivalent to an effective divisor. If D is equivalent to an effective divisor, then $l(D) =$ $\dim(X, \mathcal{O}_X(D)) \geq 1$.*

Proof. Assertion (1) is a result of the genus formula

$$p_a(C) = \frac{1}{2}(C^2) + 1.$$

Assertion (2) results from the Riemann–Roch relation

$$\chi(\mathcal{O}_D) = \frac{1}{2}D \cdot (D - K) + \chi(\mathcal{O}_X) = \frac{1}{2}(D^2) + 2 \geq 2.$$

If $H^0(X, \mathcal{O}(D)) \neq 0$, then there is an effective E equivalent to D, and if

$$H^0(X, \mathcal{O}(K - D)) \neq 0,$$

then there is an effective E equivalent to $K - D$ or equivalently $-D$ since K is equivalent to zero. Such an effective divisor is the zero locus of a nonzero section.

Assertion (3) follows from (2). This proves the proposition.

(5.3) Proposition. *If E is a smooth genus one curve on a K3 surface X, then $(E^2) = 0$ and $\ell(E) = \dim H^0(X, \mathcal{O}_X(E)) - 1 = 1$.*

Proof. the first result comes from the genus formula. For the second assertion, we use the exact sequence

$$0 \longrightarrow \mathcal{O}_X \longrightarrow \mathcal{O}_X(E) \longrightarrow \mathcal{O}_X(E)|_E \longrightarrow 0.$$

We have the restriction $\mathcal{O}_X(E)|_E = \mathcal{O}_E$ since $(E^2) = 0$. The related cohomology exact sequence is

$$0 \longrightarrow k \longrightarrow H^0(X, \mathcal{O}_X(E)) \longrightarrow k \longrightarrow 0$$

which implies that $h(E) = \dim H^0(X, \mathcal{O}_X(E)) = 2$. this proves the proposition.

Now we come to a key theorem of Pjatecň–Šapiro and Šafarev [1971].

(5.4) Theorem. *If an effective divisor D on a K3 surface X satisfies the conditions $(D^2) = 0$ and $D \cdot C > 0$ for every curve C on X, then the linear system $|D|$ contains a divisor of the form mE where $m > 0$ and E is an elliptic curve.*

Proof. We must show that $|D|$ contains a divisor with only one component.

Step 1. Let $D' \in |D|$, and consider decompositions $D' = \sum_{i=1}^{r} a_i C_i$ where the C_i are distinct and irreducible and $a_i > 0$. Assume there are at least two indices i. Then we have $D \cdot C_i \geq 0$ and $C_i \cdot C_j \geq 0$ for $i \neq j$, and this implies that $C_i^2 = 0$ by (3.8).

We show that the self intersection $C_i^2 = 0$ for some $D' \in |D|$. For this consider an embedding $X \to \mathbb{P}^N$ with general hyperplane $H \subset \mathbb{P}^N$ and hyperplane section $X \cap H$ of X. In the intersection $D \cdot H = D' \cdot H = \sum_{i=1}^{r} a_i(C_i \cdot H)$ the terms $C_i \cdot H$ are just the degrees of the embeddings $C_i \to \mathbb{P}^N$. Hence the positive integers r and a_i are bounded. Therefore, the number of $D' \in |D|$ with all $C_i \cdot C_i < 0$ is finite in number, while the linear system $|D|$ is infinite.

We have shown the existence of $D' \in |D|$ which decomposes as a sum of the form $D' = mE + D'' = mE + \sum_{i=1}^{r} a_i C_i$ where $E \cdot E = 0$ and $m \geq 0$.

Step 2. In the decompositions $D' = mE + D''$ as in step 1, we have two further intersection properties: $E \cdot D'' = 0$ and $D'' \cdot D'' = 0$. For we calculate $0 = D' \cdot D' = mE \cdot D' + D'' \cdot D'$, and by (3.8) we have $E \cdot D' = 0$ so that $D'' \cdot D' = 0$. Moreover, $E \cdot D'' \geq 0$ since they have no common components, and from the following two relations

$$0 = D' \cdot D'' = mE \cdot D'' + D'' \cdot D'' \quad \text{and} \quad 0 = D' \cdot D' = 2mE \cdot D'' + D'' \cdot D''$$

we see that $E \cdot D'' = 0$ and $D'' \cdot D'' = 0$.

Step 3. Among the $D' \in |D|$ decomposed as $D' = mE + D''$ with $E \cdot E = 0$, so that $E \cdot D'' = 0$, $D'' \cdot D'' = 0$ by step 2, we choose the one with $E'' \cdot H$ minimal as a natural number for H a hyperplane section. It remains to show that $D'' = 0$ to complete the proof, and this we do by deriving a contradiction assuming $D'' \neq 0$. If $D'' \neq 0$, then $\ell(D) \geq 2$, and we can apply the above considerations in steps 1 and 2 for $|D|$ to $|D''|$. There exists $D(1) \in |D''|$ with the decomposition

$$D'' \text{ linearly equivalent to } D(1) = nE + D(2)$$

so that D' is equivalent to $(m+n)E + D(2)$. This means that intersecting this $D(2)$ with the hyperplane H we have

$$H \cdot D(2) = H \cdot D(1) - H \cdot (mE) = H \cdot D' - H \cdot (mE) < H \cdot D''$$

which contradicts the minimal character of K''. Hence $D'' = 0$ for the minimal case and thus $D' = mH$. This proves the theorem.

(5.5) Theorem. *Let X be a K3 surface with $\rho(X) \geq 5$. Then there exists an elliptic fibration.*

Proof. The rational inner product space $\mathrm{NS}(X)_{\mathbb{Q}}$ is of rank greater than 5, and is indefinite, so that there exists $D' \in \mathrm{NS}(X)_{\mathbb{Q}}$ nonzero with

$$(D')^2 = 0$$

Then for some n we have $D = nD' \in \mathrm{NS}(X)$ and also $D \cdot D = 0$. Now we apply the previous theorem to D, and we have $mE \in |D|$ where E is an elliptic curve. Applying (2.13) to D, we have a map $\phi_D : X \to |D| = \mathbb{P}_1$ with D a fibre over a point $p \in \mathbb{P}_1$. If $f : \mathbb{P}_1 \to \mathbb{P}_1$ is the m-fold ramified covering of $\mathbb{P}_1$ ramified at p and one other point, then $\phi = f\phi_D$ is a fibration $\phi : X \to \mathbb{P}_1$ with fibre E at $f(p)$. This proves the theorem.

§6. Fibrations of 3 Dimensional Calabi–Yau Varieties

In this section we survey the results on fibrations by elliptic curves and K3 surfaces of Calabi–Yau 3-folds. the main reference is K. Oguiso [1993]. In the previous section the fibering of a K3 surface by elliptic curves started with a divisor which was a possible fibre for an elliptic fibration, that is, with an elliptic curve E with $E \cdot E = 0$.

For a 3-fold, we can look for a fibration by either surfaces or curves. In the first case, we would start with a divisor or line bundle with the intersection properties of a fibre, and in the second case we would start with a divisor with the selfintersection properties of a fibre of curves. The divisors in this picture would be numerically effective divisors.

(6.1) Definition. A divisor D on a variety is numerically effective, or nef, provided $D \cdot C \geq 0$ for all curves C on X. A line bundle $\mathcal{L}$ is numerically effective provided it is of the form $\mathcal{L} = \mathcal{O}(D)$ where D is a numerically effective divisor.

While there are many refinements of the following result in terms of singularities, the main result in the simple case reduces to the following assertion.

(6.2) Theorem. *Let X be a smooth Calabi–Yau variety of dimension 3, and let D be a numerically effective divisor on X. If the triple intersection $D \cdot D \cdot D' = 0$ for all divisors D', then X has a fibration over $\mathbb{P}^1$ by K3 surfaces. If the triple intersection $D \cdot D \cdot D = 0$ and another triple intersection $D \cdot D \cdot D' \neq 0$ for some divisor D', then X has a fibration by elliptic curves.*

The reference for this result and following related remarks is Oguiso [1993]. He uses the mapping associated with a line bundle or divisor, see (2.2). Under such mappings sub-varieties can map to a point, and the image of such a variety can have singularities. Thus we have the concept of a minimal Calabi–Yau variety which must include varieties with very mild singularities.

(6.3) Definition. A minimal Calabi–Yau 3-fold is a normal projective complex 3 dimensional variety X with K_X and $\mathcal{O}_X$ isomorphic, $\pi_1^{\text{alg}}(X) = 0$, and $c_2(X) \neq 0$. There is also a $\mathbb{Q}$-factorial condition on the singularities which we do not go into..

This generalization is related to a possible minimal model theory of 3-folds, that is, a 3-dimensional normal $\mathbb{Q}$-factorial projective complex variety.

Associated with a divisor or line bundle $\mathcal{L}$ of X is a sequence of maps $\phi_{n,\mathcal{L}} = \phi_n : X \to \mathbb{P}\left(\Gamma\left(X \cdot \mathcal{L}^{n\otimes}\right)\right)$ which are everywhere defined when for each point $x \in X$ there is a section of $\mathcal{L}^{n\otimes}$ which is nonzero at this point. In terms of a related divisor D this is just the condition of being base point free.

In his article Oguiso organizes the proof in terms of the following numerical invariants, following various notations of Kawamata.

(6.4) Notation. Consider a minimal 3-fold X with only terminal singularities and whose canonical divisor is numerically effective. The D-dimension $\varkappa(X, D)$ of a $\mathbb{Q}$-Cartier divisor D on X is

$$\varkappa(X, D) = \begin{cases} \max_{m>0} \dim \phi_{|mD|}(X) & \text{if } |mD| \text{ is nonempty for some } m > 0, \\ -\infty & \text{otherwise.} \end{cases}$$

The numerical D-dimension $\nu(X, D)$ of a numerically effective $\mathbb{Q}$-Cartier divisor D is

$$\nu(X, D) = \text{maximum natural } n \text{ with } nD \text{ not numerically equivalent to } 0.$$

(6.5) Table of Possible Fibrations.

Type	$\nu(X, \mathcal{L})$	$\mathcal{L} \cdot c_2(X)$	—structure of $\phi : X \to W$
0	0	0	—$\mathcal{L}$ and $\mathcal{O}_X$ are isomorphic and im(ϕ) is a point.
I_+	1	+	—$\phi : X \to \mathbb{P}_1$ with general fibre a K3 surface.
I_0	1	0	—$\phi : X \to \mathbb{P}_1$ with general fibre an abelian suface.
II_+	2	+	—a general fibre of ϕ is an elliptic curve and W is a rational surface with only quotient singularities.
II_0	2	0	—a general fibre of ϕ is an elliptic curve and W is a non-Gorenstein rational surface with only quotient singularities.
III	3	≥ 0	—ϕ is birational and W is a projective normal 3-fold with K_W equivalent to zero, so W is canonical of index 1.

Related assertions to the above table:

(1) the rational surface W for type II_+ has nK_W equivalent to $-\Delta$ for some effective Cartier divisor $\Delta > 0$ on W and some positive integer n.
(2) The rational surface W for type II_0 has at least one quotient singularity other than Du Val singularity, and nK_W is equivalent to zero for some positive integer n. Moreover, ϕ is a smooth fibration in codimension one on W, and there is a non-Gorenstien point $w \in W$ such that $\dim\phi^{-1}(w) = 2$. In particular, ϕ is never equidimensional in this case.
(3) All six types occur.
(4) There exists a smooth Calabi–Yau 3-fold which has a type II_0 fibration and infinitely many different fibrations of types I_+, I_0, and II_+.

An important step is to relate a singular minimal Calabi–Yau 3-fold with its minimal resolution, which we sketch now.

(6.6) Remark. Let X be a minimal Calabi–Yau 3-fold.

(1) $\dim(\mathrm{Sing}(X)) = 0$ consists of isolated points.
(2) X is factorial so every Weil divisor on X is a Cartier divisor.
(3) $\chi(\mathcal{O}_X) = 0$ by Serre duality.
(4) $h^1(\mathcal{O}_X) = 0$, for if it were strictly positive, it would be incompatible with $\pi_1^{\mathrm{alg}}(X)$ being zero.

(6.7) Remark. We consider minimal Calabi–Yau 3-folds and resolutions $\nu : Y \to X$. It follows that the numerical class $\nu_* c_2(Y)$ is independent of ν from the following formula

$$\chi(\mathcal{O}_X(D)) = \frac{1}{6}D^3 + \frac{1}{12}D \cdot \nu_* c_2(Y).$$

This can be used as a definition of $c_2(X) = \nu_* c_2(Y)$. Further, if D is a numerically effective divisor with $\nu(X, D) \geq 0$, then we have the same formula

$$\chi(\mathcal{O}_X(D)) = \frac{1}{6}D^3 + \frac{1}{12}D \cdot \nu_* c_2(X)$$

where the vanishing theorem in the following is used: Y. Kawamata, K. Matsuda, and K. Matsuki, *Introduction to the minimal model problem*, Adv. St. Pure Math. *10* (1987), pp. 283–360.

We close by indicating some references besides the main reference by Keiji Oguiso, *On algebraic fiber space structures on a Calabi–Yau 3-fold*, International Journal of Mathematics, Vol. 4, No. 3 (1993) 439–465. Article of Y. Kawamata, *Abundance Theorem for minimal Threefolds*, Invent. Math. *108* (1992) pp. 229–246 was an important background article.

A slightly later paper than that of Oguiso is: P. M. H. Wilson, *The existence of elliptic fibre space structure on Calabi–Yau threefolds*, Mathematische Annalen, 1994.

See also: N. Nakayama, *Local Weierstrass models*, Algebraic Geometry and Commutative Algebra in Honor of M. Nagata, vol II, Kinokuniya and North-Holland, 1987, pp. 403–431 is used. U. Persson, *On degenerations of algebraic surfaces*, Memoirs Amer. Math. Soc. *189* (1977).

§7. Three Examples of Three Dimensional Calabi–Yau Hypersurfaces in Weight Projective Four Space and Their Fibrings

In two early papers on mirror symmetry with authors P. Candelas, X. de la Ossa, A. Font, S. Katz, and D. Morrison [1994] there are careful analyses of particular three dimensional hypersurfaces in weighted projective spaces. We give a short sketch of three of these examples and refer to the papers where the lively interest in fibred Calabi–Yau manifolds arose among string theorists.

(7.1) Degree of the Hypersurface in the Weighted Projective Spaces Under Consideration.

Case 1. Degree 8 hypersurface in $\mathbb{P}_4(1, 1, 2, 2, 2)$ with equation

$$X : y_0^8 + y_1^8 + y_2^4 + y_3^4 + y_4^4 = 0.$$

The degree $8 = 1 + 1 + 2 + 2 + 2$ so the hypersurface in this degree is a Calabi–Yau manifold.

The intersection with the codimension 2 linear space given by equations $y_0 = y_1 = 0$ is quartic curve $C : y_2^4 + y_3^4 + y_4^4 = 0$. Let L be the linear system generated by the polynomials of degree one in y_0 and y_1. The divisors in L are the proper transforms of the zero locus under a blow up of C of the hypersurface X. These divisors can be described by $y_0 = \lambda y_1$ for $\lambda \in \mathbb{P}_1$ and denoted $D(\lambda)$. Setting $y_1^2 = x_1$ and $y_i = x_i$ for $i > 1$, we obtain a surface of degree four in $\mathbb{P}_3$, and this means that

L defines a pencil of K3-surfaces as fibres of $\phi_L : X \to \mathbb{P}_1$. Since any two distinct members of L are disjoint, the intersection product $L.L = 0$.

Case 2. Degree 12 hypersurface in $\mathbb{P}_4(1, 1, 2, 2, 6)$ with equation

$$y_0^{12} + y_1^{12} + y_2^6 + y_3^6 + y_4^2 = 0.$$

The degree $12 = 1+1+2+2+6$ so the hypersurface in this degree is a Calabi–Yau manifold.

The intersection with codimension 2 linear space given by the equations $y_0 = y_1 = 0$ is the singular curve $C : y_2^6 + y_3^6 + y_4^2 = 0$ of genus 2. When the singularity is blown up, we obtain a ruled surface over C. Let L be the linear system generated by the polynomials of degree one in y_0 and y_1. The divisors in L are the proper transforms of the zero locus under a blow up of C on the hypersurface X. These are $K3$ surfaces in $\mathbb{P}_3(1, 2, 2, 6)$ which are isomorphic to surfaces of degree 6 in $\mathbb{P}_3(1, 1, 1, 3)$.

Case 3. Degree 18 hypersurface in $\mathbb{P}_4(1, 1, 1, 6, 9)$ with equation

$$y_0^{18} + y_1^{18} + y_2^{18} + y_3^3 + y_4^2 = 0.$$

The degree $18 = 1 + 1 + 1 + 6 + 9$ so that the hypersurface in this degree is a Calabi–Yau manifold. For this equation the singularities occur along $y_1 = y_2 = y_3 = 0$ which is a curve in $\mathbb{P}_4(1, 1, 1, 6, 9)$ intersecting the hypersurface at one point $0 : 0 : 0 : -1 : 1$. Also, this weighted projective space has a quotient singularity of order 3 along this curve from the weighted projective space. Next, the polynomial $y_0 y_1 y_2$, being invariant under this group of order 3, defines a divisor with a simple intersection along the quotient curve.

The linear system L generated by y_0, y_1, y_2 maps the hypersurface to $\mathbb{P}_2$ with fibres $y_3^3 + y_4^2 =$ constant, namely elliptic curves.

For further details and implications in string theory, see the articles in *Nuclear Physics*, B416 (1994) 481–538 and B429 (1994) 626–674.

Appendix I: Calabi–Yau Manifolds and String Theory

Stefan Theisen

Calabi–Yau manifolds are not only of interest to mathematicians. They also play a central role in an area of modern theoretical physics called string theory. This common interest in Calabi–Yau manifolds, both by mathematicians and by physicists, has led to interesting new developments in both fields and has initiated fruitful collaborations. The aim of this appendix is to explain the physicists' interest in string theory and in particular how Calabi–Yau manifolds enter. This touches upon a vast area of current research and we can merely give a brief and qualitative overview of some aspects of string theory. A detailed reference, for all aspects which we will touch upon and much more, is 'Quantum Fields and Strings: A Course for Mathematicians', Vols. I & II, P. Deligne et al. (eds.), AMS publication 1999, where references to the original literature can be found.

Why String Theory?

String theory is part of a wider area of theoretical physics, the theory of elementary particles, whose main goal is to find a unifying theoretical framework to describe matter and its interactions. Three of the known interactions—the electromagnetic, and the weak and strong forces—are very successfully described by the so-called standard model of elementary particle physics within the framework of a quantum field theory with Yang–Mills gauge group $U(1) \times SU(2) \times SU(3)$. These interactions dominate at the atomic and subatomic level and are studied in present day accelerator experiments. The fourth interaction—gravity—is much weaker in the microscopic realm and can be safely neglected. However, the dynamics of stars, galaxies and the universe as a whole are governed by the laws of gravity. The gravitational interaction, at the classical (versus quantum) level, is described by Einstein's theory of general relativity. One of the hallmarks of this theory is the coupling of matter and energy to the space–time geometry, described quantitatively by Einstein's field equations. However, at very small distances, much smaller than atomic scales, or, equivalently, at very high energies, one expects that the classical concept of space-time requires modification. Gravity, like matter and the other three interactions to which it couples, must obey the laws of quantum mechanics. The familiar concept of a (four-dimensional) space–time continuum with a smooth metric is only meaningful in an appropriate classical limit. All attempts to unify gravity with the other three interactions within the framework of quantum field theory have so far been futile. The nature of the problem can be heuristically understood as follows.

The basic constituents of matter are conventionally viewed as point-particles which propagate through space–time along trajectories called world-lines. In this picture interactions between elementary particles are described by graphs, the so-called Feynman diagrams, which fail to be manifolds at the interaction points. This is the reason why in the quantum theory the concept of point-particles leads to divergent expressions when one computes physical quantities. In the case of Yang–Mills interactions these divergences can be consistently dealt with in a well-defined procedure called renormalization. In this way, the standard model allows us to predict the result of a large number of experiments after fixing the parameter (coupling constants, masses, etc.) appearing in the theory by fitting the data of few experiments. As such, it has proven very successful. However, once the gravitational interaction is included, the renormalization program fails; the infinities cannot be dealt with in a consistent way.

An intuitive cure of the infinities is to 'thicken' the Feynman diagrams into two-dimensional surfaces with smooth junctions, as indicated by this simple diagram which describes the decay of a particle into two particles:

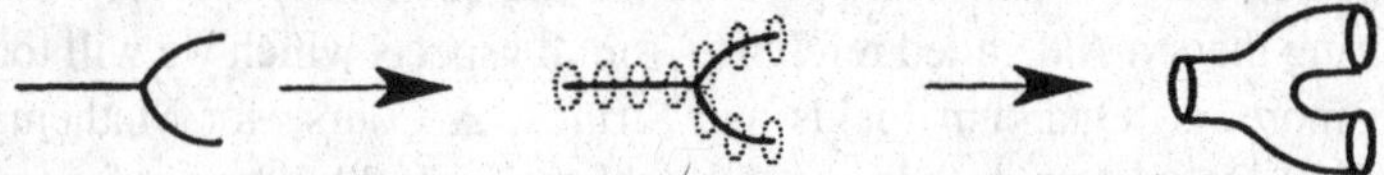

The diagrams become manifolds but no longer represent the propagation and interaction of point-particles, but rather of one-dimensional objects.

Basic Properties

This is precisely the approach taken by string theory which departs from the notion of point-like elementary constituents of matter and replaces them by a one-dimensional object: a string. The world-line of a point-particle is now replaced by the world-sheet of the string. The observed elementary particles are harmonic oscillations of the string of which there are an infinite number with ever growing mass. The analogy with a vibrating string of a harp—the overtones with increasing frequency corresponding to the exitations of increasing mass—is useful. In both cases the frequencies are determined by the tension of the string, denoted by T. But the analogy is limited as the mass is only one of the several characteristic properties of an excited fundamental string. Also, one distinguishes between open and closed strings. A closed string has two kinds of excitations, left-moving and right-moving waves, while on an open string a left-moving wave is reflected at the boundary and returns as a right-moving wave. An open string can close if its two ends meet, turning it into a closed string. There are thus theories with open and closed strings but there are also theories where only closed strings exist. Furthermore, one also distinguishes between the bosonic and fermionic strings. The excitations of the bosonic string are bosons, i.e., they transform in tensor representations ('integer spin') of the space-time Lorentz group, while the excitation spectra of fermionic strings also contain fermions ('half-integer spin'), which transform in spinor representations. Physically bosons and fermions behave very differently and both types of particles are observed in nature. The constituents of matter (electrons, neutrinos, quarks, etc.) are fermions. They interact with each other by exchanging bosons (gauge bosons for the $U(1) \times SU(2) \times SU(3)$ gauge interactions and gravitons for the gravitational interaction.) Below we will return to the distinction between different string theories. Of course a viable string theory must eventually reproduce the precise spectrum of the observed particles and their interactions with the correct strength. While this has at present not yet been achieved, an important and encouraging result is the fact that the spectrum of the closed string always contains a massless mode with properties which allow its identification with the graviton.

The general theory of relativity contains two fundamental constants of nature: the speed of light c and Newton's gravitational constant G_N. A quantum theory of gravity must also contain Planck's constant $\hbar$. As in any quantum theory, $\hbar$ enters via a quantization procedure (path integral quantization or BRS-quantization, to mention just two). From these three constants of nature one can build combinations with the dimension of length, time and mass, called Planck length $l_P = \sqrt{G_N\hbar/c^3}$, Planck time $t_P = \sqrt{G_N\hbar/c^5}$ and Planck mass $m_P = \sqrt{c\hbar/G_N}$. One can then express all distances, time intervals and masses as multiples of these three fundamental units. In string theory the string tension T replaces G_N as the fundamental constant of nature and the latter can be expressed in terms of T, c and $\hbar$. With the help of the string tension one can define a length scale $l_s = \sqrt{\hbar/Tc}$ and an energy scale $E_s = \sqrt{\hbar c^3 T}$ which are characteristic for string theory. For example, the typical size of a string is determined by l_s and the masses of its excitations are integer multiples of E_s/c^2. The natural value for the string tension is that such that $l_s \sim l_P \sim 1.6 \times 10^{-35}$ meters (for comparison: the size of an atom is 10^{-10} meters). Within the framework of string theory it now becomes evident why the classical theory of general relativity ceases to be meaningful at distances $l < l_s \sim l_P$ or, equivalently, at energies $E > E_s$. In this length and energy range the classical concept of space–time, which is adequately described by (pseudo)-Riemannian geometry, requires modification. The reason is that a one-dimensional object, a string, probes a space very differently from a zero-dimensional object at length scales of the order of and below the size of the probe.

The transition from zero-dimensional to one-dimensional fundamental building blocks, or from world-lines to world-sheets, thus has far-reaching consequences. We have already mentioned that the divergences which seem to render a quantum-field-theoretic treatment of gravity impossible no longer appear. Furthermore, since now all elementary particles arise as vibrations of a single fundamental object, the string, one has achieved a unification of particles and their interactions which necessarily includes gravity.

Any complete theory of quantum gravity should *explain* the structure of space-time. In other words, its dimension, topology, Riemannian structure, etc., which we observe at length scales much bigger than the Planck length, should be derivable consequences of the theory rather than inputs. In particular, the concept of a space–time metric should only emerge in an appropriate limit. This ambitious goal has not been reached yet. But even at the present stage of development there are many aspects in which string theory differs significantly from quantum field theory in an external gravitational field.

String Theories in Ten Dimensions

One indication of this incompleteness of string theory is the fact that its quantization is presently only possible in à priori given classical space–time backgrounds through which the string propagates. This was, for example, implicit above when we appealed to the intuitive picture of the world-sheet being embedded in space–time. The simplest case is the quantization of a free string in flat Minkowski space–time, i.e., d-dimensional $\mathbb{R}^d$ with Minkowski metric. As in theories of point-particles, consistent quantization must result in a positive-definite Hilbert space. Moreover, the symmetries of the classical theory (space-time symmetries, gauge symmetries, diffeomorphism and scale invariance on the world-sheet) must survive the process of quantization. This is the requirement of the absence of anomalies. One important restriction which follows from these requirements is that string theory cannot be quantized in a Minkowski space–time of arbitrary dimension. Rather, we must be in the so-called critical dimension which is $d_{\text{crit}} = 26$ for the bosonic string and $d_{\text{crit}} = 10$ for the fermionic string.

The excitation spectrum of the bosonic string contains, in addition to a finite number of massless states and an infinite number of massive states, a so-called tachyon with $(\text{mass})^2 < 0$. Such a state signals an instability much in the same way as the instability of a particle located at the top of a potential hill. Even though there might be nearby minima of this potential into which the system could settle, this question has, for the case of the bosonic string, not yet been answered. Also, the bosonic string does not contain fermions and can thus not be considered for a realistic description of nature, but the possibility that, once it has settled in a stable minimum fermionic degrees of freedom will appear, has not been excluded yet. The spectrum of the fermionic string can, via a suitable projection (*GSO projection*), be restricted in such a way that tachyons are absent and that the resulting string theory is *supersymmetric*. The projection also ensures that all anomalies are absent and is thus required for consistency of the theory.

Supersymmetry is a symmetry between bosons and fermions. In addition to the generators of the Poincaré symmetry—translations, rotations and Lorentz-transformations—there are $\mathcal{N}$ fermionic generators, the supercharges. They transform as spinors under the Lorentz group and, when acting on a boson (fermion) they turn it into a fermion (boson). By adding additional fermionic fields, called superpartners, the action functionals which govern the dynamics of Yang–Mills theory and general relativity can be extended in such a way that they are invariant under supersymmetry transformations. The resulting theories are known as super-Yang–Mills and supergravity, respectively. Super-Yang–Mills theories and supergravity theories exist in all space–time dimensions less than or equal to ten while a (unique) supergravity theory can also be constructed in eleven dimensions. In string theory space–time supersymmetry can be traced back to a supersymmetry on the two-dimensional world-sheet.

There are five consistent string theories with space–time supersymmetry which one can construct in ten-dimensional Minkowski space-time. Non-supersymmetric theories are also known, but they are unstable. The type I string theory, which is the only one which possesses both open and closed strings, has $\mathcal{N} = 1$ supersymmetry. The massless sector of its spectrum corresponds to supersymmetric Yang–Mills theory coupled to supergravity. The gauge fields are part of the excitation spectrum of the open string while the graviton is a massless excitation of the closed string. Consistency of the type I theory—the absence of gravitational and gauge anomalies—requires that the gauge symmetry be $SO(32)$.

The remaining four theories have only closed strings. There are two type II theories with $\mathcal{N} = 2$ supersymmetry in ten dimensions. Two inequivalent GSO projections lead to the so-called type IIA and type IIB theories. In the type IIB theory one imposes the same projection for the left- and the right-moving excitation while 'opposite' projections lead to the type IIA theory. Their massless spectra are those of the ten-dimensional type IIA and type IIB supergravity theories, respectively. Finally, the other two consistent string theories in ten-dimensional Minkowski space–time also possess, as in the type I theory, $\mathcal{N} = 1$ supersymmetry. The two possible gauge groups are $SO(32)$ and $E_8 \times E_8$. These theories are called heterotic string theories. They are constructed from a combination of the bosonic string and the fermionic string, one for the left-movers and the other for the right-movers. The difference in critical dimensions, sixteen, is equal to the rank of the two possible gauge groups. The construction involves the only two even Euclidean self-dual lattices of dimension sixteen, which enter as the weight lattices of the two possible gauge groups. Again, the absence of anomalies is responsible for this very restrictive choice. The massless spectrum of the two heterotic theories is again that of supersymmetric Yang–Mills theory coupled to supergravity, but now with gauge groups $SO(32)$ or $E_8 \times E_8$. In fact, one can show that in the limit of infinite tension, in which the string contracts to a point and the masses of the excited, massive string states become infinite, the dynamics of the massless modes of each of the five theories is governed by the appropriate ten-dimensional field theory. The full string theory is, however, much richer and can, in contrast to their field theory limits, be consistently quantized.

Compactification

So far we have only discussed string theory in the simplest possible background geometry, that of Minkowski space–time M_{10}. The most striking result is that consistency of the theory fixes the space–time dimension. The fact that it came out to be ten rather than the observed four might, at first sight, render the theory useless as a physical theory. However, through the process of *compactification* one can hide the six excess dimensions. It is in this process where string theory reveals many of its characteristic features and its rich mathematical structure. This is also how Calabi–Yau manifolds will enter.

The basic idea of compactification is to replace the Minkowski space–time by a different ten-dimensional space–time which has only four dimensions of infinite extent and the remaining six dimensions of finite extent. The simplest realization of this idea is to start with the ansatz $M_4 \times K_6$ for the ten-dimensional space–time. Here K_6 is a six-dimensional compact manifold, hence the term compactification. If one makes K_6 small enough, the world effectively looks four-dimensional. The natural size of K_6 is l_s, the only length scale which enters the theory and, in fact, this value is far below the distances we can probe with our most powerful microscopes, the high-energy particle accelerators.

A compactification which respects this ansatz is specified by a metric on K_6 and by additional tensor fields which occur in the massless spectrum of the string theory which we want to compactify. Again, consistency requirements, for example, scale-invariance on the world-sheet, restricts the allowed compactifications. The resulting four-dimensional theory depends crucially on the choice of compactification; for example, the compactification breaks supersymmetry unless K_6 admits parallel spinors. This is the case for manifolds with special holonomy $\mathrm{Hol}(K_6) \subset SO(6)$. Here the spinors may be parallel with respect to a connection which is not necessarily the Riemannian connection, but it might involve other tensor fields which characterize the chosen compactification; its precise form is fixed by supersymmetry. By an appropriate choice of background fields one can thus 'engineer' various four-dimensional theories. Without further explanation we state that on physical grounds the favored K_6's are those which admit exactly one parallel spinor field. One consequence of this choice is that when compactifying the type II string the four-dimensional theory has $\mathcal{N} = 2$ supersymmetry whereas for the compactification of the heterotic string one obtains $\mathcal{N} = 1$ supersymmetry. (Compactification on a flat six-dimensional torus would lead to $\mathcal{N} = 8$ and $\mathcal{N} = 4$ supersymmetry, instead.)

If the connection is simply the Riemannian one, and if no other other background fields are present, then the existence of precisely one parallel spinor implies that K_6 is a Calabi–Yau manifold, i.e., a compact six-dimensional Kähler manifold with $SU(3)$ holonomy. For the consistent compactification of the heterotic string one also has to specify, in addition to the metric, a holomorphic stable vector bundle with $G \subset SO(32)$ or $E_8 \times E_8$. Numerous Calabi–Yau manifolds have been explicitly constructed (the simplest one being the quintic hypersurface in $\mathbb{CP}^4$) and string compactification on them has been studied in much detail. The specific choice of a Calabi–Yau manifold determines the spectrum of the compactified string theory and also its interactions. For example, the massless states can be determined from the cohomology and topology of the Calabi–Yau manifold.

While, from a physical point of view, compactifications on six-dimensional Calabi–Yau manifolds are the most relevant ones as they lead to a four-dimensional world at long distances, other dimensions have also been considered, e.g., compactification on $K3$ surfaces, tori of various dimensions, or seven-dimensional manifolds with G_2 holonomy.

We have already mentioned that the simplest compactifications of type II string theories only involves a non-trivial metric and that this is not the most general case. In fact, compared to the heterotic theories compactified on the same manifold, the type II theories always have twice as much supersymmetry. There is an interesting class of compactifications of type IIB theory which leads to $\mathcal{N} = 1$ supersymmetry in four dimensions. The type IIB theory has in its excitation spectrum two real massless scalar fields χ and ϕ which can be combined into a single complex scalar field with positive imaginary part $\tau = \chi + ie^{-\phi}$. The type IIB string theory possesses an exact $SL(2,\mathbb{Z})$ symmetry under which the graviton is inert and which acts on τ as $\tau \to (a\tau + b)/(c\tau + d)$ where $a, b, c, d \in \mathbb{Z}$ with $ad - bc = 1$. Inequivalent values of τ are thus parametrized by points in a fundamental domain of the modular group $\mathcal{F} = \mathcal{H}_+/SL(2,\mathbb{Z})$ where $\mathcal{H}_+$ is the Siegel upper half plane. We can now interpret τ as the complex structure of an (auxiliary) elliptic curve which varies over the compact manifold in a way that is dictated by the coupled equations of motion for τ and the metric on K_n. If one combines the requirements of non-constant τ and supersymmetry of the compactified theory, one finds that K_n must be a Kähler manifold with positive first Chern class and that τ must vary holomorphically in such a way that we can view K_n as the base of an elliptically fibered $2n + 2$ dimensional Calabi–Yau manifold (which has vanishing first Chern class) where the fiber is characterized by τ. The Calabi–Yau condition implies that the fiber degenerates over (complex) co-dimension one loci. When transported around these loci, τ undergoes $SL(2,\mathbb{Z})$ monodromy transformations. The simplest example is $n = 2$ with $K_2 = \mathbb{CP}^1$. The total space is then an elliptic $K3$ surface and in the generic case the fiber degenerates (in an A_1 singularity) over 24 points. These particular compactifications of the ten-dimensional type IIB string theory can thus be viewed as compactifications of a twelve-dimensional theory, which has been called F-theory. We should point out, however, that there does not exist a string theory in twelve dimensional Minkowski space–time. What F-theory does is to interpret certain compactifications of type IIB string theory, namely those with non-trivial metric and τ, in terms of twelve-dimensional geometries. It is through this interpretation that elliptically fibered Calabi–Yau manifolds enter. Again, $n = 6$ is the most interesting case for physical applications; this leads to the study of elliptically fibered four-complex-dimensional Calabi–Yau manifolds. The physics of the four-dimensional world thus constructed depends on the topology and cohomology of the particular Calabi–Yau manifold chosen.

Duality

With the explanation of the importance of Calabi–Yau manifolds for string theory, we have made contact with the main body of the book and have thus fulfilled the purpose of this appendix. Nevertheless we want to mention one last issue which presently dominates both the physical and the mathematical aspects of string theory:

duality. We start with a simple illustrative example, namely the compactification of the bosonic string on an circle. It emphasizes once more the importance of the fact that strings are one-dimensional rather than zero-dimensional objects.

The mass-spectrum of a particle on a compact Riemannian manifold is determined by the eigenvalues of the Laplacian. Manifolds with different topologies and Riemannian structures generally give different spectra. The simplest example is the compactification on a circle of radius R. One finds the so-called Kaluza–Klein excitations with $(\text{mass})^2 = (n\hbar/Rc)^2$, for all $n \in \mathbb{Z}$. In the limit $R \to \infty$ one obtains a continuous spectrum which is the signal for an additional direction of infinite extent. The spectrum of a closed string on a circle differs from that of the particle in an important way: the string not only moves on the circle, it can also wrap it. Due to tension, stretching the string requires energy, which depends on the winding number w and on the radius R. One now finds the allowed values $(\text{mass})^2 = (n\hbar/Rc)^2 + (wRT)^2$, for all $n, w \in \mathbb{Z}$. This spectrum is symmetric under the discrete $\mathbb{Z}_2$ duality transformation $R \to l_s^2/R$ and $n \leftrightarrow w$, which exchanges Kaluza–Klein excitations and winding states. One can show that this symmetry extends, beyond the spectra, to the complete theory, that is, also to its dynamics. The important conclusion we draw from this example is that geometrically distinct compactifications of string theory may correspond to physically identical theories. Both for $R \to \infty$ and for $R \to 0$, one obtains a continuum of massless states which in each of the two cases is interpreted as the decompactification of an additional dimension. That this also happens for $R \to 0$ is at first surprising. It reveals a typical 'stringy' effect and demonstrates that strings probe space–time quite differently from point-particles. If one wants to parametrize all physically distinct compactifications on a circle, one has to choose one of the two fundamental domains of the $\mathbb{Z}_2$ duality group, either $R \geq l_s$ or, equivalently, $0 \leq R \leq l_s$. Note that the parameter space of inequivalent compactifications of a point-particle is $R \geq 0$. The bosonic string compactified on a circle of radius R is *dual* to the bosonic string compactified on a circle of radius l_s^2/R. More generally, two theories are called dual if they describe exactly the same physics. In many cases, the dual theories *appear* to be quite different; this is the reason why dual pairs are difficult to find.

As a second example consider the compactification of the type II theories on a circle. The naive expectation is that the compactification of each of the two type II theories possesses its own parameter space. What one finds instead is that the two theories are different limit points in a single space of compactified theories. For instance, if we start with the ten-dimensional type IIA theory, compactifiy on a circle of radius R, and then take the limit $R \to 0$, we get the ten-dimensional type IIB theory. It is thus appropriate to simply talk about type II theories. The distinction between A and B simply refers to the two boundary points of their common parameter space of compactifications on a circle.

This example has a very interesting generalization. Consider the compactification of, say, the type IIA theory on a six-(real)dimensional CalabiYau manifold, X. Is this dual to a compactification of the type IIB theory? The answer is yes. For any X there exists a Calabi–Yau manifold X^*, such that the compactification of the type IIB theory on X^* is identical, that is, physically indistinguishable, from the compactification of the type IIA theory on X. The pair of Calabi–Yau manifolds (X, X^*) is referred to as a mirror pair and this duality is called mirror-symmetry. Here X and X^* are topologically distinct manifolds, for example, their Euler numbers χ are related via $\chi(X) = -\chi(X^*)$. As in the previous examples, mirror symmetry is a property of the full string theory and does not hold for the corresponding supergravity theories. There is a non-geometric formulation of string compactifications (in terms of conformal field theory) where mirror symmetry appears as a triviality and where it was first observed. However, when translated into geometry, it has far reaching implications for the mathematics of Calabi–Yau manifolds.

One of the main advances in string theory after 1995 was the realization that all known string theories (and their compactifications) can be parametrized by one single parameter space. For instance, the two type II theories lie at the two boundaries of a one-dimensional subspace, parametrized by R. The all-encompassing theory, even though still unknown, has been named M-theory, where M might, for the time being, stand for Mystery. The ten-dimensional string theories correspond to certain points 'close to the boundary' of this parameter space. As a surprise one has found that a particular corner of the parameter space of M-theory corresponds to a theory whose massless fields and their interactions are precisely those of the unique eleven-dimensional supergravity theory. It is not the field theory limit of a string theory but rather of a theory whose fundamental constituents are believed to be membranes.

This example shows that whatever M-theory turns out to be, it will look like a theory of strings only in some regions of its parameter space. These regions can and have been thoroughly studied. For a complete description one certainly needs, besides string, other objects, such as membranes and higher dimensional generalizations. The nature of M-theory is the big open question.

Summary

In this appendix we could only give a glimpse of some aspects of string theory. At the present stage, the status of string theory, or perhaps more appropriately, M-theory, is far from being a completely understood and developed theory. For instance, it does not provide a mechanism for the process of compactification and thus cannot explain the dimension, the topology and the signature of the metric of our universe, the observed spectrum of elementary particles and their interactions. No background independent formulation of string theory is known. One instead considers strings moving in a consistently but otherwise arbitrarily chosen (metric, etc.) background. Ideally one would like to see this background emerge, at low energies or, equivalently, at large distances—as compared to E_s and l_s, respectively—from a 'dynamical condensation process'. The precise meaning of this statement will hopefully follow from further progress in our understanding of string theory. This requires the joint effort of physicists and mathematicians.

Appendix II: Elliptic Curves in Algorithmic Number Theory and Cryptography

Otto Forster

§1. Applications in Algorithmic Number Theory

In this section we describe briefly the use of elliptic curves over finite fields for two fundamental problems in algorithmic number theory, namely factorization and proving primality of large integers.

1.1 Factorization. The elliptic curve factorization method of H. Lenstra is a generalization of the so-called $(p-1)$-factorization algorithm of Pollard. The common setup for both methods is the following: Suppose we want to find a factor of some large integer N. Let there be given a functor that associates to N a group $G(N)$ and to any prime divisor $p \mid N$ a group $G(p)$ and a group homomorphism $\beta_p : G(N) \to G(p)$ with the following property: If $x \in G(N) \setminus \{e\}$ is a nontrivial element lying in the kernel of one of the β_p (for an unknown prime divisor $p \mid N$), then a nontrivial divisor of N can be easily calculated. In the case of Pollard's $(p-1)$-method one sets $G(m) := (\mathbb{Z}/m)^*$ for all integers $m > 0$. If an element $\bar{x} = x \mod N \in (\mathbb{Z}/N)^*$ is in the kernel of the natural homomorphism $\beta_p : (\mathbb{Z}/N)^* \to (\mathbb{Z}/p)^*$ for some prime divisor $p \mid N$ and if $x \not\equiv 1 \mod N$, then

$$d := \gcd(x-1, N)$$

is a nontrivial divisor of N. But how can we find a nontrivial element in the kernel of β_p if p is unknown? This is possible provided that the order of $G(p)$ is a "smooth" number, i.e. if

$$\#G(p) = q_1^{k_1} q_2^{k_2} \cdot \ldots \cdot q_r^{k_r}$$

with small prime powers $q_i^{k_i}$, say $q_i^{k_i} \leq B$ for all i and a given (relatively small) bound B. One then calculates the number

$$Q(B) = \prod_{q \leq B} q^{\alpha(q,B)},$$

where $\alpha(q, B) := \max\{k \in \mathbb{N} : q^k \leq B\}$. By the prime number theorem, $Q(B)$ has order of magnitude $\exp(B)$. Since by assumption $\#G(p) \mid Q(B)$, for every element

$\xi \in G(p)$ we have $\xi^{Q(B)} = e$. Therefore, if we calculate $y := x^{Q(B)}$ for an arbitrary element $x \in G(N)$, then $y \in \ker(\beta_p)$, because $\beta_p(y) = \beta_p(x)^{Q(B)} = e$. If $y \neq e$, then by assumption a divisor of N can be calculated. Pollard's method is efficient if there is a prime divisor $p \mid N$ such that $p - 1$ is a smooth number. But this is not always the case. It was Lenstra's idea to replace the multiplicative group $\mathbb{F}_p^*$ in Pollard's method by an elliptic curve $G(p) = E_{a,b}(\mathbb{F}_p)$. By varying the parameters a, b of the elliptic curve, there is a better chance that the order $\#E_{a,b}(\mathbb{F}_p)$ is a sufficiently smooth number.

Lenstra's algorithm works as follows: To start, we choose random elements $a \in \mathbb{Z}/N$, $P_0 = (x_0, y_0) \in (\mathbb{Z}/N)^2$ and determine a value $b \in \mathbb{Z}/N$ such that

$$y_0^2 \equiv x_0^3 + ax_0 + b \mod N.$$

In rare cases we will have $\gcd(4a^3 + 27b^2, N) \neq 1$. Then we have either found a nontrivial divisor of N and can stop the algorithm or else $N \mid 4a^3 + 27b^2$ and we must start again with new random values a, x_0, y_0.

If $\gcd(4a^3 + 27b^2, N) = 1$, consider the equation

$$Y^2 = X^3 + aX + b.$$

For every prime divisor $p \mid N$, we define the group $G(p) := E_{a,b}(\mathbb{Z}/p)$ as the elliptic curve defined by this equation taken modulo p and set $G(N) := \prod_{p|N} E_{a,b}(\mathbb{Z}/p)$. The homomorphisms $\beta_p : G(N) \to G(p)$ are the natural projections. If we denote by $G(p)' := G(p) \setminus \{O\}$ the affine part of $G(p)$, then $G(N)' := \prod_{p|N} G(p)'$ is the complement of $\bigcup_{p|N} \ker(\beta_p)$. The points of $G(N)'$ can be represented by pairs (x, y) of integers satisfying our equation modulo N. We have already constructed a point $P_0 = (x_0, y_0)$ of $G(N)'$. By the general principle of the factorization algorithm explained above, we must now calculate the multiple $Q(B) \cdot P_0$ (for some suitable choice of B). This can be done in $O(\log Q(B))$ steps by repeated doubling and adding. The group law to add two points $P_1 + P_2 =: P_3$ is given by the formulas

$$x_3 := \lambda^2 - x_1 - x_2, \qquad y_3 := \lambda(x_1 - x_3) - y_1,$$

where the "slope" λ is defined by

$$\lambda := \frac{y_2 - y_1}{x_2 - x_1} \text{ if } x_1 \neq x_2 \quad \text{and} \quad \lambda := \frac{3x_1^2 + a}{2y_1} \text{ if } P_1 = P_2.$$

The only problem in doing these operations in $\mathbb{Z}/N$ is the calculation of the inverses of the denominators. These inverses, if they exist in $\mathbb{Z}/N$, can be calculated by using the extended Euclidean algorithm to calculate the gcd of the denominator and N. If the gcd equals 1, the inverse can be calculated and we can go on. The exceptional case is that the gcd is a number $d \neq 1$. If $d \neq N$, we are in a lucky case because we have found a divisor of N. If one of the elliptic curves $G(p)$ has an order dividing $Q(B)$, an exceptional case must necessarily occur during the calculation of $Q(B) \cdot P_0$, because then $Q(B) \cdot P_0$ cannot be an element of $G(N)'$. If we do not encounter a lucky case we are not completely lost, because we can start again with new random parameters a, x_0, y_0, i.e. with new elliptic curves $G(p)$ with different orders. A nice feature of the elliptic curve factorization algorithm is that it is easily parallelizable, because we can let many computers work on the factorization of the same number N using different elliptic curves.

1.2 Deterministic Primality Tests. There are some very efficient probabilistic primality tests for large integers. An example is the Solovay–Strassen test. This test works as follows. Let N be a large odd integer to be tested for primality. Choose a random integer a with $1 < a < N$ and check whether (1) $\gcd(a, N) = 1$, and (2) $a^{(N-1)/2} \equiv \big(a/N\big) \mod N$, where $\big(a/N\big)$ is the Jacobi symbol. Of course, if N is prime, these conditions are satisfied (condition (2) is a theorem of Euler). Hence if one of the conditions fails, we are certain that N is not prime. If both conditions hold, we can assert the primality of N only with a certain error probability. Indeed one can show that for composite N condition (2) is satisfied for less than $N/2$ values of a. Hence the error probability is less than $1/2$. (For most N the error probabilty is much less.) By repeating the test m times with independent random values a, the error probability will be less than 2^{-m}. An integer N which has passed successfully sufficiently often a probabilistic primility test is called a "probable prime". For all practical purposes we may assume that N is prime, but this is not a mathematical certainty.

If the prime decomposition of $N - 1$ is known, there is a simple deterministic primality test: N is prime if and only if there exists an integer a such that $a^{N-1} \equiv 1 \mod N$ and $a^{(N-1)/q} \not\equiv 1$ for all prime divisors $q \mid N - 1$. An a with this property is then a primitive root modulo N. If N is prime then there exist $\varphi(N-1)$ primitive roots, hence by trying out some random numbers one can be found. But in general $N - 1$ (which is the order of $(\mathbb{Z}/N)^*$ in case N is prime) is difficult to factorize. As in the case of Lenstra's factorization method one can try to replace the group $(\mathbb{Z}/N)^*$ by an elliptic curve $E_{a,b}$. By varying the coefficients a, b, the orders of the elliptic curves vary and there is a better chance that at least one of these orders can be factorized. This was the idea of Goldwasser/Kilian. Their primality test is based on the following proposition.

Proposition. *Let N be a probable prime with* $\gcd(6, N) = 1$ *and let a, b be integers with* $\gcd(4a^3 + 27b^2, N) = 1$. *Consider the elliptic curve with affine equation*

$$E = E_{a,b}: \quad Y^2 = X^3 + aX + b.$$

Suppose there exists a prime $q > (\sqrt[4]{N} + 1)^2$ and an affine point $P = (x, y)$ on $E(\mathbb{Z}/N)$ such that $q \cdot P = O$. Then N is prime.

Remark. As in 1.1 we define $E(\mathbb{Z}/N) = \prod_{p|N} E(\mathbb{Z}/p)$. All calculations are done in $\mathbb{Z}/N$. In contrast to 1.1, here an exceptional case where we encounter a denominator, which is a nonzero noninvertible element of $\mathbb{Z}/N$, will rarely occur in practice, because N is a probable prime.

Proof. Assume that N is not prime. Then there exists a prime divisor $p \mid N$ with $p \leq \sqrt{N}$. The natural homomorphism

$$E(\mathbb{Z}/N) \longrightarrow E(\mathbb{Z}/p)$$

maps P to a point $\overline{P} = (\bar{x}, \bar{y}) \in E(\mathbb{Z}/p)$ of order q. By the theorem of Hasse (Chap. 13.1, Theorem (1.2)), the order of $E(\mathbb{Z}/p)$ satisfies

$$\#E(\mathbb{Z}/p) < p + 1 + 2\sqrt{p} \leq \sqrt{N} + 1 + 2\sqrt[4]{N} = (\sqrt[4]{N} + 1)^2.$$

Therefore it would follow that $q > \#E(\mathbb{Z}/p)$, a contradiction!

The primality test of Goldwasser/Kilian uses this proposition in the following way: Choose random numbers a, b and determine the order $m := \#E_{a,b}(\mathbb{Z}/N)$ by Schoof's algorithm (cf. 2.6), assuming N is prime. By trial division of m by small primes write m as $m = f \cdot u$, where f is the factored and u the unfactored part. If $f \geq 2$ and $u > (\sqrt[4]{N} + 1)^2$, test whether $q := u$ is a probable prime. If this is not the case or if u is not of the required size, start again with new random values a, b. If q is a probable prime, it is in general easy to find a point $P = (x, y)$ on $E_{a,b}(\mathbb{Z}/N)$ of order q. Then by the proposition N is prime provided q is prime. Since $q \leq 1/2\#E_{a,b}(\mathbb{Z}/N)$, this can be tested recursively by the same method. The primality test of Goldwasser/Kilian has expected polynomial running time (polynomial in the number of bits of N), but still is too slow in practice.

Atkin/Morain have devised an improvement which makes this primality test efficient in practice. Instead of choosing random elliptic curves and calculating their order, they construct, using a complex multiplication method, elliptic curves whose order is known a priori. Let $-D$ be the discriminant of an imaginary quadratic number field. If N is prime and the equation $4N = t^2 + Ds^2$ has an integer solution (t, s), then there exists an elliptic curve E over the field $\mathbb{Z}/N$, whose endomorphism ring is the ring of algebraic integers in $\mathbb{Q}(\sqrt{-D})$, and which has $m = \#E(\mathbb{Z}/N) = N + 1 \pm t$ elements. As above, one can test whether m can be written as $m = f \cdot q$, where q is a probable prime with $m/2 \geq q > (\sqrt[4]{N} + 1)^2$. There exists an effective algorithm of Cornacchia to decide whether the Diophantic equation $4N = t^2 + Ds^2$ is solvable and to find a solution in case of existence (of course $\left(-D/N\right) = 1$ is a necessary condition). The equation of the elliptic curve E can be constructed in the following way: We first calculate the j-invariant

$$j_D := j\left(\frac{-D + i\sqrt{D}}{2}\right) \in \mathbb{C}$$

with sufficiently high precision. This is an algebraic integer of degree equal to the class number h of the field $\mathbb{Q}(\sqrt{-D})$. Its conjugates are $j(\tau_\nu)$, $\nu = 2, \ldots, h$, where the lattices $\mathbb{Z} + \mathbb{Z}\tau_\nu$ represent the non-principal ideal classes of $\mathbb{Q}(\sqrt{-D})$. By calculating also these conjugates of j_D, we get its minimal polynomial $H_D(T) \in \mathbb{Z}[T]$. This polynomial, taken modulo N, has at least one zero $j_0 \in \mathbb{Z}/N$, which is the j-invariant of the elliptic curve $E(\mathbb{Z}/N)$. From this we can calculate the equation of the elliptic curve. Up to isomorphism, there are only two possibilities, except for $D = -3$ with 6, and $D = -4$ with 4 isomorphism classes.

Incorporating further improvements, the primality test of Atkin/Morain is very efficient and has been used to prove the primality of numbers with more than 1000 decimal digits.

§2. Elliptic Curves in Cryptography

The use of elliptic curves in cryptography is based on the discrete logarithm problem. First we describe this problem in a general group.

2.1 The Discrete Logarithm. Let G be a finite abelian group (we will write it multiplicatively) and let $g \in G$ be a fixed element of known order q. Let $G_0 = \langle g \rangle$ the cyclic subgroup of G generated by g. Then we have an isomorphism of groups

$$\exp_g : \mathbb{Z}/q\mathbb{Z} \longrightarrow G_0, \quad k \mapsto g^k.$$

The inverse map of $\exp_g$ is called the *discrete logarithm* (with respect to basis g)

$$\log_g : G_0 \longrightarrow \mathbb{Z}/q\mathbb{Z}.$$

More concretely, given an element $x \in G_0 = \langle g \rangle$, the discrete logarithm of x is the unique number $k \mod q$ such that $x = g^k$.

Popular choices for the group G are the multiplicative group of a finite field or an elliptic curve over a finite field.

The crucial point for the cryptographical applications is that the exponential map can be effectively calculated, whereas the calculation of the logarithm is in general much more complicated. To give an idea of the orders of magnitude involved, the bitsize of the number q (which should be a prime for reasons that we will explain later) is typically between 160 and 1024 (i.e. $q \approx 2^{160}$ up to $q \approx 2^{1024}$). The power g^k can be calculated by the repeated squaring algorithm: If

$$k = \sum_{i=0}^{r} b_i 2^r, \quad b_i \in \{0, 1\}$$

then

$$g^k = \prod_{b_i \neq 0} g^{2^i}$$

and g^{2^i} requires i multiplications. Hence the complexity grows linearly with the number of digits of q. The complexity of the discrete logarithm depends of course on the particular group G. We will discuss this problem later, but we say at this point only that for general elliptic curves the best known algorithms have a complexity growing exponentially with the number of digits of q.

We will now describe two cryptographical applications of the discrete logarithm in the context of a general group.

2.2 Diffie–Hellman Key Exchange. Suppose that two parties, say Alice and Bob, want to take up a confidential communication over an unsecured channel like the Internet. For this purpose they send their messages encrypted with a secret key that is known only to Alice and Bob. But how can they agree on a common secret key if this information must also be exchanged over the unsecured channel? This can be done by a public key system invented by W. Diffie and M.E. Hellman. First Alice and Bob agree on a triple (G, g, q) consisting of a group G and an element $g \in G$ of order q as in (2.1). It is supposed that the discrete logarithm problem in G is intractable. This (G, g, q) is a public key that need not to be kept secret. For every particular session a new secret key is established in the following manner:

1. Alice chooses a random number $\alpha \in \mathbb{Z}/q\mathbb{Z}$, calculates $a := g^\alpha \in G$ and sends a to Bob. The number α must be kept secret, but a may be known to an adversary.
2. Bob chooses a random number $\beta \in \mathbb{Z}/q\mathbb{Z}$, calculates $b := g^\beta \in G$ and sends b to Alice. Again β must be kept secret.
3. Alice calculates $k_a := b^\alpha \in G$, and Bob calculates $k_b := a^\beta \in G$. Of course

$$k_a = g^{\alpha\beta} = k_b;$$

so they can use $k_a = k_b$ as their common secret key. An adversary knows $a = g^\alpha$ and $b = g^\beta$. To calculate $g^{\alpha\beta}$ from g^α and g^β is known as the Diffie–Hellman problem. For this no better method is known than to calculate α or β by solving the discrete logarithm problem for one of the equations $a = g^\alpha$ or $b = g^\beta$. But this was supposed to be practically impossible.

2.3 Digital Signatures. An electronic document can be easily copied and the copy is completely identical to the original. Therefore, at first sight, it seems that a digital signature can be forged even more easily than can handwritten signature. Therefore it is surprising that a secure digital signature scheme can be established using public key cryptography. The idea is to use signatures that depend on the signed document and that can only be produced using a private (secret) key, whereas verification of the signature is possible using the public key corresponding to the secret key.

There are several digital signature schemes; we will describe one that is a variant of a scheme invented by T. ElGamal. This scheme uses the discrete logarithm and can be formulated for an arbitrary finite abelian group (for example, an elliptic curve over a finite field).

So let (G, g, q) be (as above) a triple where G is a group and $g \in G$ an element of known prime order q and suppose that the discrete logarithm problem in G is intractable. Furthermore let there be given a map $\varphi : G \to \mathbb{Z}/q\mathbb{Z}$. (For example, if G is an elliptic curve over a prime field $\mathbb{F}_p$, for a point $A \in G$, $A \neq O$, we could define $\varphi(A) = x(A) \mod q$, where $x(A) \in \{0, 1, 2, \ldots, p-1\}$ is the x-coordinate of A.)

1. To set up a public/private key pair for digital signatures, Alice chooses a random number $\xi \in (\mathbb{Z}/q)^*$ and calculates

$$h := g^\xi \in G.$$

The public key is then (G, q, φ, g, h), whereas ξ serves as Alice's private key and must be kept secret. (An adversary can calculate ξ from the public data, provided he can solve the discrete logarithm problem in G, which we supposed to be practically impossible.)

2. To sign a particular message $m \in (\mathbb{Z}/q\mathbb{Z})^*$ (in practice m will be a so called *message digest* or *cryptographic check sum* of a longer document), Alice chooses a new random number $\alpha \in (\mathbb{Z}/q\mathbb{Z})^*$ and calculates

$$a := g^{\alpha} \in G,$$

and, using her private key ξ,

$$m' := m + \xi\varphi(a) \in \mathbb{Z}/q\mathbb{Z}.$$

If $m' = 0$ (a case which in practice will never occur, because its probability is only $1/q$), another random number α has to be chosen. Then Alice calculates

$$\beta := \alpha^{-1}m' \in (\mathbb{Z}/q\mathbb{Z})^*.$$

The signature of m is

$$\sigma := (a, \beta) \in G \times \mathbb{Z}/q\mathbb{Z},$$

and the signed message is the pair (m, σ).

3. If Bob wants to verify that (m, σ) was indeed signed by Alice, he does the following calculations (which use only the public key)

$$\gamma := m\beta^{-1} \in \mathbb{Z}/q\mathbb{Z}, \quad \delta := \varphi(a)\beta^{-1} \in \mathbb{Z}/q\mathbb{Z},$$

and

$$c := g^{\gamma}h^{\delta} \in G.$$

He accepts the signature if $c = a$. If the message m was properly signed, this is indeed the case, because

$$\begin{aligned} g^{\gamma}h^{\delta} &= g^{m\beta^{-1}}g^{\xi\varphi(a)\beta^{-1}} \\ &= g^{(m+\xi\varphi(a))\beta^{-1}} = g^{m'\beta^{-1}} = g^{\alpha} = a. \end{aligned}$$

2.4 Algorithms for the Discrete Logarithm. Let G be a cyclic group of order q with generator g and $x \in G$. We wish to determine a number $\xi \in \mathbb{Z}/q\mathbb{Z}$ such that

$$x = g^{\xi}.$$

If q is not prime, but a composite with prime factorization

$$q = \prod p_j^{r_j},$$

it is easy to see that the problem can be reduced to cyclic groups of order p_j. Therefore the discrete logarithm problem is hardest if q is prime.

The baby step/giant step (BSGS) algorithm of Shanks proceeds in the following way: Let $k := \lceil\sqrt{q}\rceil$ be the smallest integer $\geq \sqrt{q}$. The (unknown) discrete logarithm ξ can be written as

$$\xi = nk + m, \quad 0 \leq n, m < k.$$

The equation $x = g^\xi$ is equivalent to

$$xg^{-m} = g^{kn}.$$

First, the "giant steps"

$$g^{k\nu}, \quad \nu = 0, 1, \ldots, k-1$$

are calculated and stored in a hash table. Then the "baby steps"

$$xg^{-\mu}, \quad \mu = 0, 1, 2, \ldots$$

are calculated one after the other and compared with the stored values until a collision

$$xg^{-m} = g^{kn}$$

is found. The discrete logarithm is then $\xi = (kn + m) \mod q$. If efficient hashing techniques are used for storing and searching, this algorithm requires roughly $O(\sqrt{q})$ steps. The memory requirement (for the giant steps) is also $O(\sqrt{q})$. However there exist probabilistic variants (Pollard's rho and lambda method) which use only a small constant amount of memory and have the same time complexity $O(\sqrt{q})$.

Remark. The complexity $O(\sqrt{q})$ is an *exponential* complexity considering it (as customary) as a function of the number of binary digits of q.

To be safe against this algorithm (i.e. to make the discrete logarithm problem intractable), q should be by today's (2002) standards at least 2^{160}. The number of required steps would then be $> 2^{80} \approx 1.2 \cdot 10^{24}$.

For special groups there exist more efficient algorithms for the discrete logarithm. For example, for the multiplicative group $\mathbb{F}_q^*$ of a finite field there exist *subexponential* algorithms (index calculus method, number field sieve). Subexponential complexity is between polynomial and exponential complexity.

For general elliptic curves over finite fields no better algorithms for the discrete logarithm problem are known than the general purpose $O(\sqrt{q})$ algorithms. However, for elliptic curves with special properties, one can do better. For example, let E be a supersingular elliptic curve over $\mathbb{F}_p$, so that $E(\mathbb{F}_p)$ has $n = p + 1$ elements. Using the Weil pairing

$$E[n] \times E[n] \longrightarrow \mu_n$$

and the fact that $\mu_n = \mu_{p+1}$ is a subgroup of $\mathbb{F}_{p^2}^*$, one can embed $E(\mathbb{F}_p)$ into the multiplicative group $\mathbb{F}_{p^2}^*$ and use the more efficient algorithms in $\mathbb{F}_{p^2}^*$ to solve the discrete logarithm problem. For several other special classes of elliptic curves algorithms with complexity better than $O(\sqrt{q})$ are known. So the recommendation for the application of elliptic curves in cryptography is to use "random" elliptic curves (i.e. curves with random coefficients) in the hope that the special algorithms for the discrete logarithm that have been found or may be found in the future do not apply to them. As we have seen, to make the discrete logarithm problem difficult, the order of the group should be a prime number or have at least a large prime factor. So the problem arises of counting the number of points of the randomly chosen elliptic curves. If one has efficient algorithms for this purpose, one chooses random elliptic curves and determines their order. If the order is not satisfactory, the curve is thrown away and a new random curve is chosen, until a good one is found.

2.5 Counting the Number of Points. A straightforward way to determine the number of points of an elliptic curve E over the prime field $\mathbb{F}_p$, (p an odd prime), given by the equation

$$Y^2 = X^3 + aX + b = P_3(X)$$

is to use the Legendre symbol. For a given $x \in \mathbb{F}_p$, the equation $Y^2 = P_3(x)$ has 2, 1 or 0 solutions in $\mathbb{F}_p$ if $(P_3(x)/p)$ equals $+1, 0$ or -1 respectively. Therefore, taking into account also the point at infinity, it follows

$$\#E(\mathbb{F}_p) = 1 + \sum_{x \in \mathbb{F}_p} \left\{1 + \left(\frac{P_3(x)}{p}\right)\right\} = (p+1) + \sum_{x \in \mathbb{F}_p} \left(\frac{P_3(x)}{p}\right).$$

However, this method has complexity $O(p)$ and can be used only for small primes p (say up to 10^6).

A better method with complexity $O(\sqrt[4]{q})$ is an adaption of Shanks's baby step/giant step algorithm. Let E be an elliptic curve over a finite field $\mathbb{F}_q$. By the theorem of Hasse, the order of E lies in the "Hasse interval"

$$H := \{n \in \mathbb{N} : |n - (q+1)| \le 2\sqrt{q}\}.$$

One chooses a random point $P \in E(\mathbb{F}_q)$ and determines by the BSGS algorithm an integer $N \in H$ such that $N \cdot P = O$. Since H has $1 + 2\lfloor 2\sqrt{q} \rfloor$ elements, this can be done with about $2\sqrt[4]{q}$ giant and baby steps. If N is the only element of the Hasse intervall with $N \cdot P = O$, this is the order of $E(\mathbb{F}_q)$. For orders up to 10^{24}, this method is effective in practice. But the elliptic curves used in cryptography are still larger, so other methods are needed.

2.6 Schoof's Algorithm. Recall that for an elliptic curve E defined over a finite field $\mathbb{F}_q$ the Frobenius automorphism $\phi = \phi_q : E \to E$ satisfies a quadratic equation

$$\phi^2 - c\phi + q = 0,$$

where the trace c is connected to the order N of the elliptic curve by

$$N = \#E(\mathbb{F}_q) = q + 1 - c.$$

The idea of Schoof is to calculate $c_\ell := c \bmod \ell$ for various small primes ℓ by restricting the Frobenius automorphism to the group of ℓ-division points $E[\ell] \subset E$, which is invariant under ϕ. If the characteristic p of the field $\mathbb{F}_q$ is bigger than ℓ, then

$$E[\ell] \cong \mathbb{Z}/\ell\mathbb{Z} \times \mathbb{Z}/\ell\mathbb{Z} = \mathbb{F}_\ell^2$$

is a 2-dimensional vector space over $\mathbb{F}_\ell$ and the restriction $\phi \mid E[\ell]$, which we denote again by ϕ, satisfies the characteristic equation

$$\phi^2 - c_\ell \phi + q_\ell = 0,$$

with $q_\ell = q \bmod \ell$. The trace c_ℓ can be calculated by choosing a point $P \in E[\ell]\setminus\{0\}$ and solving the equation

$$\phi^2(P) + q_\ell P = c_\ell \phi(P).$$

If c_ℓ is known for all $\ell \in \{\ell_1, \ldots, \ell_r\}$, then by the Chinese remainder theorem we can calculate c modulo $L := \prod \ell_\nu$. If L is greater than the length $4\sqrt{q}$ of the Hasse interval, c and therefore $N = \#E(\mathbb{F}_q)$ is uniquely determined. Even if $L < 4\sqrt{q}$, then there are at most $\lceil 4\sqrt{q}/L \rceil$ possible values for N. Using an appropriate BSGS method, one can then determine the correct value of N in about $\sqrt{4\sqrt{q}/L}$ steps.

How can we find a point $P \in E[\ell] \setminus \{0\}$? For odd ℓ, the x-coordinates of these points are the roots of the ℓ-division polynomial $\Psi_\ell(T) \in \mathbb{F}_q[T]$, which is a polynomial of degree $(\ell^2-1)/2$ (because the ℓ^2-1 points of $E[\ell]\setminus\{0\}$ come in pairs $\pm P$ having the same x-coordinate), cf. Chap. 13.9. Using the recursion formulas, the division polynomials can be easily calculated. In general, Ψ_ℓ neither has a zero in the ground field $\mathbb{F}_q$ nor is it irreducible. Suppose we know an irreducible factor $F(T)$ of degree r of the polynomial $\Psi_\ell(T)$. Then the field $K := \mathbb{F}_q[T]/(F(T))$ is isomorphic to $\mathbb{F}_{q^r}$ and the element $t := T \bmod F(T) \in K$ is the x-coordinate of an ℓ-division point. If the element $P_3(t)$ is the square of an element $s \in K$, then $(t, s) \in K^2$ is an ℓ-division point of the elliptic curve, otherwise one has to pass to a quadratic extension of K. To avoid the case distinction it is convenient, instead of working with the curve

$$E: \quad Y^2 = X^3 + aX + b =: P_3(X),$$

to work with the twisted curve

$$\widetilde{E}: \quad P_3(t)Y^2 = P_3(X).$$

On this curve, $(t, 1)$ is an ℓ-division point. The points (ξ, η) on $\widetilde{E}$ correspond to points $(\xi, \sqrt{P_3(t)}\eta)$ on E. Therefore the Frobenius automorphism $\phi : (x, y) \mapsto (x^q, y^q)$ translates to $(\xi, \eta) \mapsto (\xi^q, P_3(t)^{(q-1)/2}\eta^q)$ on $\widetilde{E}$.

There exist standard algorithms to determine an irreducible factor F of Ψ_ℓ; essentially one has to calculate the greatest common divisor of $T^{q^r} - T$ and $\Psi_\ell(T)$ for $r = 1, 2, \ldots$. However these algorithms are too expensive compared with all other operations, so it is better to leave Ψ_ℓ unfactorized and work over the ring $R := \mathbb{F}_q[T]/(\Psi_\ell(T))$, which amounts to working simultaneously over all fields $\mathbb{F}_q[T]/(F_j(T))$, where F_j are the irreducible factors of Ψ_ℓ. Working with the ring R instead of a field can cause only problems when inverses of elements $\xi \neq 0$ have to be calculated. The calculation of an inverse is done using the extended Euclidean algorithm. If the inverse does not exist, one detects automatically a factor G of Ψ_ℓ. Hence this does not hurt but is rather useful because we can pass to the smaller ring $R' = \mathbb{F}_q[T]/(G(T))$.

The algorithm of Schoof we sketched so far was the first algorithm of polynomial complexity for the point counting problem on elliptic curves. However it is still too slow for the curves used in cryptography. Atkin, Elkies and others have contributed improvements, which make the algorithm practical. In the next section we will describe one such improvement.

2.7 Elkies Primes. As before let E be an elliptic curve defined over a finite field $\mathbb{F}_q$ of characteristic $p > 3$, with trace $c = (q+1) - \#E(\mathbb{F}_q)$ and let $\ell < p$ be a an odd prime. Recall that the Frobenius automorphism restricted to the two dimensional $\mathbb{F}_\ell$-vectorspace $E[\ell]$ of ℓ-division points of E satisfies the quadratic equation

$$\phi^2 - c_\ell \phi + q_\ell = 0,$$

where $c_\ell = c \mod \ell$ and $q_\ell = q \mod \ell$. Therefore the eigenvalues of $\phi \mid E[\ell]$ are

$$\lambda_{1,2} = \frac{1}{2}(c_\ell \pm \sqrt{c_\ell^2 - 4q_\ell}).$$

If $c_\ell^2 - 4q_\ell$ is a square in $\mathbb{F}_\ell$, which will be the case for about half of the primes ℓ, these eigenvalues belong to the field $\mathbb{F}_\ell$. Primes with this property are called Elkies primes for the given elliptic curve. For such primes an eigenvector of ϕ spans a 1-dimensional subspace $C \subset E[\ell]$ invariant under the Frobenius automorphism. C is a cyclic subgroup of E of order ℓ defined over the ground field $\mathbb{F}_q$, hence the isogeny $E \to E/C$ is also defined over $\mathbb{F}_q$. Furthermore

$$G(T) := \prod_{P \in (C \setminus 0)/\pm 1} (T - x(P)) \in \mathbb{F}_q[T]$$

is a factor of degree $(\ell - 1)/2$ of the division polynomial $\Psi_\ell(T)$. The important thing about Elkies primes is that they can be determined without having to work explicitly in $E[\ell]$. This is done using the modular polynomials $\Phi_\ell(x, y)$ introduced in Chap. 11.9. These are polynomials of degree $\ell + 1$ with integer coefficients, hence they can also be regarded as polynomials over $\mathbb{F}_q$. If $j(E)$ is the j-invariant of the elliptic curve E then the zeroes of $\Phi_\ell(j(E), y)$ are the j-invariants of curves E/C, where C runs through the cyclic subgroups of E of order ℓ. Therefore ℓ is an Elkies prime if and only if the polynomial $\Phi_\ell(j(E), y) \in \mathbb{F}_q[y]$ has a zero in $\mathbb{F}_q$; this can be checked by computing the greatest common divisor of this polynomial and $y^q - y$. When a solution $j' \in \mathbb{F}_q$ of $\Phi_\ell(j(E), j') = 0$ has been found, there is also a procedure to calculate directly the factor $G(T)$ of the division polynomial $\Psi_\ell(T)$. With this, a substantial gain in efficiency of Schoof's point counting algorithm is achieved, because for the elliptic curves used in cryptography primes ℓ up to 100 or higher are needed, so it makes a big difference whether one has to deal with polynomials of degree $(\ell - 1)/2$ or $(\ell^2 - 1)/2$. There exist still further improvements, for example replacing the modular polynomials Φ_ℓ, whose coefficients grow rapidly with ℓ, by simpler polynomials. We refer to Blake/Seroussi/Smart and the references given there. We have restricted our attention here to elliptic curves over finite fields with large prime characteristic. For curves over fields of characteristic 2, other methods exist.

References

Cohen, H., *A Course in Computational Algebraic Number Theory*. Springer-Verlag, 1996.

Blake, I., Seroussi, G., and Smart, N., *Elliptic Curves in Cryptography*. LMS Lecture Notes Series 265, Cambridge University Press, 1999.

Lenstra, H. W., Factoring integers with elliptic curves. *Ann. Math.* **126**, 649–673, 1987.

Solovay, R. and Strassen, V., A fast Monte Carlo test for primality. *SIAM J. Comp.* **6**, 84–85, 1977. Erratum Vol. **7**, 118, 1978.

Goldwasser, S. and Kilian, J., Almost all primes can be quickly certified. *18th STOC*, 316–329, 1986.

Atkin, A. O. L. and Morain, F., Elliptic curves and primality proving. *Math. Comp.* **61**, 29–67, 1993.

Schoof, R., Elliptic curves over finite fields and the computation of square roots mod *p*. *Math. Comp.* **44**, 483–494, 1985.

Schoof, R., Counting points on elliptic curves over finite fields. *J. Théorie des Nombres de Bordeaux* **7**, 219–254, 1995.

Appendix III: Elliptic Curves and Topological Modular Forms

In LN 1326 [1988] we have some of the first studies into the subject of elliptic cohomology, and this subject was considered further by Hirzebruch, Berger, and Jung [1994]. The first steps in the theory centered around the characteristic classes described by series arising either from classical elliptic functions as in Chapters 9 and 10 or classical modular function as in Chapter 11.

In his 1994 ICM address Michael Hopkins proposed the cohomology theory or spectrum, denoted tmf, of topological modular forms [1994]. Since that time, the concept has gone through a development with still many of the basic results unpublished. Now the manuscript for the 2002 ICM talk of Hopkins is available, we can see a theory which brings new methods to old problems in homotopy theory including the description of the homotopy groups of spheres.

Topological modular forms start with the Weierstrass equation and its related change of variable as considered in Chapter 3. The coefficients in the Weierstrass equation become indeterminates in a polynomial algebra, and this polynomial algebra is extended with new indeterminants corresponding to the coefficients in the change of variables. These polynomial algebras are linked by a structure called a Hopf algebroid. With this Hopf algebroid we can return and give a description of the category of elliptic curves and their isomorphisms.

Associated with an elliptic curve is a formal group law, see Chapter 12, §7. Formal groups also control the multiplicative properties of the first Chern class in generalized cohomology theories, and for the complex bordism theory MU Quillen proved that the formal group was the universal one parameter formal group, see Quillen [1969]. This led to the possibility of making generalized cohomology theories from a given formal group.

Hopf algebroids were first studied by J. F. Adams LN 99 [1969] pp. 1–138. Hopf algebroids consist of two algebras with connecting data such that one algebra corresponds to the coefficients of the theory and the other to the stable cooperations in the theory. If there is an Adams spectral sequence in the theory, then the E^2-term should be an Ext term over the operation algebra of the Hopf algebroid.

Hopf algebroids in a broader perspective are an example of a category object. Categories as an organization of mathematical systems and their maps is a well established perspective in mathematics. The mathematicians differ on to what extent the categorical framework should be made explicit, but almost everybody agrees that it can be a useful way to look at certain phenomena.

Since Mumford's study of Pic on the moduli of elliptic curves [1968], Quillen's work on algebraic K-theory [1972], and Adams' work on generalized cohomology theories [1969], people have studied categories as one studies groups, topological spaces, or Hopf algebras. This led to such ideas as the category of small categories, or more generally, the category Cat($\mathcal{C}$) of category objects in a category $\mathcal{C}$. A special case are groupoids, they are categories where all morphisms are isomorphisms. This means that there is another concept of Grpoid($\mathcal{C}$) which denotes the category of groupoids in the category $\mathcal{C}$.

In the work of Mumford and further the work of Deligne and Mumford groupoids were especially important. In the work of Adams we have the opposite concept of a cocategory in the category of commutative algebras which is also called a Hopf algebroid. Many of the considerations coming into the work of Mumford and Quillen were already anticipated by Grothendieck.

The aim of this appendix is to reexamine the Weierstrass polynomial and its change of variables introduced in Chapter 3, and to put the data into a Hopf algebroid. This example becomes an algebraic motivation for the definition of Hopf algebroid. We show how to describe the category of elliptic curves and isomorphisms and how to determine the ring of modular forms in terms of the Weierstrass Hopf algebroid.

This Hopf algebroid plays a basic role in describing the new cohomology theory tmf called topological modular forms. There is a spectral sequence with $E_2^{0,*}$ the ring of modular forms, and it converges to $\pi_*(\mathrm{tmf})$. Under the edge morphism to $E_2^{0,*}$ the torsion in $\pi_*(\mathrm{tmf})$ goes to zero and certain modular forms like Δ are not in the image, but 24Δ is in the image. The edge morphism is a rational isomorphism. Since the polynomials in coefficients of the Weierstrass equation generate the ring of modular forms, we see the motivation for the term topological modular forms as a name for the cohomology theory.

With this discussion the reader has some background for the study of all of these new developments in homotopy theory related to topological modular forms.

This appendix is an elementary introduction to the bridge between elliptic curves as defined in Chapter 3 by Weierstrass equations with changes of variable and the Hopf algebroid picture used in studying cohomology theories. We begin with a discussion of categories and groupoids in a cateory and then consider the concept of Hopf algebroid which is a cogroupoid in the category of commutative algebras. This elementary material is included so to fix notation and the basic definitions. Then the Weierstrass Hopf algebroid is introduced, and its relation to isomorphism classes of elliptic curves is considered. The Weierstrass Hopf algebroid is used to compute the homotopy of the spectrum topological modular forms, denoted tmf. The constructions of tmf is still work in progress at this time.

Apart from a general sketch of ideas and suitable references to the literature, any real development of this theory would be beyond the scope of this appendix.

I wish to thank Tilman Bauer, Michael Joachim, and Stefan Schwede for the help and encouragement in preparing this appendix.

§1. Categories in a Category

We begin by a short introduction to the concept of category object and consider the axiomatic framework of categories in a category.

(1.1) Small Categories as Pairs of Sets. Let $\mathcal{C}$ be a small category which means that the class of objects $C(0)$ is a set. Form the set $C(1)$ equal to the disjoint union of all $\mathrm{Hom}(X, Y)$ for $X, Y \in C(0)$. These two sets are connected by several functions. Firstly, we have the domain (left) and range (right) functions $\mathfrak{l}, \mathfrak{r} : C(1) \to C(0)$ defined by the requirement that

$$\mathfrak{l}(\mathrm{Hom}(X, Y)) = \{X\} \quad \text{and} \quad \mathfrak{r}(\mathrm{Hom}(X, Y)) = \{Y\}$$

on the disjoint union. For $f \in C(1)$ we have $f : \mathfrak{l}(f) \to \mathfrak{r}(f)$ is a notation for the morphism f in $\mathcal{C}$.

Secondly, we have an identity morphism for each object of $\mathcal{C}$ which is a function $e : C(0) \to C(1)$ having the property that $\mathfrak{l}e$ and $\mathfrak{r}e$ is the identity on $C(0)$.

Thirdly, we have composition gf of the two morphisms f and g, but only in the case where $\mathfrak{r}(f) = \mathfrak{l}(g)$. Hence composition is not defined in general on the entire product $C(1) \times C(1)$, but it is defined on all subsets of the form $\mathrm{Hom}(X, Y) \times \mathrm{Hom}(Y, Z) \subset C(1) \times C(1)$. This subset is called the fibre product of $\mathfrak{r} : C(1) \to C(0)$ and $\mathfrak{l} : C(1) \to C(0)$ consisting of pairs $(f, g) \in C(1) \times C(1)$ where $\mathfrak{r}(f) = \mathfrak{l}(g)$. The fiber product is denoted $C(1) \times_{\mathfrak{r} C(0) \mathfrak{l}} C(1)$ with two projections $\mathfrak{r}, \mathfrak{l} : C(1) \times_{\mathfrak{r} C(0) \mathfrak{l}} C(1) \to C(0)$ defined by

$$\mathfrak{l}(f, g) = \mathfrak{l}(f) \quad \text{and} \quad \mathfrak{r}(f, g) = \mathfrak{r}(g).$$

Then composition is defined $m : C(1) \times_{\mathfrak{r} C(0) \mathfrak{l}} C(1) \to C(1)$ satisfying $\mathfrak{l}m(f, g) = \mathfrak{l}(f, g) = \mathfrak{l}(f)$ and $\mathfrak{r}m(f, g) = \mathfrak{r}(f, g) = \mathfrak{r}(g)$. Now the reader can supply the unit and associativity axioms.

Fourthly, the notion of opposite category $\mathcal{C}^{\mathrm{op}}$ where $f : X \to Y$ in $\mathcal{C}$ becomes $f^{\mathrm{op}} : Y \to X$ in $\mathcal{C}^{\mathrm{op}}$ and $(gf)^{\mathrm{op}} = f^{\mathrm{op}} g^{\mathrm{op}}$ can be described as $\mathcal{C}^{\mathrm{op}} = (C(0), C(1), e, \mathfrak{l}^{\mathrm{op}} = \mathfrak{r}, \mathfrak{r}^{\mathrm{op}} = \mathfrak{l}, m^{\mathrm{op}} = m\tau)$ where τ is the flip in the fibre product $\tau : C(1) \times_{\mathfrak{r} C(0) \mathfrak{l}} C(1) \to C(1) \times_{\mathfrak{l} C(0) \mathfrak{r}} C(1)$.

(1.2) Definition. Let $\mathcal{C}$ be a category with fibre products. A category object $C(*)$ in $\mathcal{C}$ is a sextuple $(C(0), C(1), \mathfrak{l}, \mathfrak{r}, e, m)$ consisting two objects $C(0)$ and $C(1)$ and four morphisms

(1) $\mathfrak{l}, \mathfrak{r} : C(1) \to C(0)$ called domain (left) and range (right),

(2) $e : C(0) \to C(1)$ a unit morphism, and

(3) $m : C(1) \underset{\mathfrak{r}C(0)\mathfrak{l}}{\times} C(1) \to C(1)$ called multiplication or composition satisfying the following axioms:

(Cat 1) The compositions $\mathfrak{l}e$ and $\mathfrak{r}e$ are the identities $C(0)$.

(Cat 2) Domain and range are compatible with multiplication.

$$
\begin{array}{ccc}
C(1) \underset{\mathfrak{r}C(0)\mathfrak{l}}{\times} C(1) & \xrightarrow{m} & C(1) \\
{\scriptstyle pr_1}\downarrow & & \downarrow{\scriptstyle \mathfrak{l}} \\
C(1) & \xrightarrow{\mathfrak{l}} & C(0)
\end{array}
\qquad
\begin{array}{ccc}
C(1) \underset{\mathfrak{r}C(0)\mathfrak{l}}{\times} C(1) & \xrightarrow{m} & C(1) \\
{\scriptstyle pr_2}\downarrow & & \downarrow{\scriptstyle \mathfrak{r}} \\
C(1) & \xrightarrow{\mathfrak{r}} & C(0)
\end{array}
$$

(Cat 3) (associativity) The following diagram is commutative

$$
\begin{array}{ccc}
C(1) \underset{\mathfrak{r}C(0)\mathfrak{l}}{\times} C(1) \underset{\mathfrak{r}C(0)\mathfrak{l}}{\times} C(1) & \xrightarrow{m \times C(1)} & C(1) \underset{\mathfrak{r}C(0)\mathfrak{l}}{\times} C(1) \\
{\scriptstyle C(1)\times m}\downarrow & & \downarrow{\scriptstyle m} \\
C(1) \underset{\mathfrak{r}C(0)\mathfrak{l}}{\times} C(1) & \xrightarrow{m} & C(1)
\end{array}
$$

(Cat 4) (unit property of e) $m(C(1), e\mathfrak{r})$ and $m(e\mathfrak{l}, C(1))$ are each the identities on $C(1)$.

(1.3) Definition. A morphism $u(*) : C'(*) \to C''(*)$ from the category object $C'(*)$ in $\mathcal{C}$ to the category object $C''(*)$ in $\mathcal{C}$ is a pair of morphisms $u(0) : C'(0) \to C''(0)$ and $u(1) : C'(1) \to C''(1)$ commuting with the four structure morphisms of $C'(*)$ and $C''(*)$. The following diagrams are commutative

$$
\begin{array}{ccc}
C'(0) & \xrightarrow{e'} & C'(1) \\
{\scriptstyle u(0)}\downarrow & & \downarrow{\scriptstyle u(1)} \\
C''(0) & \xrightarrow{e''} & C''(1)
\end{array}
\qquad
\begin{array}{ccc}
C'(1) & \xrightarrow{\mathfrak{l}'} & C'(0) \\
{\scriptstyle u(0)}\downarrow & & \downarrow{\scriptstyle u(0)} \\
C''(1) & \xrightarrow{\mathfrak{l}''} & C''(0)
\end{array}
\qquad
\begin{array}{ccc}
C'(1) & \xrightarrow{\mathfrak{r}'} & C'(0) \\
{\scriptstyle u(1)}\downarrow & & \downarrow{\scriptstyle u(0)} \\
C''(1) & \xrightarrow{\mathfrak{r}''} & C''(0)
\end{array}
$$

and

$$
\begin{array}{ccc}
C'(1) \underset{\mathfrak{r}C'(0)\mathfrak{l}}{\times} C'(1) & \xrightarrow{m'} & C'(1) \\
{\scriptstyle u(1) \underset{u(0)}{\times} u(1)}\downarrow & & \downarrow{\scriptstyle u(1)} \\
C''(1) \underset{\mathfrak{r}C''(0)\mathfrak{l}}{\times} C''(1) & \xrightarrow{m''} & C''(1).
\end{array}
$$

(1.4) Example. A category $C(*)$ in the category of sets (set) is a small category as in (1.1).

(1.5) Example. A category object $C(*)$ with $C(0)$ the final object in $\mathcal{C}$ is just a monoidal object $C(1)$ in the category $\mathcal{C}$.

(1.6) Example. The pair of morphisms consisting of the identities $u(0) = C(0)$ and $u(1) = C(1)$ is a morphism $u(*) : C(*) \to C(*)$ of a category object called the identity morphism on $C(*)$. If $u(*) : C'(*) \to C''(*)$ and $v(*) : C''(*) \to C(*)$ are two morphisms of category objects in $\mathcal{C}$, then $(vu)(*) : C'(*) \to C(*)$ defined by $(vu)(0) = v(0)u(0)$ and $(vu)(1) = v(1)u(1)$ is a morphism of category objects in $\mathcal{C}$.

(1.7) Definition. With the identity morphisms and the composition of morphisms in (1.5) we see that the category objects and the morphisms of category objects form a category called cat($\mathcal{C}$). There is a full subategory mon($\mathcal{C}$) of cat($\mathcal{C}$) consisting of those categories $C(*)$ where $C(*)$ is the final object in $\mathcal{C}$. This is the category of monoids over the category $\mathcal{C}$.

(1.8) Remark. The category cat(set) of category objects over the category of sets is the just the category of small categories where morphisms of categories are functors between the categories and composition is composition of functors. Also mon(set) is just the category of monoids. There is an additional structure of equivalence between morphisms as natural transform of functors, and this leads to the notion of 2-category which we will not go into here.

§2. Groupoids in a Category

In the case of a groupoid where each f is an isomorphism with inverse $\mathfrak{i}(f) = f^{-1}$, this formula defines a map $\mathfrak{i} : C(1) \to C(1)$ with domain and range interchanged $\mathfrak{r}(\mathfrak{i}(f)) = \mathfrak{i}l(f)$ and $\mathfrak{l}(\mathfrak{i}(f)) = \mathfrak{r}(f)$. There is also the inverse property $m(f, \mathfrak{i}(f)) = e(\mathfrak{l}(f))$ and $m(\mathfrak{i}(f), f) = e(\mathfrak{r}(f))$. In the next sections we have the axioms.

(2.1) Definition. Let $\mathcal{C}$ be a category with fibre products. A groupoid $G(*)$ in $\mathcal{C}$ is a septuple $(G(0), G(1), \mathfrak{l}, \mathfrak{r}, e, m, \mathfrak{i})$ where the sextuple $(G(0), G(1), \mathfrak{l}, \mathfrak{r}, e, m)$ is a category object and inverse morphism $\mathfrak{i} : G(1) \to G(1)$ satisfying in addition to axioms (cat1)–(cat4) the following axioms:

(grpoid1) The following compositions hold $\mathfrak{l}\mathfrak{i} = \mathfrak{r}$ and $\mathfrak{r}\mathfrak{i} = \mathfrak{l}$.

(grpoid2) The following commutative diagrams give the inverse property of $\mathfrak{i} : G(1) \to G(1)$

$$\begin{array}{ccc} G(1) & \xrightarrow{(G(1),\mathfrak{i})} & G(1) \underset{\mathfrak{r}G(0)\mathfrak{l}}{\times} G(1) \\ \mathfrak{l}\downarrow & & \downarrow m \\ G(0) & \xrightarrow{e} & G(1) \end{array} \qquad \begin{array}{ccc} G(1) & \xrightarrow{(\mathfrak{i},G(1))} & G(1) \underset{\mathfrak{r}G(0)\mathfrak{l}}{\times} G(1) \\ \mathfrak{r}\downarrow & & \downarrow m \\ G(0) & \xrightarrow{e} & G(1). \end{array}$$

Here the same symbol is used for an object and the identity on an object.

(2.2) Example. In the category of sets a category object $G(*)$ is a groupoid when every morphism $u \in G(1)$, which is defined $u : \mathfrak{l}(u) \to \mathfrak{r}(u)$, is an isomorphism, and in this case the morphism i is the inverse given by $\mathfrak{i}(u) = u^{-1} : \mathfrak{r}(1) \to \mathfrak{l}(u)$.

(2.3) Remark. If $\mathfrak{i}'$ and $''$ are two groupoid structures on a category $(C(0), C(1), \mathfrak{l}, \mathfrak{r}, e, m)$ then $\mathfrak{i}' = \mathfrak{i}''$. To see this, we calculate as with groups using the associative law

$$\mathfrak{i}'(u) = m(\mathfrak{i}'(u), m(u, \mathfrak{i}''(u))) = m(m(\mathfrak{i}'(u), u), \mathfrak{i}''(u)) = \mathfrak{i}''(u).$$

This means that a groupoid is not a category with an additional structure, but a category satisfying an axiom, namely $\mathfrak{i}$ exists. Also we have $\mathfrak{i}\mathfrak{i} = G(1)$, the identity on $G(1)$ by the same argument.

In general we can think of groupoids as categories where every morphism is an isomorphism, and the process of associating to an isomorphism its inverse is an isomorphism of the category to its opposite category which is an involution when the double opposite is identified with the original category.

(2.4) Example. A groupoid $G(*)$ with $G(0)$ the final object in C is just a group object in the category C.

(2.5) Definition. A morphism $u(*) : G'(*) \to G''(*)$ of groupoids in C is a morphism of categories in C.

(2.6) Remark. A morphism of groupoids has the additional groperty that $\mathfrak{i}''u(1) = u(1)\mathfrak{i}'$. This is seen as with groups from the relation $m''(\mathfrak{i}''u(1), u(1)) = \mathfrak{l}''e'' = m''(u(1)\mathfrak{i}', u(1))$. We derive (2.3) by applying this to the identity functor.

(2.7) Definition. Let grpoid(C) denote the full subcategory of cat(C) determined by the groupoids.

There is a full subcategory grp(C) of grpoid(C) consisting of those groupoids $G(*)$ where $G(0)$ is the final object in C. This is the category of groups over the category C.

The category grpoid(set) of groupoids over the category of sets is the just the category of small categories with the property that all morphisms are isomorphisms. Also grp(set) is just the category of groups (grp) of sets.

(2.8) Example. Let G be a group object in a category C with fibre products. An action of G on an object X of C is a morphism $\alpha : G \times X \to X$ satisfy two axioms given by commutative diagrams

(1) (associativity)

$$\begin{array}{ccc} G \times G \times X & \xrightarrow{G\times\alpha} & G \times X \\ {\scriptstyle \mu\times X}\downarrow & & \downarrow{\scriptstyle \alpha} \\ G \times X & \xrightarrow{\alpha} & X \end{array}$$

where $\mu : G \times G \to G$ is the product on the group object G, and

(2) (unit)

$$\begin{array}{ccc} X = \{*\} \times X & \xrightarrow{e \times X} & G \times X \\ & \searrow^{\text{id}} & \downarrow \alpha \\ & & X. \end{array}$$

(2.9) Remark. The related groupoid $X\langle G\rangle(*)$ is defined by $X\langle G\rangle(0) = X$ and $X\langle G\rangle(1) = G \times X$ with structure morphisms

$$e = e_G \times X : X\langle G\rangle(0) \to X\langle G\rangle(1) = G \times X,$$
$$\mathfrak{l} = pr_2 : X\langle G\rangle(1) = G \times X \to X = X\langle G\rangle(0) \quad \text{and}$$
$$\mathfrak{r} = \alpha : X\langle G\rangle(1) \to X = X\langle G\rangle(0).$$

For the composition we need the natural isomorphism

$$\theta : X\langle G\rangle(1) \underset{\mathfrak{r}X\langle G\rangle(0)\mathfrak{l}}{\times} X\langle G\rangle(1) \to G \times G \times X$$

given by $pr_1\theta = pr_G pr_1$, $pr_2\theta = p_G pr_2$, and $pr_3\theta = p_X pr_1$. Then composition m is defined by the following commutative diagram

$$\begin{array}{ccc} X\langle G\rangle(1) \underset{\mathfrak{r}X\langle G\rangle(0)\mathfrak{l}}{\times} X\langle G\rangle(1) & \xrightarrow{\theta} & G \times G \times X \\ m \downarrow & & \downarrow \mu \times X \\ X\langle G\rangle(1) & \longleftrightarrow & G \times X. \end{array}$$

The unit and associativity properties of m come from the unit and associativity properties of $\mu : G \times G \to G$.

(2.10) Definition. With the above notation $X\langle G\rangle(*)$ is the groupoid associated to the G-action on the object X in $\mathcal{C}$. It is also called the translation category.

§3. Cocategories over Commutative Algebras: Hopf Algebroids

(3.1) Notation. Let $(c\backslash \text{alg}/R)$ be the category of commutative algebras over a commutative ring R. The coproduct in this category is the tensor product over R and the initial object is R. Let $g : A \to A'$ and $f : A \to A''$ be two morphisms in $(c\backslash \text{alg}/R)$. The cofibre coproduct in the category $(c\backslash \text{alg}/R)$ is $A' \xrightarrow{q'} A' \otimes_A A'' \xleftarrow{q''} A''$ where $q'(x') = x' \otimes 1$ and $q''(x'') = 1 \otimes x''$. The tensor product over A defining the cofibre coproduct is formed with the right A-module structure $x'a = x'g(a)$ and the left A-module structure $ax'' = f(a)x''$ for $a \in A$, $x' \in A'$, and $x'' \in A''$.

(3.2) Definition. A cocategory object in this category $(C \text{ alg } R)$ is a sextuple $(A, \Gamma, \eta_L, \eta_H, \varepsilon, \Delta)$ such that the sextuple $(A, \Gamma, \eta_L^{\mathrm{op}}, \eta_R^{\mathrm{op}}, \varepsilon^{\mathrm{op}}, \Delta^{\mathrm{op}})$ is a category object in the dual category $(C \text{ alg } R).^{\mathrm{op}}$

In particular the category axioms (Cat1)–(Cat4) will correspond to commutative diagrams of commutative algebras.

(3.3) Remark. In particular the objects A and Γ in the cocategory object $(A, \Gamma, \eta_L, \eta_H, \varepsilon, \Delta)$ are commutative algebras over R. The first two structure morphisms $\eta_L, \eta_R : A \to \Gamma$ define a left and right A-module structure on Γ by $xa = x\eta_R(a)$ and $ax = \eta_L(a)x$ in Γ for $a \in A$ and $x \in \Gamma$. The third structure morphism $\varepsilon : \Gamma \to A$ satisfies the augmentation relation $\varepsilon_L(a) = a$ and $\varepsilon_R(a) = a$ for $a \in A$. In particular the following diagram is commutative

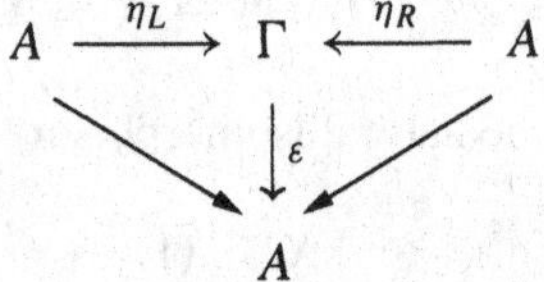

where the diagonal morphisms are identities. This is the dual to (Cat1).

The cocategory cocomposition, called the comultiplication on Γ, $\Delta : \Gamma \to \Gamma \otimes_A \Gamma$ is a morphism of A-bimodules. Using the morphisms $\Gamma \xrightarrow{q'} \Gamma \otimes_A \Gamma \xleftarrow{q''} \Gamma$, we see that the composites $q'\eta_L : A \to \Gamma \otimes_A \Gamma$ and $q''\eta_R : A \to \Gamma \otimes_A \Gamma$ define left and right A-module structures on $\Gamma \otimes_A \Gamma$.

In terms of elements $z \in \Gamma$ we can write $\Delta(z) = \sum_i z_i' \otimes z_i''$ and for $a \in A$ the left and right linearity has the form

$$\Delta(az) = a\Delta(z) \sum_i (az_i') \otimes z_i'' \quad \text{and} \quad \Delta(za) = \Delta(z)a = \sum_i z_i' \otimes (z_i''a).$$

The opposite to (Cat3) is the associativity of comultiplication which satisfies the following commutative diagram

$$\begin{array}{ccc}
\Gamma & \xrightarrow{\Delta} & \Gamma \otimes_A \Gamma \\
{\scriptstyle \Delta}\downarrow & & \downarrow{\scriptstyle \Delta\otimes\Gamma} \\
\Gamma \otimes_A \Gamma & \xrightarrow{\Gamma\otimes\Delta} & \Gamma \otimes_A \Gamma \otimes_A \Gamma
\end{array}$$

The dual to (Cat4) is counit, called the counit ε of the Hopf algebroid which satisfies the following commutative diagram

$$\begin{array}{ccccc}
 & \swarrow & \Gamma & \searrow & \\
 & & \downarrow{\scriptstyle \Delta} & & \\
A \otimes_A \Gamma & \xleftarrow[\varepsilon\otimes_A\Gamma]{} & \Gamma \otimes_A \Gamma & \xrightarrow[\Gamma\otimes_A\varepsilon]{} & \Gamma \otimes_A A
\end{array}$$

where diagonal morphisms are natural isomorphisms. A cocategory (A, Γ) is a cogroupoid provided there is a fifth structure morphism $c : \Gamma \to \Gamma$ satisfying the additional commutative diagrams

$$\text{(4)} \qquad \begin{array}{ccc} \Gamma \otimes_A \Gamma & \xrightarrow{c \otimes \Gamma} & \Gamma \otimes_A \Gamma \\ {\scriptstyle \Delta}\uparrow & & \downarrow{\scriptstyle \phi(\Gamma)} \\ \Gamma & \xrightarrow{\varepsilon \eta_L} & \Gamma \end{array} \qquad \begin{array}{ccc} \Gamma \otimes_A \Gamma & \xrightarrow{\Gamma \otimes c} & \Gamma \otimes_A \Gamma \\ {\scriptstyle \Delta}\uparrow & & \downarrow{\scriptstyle \phi(\Gamma)} \\ \Gamma & \xrightarrow{\varepsilon \eta_R} & \Gamma. \end{array}$$

Another name for such a cocategory or cogroupoid object (A, Γ) is Hopf algebroid. We also speak of Γ as a Hopf algebroid over A. Usually we would require that Γ is flat over A in order that the categories of comodules are abelian.

(3.4) Remark. We have defined a cocategory object in the category $(c\backslash\mathrm{alg}/R)$, however the same definition applies in any category $\mathcal{C}$ with finite colimits, in particular, an initial object is used and the cofibre coproduct construction.

A single cocategory object in a category $\mathcal{C}$ produce a category for each object in $\mathcal{C}$.

(3.5) Remark. Let (A, Γ) be a cocategory object in a category $\mathcal{C}$, and let Z be any object in $\mathcal{C}$. Then the cocategory object defines a category (in the category of sets) with objects $\mathrm{Hom}_{\mathcal{C}}(A, Z)$ and morphism set $\mathrm{Hom}_{\mathcal{C}}(\Gamma, Z)$.

(1) The function assigning to each object its identity is

$$e = \mathrm{Hom}_{\mathcal{C}}(\varepsilon, Z) : \mathrm{Hom}_{\mathcal{C}}(A, Z) \to \mathrm{Hom}_{\mathcal{C}}(\Gamma, Z).$$

(2) The domain and range morphisms are given by

$$\mathfrak{l} = \mathrm{Hom}_{\mathcal{C}}(\eta_L, Z) : \mathrm{Hom}_{\mathcal{C}}(\Gamma, Z) \to \mathrm{Hom}_{\mathcal{C}}(A, Z) \quad \text{and}$$
$$\mathfrak{r} = \mathrm{Hom}_{\mathcal{C}}(\eta_R, Z) : \mathrm{Hom}_{\mathcal{C}}(\Gamma, Z) \to \mathrm{Hom}_{\mathcal{C}}(A, Z).$$

(3) composition morphism in the category corresponding to the object Z on the morphism sets $\mathrm{Hom}_{\mathcal{C}}(\Gamma, Z)$ is the inverse of the following isomorphism used to define the cofibre coproduct

$$\mathrm{Hom}_{\mathcal{C}}(\Gamma, A) \underset{{}^{\mathfrak{r}}\mathrm{Hom}_{\mathcal{C}}(A,Z)^{\mathfrak{l}}}{\times} \mathrm{Hom}_{\mathcal{C}}(\Gamma, Z) \leftarrow \mathrm{Hom}_{\mathcal{C}}(\Gamma \otimes_A \Gamma, A),$$

and the morphism $\mathrm{Hom}_{\mathcal{C}}(\Delta, Z) : \mathrm{Hom}_{\mathcal{C}}(\Gamma \otimes_A \Gamma, Z) \to \mathrm{Hom}_{\mathcal{C}}(\Gamma, Z)$ induced by Δ.

In the case where (A, Γ) is a cogroupoid with additional structure morphism $c : \Gamma \to \Gamma$ the resulting category with morphisms $\mathrm{Hom}_{\mathcal{C}}(\Gamma, Z)$ is a groupoid with the inverse map

$$\mathfrak{i} : \mathrm{Hom}_{\mathcal{C}}(\Gamma, Z) \to \mathrm{Hom}_{\mathcal{C}}(\Gamma, Z)$$

given by $\mathfrak{i}(f) = fc$.

(4) The above construction defines a functor $C \to$ (Grpoid) where for a morphism $Z' \to Z''$ composition on the left induces a morphism between groupoids.

The main example of these ideas will be the category of Weierstrass equations and their transformations which leads to the category of elliptic curves over a field k. This is discussed in the next sections.

§4. The Category *WT*(*R*) and the Weierstrass Hopf Algebroid

We consider a category $WT(R)$ for each commutative ring R whose objects are Weierstrass polynomials and morphisms are triangular changes of variables over the ring R. The relation to elliptic curves is explained in 3(2.3)–3(2.7) where a Weierstrass polynomial is called a cubic in normal form.

(4.1) Definition. Let R be a commutative ring. The objects of the category $WT(R)$ are polynomials $F(a)(x, y)$ in $R[x, y]$ of the form

$$F(a)(x, y) = y^2 + a_1xy + a_3y - x^3 - a_2x^2 - a_4x - a_6$$

where $(a) = (a_1, a_2, a_3, a_4, a_6) \in R^5$.

The morphisms $\phi_{r,s,t,v} : F(a') \to F(a)$ are given by $r, s, t \in R$ and $v \in R^*$ where the follow relation is satisfied

$$F(a')(x', y') = F(a)(v^2x' + r, v^3y' + v^2sx' + t).$$

This is the triangular change of variable morphism where

$$x = v^2x' + r \quad \text{and} \quad y = v^3y' + v^2sx' + t$$

so that $F(a')(x', y') = F(a)(x, y)$ is the previous condition. Composition is given by substitution of these variables.

(4.2) Remark. We can describe composition of two morphisms $\phi_{r,s,t,v} : F(a') \to F(a)$ and $\phi_{r',s',t',v'} : F(a'') \to F(a')$ explicitly. It is given by a substitution within a substitution which has the same triangular change of variable morphism

$$\phi_{r'',s'',t'',v''} = \phi_{r,s,t,v}\,\phi_{r',s',t',v'} : F(a'') \to F(a)$$

To derive the rule of composition, we consider a substitution within a substitution starting with the following two expressions

$$F(a')(x', y') = F(a)(v^2x' + r, v^3y' + v^2sx' + t) = F(a)(x, y),$$

and

$$F(a'')(x'', y'') = F(a)(v'^2x'' + r', v'^3y'' + v'^2s'x'' + t').$$

Substituting, we have

$$x = v^2x' + r = v^2(v'^2x'' + r') + r = (vv')^2x'' + (v^2r' + r),$$

and

$$\begin{aligned} y &= v^3y' + v^2sx' + t = v^3(v'^3y'' + v'^2s'x'' + t') + v^2s(v'^2x'' + r') + t \\ &= (vv')^3y'' + (v^3v'^2s' + (vv')^2s)x'' + (v^3t' + v^2sr' + t). \end{aligned}$$

If we introduce the following three by three matrix

$$M(\phi_{r,s,t,v}) = \begin{pmatrix} 1 & r & t \\ 0 & v^2 & v^2s \\ 0 & 0 & v^3 \end{pmatrix}$$

with variables $r, s, t, v \in R$, then the substitution rule takes the form of the following matrix identity

$$\begin{pmatrix} 1 & r' & t' \\ 0 & v'^2 & s'v'^2 \\ 0 & 0 & v'^3 \end{pmatrix} \begin{pmatrix} 1 & r & t \\ 0 & v^2 & sv^2 \\ 0 & 0 & v^3 \end{pmatrix} = \begin{pmatrix} 1 & r'v^2 + r & t'v^3 + r'sv^2 + t \\ 0 & (vv')^2 & s''(vv')^2 \\ 0 & 0 & (vv')^3 \end{pmatrix}.$$

where $v'' = v'v$ and three other relations

$$r'' = v^2r' + r, \qquad v''^2s'' = v^3v'^2s' + (vv')^2s, \quad \text{and} \quad t'' = v^3t' + v^2sr' + t.$$

Here the matrix multiplication is in the opposite order from composition, but with the transpose matrices we have a matrix formula for composition

$$M^{tr}(\phi_{r,s,t,v})M^{tr}(\phi_{r',s',t',v'}) = M^{tr}(\phi_{r'',s'',t''}).$$

Also the relation between the variables x, y and x', y' can be expressed by the matrix multiplication formula

$$(1, x, y) = (1, x', y')M(\phi_{r,s,t,v}) = (1, x', y') \begin{pmatrix} 1 & r & t \\ 0 & v^2 & v^2s \\ 0 & 0 & v^3 \end{pmatrix}.$$

For $v = v' = 1$ the transpose matrices satisfy the multiplicative relation in the opposite order

$$\begin{pmatrix} 1 & 0 & 0 \\ r'' & 1 & 0 \\ t'' & s'' & 1 \end{pmatrix} = \begin{pmatrix} 1 & 0 & 0 \\ r & 1 & 0 \\ t & s & 1 \end{pmatrix} \begin{pmatrix} 1 & 0 & 0 \\ r' & 1 & 0 \\ t' & s' & 1 \end{pmatrix}.$$

(4.3) Remark. If there is a morphism $\phi_{r,s,t,v} : F(a') \to F(a)$, then we can express the constants a'_i in terms of the constants a_i and r, s, t, v. For this we begin with the relation

$$
\begin{aligned}
F(a')(x', y') &= F(a)(x, y) = F(a)(v^2x' + r, v^3y' + v^2sx' + t) \\
&= (v^3y' + v^2sx' + t)^2 + a_1(v^3y' + v^2sx' + t)(v^2x' + r) \\
&\quad + a_3(v^3y' + v^2sx' + t) - (v^2x' + r)^3 - a_2(v^2x' + r)^2 \\
&\quad - a_4(v^2x' + r) - a_6. \\
&= v^6(y')^2 + v^5(a_1 + 2s)x'y' + v^3(a_3 + a_1r + 2t)y' - v^6(x')^3 \\
&\quad - v^4(a_2 + 3r - s^2 - a_1s)(x')^2 \\
&\quad - v^2(a_4 + 2ra_2 + 3r^2 - sa_3 - a_1(rs + t) - 2st)x' \\
&\quad - (a_6 + a_4r + a_2r^2 + r^3 - a_3t - a_1rt - t^2).
\end{aligned}
$$

Thus we can write $v^ia'_i = a_i + \delta_i(r, s, t, v)$ where $\delta_i(r, s, t, v)$ is a polynomial over the integers in all a_j with $j < i$. We have the relations where the polynomials $\delta_i = \delta_i(r, s, t, 1)$ are explicitly given by

$$
\begin{aligned}
a'_1 &= a_1 + 2s && \text{thus } \delta_1 = 2s \\
a'_2 &= a_2 - a_1s + 3r - s^2 && \text{thus } \delta_2 = -a_1s + 3r - s^2 \\
a'_3 &= a_3 + a_1r + 2t && \text{thus } \delta_3 = a_1r + 2t \\
a'_4 &= a_4 + 2a_2r - a_1(rs + t) - a_3s + 3r^2 - 2st \\
&\qquad \text{thus } \delta_4 = 2a_2r - a_1(rs + t) - a_3s + 3r^2 - 2st \\
a'_6 &= a_6 + a_4r + a_2r^2 + r^3 - a_3t - a_1rt - t^2 \\
&\qquad \text{thus } \delta_6 = a_4r + a_2r^2 + r^3 - a_3t - a_1rt - t^2.
\end{aligned}
$$

Given a cubic $F(a)(x, y)$ and a four tuple $(r, s, t, v) \in T^3 \times R^*$ there exists a unique cubic $F(a')(x, y)$ with the morphism

$$\phi_{r,s,t,v} : F(a') \to F(a).$$

This means that the groupoid $WT(R)$ is of the form of group object $G(R)$ acting on the set of $WP(R)$ of Weierstrass polynomials over R by substitution, or as in (2.9) we have $WT(R) = WP(R)\langle G(R)\rangle$.

(4.4) Remark. Let $w : R' \to R''$ be a morphism of commutative rings. There is an associated functor $WT(w) : WT(R') \to WT(R'')$ given by $WT(w)(F(a)(x, y)) = F(w(a))(x, y)$ and

$$WT(w)(\phi_{r,s,t,v}) = \phi_{w(r),w(s),w(t),w(v)}.$$

For a second morphism $v : R'' \to R$ of rings we have the composition of functors $WT(v)WT(w) = WR(vw)$.

We have constructed a functor from commutative rings to the category of small categories, and now we show that it is representable by a Hopf algebroid, that is, a cocategory in the category of commutative rings.

(4.5) Definition. The Weierstrass Hopf algebroid with

$$A = \mathbb{Z}[\alpha_1, \alpha_2, \alpha_3, \alpha_4, \alpha_6] \quad \text{and} \quad \Gamma = A[\rho, \sigma, \tau, \lambda, \lambda^{-1}].$$

The unit $\varepsilon : \Gamma \to A$ is defined by giving the values on generators $\varepsilon(\rho) = 0,\ \varepsilon(\sigma) = 0,\ \varepsilon(\tau) = 0$, and $\varepsilon(\lambda) = 1$. The coaugmentation η_R is the natural inclusion $A \to \Gamma$ of rings, and η_L is given by $\eta_L(\alpha_i) = \lambda^i\alpha_i + \delta_i(\rho, \sigma, \tau, \lambda)$. We return to the comultiplication in section (4.8).

(4.6) Remark. A morphism in $WT(R)$ is a ring morphism $f : \Gamma \to R$ assigning to the variables $\rho, \sigma, \tau, \lambda$ the values r, s, t, v which determine a change of variable morphism in $WT(R)$

$$\phi_f = \phi_{r,s,t,v} : F(f\eta_L(\alpha)) \to F(f\eta_R(\alpha)).$$

Also the identity morphisms correspond to $\varepsilon : \Gamma \to A$ where a morphism of rings $g : A \to R$ determines an object $F(g(\alpha))$ in $WT(R)$ and the identity morphism on this object $F(g(\alpha))$, that is,

$$\phi_{g\varepsilon(\rho), g\varepsilon(\sigma), g\varepsilon(\tau), g\varepsilon(\lambda)} : F(g(\alpha)) \to F(g(\alpha)).$$

(4.7) Remark. For the comultiplication on Γ we begin by noting that $\eta_L, \eta_R : A \to \Gamma$ define a left and right A-module structure making Γ into a bimodule by the relations

$$ax = \eta_L(a)x \quad \text{and} \quad xa = x\eta_R(s) \quad \text{for } a \in A \quad \text{and} \quad x \in \Gamma.$$

Then the tensor product $\Gamma \otimes_A \Gamma$ is defined with respect to the right module structure on the first factor and the left module structure on the second, so that $x'a \otimes x'' = x' \otimes ax''$. It is an A-bimodule with the relations

$$a(x' \otimes x'') = (ax') \otimes x'' \quad \text{and} \quad (x' \otimes x'')a = x' \otimes (x''a)$$
$$\text{for } a \in A \quad \text{and} \quad x', x'' \in \Gamma.$$

In addition we have the bimodule linearity of Δ expressed as follows. For $\Delta(z) = \sum_i z'_i \otimes z''_i$ and a scalar $a \in A$ we have

$$\Delta(az) = \sum_i (ax'_i) \otimes z''_i \quad \text{and} \quad \Delta(za) = \sum_i z'_i \otimes (z''_i a).$$

The comultiplication structure morphism $\Delta : \Gamma \to \Gamma \otimes_A \Gamma$, which we are about to define, is an algebra morphism besides being an A-bimodule morphism. To see how to define the comultiplication, we consider two morphisms $f', f : \Gamma \to R$ with $f'\eta_R = f\eta_L$. We expect that f and f' will define two morphisms in the category $WT(R)$

$$\phi_{f'} : F(f'\eta_L(\alpha)) \to F(f'\eta_R(\alpha)) \quad \text{and} \quad \phi_f : F(f\eta_L(\alpha)) \to F(f\eta_R(\alpha))$$

which compose to $\phi_f\phi_{f'} : F(f'\eta_L(\alpha)) \to F(f\eta_R(\alpha))$, that is, we require the comultiplication $\Delta : \Gamma \to \Gamma \otimes_A \Gamma$ give the relation

$$\phi_f\phi_{f'} = \phi_{(f'\otimes f)\Delta}$$

(4.8) Definition. The comultiplication $\Delta : \Gamma \to \Gamma \otimes_A \Gamma$ on generators of Γ over A is given by

$$\Delta(\lambda) = \lambda \otimes \lambda \qquad \Delta(\rho) = \rho \otimes \lambda^2 + 1 \otimes \rho,$$
$$\Delta(\sigma) = \sigma \otimes \lambda + 1 \otimes \sigma \qquad \Delta(\tau) = \tau \otimes \lambda^3 + \rho \otimes \sigma\lambda^2 + 1 \otimes \tau.$$

(4.9) Remark. It is easy to see that the relation $v'' = v'v$ translates into the relation $\Delta(\lambda) = \lambda \times \lambda$. From the matrix identity where $v'' = v'v$

$$\begin{pmatrix} 1 & r' & t' \\ 0 & v'^2 & s'v'^2 \\ 0 & 0 & v'^3 \end{pmatrix} \begin{pmatrix} 1 & r & t \\ 0 & v^2 & sv^2 \\ 0 & 0 & v^3 \end{pmatrix} = \begin{pmatrix} 1 & r + r'v^2 & t'v^3 + r'sv^2 + t \\ 0 & v''^2 & s''v''^2 \\ 0 & 0 & v''^3 \end{pmatrix}$$

we have three other relations

$$r'' = v^2 r' + r, \qquad v''^2 s'' = v^3 v'^2 s' + (vv')^2 s, \quad \text{and} \quad t'' = v^3 t' + v^2 s r' + t.$$

The matrix

$$M(\phi_{r,s,t,v}) = \begin{pmatrix} 1 & r & t \\ 0 & v^2 & v^2 s \\ 0 & 0 & v^3 \end{pmatrix}$$

has an inverse given by the formula

$$M(\phi_{r,s,t,v})^{-1} = \begin{pmatrix} 1 & -rv^{-2} & rsv^{-3} - v^{-3}t \\ 0 & v^{-2} & -v^{-3}s \\ 0 & 0 & v^{-3} \end{pmatrix}.$$

In particular each morphism $\phi_{r,s,t,v}$ is an isomorphism. The process of carrying a morphism to its inverse is represented in Γ by a morphism $I : \Gamma \to \Gamma$ with $I\eta_L = \eta_R$, $I\eta_R = \eta_L$ corresponding to interchanging the domain and range. For the isomorphism inverse properties of I we use the algebra structure $\phi(\Gamma) : \Gamma \otimes_A \Gamma \to \Gamma$ given by $\phi(\Gamma)(z' \otimes z'') = z'z''$. The inverse properties become the formulas $\phi(\Gamma)(\Gamma \otimes I)\Delta = \varepsilon\eta_L$ and $\phi(\Gamma)(I \otimes \Gamma)\Delta = \varepsilon\eta_R$. On the generators of Γ over A we have the formulas

$$I(\lambda) = \lambda^{-1}, \qquad I(\rho) = -\rho\lambda^{-2}, \qquad I(\sigma) = -\sigma\lambda^{-1},$$
$$I(\tau) = \rho\sigma\lambda^{-3} - \tau\lambda^{-3}.$$

(4.10) Assertion. The functor WT from the category of commutative rings to groupoids is represented by the Weierstrass Hopf algebroid (A, Γ) defined above in (4.5) and (4.8).

§5. Morphisms of Hopf Algebroids: Modular Forms

Additional properties of the Weierstrass Hopf algebroid result by studying morphisms, especially localizations of cocategories. We begin with some generalities on morphisms of Hopf algebroids.

(5.1) Definition. A morphism of cocategories over a category C is just a functor $(A, \Gamma) \to (B, \Lambda)$ between the category objects in the opposite category C^{op}. In particular there are two morphisms $f : A \to B$ and $g : \Gamma \to \Lambda$ in C satisfying various commutativity relations.

(5.2) Remark. A morphism $(f, g) : (A, \Gamma) \to (B, \Lambda)$ of Hopf algebroids is a pair of morphisms which satisfy the following commutative diagrams

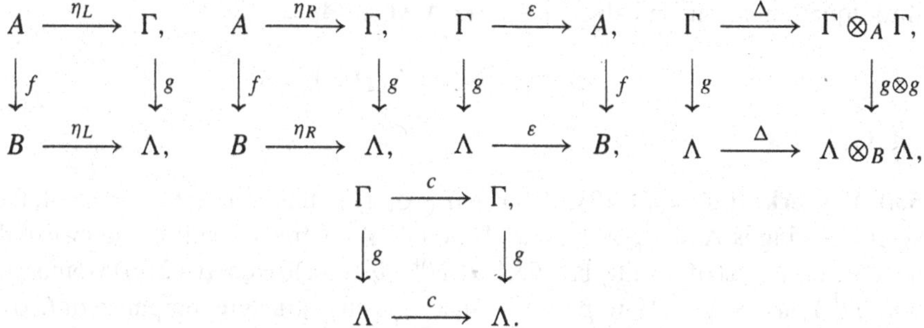

If $(f, g) : (A, \Gamma) \to (B, \Lambda)$ and $(f', g') : (B, \Lambda) \to (C, \Sigma)$ are morphism of Hopf algebroids, then the composite $(f'f, g'g) : (A, \Gamma) \to (C, \Sigma)$ is a morphism of Hopf algebroids. In particular we have a category where the objects are Hopf algebroids and the morphisms are defined in this section. We denote this category by $(h/a/oid)$. There is a subcategory $(h/a/oid)/A$ of Hopf algebroids with fixed algebra A and identity morphism on A. In this category the Hopf algebroid (A, A) with all structure morphisms equal to the identity is the final object.

(5.3) Remark. Let $(f, g) : (A, \Gamma) \to (B, \Lambda)$ be a morphism of Hopf algebroids. For each algebra E the induced morphisms

$$f^* : \mathrm{Hom}_{(c\backslash rg)}(B, E) \to \mathrm{Hom}_{(c\backslash rg)}(A, E),$$
$$g^* : \mathrm{Hom}_{(c\backslash rg)}(\Lambda, E) \to \mathrm{Hom}_{(c\backslash rg)}(\Gamma, E)$$

define a functor (f^*, g^*)

$$(\mathrm{Hom}_{(c\backslash rg)}(B, E), \mathrm{Hom}_{(c\backslash rg)}(\Lambda, E)) \to (\mathrm{Hom}_{(c\backslash rg)}(A, E), \mathrm{Hom}_{(c\backslash rg)}(\Gamma, E)).$$

Here $(c\backslash rg)$ denotes the category of commutative rings.

If a group G acts on a set or module M, then we have $H^0(G, M)$ the set or module of fixed elements $x \in M$, that is, $sx = x$ for all $s \in G$. There is a notion of fixed elements $H^0(\Gamma, M)$ for a coaction of Γ on a comodule M, and then these elements are called primitive elements.

(5.4) Remark. The Weierstrass Hopf algebroid (WT_0, WR_1) or simply (A, Γ) is $A = \mathbb{Z}[\alpha_1, \alpha_2, \alpha_3, \alpha_4, \alpha_6]$ and $\Gamma = A[\rho, \sigma, \tau, \lambda, \lambda^{-1}]$. There is the algebraic group WG with affine coordinate ring

$$K[WG] = k[\rho, \sigma, \tau, \lambda, \lambda^{-1}].$$

The coordinate ring $Z[WG]$ of this algebraic group and A can be used to construct Γ. As a groupoid we apply the construction in (2.9) to obtain $WT_1 = WR_0(\langle WG\rangle)$ functorially in R.

There is another basic construction associated with the Weierstrass groupoid which plays a role in the homotopic picture, namely modular forms. For this we use the next definition.

(5.5) Definition. The ring of multidifferential forms over the curve C with Weierstrass equation is a graded algebra $A.$ over A where $A_0 = A$,

$$A_{2n} = H^0(C, (\Omega^1_{C/A})^{n\otimes}),$$

and $A_{2n+1} = 0$.

(5.6) Remark. If $\omega = dx/(2y + a_1x + a_3) \in A_2$ is the invariant differential, then the graded ring is $A. = A[\omega^{\pm 1}]$. The A-module A_2 is free of rank 1, and is also the module of sections of the line bundle $L = H^0(\mathcal{O}_C((-1)0)/\mathcal{O}_C((-2)0))$ isomorphic to $p_*(\Omega^1_C)$ on $\mathrm{Spec}(A)$. Here $p : C \to \mathrm{Spec}(A)$ is the structure morphism of C over $\mathrm{Spec}(A)$, and $0 : \mathrm{Spec}(A) \to C$ is the section at infinity (0:0:1).

(5.7) Definition. The algebraic group WG of substitutions in the Weierstrass equations acts on $A.$, and the fixed subring under WG is $H^0(WG, A.)$ the ring of modular forms.

For the structure of this ring we return to the article of Tate on formulas in the theory of elliptic curves in LN 476. There is a version over the integers and in characteristics 2 and 3. It is in characteristics 2 and 3 where the relation to homotopy theory will be the most important.

In order to describe the ring of modular forms, we introduce the following notation from Tate LN 476 in terms of a_i instead of the indeterminants α_i.

(5.8) Notation. In terms of the Weierstrass coefficients a_i (instead of the indeterminants α_i) we have from 3(3.1)

$$b_2 = a_1^2 + 4a_2, \qquad b_4 = a_1a_3 + 2a_4, \qquad b_6 = a_3^2 + 4a_6,$$
$$b_8 = a_1^2a_6 - a_1a_3a_4 + 4a_2a_6 + a_2a_3^2 - a_4^2$$

and from 3(3.3)

$$c_4 = b_2^2 - 24b_4 \quad \text{and} \quad c_6 = -b_2^3 + 36b_2b_4 - 216b_6$$
$$\Delta = -b_2^2b_8 - 8b_4^3 - 27b_6^2 + 9b_2b_4b_6 \quad \text{and} \quad j = c_4^3/\Delta.$$

(5.9) Rings of Modular Forms. The ring of modular forms is

$$M_* = H^0(WG, A.) = \mathbb{Z}[c_4, c_6, \Delta]/(c_4^3 - c_6^2 - 12^3\Delta)$$

over the integers, and $M_*[1/6] = \mathbb{Z}[1/6][c_4, c_6]$.

In characteristic 2 the ring of modular forms is $\mathbb{F}_2[a_1, \Delta]$ where $c_4 = a_1^4$ and $c_6 = a_1^6$.

In characteristic 3 the ring of modular forms is $\mathbb{F}_3[b_2, \Delta]$ where $c_4 = b_2^2$ and $c_6 = -b_2^3$.

(5.10) Remark. The higher derived functors $H^i(WG, A.)$ were apparently of little interest in arithmetic. In December 1996, during a conversation, Michael Hopkins expressed an interest in these derived functors, and in fact they became relevant in the context of topological modular forms as we will report in §7.

§6. The Role of the Formal Group in the Relation Between Elliptic Curves and General Cohomology Theory

Recall from 12(7.3) that a formal group law over a ring R is an element $F(X, Y) \in R[[X, Y]]$ satisfying the three conditions:

$$F(X,0) = F(0, X) = X, \qquad F(X, F(Y, Z)) = F(F(X, Y), Z),$$
$$F(X, Y) = F(Y, X),$$

the formal unit, associativity, and commutativity. These one parameter commutative formal group laws $F(X, Y)$ arise in both the theory of elliptic curves, as explained in Chapter 12, §7, and in general multiplicative cohomology theories from the work of Quillen [1969] as elaborated by Adams [1974].

(6.1) Remark. For elliptic curves in Weierstrass form over k we consider functions x and y with poles of order 2 and 3, respectively, at the origin. Then the function $t = -(x/y)$ has a zero of order 1 and can be used as a local parameter at the origin. The group law on the elliptic curve E has an expansion $t_3 = F_E(t_1, t_2)$ in terms of the formal group f_E of E over k, see 12(7.2).

(6.2) Remark. A spectrum is given by a positive sequence E_n of pointed spaces together with maps $\varepsilon_n : SE_n \to E_{n+1}$ where SE_n denotes the suspension of E_n. The morphisms between spectra are certain sequences of maps, see Adams [1974]. A ring spectrum has an additional structure of pairing maps $E_m \wedge E_n \to E_{m+n}$ where $X \wedge Y = X \times Y/X \vee Y$ is called the smash product. Here $X \vee Y$ is the one-point union.

(6.3) Definition. The homology theory E_* associated with a spectrum (E_n, ε_n) is defined on a pair of spaces by

$$E_q(X, A) = \varinjlim_{m_*}[S^{m+q}, (X/A) \wedge E_m] = \pi_q^S((X/A) \wedge E_*),$$

and the cohomology theory E^* associated with a spectrum (E_n, ε_n) is defined on pairs of finite CW-complexes

$$E^q(X, A) = \varinjlim_m [S^{m-q}(X/A), E_m].$$

(6.4) Remark. In the case that the adjoint maps $\delta_n : E_n \to \Omega E_{n+1}$ to $\varepsilon_n : SE_n \to E_{n+1}$ are homotopy equivalences, as is the case for K-theory by Bott periodicity, then cohomology is just homotopy classes of map between spaces $E^q(X, A) = [(X/A), E_q]$ for $q \geq 0$.

(6.5) Example. The periodic K-theory spectrum BU (or KU) has a connected version bu where

$$\pi_i^S(bu) = \begin{cases} \pi_i(BU) & i \geq 0 \\ 0 & i < 0. \end{cases}$$

A similar relation between a modified connected theory tmf and a periodic theory TMF exists with topological modular forms in the next section.

(6.6) Remark. For suitable cohomology theories E^*, called "complex oriented", there is a first Chern class

$$c_1(L) \in E^*(\mathbb{P}_\infty(\mathbb{C})) \cong E^0[[t]]$$

associated to line bundles L. The formal group gives a formula for the tensor product of line bundles

$$c_1(L' \otimes L'') = F(c_1(L'), c_1(L'')).$$

(6.7) Example. We have a fundamental example of a cohomology theory or spectrum which is multiplicative and has a first Chern class. Moreover, the associated formal group has a universal property among formal groups, see (6.8). It is given by the complex bordism spectrum where MU_{2n} is the Thom space $T(\xi_n)$ of the universal bundle n-dimensional complex vector bundle ξ_n on the classifying space BU_n and $MU_{2n+1} = ST(\xi_n)$. Since the restriction to a subspace

$$\xi_{n+1}|BU_n = \tau \oplus \xi_n,$$

we have a natural map of Thom spaces

$$T(\tau \oplus \xi_n) = S^2T(\xi_n) \to T(\xi_{n+1})$$

giving a spectrum. This is a multiplicative spectrum and the cohomology theory, as well as K-theory, is a complex orientable theory with a canonical orientation.

(6.8) Theory of Quillen. The formal group law for MU is universal in the sense that for any formal group law F over a ring R there exists a unique morphism of rings $\phi : MU(*) \to R$ such that F has coefficients equal to the image under ϕ of the coefficients in MU of the formal group law for $c_1(L)$ in complex bordism theory.

This result led to the possibility of constructing cobordism theories and their related spectra by starting with the morphism $\phi : MU_* \twoheadrightarrow R$ and forming the tensor product $MU_*(X) \otimes_{MU_*} R$. Under suitable conditions on the morphism ϕ we obtain a homology theory, and hence, also a cohomology theory and a spectrum by Brown's representatibly theorem. The conditions are contained in the next theorem.

(6.9) The Landweber Exact Functor Theorem. This is a criterion for the functor $X \mid \to MU_*(X) \otimes_{MU_*} R = \phi_*(X)$ to be exact or equivalently a homology theory. For this we need the sequences of elements $(p, v_1, \ldots, v_n, \ldots)$ for MU localized at each prime p. If these elements form regular sequences of elements in the MU_*-module R at each prime p, then $\phi_*(X)$ is a homology theory.

(6.10) Definition. An elliptic cohomology theory is a theory which arises from the formal group of an elliptic curve by applying the Landweber exact functor theorem to the ring morphism $\phi : MU \to A$ for an elliptic curve given by a Weierstrass polynomial.

(6.11) Remark. It should be noted that not every formal group of an elliptic curve yields a homology theory via the Landweber exact functor theorem. In LN 1326, and p. 59 and also Theorem 2, p. 71 Landweber studies conditions leading to an elliptic cohomology theory. Some of these considerations were carried further in:

Baker, A., *A Supersingular congruence of modular forms*. Acta Arithmetica, LXXXVI (1998), p. 91.

§7. The Cohomology Theory or Spectrum tmf

Through the formal group, we see that certain elliptic curves give arise to a cohomology theory or spectrum, and all the cohomology theories which arise this way are called elliptic cohomology. Elliptic curves can be assembled into a space called the moduli space of elliptic curves, and the aim of current work in topology is to construct a cohomology theory from data on the moduli of elliptic curves which maps in a natural way to all elliptic cohomology theories. Due to the role of modular forms with the moduli space of elliptic curves this theory is called topological modular forms.

(7.1) Remark. The problem with constructing tmf begins with forming a representative spectrum for each elliptic cohomology theory with a higher homotopy ring structure called an E_∞-spectrum, and then organize these ring spectra into a diagram in order to take an inverse limit. The inverse limit will be a homotopy inverse limit. The result tmf, called topological modular forms, will also be a E_∞-spectrum. It is not an elliptic cohomology theory, but for each elliptic cohomology theory E there should be a morphism tmf $\to E$ inducing E.

Returning to the notation of $A.$ of multidifferential forms and $H^0(WG, A.)$ of modular forms, we have the following basic calculation.

(7.2) Basic Computation of Coefficients. There is a spectral sequence with

$$E^2_{-s,t} = H^s(WG, A_t)$$

converging to $\pi_{t-s}(\text{tmf})$. In addition, the edge morphism $\pi_*(\text{tmf}) \to H^0(WG, A.)$ has kernel and cokernel annihilated by 24.

(7.3) Remark. As of March 2003 the construction of tmf has not appeared in the mathematics literature, but it is a topic of current research in topology. There is a very sophisticated obstruction theory needed which uses the cotangent complex of the moduli space of elliptic curves over the moduli space of formal groups.

References

Adams, J.F., *Stable homotopy and generalised homology theories*, University of Chicago Press, 1974.

Ando, M., M.J. Hopkins, N.P Strickland, *Elliptic spectra, the Witten genus and the theorem of the cube*. Inventiones mathematicae, **146**, (2001), pp. 595–687.

Deligna, P. (d'après John Tate), *Modular Function of One Variable IV*, LN 476, Springer-Verlag, (1975), pp. 53–73.

Franke, Jens, *On the construction of elliptic cohomology*, Math. Nachr. **158**, (1992), pp. 43–65.

Hirzebruch, F., Thomas Berger, and Rainer Jung, *Manifolds and modular forms*, Aspects of Mathematics, E20, Friedr. Vieweg and Sohn, Braunschweig, 1992.

Hopkins, M.J., *Topological Modular Forms, the Witten Genus, and the Theorem of the Cube*, ICM 1994, pp. 554–565.

Hopkins, M.J., *Algebraic Topology and Modular Forms*, ICM 2002, Volume III.

Landweber, P., *Elliptic Cohomology and Modular Forms*, LN 1326, Elliptic curves and Modular Forms in Algebraic Topology (1988), pp. 55–68.

Landweber, P., D.C. Ravenel, R.E. Stong, *Periodic cohomology theories defined by elliptic curves*, In: Čech centennial, Boston 1993, Volume 181 of contempory Math. pp. 317–337, AMS, 1995.

Quillen, D.G., *On the formal group laws of unoriented and complex cobordism theory*, Bulletin of the American Mathematical Society, **75** (1969), pp. 1293–1298.

Ravenel, D.C., *Nilpotence and periodicity in stable homotopy theory*, Annals of Mathematics studies, **128**, 1992.

Witten, *The Index of the Dirac Operator in Loop space*, LN 1326, Elliptic curves and Modular Forms in Algebraic Topology (1988) pp. 182–215.

Appendix IV: Guide to the Exercises

Ruth Lawrence

The purpose of this appendix is to give, for each exercise, a comment, a hint, a sketch, or in a few cases, a complete solution. Pure algebra has not been worked. Exercises which are merely a matter of applying techniques given in the text to particular examples for the purpose of drill have not generally been solved fully here—the answers only being given.

When there is a batch of somewhat similar questions a representative question has been selected to be completely solved and answers only have been given to the rest.

A few questions involve rather tedious numerical calculations not reducible by technique. Readers who have carried these out can, by checking their answers against the given ones, gain confidence. They may also be comforted to discover that they have not missed some subtle point or ingeniously simple route to a solution.

In a few cases a link is suggested to connected problems which the interested reader may like to pursue.

I hope that this appendix will fulfill its purpose of helping readers who experience any difficulties with the problems to overcome them; thus gaining the maximum understanding and insight which the author intended by his carefully chosen incorporation into the text.

I wish to acknowledge the help received from Professor Husemöller whilst compiling this appendix, by way of some useful discussions and suggestions.

CHAPTER 1, §1

1. Tedious calculation gives:

$$7P = (-5/9, 8/27), \qquad 8P = (21/25, -69/125),$$
$$9P = (-20/49, -435/343), \qquad 10P = (161/16, -2065/64),$$
$$-7P = (-5/9, -35/27), \qquad 8P = (21/25, -56/125),$$
$$-9P = (-20/49, 97/343), \qquad -10P = (161/16, 2001/64).$$

Mazur's theorem then implies P has infinite order. For $7P, 8P, 9P, 10P \neq 0$ and none of the above eight points coincide, i.e., P's order does not divide any integer 14 to 20 inclusive. Thus P cannot have order $2, 3, 4, 5, 6, 7, 8, 9, 10$, or 12 (the latter since $6P \neq -6P$).

2. Very straightforward. Here $-2P = P \cdot P = (0, 2)$, i.e., $-2P = P$ as required.
3. As in Exercise 2, we work out $-2P = P \cdot P = (0, 0)$. Clearly, $(0, 0)$ has order 2, and so P has order 4.
4. It is found that $2P = (0, 1), 3P = (-1, 0)$. Thus $3P$ has order 2. So P's order is 2 or 6 (divides 6 and does not divide 3). As $2P \neq 0$, thus P's order is 6.
5. Here $-2P = (24, 108)$, and $2P = (24, -108)$. From $P, 2P$ we obtain $-3P$. It is found that $-3P = (24, -108) = 2P$. So P has order 5.
6. A suitable subgroup is $\{0, (0, 0), (1, 0), (-9, 0)\}$. The calculations for nP ($n = -7, \ldots, +7$) get very tedious! Here are the first few:

$$P = (-1, 4), \quad 2P = \left(\frac{25}{16}, \frac{195}{64}\right), \quad 3P = \left(\frac{-14161}{1681}, \frac{-466004}{68921}\right),$$

$$4P = \left(\frac{14421367921}{4090881600}, \frac{-67387812533791}{6381775296000}\right),$$

$$-P = (-1, -4), \quad -2P = \left(\frac{25}{16}, \frac{-195}{64}\right), \quad -3P = \left(\frac{-14161}{1681}, \frac{466004}{68921}\right),$$

$$-4P = \left(\frac{14421367921}{4090881600}, \frac{67387812533791}{6381775296000}\right).$$

Now use Mazur's theorem. Since P has infinite order, no finite subgroup contains P. It helps in the calculations to realize that if P_1, P_2, P_3 are collinear points on the curve, then $x_1 + x_2 + x_3 = (\text{slope of line})^2 - 8$, and thus for any Q, R on the curve $Q \cdot R$ (and hence $Q + R$) can be determined.

CHAPTER 1, §2

1. Answers:

$$6P = (-2/9, -28/27), \quad 7P = (21, -99), \quad 8P = (11/49, 20/343),$$
$$-6P = (-2/9, 1/27), \quad -7P = (21, 98), \quad -8P = (11/49, -363/343).$$

Use Mazur's theorem—very similar to §1, Exercise 1.

2. Straightforward algebra gives $u = (w/3x)(y + 4)$, $v = (w/3x)(5 - y)$.
3. For $c = 1$, the group is $\{0, (12, 36), (12, -36)\}$, i.e., $\mathbb{Z}/3\mathbb{Z}$. For $c = 2$, the group is $\{0, (3, 0)\}$, i.e., $\mathbb{Z}/2\mathbb{Z}$. The complexity increases rapidly for $c \geq 8$. The interested reader may like to consider $c = 3, 4, \ldots$.
4. Here $2P = (9, -18), 4P = (0, 0)$. So $4P = -4P$, i.e., P has order 8.
5. One easily computes $2P = (-1, -2), -3P = (3, 6), -4P = (3, -6)$. Thus P has order 7.
6. Multiples are $0, (3, 8), (-5, -16), (11, -32), (11, 32), (-5, 16), (3, -8)$ in that order. Thus $(3, 8), (11, 32)$ both have order 7, since $(11, 32) = 4(3, 8)$ and $4, 7$ are coprime.

7. Here $E(\mathbb{Q}) = \{(0,0), (0,-1)\}$ (i.e., $\mathbb{Z}/2\mathbb{Z}$).
8. The given cubic in P follows since $P = (x, y)$ gives $-P = (x, y')$ where y, y' are two roots, for Y, of

$$Y^2 + (a_1x + a_3)Y - (x^3 + a_2x^2 + a_4x + a_6) = 0$$

and so $-(y+y') = (a_1x+a_3)$. Thus $2P = 0$ iff $y = -(1/2)(a_1x+a_3)$ and this reduces to the given condition on x. Thus there are zero, one, or three solutions for x. In every case 0 is a solution of $2P = 0$.

So, the group consists of one, two or four elements. In the first two cases, the groups must be 0, $\mathbb{Z}/2\mathbb{Z}$, respectively. In the last case, there are three elements of order 2, and so we get $(\mathbb{Z}/2\mathbb{Z} \times \mathbb{Z}/2\mathbb{Z})$.

CHAPTER 1, §3

1. We get $(-1, 0)$ (order 2); $(0, 1)$,$(0, -1)$ (order 3); and $(2, 3)$, $(2, -3)$ (order 6). The latter two are possible generators.
2. (a) The condition is that there are three points of order 2. Hence we need $x^2 + ax + b$ to have two roots in k, i.e., $a^2 - 4b$ must be a square.
   (b) The condition is that there exists a point, P, on the curve, of order 4. Thus $2P$ has order 2 and is thus $(\alpha, 0)$ some α. Since $a^2 - 4b$ is not a square, so $x^2 + ax + b \neq 0$ for all $x \in k$. Therefore $\alpha = 0$, and so $2P = (0, 0)$. This gives us a condition that there exists λ such that $y = \lambda x$ intersects the cubic curve in a double point. So

$$4b = (a - \lambda^2)^2$$

   and so $b = c^2$ with $a - \lambda^2 = \pm 2c$. so one of $a \pm 2c$ must be a square.
   (c) If $a+2c$, $a-2c$ are squares (with $b = c^2$), then from (b), $(0,0) = 2P$ some P. Hence the group $E(k)$ contains a subgroup generated by P and $(\alpha_i, 0)$ where α_i are the two roots of $x^2+ax+b = 0$. Thus $\mathbb{Z}/4\mathbb{Z} \times \mathbb{Z}/2\mathbb{Z} \subseteq E(k)$.
   Conversely, if $\mathbb{Z}/4\mathbb{Z} \times \mathbb{Z}/2\mathbb{Z} \subseteq E(k)$, then $(0, 0) \in 2E(k)$ gives, from (b), that one of $a \pm 2c$ must be a square. From (a), $a^2 - 4b = (a+2c)(a-2c)$ is a square. So $a + 2c$ and $a - 2c$ are squares.
   Note that in this question, the extra conditions, which are apparently asymmetric, are required since the equation of the cubic curve has fixed $(0, 0)$ as a point of order 2.
3. Points of infinite order are $(1, 2)$, $(2, 3)$, $(-1, 1)$, $(3, 5)$, respectively. To check this, use Theorem (3.2).
4. Consider the canonical map $E(k) \xrightarrow{\theta} E(k)/2E(k)$. Then, if P, Q are linearly dependent, say $nP = mQ$ with $m, n \in \mathbb{Z}$, we get

$$n'[P] = m'[Q],$$

   where $[P]$ is the equivalence class of P (i.e., $\theta(P)$) and $n' \equiv n(\text{mod } 2)$. Thus, if $P = (3, 4)$, $Q = (15, 58)$, then P, Q have infinite order. We thus only need to check that $[P]$, $[Q]$, 0 are all distinct. In fact $Q - P = (313/36, -5491/216)$.

So, we now evaluate $2(x, y)$. The condition that $2(x, y) = (\alpha, \beta)$ on $y^2 = x^3 - 11$ is that $0 = x^4 - 4\alpha x^3 + 88x + 44\alpha$. So we use $\alpha = 3, 15, 313/36$ to show that $P, Q, Q - P \notin 2E(k)$.[1] Thus $[P], [Q], 0$ are all distinct. Hence $nP \neq mQ, \forall m, n \in \mathbb{Z}$, not both zero.

5. This is similar to Exercise 4. If $P = (0, 2), Q = (1, 0), R = (2, 0)$ then we check:
   (i) P, Q, R have infinite order.
   (ii) $[P], [Q], [R], 0$ are all distinct.
   (iii) $[P + Q + R] \neq 0$.

Chapter 1, §4

1. This is very straightforward: define $K = P' - P$ and then show that $K \in_n A$.
2. In (4.1) the conditions are that $-\alpha, -\beta$ are squares. In §3, 2(c) the conditions are

$$\alpha\beta = c^2, \quad \text{some } c,$$

and $2c - (\alpha + \beta)$ or $-2 - (\alpha + \beta)$ is a square. Here

$$\pm 2c - (\alpha + \beta) = \pm 2c - \alpha - c^2/\alpha = -(\alpha \mp c)^2/\alpha.$$

So $-\alpha$ (and hence also $-\beta$) is a square.

3. When ${}_2E(k) = 0, {}_4E(k) = 0$.
   When ${}_2E(k) = \mathbb{Z}/2\mathbb{Z} \times \mathbb{Z}/2\mathbb{Z}, {}_4E(k)$ is $\mathbb{Z}/2\mathbb{Z} \times \mathbb{Z}/2\mathbb{Z}$ or $\mathbb{Z}/4\mathbb{Z} \times \mathbb{Z}/2\mathbb{Z}$ or $\mathbb{Z}/4\mathbb{Z} \times \mathbb{Z}/4\mathbb{Z}$. When ${}_2E(k) = \mathbb{Z}/2\mathbb{Z}, {}_4E(k)$ is $\mathbb{Z}/2\mathbb{Z}$ or $\mathbb{Z}/4\mathbb{Z}$. The numbers of elements of order 4 are: 0; 0, 4, 12; 0, 2, respectively.
4. Suppose otherwise, that $(\mathbb{Z}/4\mathbb{Z})^2 \subseteq E(\mathbb{Q})$, by an inclusion map j. Then the elements of order 4 are $j((2, 0)), j((0, 2)), j((2, 2))$. Thus in normal form we get

$$y^2 = (x - \alpha)(x - \beta)(x - \gamma),$$

where $(\alpha, 0) = j((2, 0))$, etc. Hence $(\alpha, 0) = 2P$ where

$$P = j((1, 0)), \ j((1, 2)), \ j((3, 0)) \text{ or } j((3, 2)).$$

Thus $\alpha - \beta, \alpha - \gamma$ are square. Similarly $\beta - \alpha, \beta - \gamma, \gamma - \alpha, \gamma - \beta$ are squares. This leads to -1 as a square, a contradiction.

Chapter 1, §5

1. A cusp occurs if $a_2 + a_1^2/4 = 0$. A double point occurs if and only if $(a_2 + a_1^2/4) > 0$. These conditions are easily obtained from

$$y = x(-a_1/2 \pm \sqrt{x + (a_2 + a_1^2/4)}).$$

The graphs below indicate the forms of the curve for $a_1 < 0$.

[1] One checks that the quartic equation has no solution in $\mathbb{Q}$, by Eisenstein's irreducibility criterion.

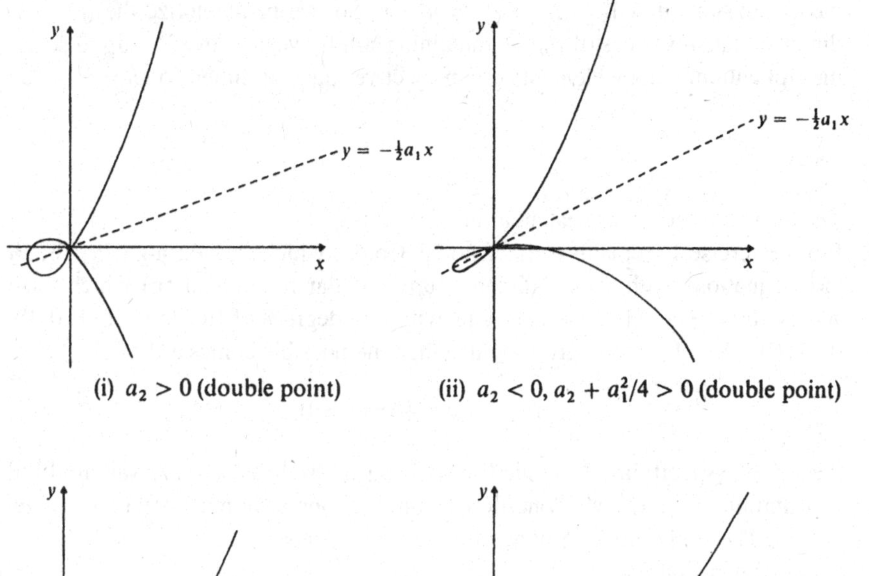

(i) $a_2 > 0$ (double point) (ii) $a_2 < 0, a_2 + a_1^2/4 > 0$ (double point)

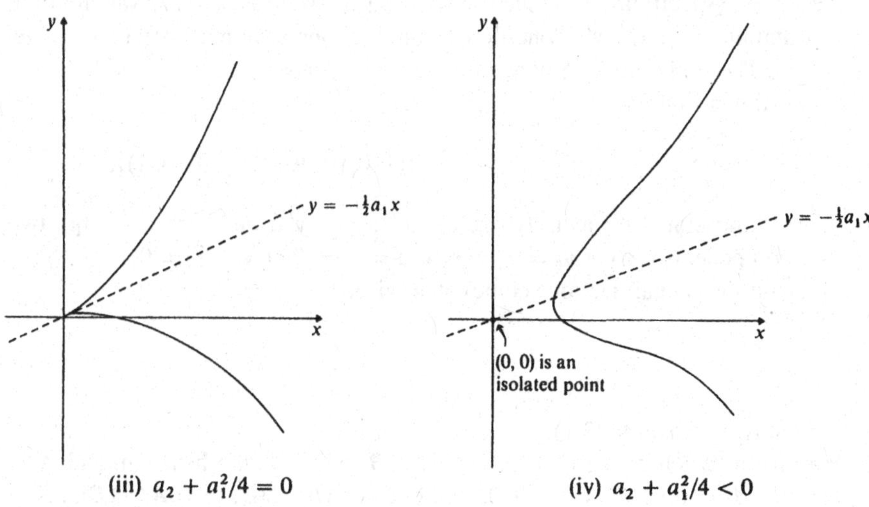

(iii) $a_2 + a_1^2/4 = 0$ (iv) $a_2 + a_1^2/4 < 0$

CHAPTER 2, §1

1. There are $(q + 1)$ points on a line and $(q^2 + q + 1)$ lines in $\mathbb{P}_2(k)$. There are

$$\frac{(q^r - 1) \cdots (q^r - q^{s-1})}{(q^s - 1) \cdots (q^s - q^{s-1})}$$

s-dimensional subspaces in $\mathbb{P}_r(k)$.
2. Any dimensional from $\max(0, s_1 + s_2 - r)$ to $\min(s_1, s_2)$ can occur.
3. Each $(r + 1)$-dimensional subspace, M, of $\mathbb{P}_n(k)$ corresponds to $M^* \subseteq k^{n+1}$ containing M_0^+ and of dimension $(r + 2)$. Such a subspace is specified by that vector $v \in M^+ \cap (M_0^+)^\perp$ (unique up to scalar multiplication). However $(M_0^+)^\perp$

has dimension $(n+1)-(r+1)=n-r$. So, we parameterize the $(r+1)$-dimensional subspaces of $\mathbb{P}_n(k)$ containing M_0 by vectors in k^{n-p} up to scalar multiplication, i.e., by elements of a projective space of dimension $n=r-1$.

CHAPTER 2, §2,

1. This is a straightforward calculation.
2. From Exercise 1, $\mathbb{H}_2^2$ is five-dimensional. Conics which pass through P_1, P_2, P_3 have equations with six coefficients (up to scalar multiplication) which must satisfy three relations. This leaves us with two degrees of freedom. If $(1,0,0)$, $(0,1,0)$, $(0,0,1)$ are the given points, then the possible conics are

$$axy+bwx+cwy=0$$

with a, b, c specifying the conic (up to scalar multiplication). The subfamily of S containing $w' : x' : y'$ is one-dimensional as long as at most one of x', y', w' is zero. The subfamily of S containing $w' : x' : y'$ and $w'' : x'' : y''$ is:

two-dimensional if

$$(w' : x' : y'), (w'' : x'' : y'') \in \{(1,0,0), (0,1,0), (0,0,1)\};$$

one-dimensional if precisely one of $(w' : x' : y')$, $(w'' : x'' : y'')$ has two components zero or $w=w'=0$ or $x=x'=0$ or $y=y'=0$;

zero-dimensional (i.e., one conic) otherwise.

CHAPTER 2, §3

1. Similar to Proposition (3.1).
2. We give a sketch of the proof. Suppose $ABCDEF$ is the hexagon, and $A=(1,0,0)$, $B=(0,1,0)$, $C=(0,0,1)$. Let $R=AB\cap DE$, $Q=AF\cap DC$, $P=BC\cap EF$. Then if $D=(d_1,d_2,d_3)$, etc.,

$$R=(d_1e_3-d_3e_1, d_2e_3-d_3e_2, 0),$$
$$P=(0, e_1f_2-e_2f_1, e_1f_3-e_3f_1),$$

and

$$Q=(d_1f_2, d_2f_2, d_2f_3).$$

Since

$$d_1f_2(e_1f_3-e_3f_1)R+d_2f_3(d_1e_3-d_3e_1)P+(d_1e_3-d_3e_1)(f_1e_3-f_3e_1)Q = 0$$

thus P, Q, R are collinear.

3. The same techniques as in Exercise 2 are used. This is somewhat more straightforward than Exercise 2.

CHAPTER 2, §4

1. Follows from the definitions in a straightforward way.
2. Family of curves through P with order $\leq r$ in $\mathbb{H}_2^m$ has dimension

$$\frac{1}{2}(r+1)(r+2).$$

The subfamily of $\mathbb{H}_2^m$ consisting of curves through P_i with order $\leq r_i$ at P_i for $i = 1, 2, \ldots, t$ has dimension at least

$$\sum_{i=1}^{t}(1/2)(r_i+1)(r_i+2) - (1/2)m(m+3).$$

3. Transform so that (w, x, y) is at $(1, 0, 0)$. Then $(1, 0, 0)$ has order r on C_f and s on C_g. Thus, a_m has terms of degree $\geq r + m' - m$ only when $m' = m - r + 1, \ldots, m$. Similarly for the b_i. So $R(f, g)$ is homogeneous in w, x and of degree mn and divisible by x^{rs}. Hence result.
4. Use Exercise 3.
5. Apply Exercise 4 to f'. For the last part, the case in which we get n lines through one point, gives only one r_i, namely n; and $m = n$.
6. Consider the family of curves of degree $m-1$ (containing f') and the subfamilies of those curves of degree r_i–1 at P_i. Thus subfamily has dimension $(1/2)r_i(r_i{+}1)$ and the original family has dimension $(1/2)(m-1)(m+2)$.
7. If f is reducible, then $f = gh$ where g, h are of degree 1. Recall the formula $\nabla f = h\nabla g + g\nabla h$. So ∇f is a linear combination of ∇g, ∇h. Note that ∇g, ∇h are constant vectors. Thus $\partial f/\partial x_1, \partial f/\partial x_2, \partial f/\partial x_3$ satisfy a linear relation with constant coefficients, namely $\nabla f \cdot (\nabla g \wedge \nabla h) = 0$. Hence $\partial^2 f/\partial x_i \partial x_j$ with $i = 1, 2, 3$ satisfy a common relation for all j, and thus $\det(\partial^2 f/\partial x_i \partial x_j) = 0$.

   Conversely, suppose $\det(\partial^2 f/\partial x_i \partial x_j) = 0$. Let $\beta_{ij} = \partial^2 f/\partial x_i \partial x_j$. Now diagonalize β_{ij} (possible since $\beta_{ij} = \beta_{ji}$). One of the diagonal elements is zero, since the determinant is zero. Thus we get a diagonal form

$$\begin{pmatrix} \beta_{11} & 0 & 0 \\ 0 & \beta_{22} & 0 \\ 0 & 0 & 0 \end{pmatrix}$$

   However, by a version of Euler's theorem for second derivatives.

$$\begin{aligned} 2f &= \sum_{i,j} \beta_{ij}x_ix_j = \beta_{11}x_1^2 + \beta_{22}x_2^2 \\ &= (\sqrt{\beta_{11}}x_1 + \sqrt{-\beta_{22}}x_2)(\sqrt{\beta_{11}}x_1 - \sqrt{-\beta_{22}}x_2). \end{aligned}$$

   This factorization shows that f is reducible.

8. The assertion does not hold in general for characteristics p unless we impose a condition on f, e.g., it has degree $\leq (p-1)$.
9. Same answer as for Exercise 8.
10. If $\mathbb{A}$ is the matrix of $\partial^2 f/\partial x_i$ then $\det \mathbb{A}$ is homogeneous of degree ≥ 3 (it cannot be identically 0 by Exercise 7). Thus $\det \mathbb{A} = 0, f = 0$ has at least one common solution, and hence we get at least one flex by Exercise 9.

APPENDIX TO CHAPTER 2

Exercise 1, 2, and 3 are straightforward verifications.

4. If f has a repeated root in an extension field k' of k then f, f' have a common root in k' and so $R(f, f') = 0$ since f, f' have a common factor in $k'(X)$. Thus $D(f) = 0$.

   Conversely, if $D(f) = 0$, then $R(f, f') = 0$ and so f, f' have a common factor $g \in k[X]$ with $\partial g > 0$. Let k' be an extension of k over which g splits completely. Then f, f' have a common linear factor over $k'[X]$. So f, g' have a common root in k', i.e., f has a repeated root in k'.
5. A straightforward calculation gives

$$D(ax^2 + bx + c) = -a(b^2 - 4ac) \quad \text{and} \quad D(x^2 + px + q) = 27q^2 + 4p^3.$$

CHAPTER 3, §1

These questions are quite tedious to do completely. They just consist of many cases. For Exercise 2, we consider $O, P, Q, PP, PQ + P, P + Q, P(P + Q), (P + P)Q$. The latter two are to be shown equal. Using Theorem (3.3), $(P + P)Q$ is one of the other eight points. We must eliminate all other possibilities. For example, $(P + P)Q = P$ implies $P + P = PQ$. Then

$$P(P + Q) = P(O \cdot PQ) = P(O \cdot (P + P)) = P(PP) = P, \quad \text{as required.}$$

CHAPTER 3, §2

Both follow by straight algebra, Substitute the relations in (2.3) into the equation in (2.4).

CHAPTER 3, §3

1. Discriminant is $27c^2 - 18abc + 4a^3c + 4b^3 - a^2b^2$ (obtained by evaluating a 5×5 determinant).
2. Discriminant is $(\alpha_1 - \alpha_2)^2(\alpha_2 - \alpha_3)^2(\alpha_3 - \alpha_1)^2$. This is obtained as follows. It is a homogeneous polynomial of degree 6 which vanished whenever two of α_i's coincide. Thus it is divisible by $(\alpha_1 - \alpha_2)(\alpha_2 - \alpha_3)(\alpha_3 - \alpha_1)$. It is zero only if two α_i's coincide, giving $D(f) = c(\alpha_1 - \alpha_2)^2(\alpha_2 - \alpha_3)^2(\alpha_3 - \alpha_1)^2$ some constant c. Use the special case $f = x^3 - x$ to get $c = 1$.

3. Evaluation of a 7×7 determinant gives discriminant

$$256c^3 - 51b^4 - 20a^3b^2 + 80ab^2c + 96a^2c^2 - 32a^4c - 16a^3c + 16ab^3c - 48a^2bc^2.$$

4. (a) $j = -4906/11$, $\Delta = -11$.
   (b) $j = 4096/43$, $\Delta = -43$.
   (c) $j = (27/37)(4096)$, $\Delta = 37$.
   (d) $j = (27/91)(4096)$, $\Delta = -91$.
   (e) $j = (27 \times 10^6/79634)$, $\Delta = -28$.
   (f) $j = 0$, $\Delta = -3^9$.
5. General formula: $16b^2(a^2 - 4b)$. So we get (a) 80; (b) -48; (c) 128; (d) 230400.

CHAPTER 3, §5

1. $\Delta = -16D$. To simplify algebra, shifting to give $a_2 = 0$ is convenient. Then $D = 27a_6^2 + 4a_4^3$ and $\Delta = 9b_2b_4b_6 - 27b_6^2 - 8b_4^3 - b_2^2b_8$ reducing to $-16D$.
2. A suitable elliptic curve is $y^2 = x^3 - x$. It has $\Delta = -1$, $j = 0$. In $\mathbb{F}_3$, it has graph:

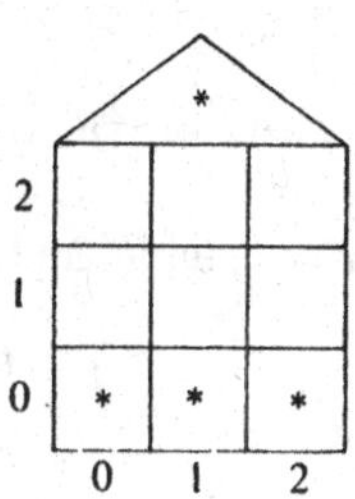

CHAPTER 3, §6

1. It is found that E_3, E_4 have four points while E_5 has eight points. So $E_3 \not\cong E^5$ while $E_3 \cong E_4$ under a shift.
2. The only part that needs checking is inverses. In $\mathbb{F}_{16}$, the relation given leads to $(a + bv)^{-1} = (a' + b'v)$ where

$$a' = (b - 1)/(wb^2 + ab - a^2),$$
$$b' = b/(wb^2 + ab - a^2),$$

and $wb^2 + ab - a^2 \neq 0$, as can be easily verified. A slightly more complicated version of this holds in $\mathbb{F}_{256}$.
3. Going back to (2.4) gives us that for an isomorphism between E_3 and E_5, $t^2 + t + v^3 = 0$ must be soluble. In $\mathbb{F}_{256}$, $t = w$ is a solution. In $\mathbb{F}_{16}$, there are no such solutions.

4. This question is best answered by writing down (2.4) with $a_i = \bar{a}_i$ in each of the cases $E_i (i = 1, 2, 3, 4, 5)$ separately. We then find that E_1, E_2 give automorphism groups $\mathbb{Z}/2\mathbb{Z}$ over any field of characteristic 2, of the identity and

$$\left.\begin{aligned} x &= \bar{x} \\ y &= \bar{x} + \bar{y} \end{aligned}\right\}.$$

However E_3, E_4, E_5 are more complicated, and the different k fields must be considered separately. In fact, E_3, E_4 give an identical set of equations and thus the same automorphism groups,

$$\begin{aligned} &\text{over } \mathbb{F}_2, \mathbb{F}_4: && \mathbb{Z}/4\mathbb{Z}, \\ &\text{over } \mathbb{F}_{16}, \mathbb{F}_{256}: && \text{24-element group.} \end{aligned}$$

Also E_5 gives automorphism groups,

$$\begin{aligned} &\text{over } \mathbb{F}_2: && \mathbb{Z}/2\mathbb{Z}, \\ &\text{over } \mathbb{F}_4, \mathbb{F}_{16}, \mathbb{F}_{256}: && \text{24-element group, } G. \end{aligned}$$

In this latter case, (2.4) gives:

$$\begin{aligned} x &= v\bar{x} + r, \\ y &= \bar{y} + r^2 v\bar{x} + t, \end{aligned}$$

where $v = u^2$, $s = r^2$. Here, $u = 1$, w, w' and either

$$r = 0, \qquad t = 0, 1$$

or

$$r \in \mathbb{F}_4^*, \qquad t = w, w'.$$

Let $\alpha_{v,r,t}$ be the above transformation. Then

$$\alpha_{v,r,t} \circ \alpha_{vv', vr'+r, t+t'+r^2r'v}.$$

When $v = 1$, we get a subgroup of eight elements isomorphic to $\mathbb{H}$ (the quaternion group $\{\pm 1, \pm i, \pm j, \pm k\}$). In this isomorphism:

$$\begin{aligned} 1 &\to \alpha_{1,0,0} & -1 &\to \alpha_{1,0,1} \\ i &\to \alpha_{1,1,w} & -i &\to \alpha_{1,1,w'} \\ j &\to \alpha_{1,w,w} & -j &\to \alpha_{1,w,w'} \\ k &\to \alpha_{1,w',w} & -k &\to \alpha_{1,w',w'} \end{aligned}$$

and

$$\alpha_{1,0,1} \circ \alpha_{v,r,t+1} = \alpha_{v,r,t+1} = \alpha_{v,r,t} \circ \alpha_{1,0,1}.$$

In fact, the group of 24 elements in the E_5 case is isomorphic to

$$\left\{\pm 1, \pm i, \pm j, \pm k, \frac{1}{2}(\pm 1 \pm i \pm j \pm k)\right\}$$

(all are units). This can be seen as follows. If

$$l = \frac{1}{2}(1 + i + j + k)$$

then l has order 6.

Also $\alpha_{u,0,0} \circ \alpha_{1,r,t} \circ \alpha_{u,0,0}^{-1} = \alpha_{1,ur,1}$. Thus the subgroup, H, of our 24-element group G isomorphic to $\mathbb{H}$, given by

$$H = \{\alpha_{1,r,t} \mid r = 0, t \in \{0, 1\} \text{ or } r \in \mathbb{F}_4^* \text{ and } t \in \{w, w'\}\}.$$

is a normal subgroup of G.

Now $l^{-1}il = -k$ can easily be verified. Thus conjugation by l^2 permutes i, j, k cyclically, and we can thus correspond

$$l^2 \leftrightarrow \alpha_{w',0,0}$$

and since $\alpha^2_{v^2,(v+1)r,vr^3}$, thus l can correspond to $\alpha_{w,0,0}$ or to $\alpha_{w,0,1}$. Thus G is a semidirect product of H with $\mathbb{Z}/3\mathbb{Z}$. There are four 3-Sylow subgroups of G, namely those generated by

$$\frac{1}{2}(-1 + i + j + k),\ \frac{1}{2}(-1 + i - j - k),\ \frac{1}{2}(-1 - i - j + k),$$
$$\frac{1}{2}(-1 - i + j - k).$$

These are corresponded to

$$\alpha_{w',0,0} \quad \alpha_{w',w,w} \quad \alpha_{w',1,w} \quad \alpha_{w',w',w}$$

and

$$\frac{1}{2}(-1 - i - j - k),\ \frac{1}{2}(-1 - i + j + k),\ \frac{1}{2}(-1 + i + j - k),$$
$$\frac{1}{2}(-1 + i - j + k)$$

are their squares (= conjugates) and are thus corresponded to

$$\alpha_{w,0,0} \quad \alpha_{w,w',w'} \quad \alpha_{w,w,w'} \quad \alpha_{w',1,w'}.$$

So, in the full correspondence,

$$1 \leftrightarrow \alpha_{1,0,0} \qquad -1 \leftrightarrow \alpha_{1,0,1} \qquad i \leftrightarrow \alpha_{1,1,w} \qquad -i \leftrightarrow \alpha_{1,1,w'}$$

$$
\begin{array}{llll}
j \leftrightarrow \alpha_{1,w,w} & -j \leftrightarrow \alpha_{1,w,w'} & k \leftrightarrow \alpha_{1,w',w} & -k \leftrightarrow \alpha_{1,w',w'}
\end{array}
$$

$$
\begin{array}{ll}
\frac{1}{2}(1+i+j+k) \leftrightarrow \alpha_{w,0,1} & \frac{1}{2}(-1-i-j-k) \leftrightarrow \alpha_{w,0,0} \\
\frac{1}{2}(1+i+j-k) \leftrightarrow \alpha_{w',1,w'} & \frac{1}{2}(-1-i-j+k) \leftrightarrow \alpha_{w',1,w} \\
\frac{1}{2}(1+i-j+k) \leftrightarrow \alpha_{w',w',w'} & \frac{1}{2}(-1-i+j-k) \leftrightarrow \alpha_{w',w',w} \\
\frac{1}{2}(1+i-j-k) \leftrightarrow \alpha_{w,w',w} & \frac{1}{2}(-1-i+j+k) \leftrightarrow \alpha_{w,w',w'} \\
\frac{1}{2}(1-i+j+k) \leftrightarrow \alpha_{w',w,w'} & \frac{1}{2}(-1+i-j-k) \leftrightarrow \alpha_{w',w,w} \\
\frac{1}{2}(1-i+j-k) \leftrightarrow \alpha_{w,1,w} & \frac{1}{2}(-1+i-j+k) \leftrightarrow \alpha_{w,1,w'} \\
\frac{1}{2}(1-i-j+k) \leftrightarrow \alpha_{w,w,w} & \frac{1}{2}(-1+i+j-k) \leftrightarrow \alpha_{w,w,w'} \\
\frac{1}{2}(1-i-j-k) \leftrightarrow \alpha_{w,0,1} & \frac{1}{2}(-1+i+j+k) \leftrightarrow \alpha_{w',0,0}
\end{array}
$$

Thus, in the case of E_5, the automorphism group (when it is not restricted by the size of the field) over characteristic 2 is isomorphic to the group

$$
\left\{\pm 1, \pm i, \pm j, \pm k, \frac{1}{2}(\pm 1, \pm i \pm j \pm k)\right\}
$$

of quaternions.

Since E_3, E_4, E_5 are all isomorphic over $\mathbb{F}_{256}$, their automorphism groups are all isomorphic over $\mathbb{F}_{256}$; i.e., the order 24 automorphism groups occurring over fields $\mathbb{F}_{16}$, $\mathbb{F}_{256}$ are isomorphic to the above group of quaternions.

The interested reader might like to consider the case of elliptic curves with $j = 0$ over fields of characteristic 3. In that case, we get a group of order 12 over $\mathbb{F}_9$.

5. The elliptic curves over $\mathbb{F}_4$ up to isomorphism over $\mathbb{F}_4$ are:

$$
\begin{aligned}
j = 0: \quad & y^2 + y = x^3, \\
& y^2 + y = x^3 + w, \\
& y^2 + y = x^3 + x + w, \\
& y^2 + y = x^3 + x, \\
& y^2 + y = x^3 + wx, \\
& y^2 + y = x^3 + wx + w, \\
& y^2 + y = x^3 + w'x, \\
& y^2 + y = x^3 + w'x + w, \\
& y^2 + wy = x^3,
\end{aligned}
$$

$$
\begin{aligned}
& & y^2 + wy &= x^3 + w, \\
& & y^2 + w'y &= x^3, \\
& & y^2 + w'y &= x^3 + w', \\
j = 1: & & y^2 + xy &= x^3 + 1, \\
& & y^2 + xy &= x^3 + wx^2 + 1, \\
j = w: & & y^2 + xy &= x^3 + w', \\
& & y^2 + xy &= x^3 + wx^2 + w', \\
j = w': & & y^2 + xy &= x^3 + w, \\
& & y^2 + xy &= x^3 + wx^2 + w.
\end{aligned}
$$

Over $\mathbb{F}_{16}$, these reduce to:

$j = 0: \quad y^2 + y = x^3$ (isomorphic to all above curves with $j = 0$, $a_3 = 1$),

$$
\begin{aligned}
y^2 + wy &= x^3, \\
y^2 + wy &= x^3 + w, \\
y^2 + w'y &= x^3, \\
y^2 + w'y &= x^3 + w'.
\end{aligned}
$$

$j = 1: \; y^2 + xy = x^3 + 1$ (isomorphic to $y^2 + xy = x^3 + wx^2 + 1$).
$j = w: \; y^2 + xy = x^3 + w'$ (isomorphic to $y^2 + xy = x^3 + wx^2 + w'$).
$j = w': \; y^2 + xy = x^3 + w$ (isomorphic to $y^2 + xy = x^3 + wx^2 + w$).

6. The elliptic curves over F_3 up to isomorphism over F_3 are:

$$
\begin{aligned}
j = 0: \quad & \left.\begin{aligned} y^2 &= x^3 + x \\ y^2 &= x^3 - x \end{aligned}\right\} \text{isomorphic over } F_9 \text{ (16 elements)}, \\
& \left.\begin{aligned} y^2 &= x^3 - x + 1 \\ y^2 &= x^3 - x - 1 \end{aligned}\right\} \text{isomorphic over } F_9 \text{ (7 elements)}, \\
j = 1: \quad & \left.\begin{aligned} y^2 &= x^3 - x^2 + 1 \\ y^2 &= x^3 + x^2 - 1 \end{aligned}\right\} \text{isomorphic over } F_9 \text{ (15 elements)}, \\
j = -1: \quad & \left.\begin{aligned} y^2 &= x^3 + x^2 + 1 \\ y^2 &= x^3 - x^2 - 1 \end{aligned}\right\} \text{isomorphic over } F_9 \text{ (12 elements)},
\end{aligned}
$$

7. We find that when $k = F_3$,

$$
\mathrm{Aut}_k(E) = \begin{cases} \mathbb{Z}/2\mathbb{Z} & \text{for curves with } j = 1, \\ \mathbb{Z}/2\mathbb{Z} & \text{for } y^2 = x^3 + x, \\ S_3 & \text{for } y^2 = x^3 - x + \alpha, (\alpha = 0, \pm 1), \end{cases}
$$

and with the notation $V_4 = \mathbb{Z}/2\mathbb{Z} \times \mathbb{Z}/2\mathbb{Z}$

$$\mathrm{Aut}_{F_9}(E) = \begin{cases} \mathbb{Z}/2\mathbb{Z} & \text{for } j \neq 0, \\ \mathbb{Z}/4\mathbb{Z} & \text{for } y^2 = x^3 + x, \\ V_4 \times \mathbb{Z}/3\mathbb{Z} & \text{for } y^2 = x^3 - x, \\ \mathbb{Z}/6\mathbb{Z} & \text{for } y^2 = x^3 - x + 1, \\ \mathbb{Z}/6\mathbb{Z} & \text{for } y^2 = x^3 - x - 1, \end{cases}$$

8. We think of F_n as a cubic extension of F_2 formed by adjoining w such that $w^3 + w + 1 = 0$. We find that noncyclic groups occur only over F_{16}, F_{256} in this example. For

$i = 1:$ F_8 gives $\mathbb{Z}/14\mathbb{Z}$
$i = 2:$ F_8 gives $\mathbb{Z}/4\mathbb{Z}$
} F_{16} gives a 16-element group with a subgroup $\mathbb{Z}/8\mathbb{Z}$ of $\left\{0, (0, 1), \left(1, \frac{w}{w'}\right), \left(2, \frac{1}{w'}\right), \left(w', \frac{1}{w}\right)\right\}$ for E.

$i = 3:$ F_8 gives $\mathbb{Z}/5\mathbb{Z}$
$i = 4:$ F_8 gives 0
} F_{16} gives a 25-element group with subgroup of order 5 given by $\{0, (0, 0), (0, 1), (1, 0), (1, 1)\}$.

$i = 5:$ F_8 gives $\mathbb{Z}/9\mathbb{Z}$; F_{16} gives $\left\{(0, 0), (0, 1), \begin{pmatrix} 1 & \\ w, & \frac{w}{w'} \\ w & \end{pmatrix}, 0\right\}$.

This exercise can become a bit tedious: we only need to check F_8, F_{16} here.

9. The elliptic curves over F_5 are:

$j = 0:$	$y^2 = x^3 + 1(\mathbb{Z}/6\mathbb{Z})$
	$y^2 = x^3 + 2(\mathbb{Z}/6\mathbb{Z})$
$j = 12^3:$	$y^2 = x^3 + x(\mathbb{Z}/2\mathbb{Z} \times \mathbb{Z}/2\mathbb{Z})$
$(= 3)$	$y^2 = x^3 + 2x(\mathbb{Z}/2\mathbb{Z})$
	$y^2 = x^3 + 3x(\mathbb{Z}/10\mathbb{Z})$
	$y^2 = x^3 + 4x(\mathbb{Z}/2\mathbb{Z} \times \mathbb{Z}/2\mathbb{Z})$
$j = 1:$	$y^2 = x^3 + x + 2(\mathbb{Z}/4\mathbb{Z})$
	$y^2 = x^3 + 4x + 1(\mathbb{Z}/8\mathbb{Z})$
$j = 2:$	$y^2 = x^3 + 4x + 2(\mathbb{Z}/3\mathbb{Z})$
	$y^2 = x^3 + x + 1(\mathbb{Z}/9\mathbb{Z})$
$j = 4:$	$y^2 = x^3 + 2x + 4(\mathbb{Z}/7\mathbb{Z})$
	$y^2 = x^3 + 3x + 2(\mathbb{Z}/5\mathbb{Z})$

CHAPTER 4, §1

These exercises are routine algebra, using the definitions of the b_i, c_i, Δ, and j.

CHAPTER 4, §2

Here p, q are clearly inverses. Thus one only needs to check that $q = 2p$. This is done by showing that the tangent at p cuts the elliptic curve at a triple point (namely p).

CHAPTER 5, §1

1. It is easy to verify that C is nonsingular over $\mathbb{Q}$ and that $\overline{C}$ is singular. A suitable pair P, P' is

$$P \equiv (p^2, p, 1), \qquad P' = (p, p^2, 1),$$

and then L is $w + x = (p + p^2)y$. Thus $\bar{l}$ is $w + x = 0$ and all the conditions are satisfied.

CHAPTER 5, §2

1. The required condition is that

$$\mathrm{ord}_p(\Delta) + \min(0, \mathrm{ord}_p(j)) < 12 + 12\delta_{2p} + 6\delta_{3p}$$

for all primes p. It is easily verified that

$$\Delta = -16(4a^3 + 27b^2),$$
$$j = (3^3 \cdot 2^8)a^3/(4a^3 + 27b^2)$$

for $y^2 = x^3 + ax + b$. Thus, for $y^2 = x^3 + ax$, $\Delta = -2^6a^3$, and $j = 3^3 \cdot 2^6$. This gives the conditions for minimality:

$$\left.\begin{array}{ll} \mathrm{ord}_2(a) < 6 & \\ \mathrm{ord}_3(a) < 6 & \\ \mathrm{ord}_p(a) < 4, & \forall p \neq 2, 3 \end{array}\right\}.$$

For $y^2 = x^3 + a$, $\Delta = -2^4 \cdot 3^3 a^2$, $j = 0$. This gives the condition for minimality

$$\left.\begin{array}{ll} \mathrm{ord}_2(a) < 10 & \\ \mathrm{ord}_3(a) < 8 & \\ \mathrm{ord}_p(a) < 6, & \forall p \neq 2, 3 \end{array}\right\}.$$

CHAPTER 5, §3

The first four questions in this section are all very similar. We therefore give only the answers, and in Exercise 4(c) we give a complete solution.

1. (a) Never get bad reduction at prime p, unless $a \equiv 0 \mod p$. So the primes p at which bad reduction occurs are those which divide a.

(b) Bad reduction occurs at primes $p \neq 3$ as long as p divides a, and occurs at $p = 3$ always.
(c) Bad reduction occurs at $p = 13$.
(d) Bad reduction occurs at $p = 3, 11$.

2. (a) $p = 5$.
(b) $p = 3$.
(c) We never get bad reduction.
(d) $p = 3, 5$.
3. (a) $p = 37$. Modulo 2 gives E_3 in 3(6.4) and modulo 3 gives $y^2 = x^3 - x + 1$.
(b) $p = 43$. Modulo 2 gives E_3' in 3(6.4) and modulo 3 gives $y^2 = x^3 + x^2 + 1$.
(c) $p = 91$. Modulo 2 gives E_3 in 3(6.4) and modulo 3 gives $y^2 = x^3 + x + 1$.
(d) $p = 3$. Modulo 2 gives E_5'.
(e) $p = 2, 53$. Modulo 3 gives $y'^2 = x^3 - x + 1$.
(f) $p = 5, 17, 31$. Modulo 2 gives E_1 and modulo 3 gives $y^2 = x^3 + 1$.
4. (a) $p = 3, 7$ give good reduction and $p = 2, 5$ give bad reduction.
(b) $p = 2, 3$ give good reduction and $p = 5, 7$ give bad reduction.
(c) Here, $y^2 + xy + y = x^3 - x^2 - 3x + 3$. Thus

$$(2y + x + 1)y' = 3x^2 - 2x - 3 - y$$

and bad reduction at an odd prime p requires

$$2y + x + 1 = 0 \quad \text{i.e.} \quad y = -\frac{1}{2}(x + 1)$$

and

$$3x^2 - 2x - 3 = y.$$

This is never satisfied for odd p, and (x, y) on the curve. So, we consider $p = 2$. This gives $y^2 + x'y = x'^3$ where $x' = x + 1$. Thus bad reduction occurs at $p = 2$ and good reduction occurs for $p = 3, 5, 7$.
(d) $p = 2, 5, 7$ give good reduction and $p = 3$ gives bad reduction.
(e) $p = 5$ gives good reduction and $p = 2, 3, 7$ give bad reduction.
5. The subgroup, G, generated by $(0, 0)$ has image under r_2 which is

$$\{(0, 0), (1, 1), (1, 0), (0, 1), 0\}.$$

6. The discriminant is 5077 and this is easily seen to be prime.

CHAPTER 5, §4

1. These follow very simply from the relation for $x + x' + x''$ given in (4.3) since $\text{ord}_p(x + x' + x'') \geq 3n$ as $a_1 = 0$.
2. It is quite clear that R_p is a maximal ideal in R iff $R_{(p)}$ forms an additive group, and thus $R_{(p)} = R$. Uniqueness of this maximal ideal follows by assuming I and $J = R_p$ to be two distinct maximal ideals. Thus $I \not\subseteq J$, $J \not\subseteq I$. So, there exists $a_0 \in I$ such that $\text{ord}_p a_0 = 0$, and then $\forall a \in k$,

$$\mathrm{ord}_p(aa_0) \geq 1 \quad \text{if and only if} \quad \mathrm{ord}_p a \geq 1.$$

But,

$$aa_0 \in I \qquad \text{for all } a \in R.$$

So, $I = R$ because a_0 is invertible, a contradiction. It is now easy to show that $I + Rp \subseteq R^*$. We then get

$$k^*/R^* \to \mathbb{Z},$$
$$a + R^* \to \max_{r \in R^*}\{\mathrm{ord}_p(a+r)\}.$$

Reduction mod p maps $R^* \to k(p)^*$ with kernel $1 + Rp$. The canonical map $R \to R/Rp$ composed with the map:

$$\left.\begin{array}{l} 1 + Rp^n \to R \\ 1 + ap^n \to a \end{array}\right\}$$

produces a map $1 + Rp^n \to R/Rp$. The kernel consists of $1 + Rp^{n+1}$. Thus $(1 + Rp^n)/(I + Rp^{n+1}) \simeq R/Rp$ by the first isomorphism theorem.

3. This is very straightforward from the definitions.

Chapter 5, §5

1. For the last part, note that good reduction occurs mod p for all odd primes other than 3. When E is reduced mod 5, we get a group of order 6. Since $E(\mathbb{Q})_{\text{tors}}$ has no elements of order 2, all elements must be of order 3. Hence $E(\mathbb{Q})_{\text{tors}} \simeq \mathbb{Z}/3\mathbb{Z}$.
2. Modulo 2 and modulo 3 give no singularities. So, there are injections from $E(\mathbb{Q})_{\text{tors}}$ into $\mathbb{Z}/3\mathbb{Z}, \mathbb{Z}/7\mathbb{Z}$. Hence $E(\mathbb{Q})_{\text{tors}} = 0$.
3. Modulo 2, E contains just 0. Modulo 3, we get no singularities and the group is $\mathbb{Z}/4\mathbb{Z}$. So, $E(\mathbb{Q})_{\text{tors}}$ is thus 0 or $\mathbb{Z}/2\mathbb{Z}$. However, $(2, -1)$ has order 2, and so $E(\mathbb{Q})_{\text{tors}} = \mathbb{Z}/2\mathbb{Z}$.
4. Here, $E(\mathbb{Q})_{\text{tors}} = \mathbb{Z}/7\mathbb{Z}$ since modulo 3 gives $\mathbb{Z}/7\mathbb{Z}$ and $(1, 0)$ has order 7.
5. Here $\mathrm{GL}_n(\mathbb{Z}$ denotes the group of $n \times n$ matrices with entries in $\mathbb{Z}$ and determinant ± 1. Thus $\mathrm{GL}_n(\mathbb{Z}$ maps into $\mathrm{GL}_n(\mathbb{Z}/q\mathbb{Z})$ under r_q. Showing that $(I_n + p^a X)^n \equiv I_n + np^a X \mod p^{a+b+1}$. Thus we find that $\binom{n}{rp^{ar}}$ is divisible by p^{a+b+1} whenever $r \geq 2$.

   If $G \subseteq \mathrm{GL}_n(\mathbb{Z}$ is finite and $A \in G \cap \ker r_p$, then A has finite order. Thus, as $A \in \ker r_p$, $A = I + X$, and $A^m = I + mX \mod p^{b+1}$. So, $X = 0$ since the A^m cannot all be distinct. This gives $G \cap \ker r_p \doteq 1$, for all $p > 2$. For $p = 2$, $A^m = I_n + 2mX \mod p^{\mathrm{ord}_p(m)+2}$ and so $G \cap \ker r_4 = 1$ for $p = 2$.
6. It is easily seen that $p = 3, 5$ give good reduction, and groups $\mathbb{Z}/7\mathbb{Z}$ $(p = 3)$ and a group of order 10 $(p = 5)$. So $E(\mathbb{Q})_{\text{tors}} = 0$.

CHAPTER 5, §6

1. It is easily computed that E has discriminant 3. So, any torsion point has $y = 0, \pm 1, \pm 3$. Thus $0, (0,0), (1,1), (1,-1)$ are the torsion points $(\mathbb{Z}/4\mathbb{Z})$. The curve has good reduction modulo 5 and $E(\mathbb{F}_5)$ is a group of eight points:

$$\{0, (0,0), (1, \pm), (2, \pm 1), (-2, \pm 2)\}.$$

2. The torsion points are $0, (0,0)$. The curve has good reduction both modulo 3 and modulo 5, and $E(\mathbb{F}_3) = \mathbb{Z}/6\mathbb{Z}$, $E(\mathbb{F}_5) = \mathbb{Z}/4\mathbb{Z}$. In fact, $E(\mathbb{F}_3) = \{0, (0,0), (\pm 1, \pm 1)\}$, $E(\mathbb{F}_5) = \{0, (0,0), (-2, \pm 1)\}$.
3. Here $E(\mathbb{F}_3) = \mathbb{Z}/6\mathbb{Z} = \{0, (0,0), (\pm 1, \pm 1)\}$. Also, $E(\mathbb{Q})_{\text{tors}} = E(\mathbb{F}_3)$ and so both are $\mathbb{Z}/6\mathbb{Z}$.
4. To prove this, we use the fact that $-2P$ is not a torsion point, so that the tangent line at P cuts the curve at an integer point. This is used to determine that the slope of this line must be integral.

CHAPTER 6, §4

1. The first part is quite easy, as $x + (f)$ generates R_f as an algebra. If f factors as distinct linear factors, $\prod_{i=1}^{n}(x - \alpha_i)$, then we can map $R_f \to k''$ by evaluating at the α_i. When f factors as $\prod_{i=1}^{k}(x - \alpha_i)^{r_i}$, we can map R_f to a direct sum of $R_{x^{r(i)}}$ where $r(i) = r_i$. This then gives the structure of R_f.
2. We use θ_i as our three maps which produce a triple for each $P \in E$. This triple is converted into an element of R_f using $g(r_i)$. Thus, $\operatorname{Im} g \subseteq R^*_{f,1}/(R^*_f)^2$.

CHAPTER 9, §1

1. The proof is by induction on n, and so we assume that it holds for $(n-1)$. For $n = 2$, it obviously holds. A discrete subgroup Γ of $\mathbb{R}^n$ gives $\Gamma' = \Gamma/(\mathbb{R}\omega_1) \subseteq \mathbb{R}^n/(\mathbb{R}\omega_2) \simeq \mathbb{R}^{n-1}$ where $\omega_1 \in \Gamma$ has minimal nonzero absolute value. (This holds unless $\Gamma = \{0\}$). Thus $\Gamma = \mathbb{Z}\omega_1 + \cdots + \mathbb{Z}\omega_r$ since $\Gamma' = \mathbb{Z}\omega_2 + \cdots + \mathbb{Z}\omega_r$ by inductive assumption, and by discreteness of Γ.

   The condition for compactness of $\mathbb{R}^n/\Gamma$ follows since $\mathbb{R}^n/\Gamma$ is $\mathbb{R}^{n-r} \times T_r$ where T_r is torus.
2. This follows from the last exercise.
3. Suppose that f is analytic, so that $f, g : \mathbb{C} \to \mathbb{C}$ and $f \circ g = g \circ f = \text{id}$. We can extend to maps $\hat{\mathbb{C}} \to \hat{\mathbb{C}}$ where $\hat{\mathbb{C}} = \mathbb{C} \cup \{\infty\}$ by

$$f(\infty) = g(\infty) = \infty.$$

   Hence $f(z) = (az + b)/(cz + d)$.

   Since $f(\infty = \infty, c = 0$, and so $f(z) = az + b$ with $a \neq 0$. Since f is a homomorphism,

$$f(z_1 + z_2) = f(z_1) + f(z_2).$$

   Thus $b = 0$, and so $f(z) = \lambda z$ for some $\lambda \in \mathbb{C}$.

CHAPTER 9, §3

1. We consider $d/dz(\zeta(z+\omega) - \zeta(z))$ and show it vanishes for all z. Thus $\zeta(z+\omega) - \zeta(z)$ is constant, independent of z, and is $\eta(\omega)$ for some function η of ω.
2. Integrate ζ around a rectangle from x_0 to $x_0+\omega_1, x_0+\omega_1+\omega_2$, and $x_0+\omega_2$. Then we get $2\pi i$ by the calculus of residues. Integrating around pairs of opposite sides will give the result (using Exercise 1).
3. To show that σ is entire, we only need to show that $\ln(1-z/\omega)+z/\omega+z^2/2\omega^2$ forms a convergent sum over $\omega \in L - \{0\}$.
4. In this exercise, $\sigma(z+\omega_i) = \sigma(z)e^{\eta_i z}A_i$ some constant A_i from Exercises 1 and 3. Also, σ is odd, and so $A_i = -e^{-\eta_i\omega_i/2}$. Hence the result.
5. Since any elliptic function has only a finite number of zeros and poles a_i, b_i, say, we have

$$f(z) = \prod_{i=1}^{n} \frac{\sigma(z-b_i)}{\sigma(z-a_i)}$$

containing no zeros or poles and also periodic. Thus we get a constant function as required.

CHAPTER 9, §4

1. Since h is a group isomorphism and $A+B = -A\cdot B$, therefore

$$(1, \wp(z_1), \wp'(z_1)), (1, \wp(z_2), \wp'(z_2)), -(1, \wp(z_1+z_2), \wp'(z_1+z_2))$$

are collinear. This gives the result.
2. To check the first result, we note that both sides, as a function of z_1 are periodic with poles when $z_1 \in L$. These poles are double poles with equal coefficients of $1/z_1^2$ near 0 (namely 1). So, they can differ at most by a constant. At $z_1 = z_2$, both sides agree. Hence the result follows. As $z_2 \to z_1$, we obtain the second part.
3. Differentiate the first result in Exercise 2.
4. Apply Exercise 3 with z_1, z_2 interchanged, and add to the result of Exercise 3. Simplifying gives the result. To derive the addition formulae, we need only derive (1).

CHAPTER 9, §5

1. We have

$$\begin{aligned}
&\Gamma(s)\Gamma(1-s)\\
&\quad = \left(\int_0^\infty x^{s-1}e^{-x}\,dx\right)\left(\int_0^\infty x^{-s}e^{-x}\,dx\right)\\
&\quad = \left(\int_0^\infty 2x^{2s-1}e^{-x^2}\,dx\right)\left(\int_0^\infty 2y^{-2s+1}e^{-y^2}\,dy\right)
\end{aligned}$$

$$= \int_0^\infty \int_0^{\pi/2} 4e^{-r^2} (\cos\theta)^{2s-1} (\sin\theta)^{1-2s} r \, d\theta \, dr \quad \text{putting } r^2 = x^2 + y^2$$

$$= 2 \int_0^{\pi/2} (\tan\theta)^{1-2s} \, d\theta$$

$$= \int_0^\infty u^{-s} \, du/(1+u) \quad (u = \tan^2\theta).$$

Integrating $u^{-s}/(1+u)$ around the contour below, since $0 < \mathrm{Re}(s) < 1$ so thus the integrals around circles of radii ε, R about 0 tend to 0 as $\varepsilon \to 0$, $R \to \infty$. The only enclosed pole is at -1, with residue $e^{-\pi is}$. Thus

$$2\pi i e^{-\pi is} = \Gamma(s)\Gamma(1-s)(1 - e^{-2i\pi s})$$

and so $\Gamma(s)\Gamma(1-s) = \pi/\sin \pi s$. Q.E.D.

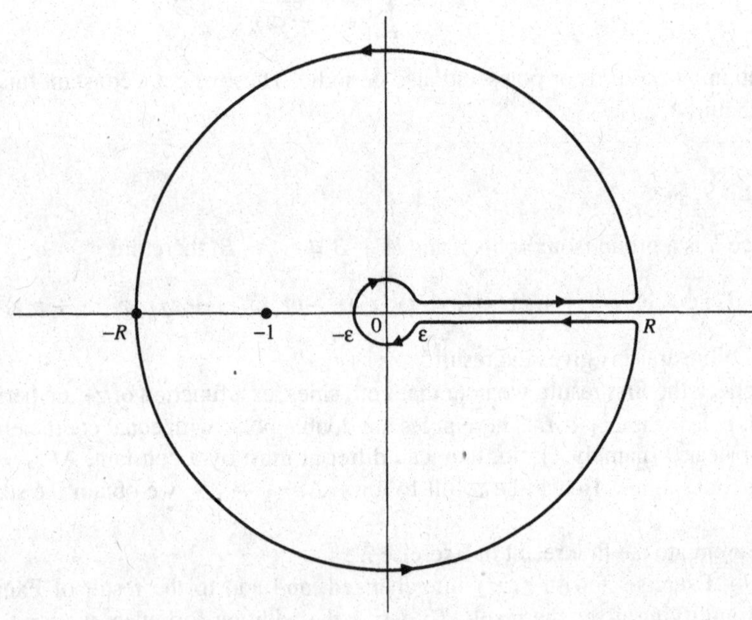

Questions 2, 3, 4 follow straight from the definitions of the hypergeometric function.

References

Before giving a more complete bibliography, we assemble those references arising frequently in the text or which have had a special impact on the development of this book.

Ahlfors, L.: *Complex Analysis*, 2nd ed., McGraw-Hill, 1996.

Birch, B., Swinnerton-Dyer, H. P. F.: Notes on ellipti̧c curves, (I), *J. Reine Angew. Math.*, **212**, 7–25 (1963).

Birch, B., Swinnerton-Dyer, H. P. F.: Notes on elliptic curves, (II), *J. Reine Angew. Math.*, **218**, 79–108 (1965).

Cassels, J. W. S.: Diophantine equations with special reference to elliptic curves, *J. London Math. Soc.*, **41**, 193–291 (1966). (This reference contains an extensive bibliography of the classical literature.)

Koblitz, N.: *Introduction to Elliptic Curves and Modular Forms*, Springer-Verlag. 1984.

Lang, S.: *Algebraic Number Theory*, Addison-Wesley, 1970.

Mumford, D.: *Abelian Varieties*, Oxford University Press, 1974.

Mumford, D.: *Tata Lectures on Theta, I* and *II*, Progress in Mathematics, Vol. 28 and 43, Birkhäuser, 1983, 1984.

Serre, J.-P.: *Abelian l-adic Representations and Elliptic Curves*, Benjamin, 1968.

Serre, J.-P.: Proprietes galoisiennes des points d'ordre fini des corbes elliptiques, *Invent. Math.* **15**, 259–331 (1972).

Sierre, J.-P.: *A course in Arithmetic*, Springer-Verlag, 1973.

Silverman, J.: *The Arithmetic of Elliptic Curves*, Graduate Texts in Mathematics, 106, Springer-Verlag, 1986.

Silverman, J.: *Advanced Topics in the Arithmetic of Elliptic Curves*, Graduate Texts in Mathematics, 155, Springer-Verlag, 1994.

Tate J.: The arithmetic of elliptic curves, *Invent. Math.* **23**, 179–206 (1974).

Ahlfors, L.: *Complex Analysis*, 2nd ed., McGraw-Hill, 1966.

Amice, Y., Vélu, J.: Distributions p-adiques associées aux séries de Hecke, *Astérisque*, **24/25**, 119–131 (1975).

Apostol, T.: *Modular Functions and Dirichlet Series in Number Theory*, Springer-Verlag, 1976.

Arnaud, B.: *Interpolation p-adique d'un Produit de Rankin*, University of Orsay, 1984 (preprint).

Arthaud, N.: On the Birch and Swinnerton-Dyer conjecture for elliptic curves with complex multiplication, *Compositio Math.* **37**, 209–232 (1978).

Artin, E.: *Galois Theory*, University of Notre Dame Press, 1942.

Artin, M., Swinnerton-Dyer, H. P. F.: The Shafarevitch-Tate conjecture for pencils of elliptic curves on $\mathbb{K}3$ surfaces, *Invent. Math.* **20**, 249–266 (1973).

Atiyah, M., Macdonald, I.: *Introduction to Commutative Algebra*, Addison-Wesley, 1969.

Atiyah, M., Wall, C.: Cohomology of groups. In *Algebraic Number Theory*, J. W. S. Cassels and A. Fröhlich, eds., Academic Press, 1967, pp. 94–115.

Atkin, A. O. L., Lehner, J.: Hecke operators on $\Gamma_0(m)$, *Math, Ann.* **185**, 134–160 (1970).

Atkin, O., Li, W.: Twists of newforms and pseudo-eigenvalues of W-operators, *Invent. Math.* **48**, 221–243 (1978).

Baker, A.: *Transcendental Number Theory*, Cambridge University Press, 1975.

Baker, A., Coates, J.: Integer points on curves of genus 1, *Proc. Camb. Phil. Soc.* **67**, 595–602 (1970).

Barsotti, I.: Analytical methods for abelian varieties in positive characteristic, *Colloque de Bruxelles*, 1962, 77–85.

Birch, B. J.: Cyclotomic fields and Kummer extensions. In *Algebraic Number Theory*, J. W. S. Cassels and A. Fröhlich, eds., Academic Press, 1967, pp. 85–93.

Birch, B. J.: Elliptic curves and modular functions, *Symp. Math. 1st. Alta Mat.* **4**, 27–32 (1970).

Birch, B. J.: Elliptic curves. A progress report, *Proceedings of the 1969 Summer Institute on Number Theory, Stony Brook, New York*, American Mathematical Society, 1971, pp. 396–400.

Birch, B. J.: Heegner points of elliptic curves, *Symp. Math.* **15**, 441–445 (1975).

Birch, B. J., Stephens, N.: Heegner's construction of points on the curve $y^2 = x^3 - 1728e^2$, *Séminaire de Théorie des Nombres*, Paris, 1981–82, Progress in Mathematics, 38, Birkhäuser, 1983, pp. 1–19.

Birch, B. J., Stephens, N.: Computation of Heegner points. In *Modular Forms*, R. A. Rankin, ed., Ellis Horwood, 1984, pp. 13–41.

Birch, B. J., Swinnerton-Dyer, H. P. F.: Notes on elliptic curves, (I), *J. Reine Angew. Math.* **212**, 7–25 (1963).

Birch, B. J., Swinnerton-Dyer, H. P. F.: Notes on elliptic curves, (II), *J. Reine Angew. Math.* **218**, 79–108 (1965).

Birch, B. J., Swinnerton-Dyer, H. P. F.: Elliptic curves and modular functions. In *Modular Functions of One Variable, IV*, B. J. Birch and W. Kuyck, eds. Lecture Notes in Mathematics, **476**, Springer-Verlag, 1975, pp. 2–32.

Bloch, S.: A note on height pairings, Tamagawa numbers, and the Birch and Swinnerton-Dyer conjecture, *Invent. Math*, **58**, 65–76 (1980).

Borevich, Z. I., Shafarevich, I. R.: *Number Theory*, Academic Press, 1966.

Bremner, A., Cassels, J. W. S.: On the equation $Y^2 = X(X^2 + p)$, *Math. Com.* **42**, 257–264 (1984).

Brumer, A., Kramer, K.: The rank of elliptic curves, *Duke Math. J.* **44**, 715–743 (1977).

Cartier, P.: Groupes formels, fonctions automorphes et functions zeta des courbes elliptiques, *Actes des Congres Intern. Math. T.* **2**, 291–299 (1970).

Cassels, J. W. S.: A note on the division values of $p(u)$, *Proc. Camb. Phil. Soc.* **45**, 167–172 (1949).

Cassels, J. W. S.: Arithmetic curves of genus 1 (III). The Tate–Shafarevitch and Selmer groups, *Proc. London Math. Soc.* **12**, 259–296 (1962).

Cassels, J. W. S.: Arithmetic on curves of genus 1 (IV). Proof of the Hauptvermutung, *J. Reine Angew. Math.* **211**, 95–112 (1962).

Cassels, J. W. S.: Arithmetic on curves of genus 1 (V). Two counter-examples, *J. London Math. Soc.* **38**, 244–248 (1963).

Cassels, J. W. S.: Arithmetic on curves of genus 1 (VI). The Tate–Shafarevitch group can be arbitrarily large, *J. Reine Angew. Math.* **214/215**, 65–70 (1964).

Cassels, J. W. S.: Arithmetic on curves of genus 1 (VII). The dual exact sequence, *J. Reine Angew. Math.* **216**, 150–158 (1964).

Cassels, J. W. S.: Arithmetic on curves of genus 1 (VIII). On the conjectures of Birch and Swinnerton-Dyer, *J. Reine Angew. Math.* **217**, 180–189 (1965).

Cassels, J. W. S.: Diophantine equations with special reference to elliptic curves, *J. London Math. Soc.* **41**, 193–291 (1966).

Cassels, J. W. S.: Global fields. In *Algebraic Number Theory*, J. W. S. Cassels and A. Fröhlich, eds., Academic Press, 1967, pp. 42–84.

Cassels, J. W. S., Ellison, W. J., Pfister, A.: On sums of squares and on elliptic curves over function fields, *J. Number Theory* **3**, 125–149 (1971).

Chinberg, T.: An introduction to Arakelov intersection theory. In *Arithmetical Geometry*, G. Cornell and J. Silverman, eds., Springer-Verlag, 1986.

Clemens, C. H.: *A Scrapbook of Complex Curve Theory*, Plenum, 1980.

Coates, J.: Construction of rational functions on a curve, *Proc. Camb. Phil. Soc.* **68**, 105–123 (1970).

Coates, J.: An effective p-adic analog of a theorem of Thue, III. The Diophantine equation $y^2 = x^3 + k$, *Acta Arith.* **XVI**, 425–435 (1970).

Coates, J., Wiles, A.: On the conjecture of Birch and Swinnerton-Dyer, *Invent. Math.* **39**, 223–251 (1977).

Cox, D.: The arithmetic-geometric mean of Gauss, *Enseign. Math.* **30**, 275–330 (1984).

Danilov, L. V.: The Diophantine equation $x^3 - y^2 = k$ and Hall's conjecture, *Math. Notes* **32**, 617–618 (1982).

Davenport, H.: On $f^3(t) - g^2(t)$, *Norske Vid. Selsk. Forh. (Trondheim)* **38**, 86–67 (1965).

Deligne, P.: La conjecture de Weil, *Inst. Hautes Etudes Sci. Publ. Math.* **43**, 273–307 (1974).

Deligne, P.: Valeurs de fonctions L et périodes d'intégrales, *Proc. Sympos. Pure Math.* **33**, 313–346 (1979).

Deligne, P., Rapoport, M.: Les schémes de modules de courbes elliptiques. In *Modular Functions of One Variable, II*, P. Deligne and W. Kuyck, eds., Lecture Notes in Mathematics, 349, Springer-Verlag, 1973, pp. 143–316.

Demjanenko, V. A.: On the torsion of elliptic curves, *Izv. Acad. Nauk. SSSR Ser. Mat.* **35**, 280–307 (1971) (in Russina). (English trans. *Math. USSR-Izv.*)

Demjanenko, V. A.: On the uniform boundedness of the torsion of elliptic curves over algebraic number fields, *Izv. Akad. Nauk. SSSR Ser. Mat.* **36** (1972) (in Russian).

Deuring, M.: Die Typen der Multiplikatorenringe elliptischer Funktionenkörper, *Abh. Math. Sem. Univ. Hamburg* **14**, 197–272 (1941).

Deuring, M.: Invarianten und Normalformen elliptischer Funktionenkörper, *Math. Z.* **47**, 47–56 (1941).

Deuring, M.: Die Typen der Multiplikatorenringe elliptischer Funktionenkörper, *Abh. Math. Sem. Univ. Hamburg* **16**, 32–47 (1949).

Deuring, M.: *Die Zetafunktion einer algebraischen Kurve vom Geschlechte Eins, I. II, III, IV*, Gott. Nach., 1953, 1955, 1956, 1957.

Deuring, M.: *Die Klassenkörper der komplexen Multiplikation*, Enz. Math. Wiss., 12, 23. Teubner, 1958.

Drinfeld, V. G.: Elliptic modules, *Mat. Sb. (N.S.)* **94**, 596–627 (1974) (in Russian). (English trans, *Math USSR-Sb.* **23** (4), 1973.)

Drinfeld, V. G.: Coverings of p-adic symmetric regions, *Functional Anal. Appl.* **10**, 29–40 (1976).

Dwork, B.: On the rationality of the zera function of an algebraic variety, *Amer. J. Math.* **82**, 631–648 (1960).

Eichler, M.: Quaternäre quadratische Formen und die Riemannsche Vermutung für die Kongruenzzetafunktion, *Arch. Math. (Basel)* **5**, 355–366 (1954).

Eichler, M.: *Lectures on Modular Correspondences*, Lecture Notes of the Tata Institute, 9 (1956).

Evertse J. H.: On equations in S-units and the Thue-Mahler equation, *Invent. Math.* **75**, 561–584 (1984).

Faltings, G.: Endlichkeitssätze für abelsche Varietäten über Zahenkörpern, *Invent. Math.* **73**, 349–366 (1983).

Faltings, G.: Calculus on arithmetic surfaces, *Ann. of Math.* **119**, 387–424 (1984)

Fröhlich, A.: Local fields. In *Algebraic Number Theory*, J. W. S. Cassels and A. Fröhlich, eds., Academic Press, 1967, 1–41.

Fröhlich, A.: *Formal Groups*, Lecture Notes in Mathematics, 74, Springer-Verlag, 1968.

Fueter, R.: Über kubische diophantische Gleichungen, *Comment. Math. Helv.* **2**, 69–89 (1930).

Fulton, W.: *Algebraic Curves*, Benjamin, 1969.

Goldfeld, D.: The class numbers of quadratic fields and the conjectures of Birch and Swinnerton-Dyer, *Ann. Scuola Norm. Sup. Pisa* **3**, 623–663 (1976).

Greenberg, R.: On the Birch and Swinnerton-Dyer conjecture, *Invent. Math.* **72**, 241–265 (1983).

Griffith, P., Harris, J.: *Principles of Algebraic Geometry*, Wiley, 1978.

Gross, B.: Heegner points on $X_0(N)$. In *Modular Forms*, R. A. Rankin, ed., Ellis Horwood, 1984, pp. 87–106.

Gross, B.: Local heights on curves, *Proceedings of the Conference on Arithmetic Algebraic Geometry, Storrs*, Springer-Verlag.

Gross, B.: On canonical and quasi-canonical liftings, *Invent. Math.* **84**, 321–326 (1986).

Gross, B.: Heights and the special values of L-series, *Conference Proceedings of the CMS*, Vol. 7, American Mathematical Society, 1986.

Gross, B., Zagnier, D.: Points de Heegner et dérivées de fonctions, *L.C.R. Acad. Sci. Paris* **297**, 85–87 (1983).

Gross, B., Zagier, D.: On singular moduli, *J. Reine Angew. Math.* 355, 191–219 (1985).

Gross, B., Zagier, D.: Heegner points and derivates of L-series, *Invent. Math.* **84**, 225–320 (1986).

Gruenberg, K.: Profinite groups. In *Algebraic Number Theory*, J. W. S. Cassels and A. Fröhlich, eds., Academic Press, 1967, pp. 116–127.

Hall, M.: The Diophantine equation $x^3 - y^2 = k$. In *Computers in Number Theory*, A. Atkins and B. Birch, eds., Academic Press, 1971.

Hardy, G., Wright, E.: *An Introduction to the Theory of Numbers*, Oxford Uiversity Press, 1960.

Hartshorne, R.: *Algebraic Geometry*, Springer-Verlag, 1977.

Hazewinkel, M.: *Formal Groups and Applications*, Academic Press, 1978.

Hecke, E.: *Vorlesungen über die Theorie der algebraische Zahlen*, Chelsea, 1948.

Hecke, E.: Analytische Arithmetik der positiven Formen. In *Mathematische Werke*, Vandenhoeck und Rupecht, 1959, pp. 789–918.

Hejhal, D.: *The Selberg Trace Fromula for PSL (2, $\mathbb{R}$)*. Lecture Notes in Mathematics, 1001, Springer-Verlag, 1983.

Honda, T.: Formal groups and zeta functions, *Osaka J. Math.* **5**, 199–213 (1968).

Honda, T.: On the theory of commutative formal groups, *Math. Soc. Japan* **22**, 213–246 (1970).

Hurwitz, A.: Über tenäre diophatische Gleichungen dritten Grades, *Vieteljahrschrift d. Naturf. Ges. Zürich* **62**, 207–229 (1917).

Igusa, J.-I.: Class number of a definite quaternion with prime discriminant, *Proc. Nat. Acad. Sci. U.S.A.* **44**, 312–314 (1958).

Igusa, J.-I.: On the transformation theory of elliptic functions, *Amer. J. Math.* **81**, 436–452 (1959).

Isuga, J.-I.: Kroneckerian model of fields of elliptic modular functions, *Amer. J. Math.* **81**, 561–577 (1959).

Iwasawa, K.: *Lectures on p-adic L-functions*, Annals of Mathematical Studies, 74, Princeton University Press, 1972.

Katz, N.: p-adic properties of modular schemes and forms. In *Modular Functions of One Variable, III*, Lecture Notes in Mathematics, 350, Springer-Verlag, 1973, pp. 69–191.

Katz, N.: An overview of Deligne's proof, *Proc. Sympos. Pure Math.* **28** (1976).

Katz, N., Mazur, B.: *Arithmetic Moduli of Elliptic Curves*, Annals of mathematical Studies, 108, Princeton University Press, 1985.

Kenku, M.: On the number of $\mathbb{Q}$-isomorphism class of elliptic curves in each $\mathbb{Q}$-isogeny class, *J. Number Theory* **15**, 199–202 (1982).

Koblitz, N.: *Introduction to Elliptic Curves and Modular Forms*, Springer-Verlag, 1984.

Kodaira, K.: *On Compact Analytic Surfaces*, Annals of Mathematical Studies, 24, Princeton University Press, 1960, pp. 121–135.

Kotov, S. V., Trelina, L. A.: S-ganze Punkte auf elliptischen Kurven, *J. Reine Angew. Math.* **306**, 28–41 (1979).

Kramer, K.: Arithmetic of elliptic curves upon quadratic extension, *Trans. Amer. Math. Soc.* **264**, 121–135 (1981).

Kubert, D.: Universal bounds on the torsion and isogenies of elliptic curves. Ph.D. thesis, Harvard, 1973.

Kubert, D.: Universal bounds on the torsion of elliptic curves, *Proc. London Math. Soc.* **33**, 193–237 (1976).

Lai, M., Jones, M. F., Blundon, W. J.: Numerical solutions of the Diophantine equation $y^3 - x^2 = k$, *Math. Comp.* **20**, 322–325 (1966).

Lang, S.: *Diophantine Geometry*, Interscience, 1962.

Lang, S: *Les Formes Bilinéaires de Néron et Tate*, Séminaire Bourbaki, 274 (1964).

Lang, S.: *Algebraic Number Theory*, Addison-Wesley, 1970, and Graduate Texts in Mathematics, 110, Springer-Verlag, 1986.

Lang, S.: Division points of elliptic curves and abelian functions over number fields, *Amer. J. Math.* **97**, 124–132 (1972).

Lang, S.: *Elliptic Functions*, Addison-Wesley, 1973.

Lang, S.: *Elliptic Curves: Diophantine Analysis*, Springer-Verlag, 1978.

Lang, S.: *Introduction to Algebraic and Abelian Functions*, 2nd ed., Springer-Verlag, 1982.

Lang, S.: *Fundamentals of Diophantine Geometry*, Springer-Verlag, 1983.

Lang, S.: *Conjectured Diophantine Estimates on Ellptic Curves*, Progress in Mathematics, 35, Birkhäuser, 1983.

Lang, S.: *Complex Multiplication*, Springer-Verlag, 1983.

Lang, S.: *Algebra*, 2nd ed., Addison-Wesley, 1984.

Lang, S., Néron, A.: Rational points of abelian varieties over function fields, *Amer. J. Math.* **81**, 95–118 (1959).

Lang, S., Tate, J.: Principal homogeneous spaces over abelian varieties, *Amer. J. Math.* **80**, 659–684 (1958).

Lang, S., Trotter, H.: *Frobenius Distributions in* GL_2*-Expensions*, Lecture Notes in Mathematics, 504, Springer-Verlag, 1976.

Laska, M.: An algorithm for finding a minimal Weierstrass equation for an elliptic curve, *Math. Comp.* **38**, 257–260 (1982).

Laska, M.: *Elliptic Curves over Number Fields with Prescribed Reduction Type*, Aspects of Mathematics, 4, Friedr. Vieweg & Sohn, 1983.

Laurent, M.: Minoration de la hauteur de Néron–Tate, *Séminaire de Théorie des Nombres, Paris*, 1981–82, Progress in Mathematics, 38, Birkhäuser, pp. 137–152.

Lazard, M.: Sur des groupes de Lie formels à un paramètre, *Bull. Soc. Math. France* **83**, 251–274 (1955).

Li, W.: Newforms and functional equations, *Math. Ann* **212**, 285–315 (1975).

Ligozat, G.: Fonctions L des courbes modulaires, *Séminaire Delange–Pisot–Poitou*, 1970.

Lind, C.-E.: Untersuchungen über die rationalen Punkte der ebenen Kubischen Kurven vom Geschlecht Eins. Thesis, Uppsala.

Liouville, J.: Sur des classes trés-étendues de quantitiés dont la irrationelles algébriques, *C. R. Acad. Sci. Paris* **18**, 883–885, 910–911 (1844).

Lubin, J.: Finite subgroups and isogenies of one-parameter formal Lie groups, *Ann. of Math.* **85**, 296–302 (1967).

Lubin, J.: Canonical subgroups of formal groups, *Trans. Amer. Math. Soc.* **251**, 103–127 (1979).

Lubin, J., Serre, J.-P., Tate, J.: *Seminar at Woods Hole Institute on Algebraic Geometry*, 1964.

Lubin, J., Serre, J.-P., Tate, J. T.: Elliptic curves and formal groups, *A.M.S. Summer Institute on Algebraic Geometry, Woods Hole*, American Mathematical Society, 1964.

Lubin, J., Tate, J.: Formal complex multiplication in local fields, *Ann. Math.* **81**, 380–387 (1965).

Lubin, J., Tate, J.: Formal moduli for one-parameter formal Lie groups, *Bull. Soc. Math. France* **94**, 49–60 (1966).

Lutz, E.: Sur l'equation $y^2 = x^3 - Ax - B$ dans les corps p-adic, *J. Reine Angew. Math.* **177**, 237–247 (1937).

Mahler, K.: On the lattice points on curves of genus 1, *Proc. London Math. Soc.* **39**, 431–466 (1935).

Manin, Yu. I.: *The Hasse-Witt Matrix of an Algebraic Curve*, AMS Translations, 45, 1965, pp. 245–264.

Manin, Yu. I.: The p-torsion of elliptic curves is uniformly bounded, *Izv. Akad, Nauk SSSR Ser. Mat.* **33**, 433–438 (1969).

Manin, Yu. I.: Parabolic points and zeta functions of modular curves, *Izv. Akad. Nauk SSSR Ser. Mat.* **36**, 19–64 (1972).

Manin, Yu. I.: Uniform bound for the p-torsion of elliptic curves, *Izv. Akad. Nauk. SSSR Ser. Mat.* **33**, 459–465 (1969) (in Russian). (English trans, *Math. USSR-Izv.*).

Manin, Yu. I.: Le group de Brauer–Grothendieck en géométric diophantienne, *Actes, Congres Intern. Math.* **1**, 401–411 (1970).

Manin, Yu. I.: Cyclotomic fields and modular curves, *Uspekhi Mat. Nauk* **26**, 7–71 (1971) (in Russian).

Manin, Yu. I., Zarkin, Y. G.: Heights on families of abelian varieties, *Mat. Sb. (N.S.)* **89**, 171–181 (1972) (in Russian).

Masser, D.: *Elliptic Functions and Transcendence*, Lecture Notes in Mathematics, 437, Springer-Verlag, 1975.

Masser, D., Wüstholz, G.: Fields of large transcendence degree generated by the values of elliptic functions, *Invent. Math.* **72**, 407–464 (1983).

Matsumura, H.: *Commutative Algebra*, 2nd ed., Benjamin/Cummings, 1980.

Mazur, B.: *Courbes Elliptiques et Symboles Modulaires*. Séminaire Bourbaki, 414, 1972.

Mazur, B.: Rational points of abelian varieties with values in towers of number fields, *Invent. Math.* **18**, 183–266 (1972).

Mazur, B.: *Courbes Elliptiques et Symboles Modularies*, Séminaire Bourbaki, 414, Lecture Notes in Mathematics, 317, Springer-Verlag, 1973.

Mazur, B.: Modular curves and the Eisenstein ideal, *Inst. Hautes Études Sci. Publ. Math.* **47**, 33–186 (1977).

Mazur, B.: Rational isogenies of prime degree, *Invent. Math.* **44**, 129–162 (1978).

Mazur, B., Swinnerton-Dyer, H. P. F.: Arithmetic of Weil curves, *Invent. Math.* **25**, 1–61 (1974).

Mazur, B., Tate, J.: Canonical height pairings via biextensions. In *Arithmetic and Geometry*, Progress in Mathematics, 35, Birkhäuser, 1983, pp. 195–237.

Mazur, B., Wiles, A.: Class fields of abelian extensions of $\mathbb{Q}$, *Invent. Math.* **76**, 179–330 (1984).

Mazur, B., Tate, J., Teitelbaum, J.: On p-adic analogues of the conjectures of Birch and Swinnerton-Dyer, *Invent. Math.* **84**, 1–48 (1986).

McCabe, John: p-adic theta functions. Ph.D. Thesis, Harvard, 1968, pp. 1–222.

Mestre, J.-F.: Courbes elliptiques et formules explicités, *Séminaire de Théorie des Nombres, Paris*, 1981–82, Progress in Mathematics, 38, Birkhäuser, pp. 179–188.

Mestre, J.-F.: Construction of an elliptic curve of rank ≥ 12, *C. R. Acad. Sci. Paris* **295**, 643–644 (1982).

Mestre, J.-F.: Courbes de Weil de conducteur 5077, *C. R. Acad. Sci. Paris* **300**, 509–512 (1985).

Mignotte, M.: Quelques remarques sur l'approximation rationelle des nombres algébriques, *J. Reine Angew. Math.* **268/269**, 341–347 (1974).

Milne, J. S.: The Tate–Šhafarevitch group of a constant abelian variety, *Invent. Math.* **6**, 91–105 (1968).

Milne, J. S.: Weil–Châtelet groups over local fields, *Ann. Sci. École. Norm. Sup.*, **3**, 273–284 (1970); Addendum, ibid. **5**, 261–264 (1972).

Milne, J. S.: On the arithmetic of abelian varieties, *Invent. Math.* **17**, 177–190 (1972).

Mordell, L. J.: On the rational solutions of the indeterminate equations of the third and fourth degrees, *Proc. Camb. Phil. Soc.* **21**, 179–192 (1922).

Mordell, L. J.: *A Chapter in the Theory of Numbers*, Cambridge University Press, 1947.

Mordell, L. J.: The Diophantine equation $x^4 + my^4 = z^2$, *Quart. J. Math. Oxford Ser.* (2) **18**, 1–6 (1967).

Mordell, L. J.: *Diophantine Equations*, Academic Press, 1969.

Morikawa, H.: On theta functions and abelian varieties over valuation fields of rank one, I and II, *Nagoya Math. J.* **20**, 1–27, 231–250 (1962).

Mumford, D.: An analytic construction of degenerating curves over complete local rings, *Composition Math.* **24**, 129–174 (1972).

Mumford, D.: *Abelian Varieties*, Oxford University Press, 1974.

Mumford, D., Fogerty, J.: *Geometric Invariant Theory*, 2nd ed., Springer-Verlag, 1982.

Mumford, D.: *Tate Lectures on Theta, I* and *II*. Progress in Mathematics, 28 and 43, Birkhäuser, 1983, 1984.

Nagell, T.: Solution de quelque problémes dans la théorie arithmétique des cubiques planes du premier genre, *Wid. Akad. Skrifter Oslo* **I**, No. 1 (1935).

Néron, A.: Problémes arithmétiques et géométriques rattachés á la notion de rang d'une courbe algébrique dans un corps, *Bull. Soc. Math. France* **80**, 101–166 (1952).

Néron, A.: Propriétés arithmétiques de certaines families de courbes algébriques, *Proceedings of an International Congress, Amsterdam*, Vol. **III**, 481–488 (1954).

Néron, A.: Modéles minimaux des variétés abéliennes sur les corps loceax et globeax, *Inst. Hautes Études Sci. Publ. Math.* **21**, 361–482 (1964).

Néron, A.: Quasi-fonctions et hauteurs sur les variétés abéliennes, *Ann. of Math.* **82**, 249–331 (1965).

Neumann, O.: Zur Reduktion der elliptischen Kurven, *Math. Nachr.* **46**, 285–310 (1970).

Norman, P.: p-adic theta functions, *Amer. J. Math.* **106**, 617–661 (1985).

Oesterlé, J.: *Nombres de Classes des Corps Quadratiques Imaginaires*, Séminaire Bourbaki, 631, 1984.

Ogai, S. V.: On rational points on the curve $y^2 = x(x^2 + ax + b)$, *Trudy Mat. Inst. Steklov* **80** (1965) (in Russian).

Ogg, A. P.: Cohomology of abelian varieties over function fields, *Ann. of Math.* **76**, 185–212 (1962).

Ogg, A. P.: Abelian curves of 2-power conductor, *Proc. Camb. Phil. Soc.* **62**, 143–148 (1966).

Ogg, A. P.: Abelian curves of small conductor, *J. Reine Angew. Math.*, **226**, 204–215 (1967).

Ogg, A. P.: Elliptic curves and wild ramification, *Amer. J. Math.* **89**, 1–21 (1967).

Ogg, A. P.: *Modular Forms and Dirichlet Series*, Benjamin, 1969.

Ogg, A. P.: On the eigenvalues of Hecke operators, *Math. Ann.* **179**, 101–108 (1969).

Ogg, A. P.: Rational points of finite order on elliptic curves, *Isv. Math.* **12**, 105–111 (1971).

Ogg, A. P.: Rational points on certain elliptic modular curves. (A talk given in St. Louis on 29 March 1972, at the A.M.S. Symposium on Analytic Number Theory and Related Parts of Analysis.)

Olson, L.: Torsion points on elliptic curves with given j-invariant, *Manuscripta Math.* **16**, 145–150 (1975).

Panciskin, A.: Le prolongement p-adique analytiques des fonctions L de Rankin, *C. R. Acad. Sci. Paris* **295**, 51–53, 227–230 (1982).

Parshin, A. N.: Algebraic curves over function fields, *Math. USSR-Izv.* **2**, 1145–1170 (1968).

Perrin-Riou, B.: Descente infinie et hauteur p-adique sur les courbes elliptiques à multiplication complexe, *Invent. Math.* **70**, 369–398 (1983).

Pinch, R. G. E.: Elliptic curves with good reduction away from 2, *Proc. Camb. Phil. Soc.* **96**, 25–38 (1984).

Rajwade, A. R.: Arithmetic on curves with complex multiplication by $\sqrt{2}$, *Proc. Camb. Phil. Soc.* **64**, 659–672 (1968).

Raynaud, M.: *Caractéristique d'Euler-Poincaré d'un Faisceau et Cohomologie des Variétés Abéliennes*. Séminaire Bourbaki, 286, 1965.

Raynaud, M.: Spécialisation du foncteur de Picard, *Inst. Hautes Études Sci. Publ. Math.* **38**, 27–76 (1970).

Raynaud, M.: Variëtés abéliennes et géométrie rigide, *Procedings of the International Congress, Nice*, Vol. I, 1970, pp. 473–477.

Razar, M.: The non-vanishing of $L(1)$ for certain elliptic curves. Ph.D. Thesis, Harvard, 1971.

Reichardt, H.: Einige im Kleinen überall lösbare, im Großen Unlösbare diophantische Gleichungen, *J. Reine Angew. Math.* **184**, 12–18 (1942).

Robert, A.: *Elliptic Curves*, Lecture Notes in Mathematics, 326, Springer-Verlag, 1973.

Rohrlich, D.: On L-functions of elliptic curves and anti-cyclotomic towers, *Invent. Math.* **75**, 383–408 (1984).

Roquette, P.: *Analytic Theory of Elliptic Functions Over Local Fields*, Vandenhoeck & Ruprecht, 1970.

Roquette, P. Analytic theory of elliptic functions over local fields, *Hamb. Math. Einzelschriften. Neue Folge*, Heft 1, Göttingen, 1970.

Rubin, K.: Elliptic curves with complex multiplication and the conjecture of Birch and Swinnerton-Dyer, *Invent. Math.* **64**, 455–470 (1981).

Satge, P.: Une généralisation du calcul de Selmer, *Séminaire de Théorie des Nombres*, Paris, 1981–82, Progress in Mathematics, 38, Birkhäuser, pp. 245–266.

Schanuel, S.: Heights in number fields, *Bull. Soc. Math. France* **107**, 443–449 (1979).

Schmidt, W.: Thue's equation over function fields, *J. Austral. Math. Soc.* **25**, 385–422 (1978).

Schmidt, W.: *Diophantine Approximation*, Lecture Notes in Mathematics, 785, Springer-Verlag, 1980.

Schneider, P: *p*-adic heights, *Invent. Math* **69**, 401–409 (1982).

Selmer, E: The diophantine equation $ax^3 + by^3 + cz^3 = 0$, *Acta. Math.* **85**, 203–362, (1951); **92**, 191–197 (1954).

Selmer, E.: A conjecture concerning rational points on cubic surfaces, *Math. Scand.* **2**, 49–54 (1954).

Serre, J.-P.: Géométrie algébrique et géométrie analytique, *Ann. Inst. Fourier (Grenoble)* **6**, 1–42 (1956).

Serre, J.-P.: *Groupes Algébriques et Corps de Class*, Hermann, 1959.

Serre, J.-P.: Groupes de Lie *l*-adiques attachés aux courbes elliptiques, *Colloque de Clermont-Ferrand*, 1964.

Serre, J.-P.: Sur les groups de congruences des variétés abélinennes, *Izv. Akad. Nauk SSSR Ser. Mat.* **28**, 3–20 (1964).

Serre, J.-P.: Groups de Lie *l*-adiques attachés aux courbes elliptiques, *Coll. Internat. du C.N.R.S.*, No. 143, Editions du C.N.R.S., 1966.

Serre, J.-P.: *Zeta- and L-functions in Arithmetical Algebraic Geometry*. Harper and Row, 1966.

Serre, J.-P.: Une interprétation des congruences relatives à la fonction τ de Ramanujan, *Séminaire Delange-Pisot-Poitou: Théorie des Nombres*, Exposé 14, 1967/68.

Serre, J.-P.: Complex multiplication. In *Algebraic Number Theory*, J.W.S. Cassels and A. Fröchlich, eds., Academic Press, 1967, pp. 292–296.

Serre, J.-P.: *Abelian l-adic Representations and Elliptic Curves*, Benjamin, 1968.

Serre, J.-P.: Facteurs locaux des fonctions zêta des variétés algébriques (définitions et conjectures), *Séminaire Delange-Pisot-Poitou*, 1970.

Serre, J.-P.: *p-Torsion des Courbes Elliptiques (d'après Y. Manin)*, Séminaire Bourbaki, 380, 1970.

Serre, J.-P.: Properiétés galoisienne des points d'ordre fini des courbes elliptiques, *Invent. Math.* **15**, 259–333 (1972).

Serre, J.-P.: *Congruences et Formes Modular (d'après H. P. F. Swinnerton-Dyer)*, Séminaire Bourbaki, 416, 1972.

Serre, J.-P.: *A Course in Arithmetic*, Springer-Verlag, 1973.

Serre, J.-P.: *Cohomologie Galoisienne*, Lecture Notes in Mathematics, 5, Springer-Verlag, 1973.

Serre, J.-P.: *Local Fields*, Springer-Verlag, 1973.

Serre, J.-P.: Diophantine geometry. Unpublished notes from a course given at Harvard University, 1979.

Serre, J.-P.: Quelques application du théorème de densité de Chebotarev, *Inst. Hautes Études Sci. Publ. Math* **54**, 123–202 (1981).

Serre, J.-P.: Sur la lacunarité des puissances de η, *Glasgow Math. J.* **27**, 203–221 (1985).

Serre, J.-P., Tate, J.: Mimeographed noes from the A.M.S. Summer Institute at Woods Hole, 1964.

Serre, J.-P., Tate, J.: Good reduction of abelian varieties, *Ann. of Math.* **88**, 492–517 (1968).

Setzer, C. B.: Elliptic curves of prime conductor. Ph.D. Thesis, Harvard, 1972.

Setzer, B.: Elliptic curves over complex quadratic fields, *Pacific J. Math.* **74**, 235–250 (1978).

Shafarevitch, I. R.: Algebraic number fields, *Proceedings of an International Congress on Mathematics, Stockholm*, 1962, pp. 163–176. (A.M.S. Translations, Series 2, Vol. 31, pp. 25–39.)

Shafarevitch, I. R.: *Principal Homogeneous Spaces Defined over a Function-field.* A.M.S. Translations, Vol. 37, 1964, pp. 85–114.

Shafarevitch, I. R.: *Basic Algebraic Geometry*, Springer-Verlag, 1977.

Shafarevitch, I. R., Tate, J. T.: The rank of elliptic curves, *Dokl. Akad. Nauk. USSR.* **175**, 770–773 (1967) (in Russian). (A.M.S. Translations, Vol. 8, 1967, pp. 917–920.)

Shimura, G.: On the zeta functions of the algebraic curves uniformized by certain automorphic functions, *J. Math. Soc. Japan* **13**, 275–331 (1961).

Shimura, G.: A reciprocity law in non-solvable extensions, *J. Reine Angew. Math.* **221**, 209–220 (1966).

Shimura, G.: Construction of class fields and zeta functions of algebraic curves, *Ann. of Math.* **85**, 58–159 (1967).

Shimura, G.: On the zeta-function of an abelian variety with complex multiplication, *Ann. of Math.* **94**, 504–533 (1971).

Shimura, G.: On elliptic curves with complex multiplication as factors of the jacobians of modular function fields, *Nagoya Math. J.* **43**, 199–208 (1971).

Shimura, G.: *Introduction to the Arithmetic Theory of Automorphic Forms*, Publications of the Mathematical Society of Japan, Vol. 11, Iwanami Shoten and Princeton University Press, 1971.

Shimura, G., Taniyama, Y.: *Complex Multiplication of Abelian Varieties and Its Application to Number Theory*, Publications of the Mathematical Society of Japan, 6, 1961.

Shioda, T.: On elliptic modular surfaces, *J. Math. Soc. Japan* **24**, 20–59 (1972).

Siegel, C. L.: Über einige Anwendungen diophantischer Approximationen (1929). In *Collected Works*, Springer-Verlag, 1966, pp. 209–266.

Silverman, J.: Lower bound for the canonical height on elliptic curves, *Duke. Math. J.* **48**, 633–648 (1981).

Silverman, J.: Integer points and the rank of Thue elliptic curves, *Invent. Math.* **66**, 395–404 (1982).

Silverman, J: Heights and the specialization map for families of abelian varieties, *J. Reine Angew. Math.* **342**, 197–211 (1983).

Silverman, J.: Integer points on curves of genus 2, *J. London Math. Soc.* **28**, 1–7 (1983).

Silverman, J.: Weierstrass equations and the minimal discriminant of an elliptic curve, *Mathematika* **31**, 245–251 (1984).

Silverman, J.: Divisibility of the specialization map for families of elliptic curves, *Amer. J. Math.* **107**, 555–565 (1985).

Silverman, J.: *The Arithmetic of Elliptic Curves*, Graduate Texts in Mathematics, 106, Springer-Verlag, 1986.

Stark, H.: Effective estimates of solutions of some Diophantine equations, *Acta Arith.* **24**, 251–259 (1973).

Stephens, N. M.: The diophantine equation $x^3 + y^3 = DZ^3$ and the conjectures of Birch and Swinnerton-Dyer, *J. Reine Angew. Math.* **231**, 121–162 (1968).

Stevens, G.: *Arithmetic on Modular Curves*, Progress in Mathematics 20, Birkhäuser, 1982.

Sturm, J.: Projections of C^{∞} automorphic forms, *Bull. Amer. Math. Soc.* **2**, 435–439 (1980).

Swinnerton-Dyer, H. P. F.: The conjectures of Birch and Swinnerton-Dyer, and of Tata, *Proceedings of a Conference on Local Fields*, Springer-Verlag, 1967, pp. 132–157.

Swinnerton-Dyer, H. P. F., Birch, B. J.: Elliptic curves and modular functions. In *Modular Functions of One Variable, IV*, B. J. Birch and W. Kuyck, eds., Lecture Notes in Mathematics, 476, Springer-Verlag, 1975, pp. 2–32.

Tate, J.: *W. C. Groups over P-adic Fields*. Séminaire Bourbaki, 156, 1957.

Tate, J.: Duality theorems in Galois cohomology over number fields, *Proceedings of an International Congress on Mathematics Stockholm,* Institut Mittag-Leffer, 1962, pp. 288–295.

Tate, J.: Algebraic cycles and poles of zeta functions. In *Arithmetical Algebraic Geometry*, Harper and Row, 1966.

Tate, J.: Endomorphisms of abelian varieties over finite fields, *Invent. Math.* **2**, 134–144 (1966).

Tate, J.: *On the Conjecture of Birch and Swinnerton-Dyer and a Geometric Analog*. Séminaire Bourbaki, 306, 1966.

Tate, J.: Global classified theory. In *Algebraic Number Theory*, J. W. S. Cassels and A Fröhlich, eds., Academic Press, 1967, pp. 162–203.

Tate, J.: Residues of differentials on curves, *Ann. Sci. de École Norm. Sup.* (4) **1**, 149–159 (1968).

Tate, J.: Letter to J.-P. Serre, 1968.

Tate, J.: *Classes d'Isogénie des Variétés Abéliennes sur un Corps Fini (d'après T. Honda)*. Séminaire Bourbaki, 352, 1968.

Tate, J.: The arithmetic of elliptic curves, *Invent. Math.* **23**, 179–206 (1974).

Tate, J.: Algorithm for determining the type of a singular fiber in an elliptic pencil. In *Modular Functions of One Variable, IV*, B. J. Birch and W. Kuyck, eds., Lecture Notes in Mathematics, 476, Springer-Verlag, 1975, pp. 33–52.

Tate, J.: Variation of the canonical height of a point depending on a parameter, *Amer. J. Math.* **105**, 287–294 (1983).

Van der Waerden, B. L.: *Algebra*, 7th ed., Ungar, 1970.

Vélu, J: Isogénies entre courbes elliptiques, *C. R. Acad. Sci. Paris* 238–241 (1971).

Vélu, J.: Courbes elliptiques sur $\mathbb{Q}$ ayant bonne réduction en dehors de (11), *C. R. Acad. Sci. Paris*, 73–75 (1971).

Vignèras, M.-F.: Valeur au centre de symétrie des fonctions L associées aux formes modulaires, *Séminaire de Théorie des Nombres, Paris*, 1979–80. Progress in Mathematics, 12, Birkhäuser, 1981, pp. 331–356.

Vishik, M.: Nonarchimedean measures connected with Dirichlet series, *Math. USSR-Sb.* **28**, 216–228 (1976).

Vojta, P.: A higher dimensional Mordell conjecture. In *Arithmetic Geometry*, G. Cornell and J. Silverman, eds., Springer-Verlag, 1986.

Walker, R.J.: *Algebraic Curves*, Dover, 1962.

Waterhouse, W. C., Milne, J. S.: Abelian varieties over finite fields, *A.M.S. Summer Institute on Number Theory, Stony Brook*, 1969, Proceedings of Symposia in Pure Mathematics, Vol. XX, American Mathematical Society, 1971.

Weil, A.: L'arithmétique sur les courbes algébriques, *Acta Math.* **52**, 281–315 (1928).

Weil, A.: Sur un théorème de Mordell, *Bull. Sci. Math.* **54**, 182–191 (1930).

Weil, A.: Number of solutions of equations in finite fields, *Bull. Amer. Math. Sci.* **55**, 497–508 (1949).

Weil, A.: Jacobi sums as Größencharaktere, *Trans. Amer. Math. Soc.* **75**, 487–495 (1952).

Weil, A.: On algebraic groups and homogeneous spaces, *Amer. J. Math.* **77**, 493–512 (1955).

Weil, A.: Über die Bestimmung Dirichletscher Reihen durch Funktionalgleichungen, *Math. Ann.* **168**, 149–156 (1967) (*Oeuvres*, Vol. III, pp. 165–172).

Weil, A.: *Dirichlet Series and Automorphic Forms*. Lecture Notes in Mathematics, 189, Springer-Verlag, 1971.

Whittaker, E. T., Watson, G. N.: *A Course in Modern Analysis*, 4th ed., Cambridge University Press, 1927.

Wüstholz, G.: *Recent Progress in Transcendence Theory*, Lecture Notes in Mathematics, 1068, Springer-Verlag, 1984, pp. 280–296.

Zagier, D.: L-series of elliptic curves, the Birch–Swinnerton-Dyer conjecture, and the class number problem of Gauss, *Notices Amer. Math. Soc.* **31**, 739–743 (1984).

Zimmer, H.: On the difference of the Weil height and the Néron–Tate height, *Math. Z.* **147**, 35–51 (1976).

Additional Bibliography for the Second Edition

Baker, A.: A supersingular congruence for modular forms, *Acta Arithmetica*, **LXXXVI**, 91–100 (1998).

Barth, Peters, and van de Ven: *Compact complex surfaces*, Springer-Verlag, Ergebnesse der Mathematik und ihrer Grenzgebiete, 3 Folge, Band 4.

Bosch, S., Lütkebohmert, W., Raynaud, M.: *Néron Models*, Springer-Verlag, 1987.

Breuil, C., Conrad, B. Diamond, F. and Taylor, R.: On the modularity of elliptic curves over $\mathbb{Q}$, or wild 3-adic exercies, *Journal of the American Math. Soc.* **14**, 843–940 (2001).

Carayol, H.: Sur les repésentions ℓ-adiques associées aux formes modularies de Hilbert, *Ann. Sci. Éc. Norm. Sup.* **19**, 409–468 (1986).

Carayol, H.: Sur les représentations galoisiennes modulo l attachées aus formes modulaires, *Duke Math. J.* **59**, 785–801 (1989).

Coates, J., Sujatha, R.: Galois cohomology of elliptic curves, *TIFR, Mumbai*, Narosa (2000).

Conrad, B.: Finite group schemes over bases with low ramificaiton, *Compositio Mathematica*.

Conrad, B.: Ramified deformation problems, *Duke Math. J.* **97** (1999), 439–514.

Conrad, B., Diamond, F. and Taylor, R.: Modularity of certain protentially Barsotti–Tate Galois representations, *Journal of the American Mathematical Society* **12**, 521–567 (1999).

Cox, D. A., Katz, S.: *Mirror Symmetry and Algebraic Geometry*, Math. Surveys and Monographs, Vol. 68, AMS, 1999.

Cremona, J. E.: *Algorithms for Modular Elliptic Curves*, Cambridge Univ. Press, 1992.

Curtis, C., Reiner, I.: *Methods of Representation Theory*, Wiley & Sons, New York, 1981.

Darmon, H., Diamond, F., Taylor, R.: Fermat's Last Theorem, in *Current Developments in Mathematics* 1995, International Press, 1996, pp. 1–154.

de Jong, A. J., Finite locally free group schemes in characteristic p and Dieudonné modules, *Inv. Math.* **114**, 89–138 (1993).

de Smit, B., Lenstra, H.: Explicit construciton of universal deformation rings, in *Modular Forms and Fermat's Last Theorem* (Boston 1995), Springer-Verlag 1997, pp. 313–326.

Diamond, F.: The refined conjecture of Serre, in *Elliptic Curves, Modular Forms and Fermat's Last Theorem* (Hong Kong, 1993), International Press, 1995, pp. 22–37.

Diamond, F.: On deformation rings and Hecke rings, *Ann. Math.* **133**, 137–166 (1996).

Diamond, F.: The Taylor–Wiles construction and multiplicity one, *Inv. Math.* **128**, 379–391 (1997).

Diamond, F., Taylor, R.: Non-optimal levels for mod ℓ modular represenations of $\mathrm{Gal}(\overline{\mathbb{Q}}/\mathbb{Q})$, *Inv. Math.* **115**, 435–462 (1994).

Diamond, F., Taylor, R.: Lifting modular mod ℓ representations, *Duke Math. J.* **74**, 253–269 (1994).

Ekedahl, T.: An effective version of Hilbert's irreducibility theorm, *Séminare de Théorie des Nombres, Paris 1988–89*, Birkhäuser, 1990.

Edixhoven, B.: The weight in Serre's conjectures on modular forms, *Inv. Math.* **109**, 563–594 (1992).

Elkies, N.: Elliptic and modular curves over finite fields, and related computational issures, *Computational Perspectives on Number Theory* (J. Teitebaum, ed.).

Faltings, G.: The Proof of Fermat's Last Theorem by R. Taylor and A. Wiles, *Notices of the American Mathematical Society* **42**, 743–746 (1995).

Flach, M.: A finiteness theorem for the symmetric square of an elliptic curve, *Invent. Math.* **109**, 307–327 (1992).

Fontaine, J.-M.: Groupes p-divisibles sur les corps locaux, *Astérisque* **47–48**, Société mathématique de France, Paris, 1977.

Fontaine, J.-M.: Le corps des périodes p-adiques in Périodes p-adiques, *Astérisque* **223**, 59–111.

Fontaine, J.-M.: Représentations p-adiques semi-stables, Périodes p-adiques, *Astérisque* **223**, 113–184.

Fontaine, J.-M.: Sur certains types de représentations p-adiques du groupe de Galois d'un corps loca: construction d'un anneau de Barsotti–Tate, *Annals of Mathematics*, **115**, 529–577 (1982).

Fontaine, J.-M.: Il n'y a pas de variété abélienne sur $\mathbb{Z}$, *Invent. Math.* **81**, 515–538 (1985).

Fontaine, J.-M., Mazur, B.: Geometric Galois representations, in *Elliptic Cruves, Modular Forms and Fermat's Last Theorem*, (Hong Kong, 1993), International Press, 1995, 41–78.

Frey, G.: Links between stable elliptic curves and certain diophantine equation, *Ann. Univ. Sarav. ser. Math.* **1**, 1–40 (1986).

Frey, G.: Links between elliptic curves and solutions of $A - B = C$, *J. Indian Math. Soc.* **51**, 117–145 (1987).

Garcia-Selfa, Olalla, M. A., Tornero, J. M.: Computing the rational torsion of an elliptic curve using Tate normal form, *Journal of Number Theory* **96**, 78–88 (2002).

Greenberg, R.: Iwasawa theory of p-adic representation, *Adv. Studies in Pure Math.* **17**, pp. 97–137 (1989).

Greenberg, R.: The structure of Sellmer groups, *Proc. Nat. Acad. Sc.* **94**, 11125–11128 (1997).

Griffiths, P., Harris, J.: *Principles of Algebraic Geometry*, Wiley, 1978.

Gross, B. H.: Kolyvagin's work on modular elliptic curves, *L-functions and arithmetic* (Durham 1989) 235–256, London Math. Soc. Lecture Note Ser., 153, Cambridge Univ. Press, 1991.

Grothendieck, A.: (SGA 3), Lecture Notes, 151, Springer-Verlag, 1970.

Grothendieck, A.: (SGA 7), Lecture Notes, 288, Springer-Verlag, 1972.

Ihara, Y.: On modular curves over finite fields, *Discrete Subgroups of Lie Groups and Applicaitons to Moduli* (Bombay 1973) Oxford University Press, 1975, pp. 161–202.

Katz, N., Messing, W.: Some consequences of the Riemann Hypothesis for varieties over finite fields, *Inv. Math.* **23**, 73–77 (1974).

Kobayashi, S., K. Nomizu: *Foundations of Differential Geometry*, Interscience Publishers, 1969.

Kolyvagin, V. A.: *The Grothendieck Festschrift*, Vol. II, 435–483, Progress in Mathathematics, 87, Birkhäuser Boston, 1990.

Kolyvagin, V. A., Logachev, D. Yu.: *Algebra i Analiz* **1** (1), 522–540 (1989).

Mazur, B.: Deforming Galois representations, in *Galois Groups over* $\mathbb{Q}$, Springer-Verlag, 1989, pp. 385–437.

MaCallum, W. G.: Kolyvagin's work on Shafarevich–Tate groups, *L-functions and arithmetic* London Math. Soc. Lecture Note Ser., 153, Cambridge Univ. Press, 1991, pp. 295–316.

Perrin-Riou, B.: Travaux de Kolyvagin et Rubin, Séminaire Bourbaki, Vol. 1989/90, *Astérisque* No. 189–190 (1990), Exp. No. 717, 69–106.

Ramakrishna, R.: On a variation of Mazur's deformation functor, *Comp. Math.* **87**, 269–286 (1993).

Raynaud, M.: Schémas en groupes de type $(p, p, \ldots, p)$, *Bull. Soc. Math. France* **102**, 241–280 (1974).

Ribet, K. A.: On modular representations of $\mathrm{Gal}(\overline{\mathbb{Q}}/\mathbb{Q})$ arising from modular forms, *Invent. Math.*, **100**, 431–471 (1990).

Rubin, K.: Global units and ideal class groups, *Invent. Math.* **89**, 511–526 (1987).

Rubin, K.: Tate–Shafarevich groups and L-functions of elliptic curves with complex multiplication, *Invent. Math.* **89**, 527–560 (1987).

Rubin, K.: *Invent. Math.* **103**, 25–68 (1991).

Rubin, K.: The work of Kolyvagin on the arithmetic of elliptic curves, *Arithmetic of complex manifolds* (Erlangen, 1988) SLN, 1989, pp. 128–136.

Rubin, K., Silverberg, A.: Ranks of elliptic curves, *Bulletin of the AMS* **39**, 455–474 (2002).

Saito, T.: Modular forms and p-adic Hodge theory, *Invent. Math.* **129**, 607–620 (1997).

Serre, J.-P.: Le problème des groupes de congruence pur SL_2, *Ann. Math.* **92**, 489–527 (1970).

Serre, J.-P.: *Linear Representations of Finite Groups*, Springer-Verlag, 1977.

Serre, J.-P.: *Trees*, Springer-Verlag, 1980.

Serre, J.-P.: Sur les représentations modulaires de degré 2 de $\mathrm{Gal}((\overline{\mathbb{Q}}/\mathbb{Q}))$, *Duke Math. J.* **54**, 179–230 (1987).

Shimura, G.: Algebraic number fields and symplectic discontinuous groups, *Ann. Math.* **87**, 503–592 (1967).

Silverman, J.: Advanced topic in the arithmetic of elliptic curves, GTM 151, 1994.

Tate, J.: p-divisible groups, in *Proceedings of a Conference on Local Fields* (Dreibergen, 1966), Springer, 1967, pp. 158–183.

Tate, J.: A review of non-archimedean elliptic functions, in *Elliptic Curves, Modular Forms and Fermat's Last Theorem*, 2nd ed., International Press, 1997, pp. 310–314.

Taylor, R., Wiles, A.: Ring theoretic properties of certain Hecke algebras, *Ann. Math.* **141**, 553–572 (1995).

Voisin, C.: *Mirror Symmetry*, SMF/AMS Texts and Monographs, Vol. 1, 1996.

Wiles, A.: Modular elliptic curves and Fermat's last Theorem, *Ann. Math.* **141**, 445–551 (1995).

List of Notation

Index

(continued from page ii)

124 DUBROVIN/FOMENKO/NOVIKOV. Modern Geometry—Methods and Applications. Part III
125 BERENSTEIN/GAY. Complex Variables: An Introduction.
126 BOREL. Linear Algebraic Groups. 2nd ed.
127 MASSEY. A Basic Course in Algebraic Topology.
128 RAUCH. Partial Differential Equations.
129 FULTON/HARRIS. Representation Theory: A First Course.
Readings in Mathematics
130 DODSON/POSTON. Tensor Geometry.
131 LAM. A First Course in Noncommutative Rings. 2nd ed.
132 BEARDON. Iteration of Rational Functions.
133 HARRIS. Algebraic Geometry: A First Course.
134 ROMAN. Coding and Information Theory.
135 ROMAN. Advanced Linear Algebra.
136 ADKINS/WEINTRAUB. Algebra: An Approach via Module Theory.
137 AXLER/BOURDON/RAMEY. Harmonic Function Theory. 2nd ed.
138 COHEN. A Course in Computational Algebraic Number Theory.
139 BREDON. Topology and Geometry.
140 AUBIN. Optima and Equilibria. An Introduction to Nonlinear Analysis.
141 BECKER/WEISPFENNING/KREDEL. Gröbner Bases. A Computational Approach to Commutative Algebra.
142 LANG. Real and Functional Analysis. 3rd ed.
143 DOOB. Measure Theory.
144 DENNIS/FARB. Noncommutative Algebra.
145 VICK. Homology Theory. An Introduction to Algebraic Topology. 2nd ed.
146 BRIDGES. Computability: A Mathematical Sketchbook.
147 ROSENBERG. Algebraic *K*-Theory and Its Applications.
148 ROTMAN. An Introduction to the Theory of Groups. 4th ed.
149 RATCLIFFE. Foundations of Hyperbolic Manifolds.
150 EISENBUD. Commutative Algebra with a View Toward Algebraic Geometry.
151 SILVERMAN. Advanced Topics in the Arithmetic of Elliptic Curves.
152 ZIEGLER. Lectures on Polytopes.
153 FULTON. Algebraic Topology: A First Course.
154 BROWN/PEARCY. An Introduction to Analysis.
155 KASSEL. Quantum Groups.
156 KECHRIS. Classical Descriptive Set Theory.
157 MALLIAVIN. Integration and Probability.
158 ROMAN. Field Theory.
159 CONWAY. Functions of One Complex Variable II.
160 LANG. Differential and Riemannian Manifolds.
161 BORWEIN/ERDÉLYI. Polynomials and Polynomial Inequalities.
162 ALPERIN/BELL. Groups and Representations.
163 DIXON/MORTIMER. Permutation Groups.
164 NATHANSON. Additive Number Theory: The Classical Bases.
165 NATHANSON. Additive Number Theory: Inverse Problems and the Geometry of Sumsets.
166 SHARPE. Differential Geometry: Cartan's Generalization of Klein's Erlangen Program.
167 MORANDI. Field and Galois Theory.
168 EWALD. Combinatorial Convexity and Algebraic Geometry.
169 BHATIA. Matrix Analysis.
170 BREDON. Sheaf Theory. 2nd ed.
171 PETERSEN. Riemannian Geometry.
172 REMMERT. Classical Topics in Complex Function Theory.
173 DIESTEL. Graph Theory. 2nd ed.
174 BRIDGES. Foundations of Real and Abstract Analysis.
175 LICKORISH. An Introduction to Knot Theory.
176 LEE. Riemannian Manifolds.
177 NEWMAN. Analytic Number Theory.
178 CLARKE/LEDYAEV/STERN/WOLENSKI. Nonsmooth Analysis and Control Theory.
179 DOUGLAS. Banach Algebra Techniques in Operator Theory. 2nd ed.
180 SRIVASTAVA. A Course on Borel Sets.
181 KRESS. Numerical Analysis.
182 WALTER. Ordinary Differential Equations.

183 MEGGINSON. An Introduction to Banach Space Theory.
184 BOLLOBAS. Modern Graph Theory.
185 COX/LITTLE/O'SHEA. Using Algebraic Geometry.
186 RAMAKRISHNAN/VALENZA. Fourier Analysis on Number Fields.
187 HARRIS/MORRISON. Moduli of Curves.
188 GOLDBLATT. Lectures on the Hyperreals: An Introduction to Nonstandard Analysis.
189 LAM. Lectures on Modules and Rings.
190 ESMONDE/MURTY. Problems in Algebraic Number Theory. 2nd ed.
191 LANG. Fundamentals of Differential Geometry.
192 HIRSCH/LACOMBE. Elements of Functional Analysis.
193 COHEN. Advanced Topics in Computational Number Theory.
194 ENGEL/NAGEL. One-Parameter Semigroups for Linear Evolution Equations.
195 NATHANSON. Elementary Methods in Number Theory.
196 OSBORNE. Basic Homological Algebra.
197 EISENBUD/HARRIS. The Geometry of Schemes.
198 ROBERT. A Course in p-adic Analysis.
199 HEDENMALM/KORENBLUM/ZHU. Theory of Bergman Spaces.
200 BAO/CHERN/SHEN. An Introduction to Riemann–Finsler Geometry.
201 HINDRY/SILVERMAN. Diophantine Geometry: An Introduction.
202 LEE. Introduction to Topological Manifolds.
203 SAGAN. The Symmetric Group: Representations, Combinatorial Algorithms, and Symmetric Functions.
204 ESCOFIER. Galois Theory.
205 FÉLIX/HALPERIN/THOMAS. Rational Homotopy Theory. 2nd ed.
206 MURTY. Problems in Analytic Number Theory.
Readings in Mathematics
207 GODSIL/ROYLE. Algebraic Graph Theory.
208 CHENEY. Analysis for Applied Mathematics.
209 ARVESON. A Short Course on Spectral Theory.
210 ROSEN. Number Theory in Function Fields.
211 LANG. Algebra. Revised 3rd ed.
212 MATOUŠEK. Lectures on Discrete Geometry.
213 FRITZSCHE/GRAUERT. From Holomorphic Functions to Complex Manifolds.
214 JOST. Partial Differential Equations.
215 GOLDSCHMIDT. Algebraic Functions and Projective Curves.
216 D. SERRE. Matrices: Theory and Applications.
217 MARKER. Model Theory: An Introduction.
218 LEE. Introduction to Smooth Manifolds.
219 MACLACHLAN/REID. The Arithmetic of Hyperbolic 3-Manifolds.
220 NESTRUEV. Smooth Manifolds and Observables.
221 GRÜNBAUM. Convex Polytopes. 2nd ed.
222 HALL. Lie Groups, Lie Algebras, and Representations: An Elementary Introduction.
223 VRETBLAD. Fourier Analysis and Its Applications.
224 WALSCHAP. Metric Structures in Differential Geometry.
225 BUMP: Lie Groups
226 ZHU. Spaces of Holomorphic Functions in the Unit Ball.
227 MILLER/STURMFELS. Combinatorial Commutative Algebra.
228 DIAMOND/SHURMAN. A First Course in Modular Forms.
229 EISENBUD. The Geometry of Syzygies.
230 STROOCK. An Introduction to Markov Processes.